Progress in Probability

Volume 36

Series Editors
Thomas Liggett
Charles Newman
Loren Pitt

Seminar on Stochastic Analysis, Random Fields and Applications

Centro Stefano Franscini, Ascona, 1993

Erwin Bolthausen
Marco Dozzi
Francesco Russo
Editors

Springer Basel AG

Editors:

Erwin Bolthausen
Institut für Mathematik
Abteilung Angewandte Mathematik
Universität Zürich-Irchel
Winterthurerstr. 190
CH–8057 Zürich

Marco Dozzi
Département de Mathématiques
Université de Nancy 1
B.P. 239
F–54506 Vandoeuvre-Les-Nancy Cedex

Francesco Russo
Université Paris-Nord
Département de Mathématiques
Institut Galilée
Av. Jean-Baptiste Clément
F–93430 Villetaneuse and
Universität Bielefeld
BIBOS
D–33615 Bielefeld 1

A CIP catalogue record for this book is available from the Library of Congress,
Washington D.C., USA

Deutsche Bibliothek Cataloging-in-Publication Data

**Seminar on Stochastic Analysis, Random Fields and
Applications <1993, Ascona>:**
Seminar on Stochastic Analysis, Random Fields and
Applications : Centro Stefano Franscini, Ascona, 1993 / Erwin
Bolthausen ... ed. – 1995 Springer Basel AG
  (Progress in probability ; Vol. 36)
    ISBN 978-3-0348-7028-3   ISBN 978-3-0348-7026-9 (eBook)
    DOI 10.1007/978-3-0348-7026-9
NE: Bolthausen, Erwin [Hrsg.]; GT

© 1995 Springer Basel AG
Originally published by Birkhäuser Verlag Basel in 1995
Softcover reprint of the hardcover 1st edition 1995
Printed on acid-free paper produced from chlorine-free pulp ∞

ISBN 978-3-0348-7028-3
9 8 7 6 5 4 3 2 1

CONTENTS

Financial models

PREFACE

This volume contains the proceedings of a six-day *Seminar on Stochastic Analysis, Random Fields and Applications* which took place at the Centro Stefano Franscini (Monte Verità) in Ascona, Switzerland, from Monday June 7 to Saturday June 12, 1993. It was financially supported by ETH Zürich and sponsored by Banca della Svizzera Italiana. The seminar focused on four topics, namely stochastic partial differential equations, analysis of Wiener functionals and non-causal stochastic calculus, random media and financial models. The fourth topic was the subject of a special *Minisymposium*.

The purpose of the seminar was to promote the interaction between specialists in these areas and young scientists. We tried to provide an up-to-date picture of current issues and outstanding problems. We hope that these proceedings will convey this picture to a larger audience. Several lecturers had been asked to present a review of their research areas. Two public lectures were given by Prof. Nicole El Karoui (Paris 6) and Dott. Riccardo Cesari (Banca d'Italia, Bologna). The titles were:

- Modèles stochastiques des taux d'intérêts,
- Il mercato finanziario in Italia. Istituzioni e strumenti.

Let us briefly discuss the state of the art in the topics of the seminar:

Stochastic partial differential equations basically include nonlinear partial differential equations perturbed by Gaussian white or coloured noise. In principle, there are two classes of stochastic partial differential equations: those which admit continuous multi-parameter processes as solutions and those which involve non-linearities of Schwartz distributions. The first case arises in low space dimension d, say $d = 1$ for hyperbolic and parabolic stochastic partial differential equations and $d \leq 3$ in the elliptic case. Problems of existence, uniqueness and related questions have been considered since the '70s. The hyperbolic case constitutes one extension of the classical theory of two-parameter processes, whose index set is the first quarter of the plane; this field was very fashionable in the '70s. Since the mid '80s the Malliavin calculus has been applied to establish and discuss existence, regularity, positivity and support of the law density. Particular interest has been devoted to support theorems in the case of a stochastic heat equation with nonlinearities of power type; these equations are strongly connected with limit theorems for interacting particle systems. The case where no continuous solutions exist is partially motivated by mathematical physics and more specifically by quantum field theory. One of the basic problems in this area is stochastic quantization which is a particular heat equation perturbed by white noise and involving nonlinearities of polynomial or trigonometrical type. The polynomial case has recently been treated with the help of Dirichlet form techniques, which have had a big development in the past years.

Non-causal stochastic calculus has mainly been developed in the last ten years after the rediscovery of the Skorohod integral. In the framework of stochastic calculus of variations, this integral can be seen as a divergence, that is, the dual operator of the derivative on the Wiener space. From this starting point, a stochastic calculus has been developed, sometimes also in connection with two-parameter pocesses and manifold-valued processes, and at the same time, a large class of anticipating stochastic differential equations has been discussed. Extensions from the Brownian case to some jump processes (Poisson processes, queuing processes and so on) deserve particular interest. Under the influence of Malliavin, connections with quasi sure analysis have been established. There are other approaches, not directly related to the Skorohod calculus, for instance pathwise approaches.

Random media is a rapidly developing field with many successes in the last years and many more unsolved problems. Generally speaking, one is interested in the behaviour of some quantity which is usually well understood in a completely homogeneous medium but which one now would like to understand in a disordered one. This quantity may be connected with some second random mechanism, e. g. the diffusive behaviour of a random walk, which of course is well known in a nonrandom medium but is very challenging for some types of random media. Some of the problems are directly connected with stochastic partial differential equations. The presentations at the conference in this field were naturally restricted to a few highlights.

The success of mathematically oriented *financial models* has been possible through a parallel development of mathematics, computing power and new financial tools (basically options). The fundamental problems are pricing and hedging. The starting point was of course the famous Black–Scholes formula. Later, many generalizations were studied, for instance involving jump processes. The field has much profited from very recent developments in stochastic analysis, like forward–backward stochastic differential equations in connection with control theory, models involving fractional Brownian motion, stable processes and so on. On the computational side, different classes of Monte Carlo methods have been exploited for simulation.

We would like to thank the Centro Stefano Franscini and especially its director Prof. Konrad Osterwalder for making this conference possible. The Banca della Svizzera Italiana generously sponsored the seminar. We wish especially to thank Dr. Grandi and Dr. Gysi. We are particularly grateful to Ms. Gerda Schacher and Dr. Uwe Schmock for the tremendous work involved in preparing this volume for publishing; concerning the organization of the seminar we are grateful for the help which we had from Ms. Katia Bastianelli, Ms. Aline Cossu and Ms. Schkölziger.

Zürich, March 1995

Erwin Bolthausen
Marco Dozzi
Francesco Russo

LIST OF PARTICIPANTS

Robert Aebi. Universität Bern. Switzerland
Sergio Albeverio. Ruhr-Universität Bochum. Germany
Renzo G. Avesani. Milano, Italy
Paolo Baldi. Università di Roma 2. Italy
Catherine Bandle. Universität Basel. Switzerland
Andrew D. Barbour. Universität Zürich. Switzerland
Erwin Bolthausen. Universität Zürich. Switzerland
Ulriche Bucher. Basel, Switzerland
Rainer Buckdahn. Humboldt-Universität. Berlin, Germany
René Carmona. University of California. Irvine, U.S.A.
Fabienne Castell. Université Paris-Sud. France
Riccardo Cesari. Banca d'Italia. Bologna, Italy
Mireille Chaleyat-Maurel. Université Paris VI. France
Robert Dalang. Ecole polytechnique fédérale de Lausanne. Switzerland
Giuseppe Da Prato. Scuola normale superiore. Pisa, Italy
Jean-Dominique Deuschel. Technische Universität Berlin. Germany
Markus Dozzi. Université de Nancy 1. France
Nicole El Karoui. Université Paris VI. France
Franco Flandoli. Scuola normale superiore. Pisa, Italy
Hans Föllmer. Universität Bonn. Germany
Jean-Pierre Fouque. Ecole Polytechnique. Palaiseau, France
Christoph Gallus. Universität Erlangen-Nürnberg. Germany
Nina Gantert. Technische Universität Berlin. Germany
Hélyette Geman. Essec. Cergy, France
Barbara Gentz. Universität Zürich. Switzerland
Axel Grorud. Université de Provence. France
Zbigniew Haba. University of Wroclaw. Poland
Peter Imkeller. Université de Besançon. France
Uwe Küchler. Humboldt-Universität. Berlin, Germany
Damien Lamberton. Université de Marne-la-Vallée. Noisy-le-Grand, France
Rémi Léandre. Université de Nancy 1. France
C. W. Li. City Polytechnic of Hong Kong. Hong Kong
Terry Lyons. Imperial College. London, Great Britain
Paul Malliavin. Paris, France
Christiane Mathis. Patria Life Insurance. Basel, Switzerland
Franco Moriconi. Università di Perugia. Italy
Masao Nagasawa. Universität Zürich. Switzerland

John Noble. University College. Cork, Ireland
David Nualart. Universitat de Barcelona. Spain
Michael Oberguggenberger. Universität Innsbruck. Austria
Bernt Øksendal. University of Oslo. Norway
Etienne Pardoux. Université de Provence. France
Monique Pontier. Université d'Orléans. France
Jürgen Potthoff. Universität Mannheim. Germany
Nicolas Privault. Université d'Evry. France
Michael Röckner. Universität Bonn. Germany
Bernard Roynette. Université de Nancy 1. France
Wolfgang Runggaldier. Università di Padova. Italy
Francesco Russo. Université Paris-Nord. France
Marta Sanz-Solé. Universitat de Barcelona. Spain
Bruno Scarpellini. Universität Basel. Switzerland
Christian Schwab. Universität Fribourg. Switzerland
Martin Schweizer. Universität Göttingen. Germany
Josep Lluis Sole I Clivilles. Barcelona, Spain
Richard Sowers. University of Southern California. Los Angeles, U.S.A.
Christophe Stricker. Université de Besançon. France
Theo Sturm. Universität Erlangen-Nürnberg. Germany
Alain-Sol Sznitman. ETH. Zürich, Switzerland
Angès Tourin. Université de Paris IX. France
Luciano Tubaro. Università di Trento. Italy
Ali Sûleyman Ustunel. ENST. Paris, France
Walter Willinger. Bellcore. Morristown, NJ, U.S.A.
Marc Yor. Université de Paris VI. France
Jerzy Zabczyk. Polish Academy of Sciences. Poland
Bijan Z. Zangeneh. Sharif University of Technology. Teheran, Iran
Boguslaw Zegarlinski. Imperial College. London, Great Britain
Ofer Zeitouni. Technion Haifa. Israel
Xialiau Zhang. Société générale. Paris, France

Progress in Probability, Vol. 36
© 1995 Birkhäuser Verlag Basel/Switzerland

PROPAGATION OF CHAOS – THE INVERSE PROBLEM

ROBERT AEBI

ABSTRACT. The phenomenon propagation of chaos describes a cloud of identical particles which behave asymptotically, as their number tends to infinity, independently and identically according to a limiting distribution. We introduce a coherent definition in terms of relative entropy in the tradition of statistical mechanics.

Investigating the inverse problem raised in Nagasawa [38] means to start with certain limiting distributions and to furnish them with microscopical systems which perform propagation of chaos.

The considered limiting distributions are diffusion processes arising from the statistically relevant situation of given particle dynamics and prescribed marginal distributions at initial and final time. They actually describe a wide range of observed phenomena to which the fundamental hypothesis of statistical mechanics applies. We will give an application to diffusions related to wave functions in quantum mechanics.

§1 INTRODUCTION

Let us consider diffusion processes which are "most probable" in the sense of large deviations among those which possess prescribed marginal distributions at finite initial and final time, respectively. We show that they can be understood as limiting processes of naturally given microscopical systems which are shown to perform propagation of chaos. Accordingly, the investigated limiting processes, which we will call Schrödinger processes based on a crucial example in Schrödinger [43], may be seen as Gibbs states.

Starting with a given limiting distribution Q, we intend to furnish it with a microscopical system $\mathbb{Q}^{(n,k)}$, $n, k \in \mathbb{N}$. Hence, what we actually do is solving the inverse problem of propagation of chaos in case of Schrödinger processes Q. This paper deals with the existence of microscopical systems $\mathbb{Q}^{(n,k)}$, $n, k \in \mathbb{N}$, which converge in entropy to Q performing propagation of chaos. For a discussion of a kind of uniqueness, a further characterization of the wanted microscopical system would be required. However, the proposed $\mathbb{Q}^{(n,k)}$, $n, k \in \mathbb{N}$, can truly be considered as the most "natural" one.

Our approach clearly depends essentially on the meaning of the investigated limiting processes. Schrödinger processes can be met on various occasions. Actually,

1991 *Mathematics Subject Classification.* primary 82C22; secondary 60J60, 70K50.

Key words and phrases. propagation of chaos; relative entropy; diffusion with creation and killing.

as shown in Aebi [7], any diffusion process in its Schrödinger–Nagasawa representation expressing the time-reversible nature of diffusions, is a Schrödinger process. Following Schrödinger's original intention [43], Nagasawa [37, 39] and Aebi [7] establish Schrödinger processes as real-valued counterparts to wave functions in quantum mechanics. For the purpose of limiting distributions in propagation of chaos, we adopt the large deviation approach of Aebi–Nagasawa [2] which is based on Föllmer [18] using tools developed by Csiszár [14, 15].

With Schrödinger [43], we find Schrödinger processes as the solutions of a statistically motivated problem. Let us watch a cloud of independent and identical particles where their dynamics is given by a transition density p possessing in general even singular creation and killing. Fixing a finite time horizon $-\infty < a < b < \infty$, an observer reports the distribution density q_a of the particle cloud at time a and the distribution density q_b at time b. What is the "most probable" distribution density ρ_t of the particle cloud at intermediate times $t \in [a, b]$? The triple (p, q_a, q_b) determines uniquely a so-called Schrödinger process Q, $dQ(X_t) = \rho_t \, dt$, as the "most probable" diffusion process with marginal distribution densities q_a and q_b at initial time a and final time b, respectively. "Most probable" is defined by means of a large deviation principle where the rate function is the relative entropy with respect to the dynamics p.

In case that only q_a is observed, the "most probable" diffusion is simply the diffusion process with initial distribution q_a and transition density p. If q_b at time b is the only observation, it is the final distribution density of the "most probable" diffusion given by the time-reversed transition density of p. Installing three observation times $-\infty < a < b_1 < b_2 < \infty$, the solution for the "most probable" diffusion with observed marginal distribution densities q_a, q_{b_1} and q_{b_2} can obviously be decomposed in two Schrödinger processes $(X_t, a \leq t \leq b_1, Q_1)$ and $(X_t, b_1 \leq t \leq b_2, Q_2)$. These simple remarks are intended to suggest that Schrödinger processes cover more general situations than it might look at first sight.

The phenomenon propagation of chaos describes a cloud consisting of identical particles which behave asymptotically, as their number tends to infinity, independently and identically according to a limiting distribution. It is introduced by McKean [32, 33] and investigated by Nagasawa–Tanaka [34, 35, 36] in connection with Schrödinger processes. Various approaches to propagation of chaos can be found in Gutkin–Kac [20], Tanaka [46], Kusuoka–Tamura [30], Ölschläger [42], Sznitman [45] and Dawson–Gärtner [16].

In the present context we continue developing the notion of propagation of chaos in entropy introduced in Aebi–Nagasawa [2] to a coherent form. It is our intention to have convergence in entropy to the limiting distribution in the tradition of statistical mechanics.

Definition 1.1. A sequence of systems $((X_1, \ldots, X_n),\ \mathbb{Q}^{(n,k)})_{n,k \in \mathbb{N}}$ of interacting diffusion processes performs *propagation of chaos in entropy* with limiting distribution Q as n and k tend to infinity, if for each $m \in \mathbb{N}$, the marginal distribution $\mathbb{Q}_m^{(n,k)}$ of $\mathbb{Q}^{(n,k)}$ on Ω^m and the empirical distribution $L_n = (1/n)\sum_{i=1}^n \delta_{\omega_i}$, $\omega_i \in \Omega$, $i \in \mathbb{N}$, satisfy

$$\lim_{k\nearrow\infty}\lim_{n\nearrow\infty} H(Q^m \,|\, \mathbb{Q}_m^{(n,k)}) = 0 \tag{1}$$

$$\lim_{k\nearrow\infty}\lim_{n\nearrow\infty} H(Q \,|\, \mathbb{Q}^{(n,k)}[L_n]) = 0 \tag{2}$$

where k is called a modeling parameter.

This definition allows general microscopical systems represented by $\mathbb{Q}^{(n,k)}$ as it is required in the situation of Section 3. In fact, while $Q^n \ll \mathbb{Q}^{(n,k)}$ seems to be quite reasonable, the equivalence of measures does restrict the choice of microscopical systems too much. As a consequence, our limiting distributions Q appear as the first argument in the non-symmetrical relative entropy formula $H(\cdot|\cdot)$ at (23). This will cause the main technical difficulties we have to deal with in Section 4.

The relative entropy has been proved to be a natural indicator for the mutual randomness of probability measures. See for example Csiszár [14, 15] in connection with large deviations, Boltzmann [12] and Lanford [31] who deal with statistical mechanics and Khinchin [26] for an approach to information theory. Different types of convergence will be compared in Remark 4.2 where convergence in entropy will turn out to be the strongest one.

Remark 1.1. The definition of propagation of chaos in entropy of Aebi–Nagasawa [2] is refined in Section 8.2 of Nagasawa [39]. Caused by the non-symmetrical relative entropy formula at (23), each of the two limit properties in Nagasawa's definition [39] consists of two distinct limits. First, $H(\mathbb{Q}_m^{(n,k)}\,|\,Q_k^m)$, $m \in \mathbb{N}$, as well as $H(\mathbb{Q}^{(n,k)}[L_n]\,|\,Q_k)$ are required to vanish as n tends to infinity where $(Q_k)_{k\in\mathbb{N}}$ is any family of approximative limiting distributions indexed by the modeling parameter k. Second, $H(Q\,|\,Q_k)$ is required to vanish as k tends to infinity. Our definition comes back to a direct convergence in entropy.

Let us reveal the meaning of Definition 1.1. In one sentence we can say that (1) guarantees the existence of a "typical particle" and (2) ensures that asymptotically it is the "typical particle" which is observed. More precisely, (1) describes m arbitrary of the interacting diffusion processes in terms of their joint m-dimensional distribution $\mathbb{Q}_m^{(n,k)}$ which converges in entropy to Q^m. This means that any m interacting diffusion processes are asymptotically independent and that Q is the asymptotical distribution of each interacting diffusion process. Property (2) deals with a mixing of the involved interacting diffusion processes by means of L_n. It represents the observed distribution on Ω under $\mathbb{Q}^{(n,k)}$ which tends in entropy to Q as the number n of interacting diffusion processes and the modeling parameter k increase to infinity.

The purpose of this paper is to furnish Schrödinger processes Q with systems of interacting diffusion processes $\mathbb{Q}^{(n,k)}$ which perform propagation of chaos in entropy in the sense of Definition 1.1. The proof will be given in Section 4. Section 2 is devoted to the construction and the properties of Schrödinger processes Q. In Section 3 we will deduce naturally associated systems of interacting diffusion processes $\mathbb{Q}^{(n,k)}$ playing the role of microscopical systems.

§2 SCHRÖDINGER PROCESSES

2.1 Motivation and interpretation. For a finite time horizon $-\infty < a < b < \infty$, let us consider partitions $(\Delta x_k)_{1 \leq k \leq m}$ and $(\Delta z_l)_{1 \leq l \leq n}$ of copies of the state space $\mathbb{R}^d$ at initial time a and final time b, respectively, as well as positions $x_k \in \Delta x_k$ for $1 \leq k \leq m$ and $z_l \in \Delta z_l$ for $1 \leq k \leq n$.

Suppose that a large number N of independent and identical *particles* perform a *journey* starting at time a and ending at time b. The *observation* of the particle cloud at time a and b, respectively, can be registered in terms of probability densities q_a and q_b on the state space $\mathbb{R}^d$. Schrödinger comments that this only observation (q_a, q_b) might be more or less exceptional, which should not bother us. He takes the viewpoint that the distributions q_a and q_b represent the reality on which further conclusions have to be based. What we actually observe is the result of *N-particle journeys* and it might be interesting to know the numbers Γ_{kl} of those particles leaving from cell Δx_k which finally arrive at cell Δz_l, $1 \leq k \leq m$, $1 \leq l \leq n$. Hence we are going to seek for classes of N-particle journeys

$$(\Gamma_{kl})_{1 \leq k \leq m, 1 \leq l \leq n} \tag{3}$$

which satisfy

$$N q_a(x_k) \Delta x_k = \sum_{l=1}^{n} \Gamma_{kl}, \quad 1 \leq k \leq m \tag{4}$$

$$N q_b(z_l) \Delta z_l = \sum_{k=1}^{m} \Gamma_{kl}, \quad 1 \leq l \leq n. \tag{5}$$

Schrödinger's intention is to predict the distribution density of an N-particle cloud as N tends to infinity at *intermediate* times $t \in (a, b)$ for given *particle dynamics* p under prescribed *probability densities* q_a and q_b at initial time a and final time b, respectively. Let $\rho_t(dy \mid q_a, q_b)$ denote the distribution density of an N-particle cloud at time $t \in (a, b)$ for N tending to infinity. Its construction is going to be decomposed in a so-called *Schrödinger bridge* at (8) and a *"most probable"* Γ at (9).

(Travel-)dynamics of the observed kind of particles can be described in analytical terms by a parabolic differential operator L or in probabilistical terms by a *transition density* $p(s, x; t, y)$, $a \leq s < t \leq b$, $x, y \in \mathbb{R}^d$, where in this case p is a (weak) fundamental solution of the *Fokker–Planck equation* $L = 0$. Hence, for

any departure density d, the probability of an N-*particle journey* belonging to a certain Γ at (3) is given by

$$\prod_{k,l=1}^{m,n} (d_k \, p_{kl} \, \Delta x_k \, \Delta z_l)^{\Gamma_{kl}} \tag{6}$$

where we denote $p_{kl} = p(a, x_k; b, z_l)$ and $d_k = d(x_k)$ for $1 \leq k \leq m$, $1 \leq l \leq n$. Since the number of N-particle journeys possible under a fixed Γ is

$$\frac{N!}{\prod_{k,l=1}^{m,n} \Gamma_{kl}!}$$

we conclude by means of (6) that Γ at (3) has the probability

$$P_N(\Gamma) = N! \prod_{k,l=1}^{m,n} \frac{(d_k \, p_{kl} \, \Delta x_k \, \Delta z_l)^{\Gamma_{kl}}}{\Gamma_{kl}!}. \tag{7}$$

Obviously there are many Γ which obey the continuity conditions (4) and (5). Hence the *fundamental hypothesis of statistical mechanics* will serve by means of (9) as criterion to determine the relevant Γ uniquely.

The situation of a Schrödinger bridge deals with the degenerate case of Dirac marginal distributions

$$q_a = \delta_{x_a}, \quad q_b = \delta_{z_b}$$

for fixed x_a, $z_b \in \mathbb{R}^d$. Considering the proportion of particles which visit $(y, y + dy)$ at time $t \in (a, b)$ on the journey from x_a to z_b, we obtain

$$\rho_t(dy \,|\, \delta_{x_a}, \delta_{z_b}) = \frac{p(a, x_a; t, y) \, dy \, p(t, y; b, z_b)}{p(a, x_a; b, z_b)} \tag{8}$$

which is a *conditional distribution density*.

Seeking for the "most probable" Γ at (3) as N tends to infinity, we write

$$\gamma_{kl} \, \Delta x_k \, \Delta z_l = \lim_{N \nearrow \infty} \Gamma_{kl} / N, \quad 1 \leq k \leq m, \; 1 \leq l \leq n$$

and obtain by means of Stirling's formula that

$$\lim_{N \nearrow \infty} \frac{1}{N} \log P_N(\Gamma) = -H_\Delta(\gamma_{..} \,|\, d_. \, p_{..}). \tag{9}$$

Accordingly, the probability (7) vanishes exponentially fast as N tends to infinity with the rate function

$$H_\Delta(\gamma_{..} \,|\, d_. \, p_{..}) = \sum_{k,l=1}^{m,n} \log\left(\frac{\gamma_{kl}}{d_k \, p_{kl}}\right) \gamma_{kl} \, \Delta x_k \, \Delta z_l$$

which is known as relative entropy. What we meet here is a so-called large deviation phenomenon where "most probable" under the continuity conditions (4) and (5) means "vanishing with smallest rate function" among those which satisfy (4) and (5). As a matter of fact, $(d_k\,p_{kl})$ does in general not satisfy these continuity conditions which result from an observation. Applying the Lagrangeian procedure in the limit of N tending to infinity, we consider

$$L_\Delta(\gamma_{kl}, \lambda_{k\cdot}, \lambda_{\cdot l}) = \sum_{k,l=1}^{m,n} \log\left(\frac{\gamma_{kl}}{d_k\,p_{kl}}\right) \gamma_{kl}\,\Delta x_k\,\Delta z_l$$

$$-\sum_{k=1}^{m}\lambda_{k\cdot}\left(\sum_{l=1}^{n}\gamma_{kl}\,\Delta z_l - q_a(x_k)\right)\Delta x_k$$

$$-\sum_{l=1}^{n}\lambda_{\cdot l}\left(\sum_{k=1}^{m}\gamma_{kl}\,\Delta x_k - q_b(z_l)\right)\Delta z_l$$

for $m+n$ Lagrangeian multipliers $\lambda_{k\cdot}$ and $\lambda_{\cdot l}$. The first $m\,n$ Lagrangeian equations $\partial L_\Delta/\partial\gamma_{ij}=0$ yield

$$e\,\frac{\gamma_{ij}}{d_i\,p_{ij}} = \exp\{\lambda_{i\cdot}\}\,\exp\{\lambda_{\cdot j}\}$$

for $1 \le i \le m$, $1 \le j \le n$. The $m+n$ Lagrangeian equations $\partial L_\Delta/\partial\lambda_{k\cdot}=0$ and $\partial L_\Delta/\partial\lambda_{\cdot l}=0$ reproduce the continuity conditions (4) and (5), respectively, in terms of (γ_{kl}). Hence the "most probable" Γ expressed in term of the density q_{ij}, $1 \le i \le m$, $1 \le j \le n$, occurs as

$$q_{ij} = \widehat{\varphi}_{ai}\,d_i\,p_{ij}\,\varphi_{bj} \tag{10}$$

where

$$\widehat{\varphi}_{ai} = \exp\{\lambda_{i\cdot}\}, \quad 1 \le i \le m$$

depends exclusively on *the past at time a* and

$$\varphi_{bj} = \exp\{\lambda_{\cdot j} - 1\}, \quad 1 \le j \le n$$

depends exclusively on *the future at time b*. We call the non-negative weights $(\widehat{\varphi}_{ai})_{1\le i\le m}$ and $(\varphi_{bj})_{1\le j\le n}$ Schrödinger multipliers and notice that only their product $(\widehat{\varphi}_{ai}\,\varphi_{bj})_{1\le i\le m,1\le j\le n}$ is uniquely determined by the Lagrangeian procedure. Accordingly, the maximizing problem based on (9) is reduced to solving the pair of continuity conditions (4) and (5) expressed in terms of (10). Schrödinger himself considered neither existence nor uniqueness of solutions of these so-called *Schrödinger systems*. These questions have been investigated by Bernstein [9], Fortet [19], Beurling [10], Jamison [23, 24], Föllmer [18], Nagasawa [37, 38] and Aebi–Nagasawa [2] where the latter deals with more general dynamics. Schrödinger rather prefers to give a justification of his approach which should not be missed here. An attempt to translate Schrödinger's explanations into English yields:

If considered from a statistical point of view, the so-called irreversible natural laws do actually not determine a specific direction of time. In fact, their statements only depend on the cross-sections in space at two points of time a and b and they are entirely symmetrical with respect to the conditions at time a and b. This is slightly covered by our habit of observing only one cross-section in space. The second cross-section in space we usually transfer into a sufficiently large distance of time. There, it holds the reliable rule that the state of biggest randomness, i. e. of maximal entropy, is achieved. Basically it is very strange that this rule should be true. I believe that it cannot be deduced logically. Anyway, this rule does not determine a specific direction of time either. It holds uniformly for both directions of time as soon as the second cross-section in space is in a sufficiently large distance of time from the first one.

We realize that Schrödinger propagates another understanding of diffusions where their *time-reversals* are always considered simultaneously.

In our investigations of diffusion processes we would like to have a continuous version of (9). It is provided by Theorem 3.1 in terms of a large deviation principle where (10) becomes (13) which is (28) as a consequence of substantial results in Csiszár [14, 15]. Since the product form of (10) is crucial for an interpretation of these so-called Schrödinger processes, we prefer the Lagrangeian approach for illustrative reasons. The switch to the continuous situation is conveniently realized on the level of Schrödinger systems. In fact, carrying out a refinement of the discretization, the equations (4) and (5) in terms of (10) yield

$$\widehat{\varphi}_a(x)\, d(x) \int p(a, x; b, z)\, dz\, \varphi_b(z) = q_a(x) \tag{11}$$

and

$$\int \widehat{\varphi}_a(x)\, d(x)\, dx\, p(a, x; b, z)\, \varphi_b(z) = q_b(z) \tag{12}$$

for λ-almost all $x, z \in \mathbb{R}^d$. The conditions in Lemma 2.1, which is going to be formulated for a more general situation, guarantee a unique solution $\widehat{\varphi}_a(x)\, \varphi_b(z)$, $(x, z) \in \mathbb{R}^d \times \mathbb{R}^d$, of the Schrödinger system (11) and (12). In analogy to (10) we set

$$q(x, z) = \widehat{\varphi}_a(x)\, d(x)\, p(a, x; b, z)\, \varphi_b(z) \tag{13}$$

which is "most probable" in the inherited sense from the discrete situation. It coincides with (28) obtained by the large deviation principle established in Theorem 3.1, where "most probable" has a meaning of its own.

Schrödinger's program has provided (8) and (13). Accordingly, the distribution density ρ_t of a cloud with infinitely many particles at intermediate times $t \in (a, b)$

is given as

$$\rho_t(dy \,|\, q_a, q_b)$$

$$= \int dx \int dz\, \rho_t(dy \,|\, \delta_x, \delta_z)\, q(x, z)$$

$$= \int dx \int dz\, \frac{p(a, x; t, y)\, dy\, p(t, y; b, z)}{p(a, x; b, z)}\, \widehat{\varphi}_a(x)\, d(x)\, p(a, x; b, z)\, \varphi_b(z) \qquad (14)$$

$$= \widehat{\varphi}_t(y)\, \varphi_t(y)$$

where

$$\varphi_t(y) = \int p(t, y; b, z)\, dz\, \varphi_b(z), \quad t \in [a, b) \qquad (15)$$

$$\widehat{\varphi}_t(y) = \int \widehat{\varphi}_a(x)\, d(x)\, dx\, p(a, x; t, y), \quad t \in (a, b] \qquad (16)$$

for $y \in \mathbb{R}^d$, are *space-time harmonic* and *co-harmonic* functions of $p(s, x; t, y)$, $a \le s < t \le b$, respectively. The *structures of the distribution density* ρ_t at (14) in terms of (15) and (16) are characteristics of Schödinger processes. Their interpretation becomes evident by considering $\widehat{\varphi}_a$ and φ_b as the *"history" from past and future*, respectively, which is determined by the observation (q_a, q_b) through the Schrödinger system (11) and (12). According to (16), $\widehat{\varphi}_t$ is the *prediction from the past* and according to (15), φ_t is the *prediction from the future*, i. e. time-reversed prediction. Hence it follows by means of (14) that the distribution density ρ_t at intermediate times $t \in (a, b)$ is the *product of prediction from the past and prediction from the future*. This key point of the propagated way of thinking has already been claimed by Boltzmann in [13]. An attempt to translate his comment yields:

> There is no doubt that a world could also be imagined in which all natural laws happen in time-reversed order. A human being living in such a time-reversed world would not feel differently from us. He would simply call past what we call future and vice versa.

Remark 2.1. Schrödinger stresses in [43] that for him, the most interesting aspect of the distribution density ρ_t, $t \in [a, b]$, at (14) in terms of (15) and (16) is its striking formal analogy to distribution densities $\overline{\Psi}_t\, \Psi_t$ formed out of solutions Ψ of Schrödinger equations in quantum mechanics. For example, let us consider the free (Brownian) particle according to $L = 0$ with

$$L = \frac{\partial}{\partial t} + \frac{1}{2}\triangle$$

which is described by the Schrödinger equation

$$\sqrt{-1}\, \frac{\partial \Psi}{\partial t} + \frac{1}{2}\triangle\, \Psi = 0. \qquad (17)$$

In diffusion theory, the distribution density itself satisfies a linear differential equation, e. g. $L = 0$. On the other hand, in quantum mechanics, probability densities are obtained in a bilinear way from wave functions Ψ which are solutions of complex linear differential equations, see e. g. (17). Both differential equations $L = 0$ and (17) are of first order in time t, but the Fokker–Planck equation $L = 0$ is of a parabolic irreversible nature and the Schrödinger equation shows a hyperbolic reversible behavior caused by the factor $\sqrt{-1}$. Concerning both of these two differences, the distribution density ρ_t, $t \in [a, b]$, at (14) as a product of (15) and (16) shows a much closer analogy to distribution densities $\bar{\Psi}_t \Psi_t$ obtained from solutions Ψ of the Schrödinger equation (17) than to distribution densities which satisfy one of the irreversible differential equations $L = 0$ or $L^* = 0$, where L^* denotes the adjoint of L. In fact, ρ_t, $t \in [a, b]$, is not solution of a certain Fokker–Planck equation, but it is the product of two solutions of adjoint Fokker–Planck equations. As a consequence, ρ_t, $t \in [a, b]$, does not evolve in a specified direction of time. If q_a and q_b are interchanged, then ρ_t, $t \in [a, b]$, simply develops in time-reversed order between a and b.

2.2 Construction in terms of Csiszár's projection. Let $\Omega = C([a, b], \mathbb{R}^d)$, $-\infty < a < b < \infty$, be the space of continuous paths taking values in $\mathbb{R}^d$, d fixed, with the Borel σ-field $\sigma(\Omega)$. The state of a path $\omega \in \Omega$ at time t, $a \le t \le b$, is denoted by the projection $X_t(\omega) = X(t, \omega) = \omega(t)$. Let $((t, X_t), s \le t \le b$, $P_{(s,x)})$, $(s, x) \in [a, b] \times \mathbb{R}^d)$ be a space-time diffusion process and let $c(s, x)$ be a measurable extended real-valued function on $[a, b] \times \mathbb{R}^d$ with positive and negative parts $c^+(s, x)$ and $c^-(s, x)$, respectively. We intend to consider space-time diffusion processes with possibly singular creation and killing $c(s, x) = c^+(s, x) - c^-(s, x)$ as it is required by Corollary 4.1 when associating Schrödinger processes to Schrödinger equations. In terms of the killing part $c^-(s, x)$, we set

$$T_s = \inf\left\{ t > s : \int_s^t c^-(v, X_v) \, dv = \infty \right\}$$

and determine our state space $D \subset [a, b] \times \mathbb{R}^d$ as

$$D = \{ (s, x) : P_{(s,x)}[b < T_s] > 0 \}.$$

Let us define first of all the measure $P^-_{(s,x)}$ with killing $c^-(s, x)$ by

$$P^-_{(s,x)}[F] = P_{(s,x)}\left[\exp\left\{ -\int_s^b c^-(v, X_v) \, dv \right\} F \, 1_{\{b < T_s\}} \right] \qquad (18)$$

for any non-negative measurable function F. We notice that the exponential function on the right-hand side of (18) is positive on the set $\{b < T_s\}$, i. e. $P^-_{(s,x)}[1] > 0$ holds for all $(s, x) \in D$. Hence the probability that a particle starting at a point in D survives until the final time b is positive.

Dealing with the creation part $c^+(s, x)$, we establish the basic integrability condition

$$\xi(s, x) = P^-_{(s,x)}\left[\exp\left\{\int_s^b c^+(v, X_v)\, dv\right\}\right] < \infty \qquad (19)$$

for all $s \in [a, b]$ and λ-almost all $x \in \mathbb{R}^d$ with $(s, x) \in D$, where $P^-_{(s,x)}$ is the measure defined at (18). It is immediate from the definition of the state space D that $\xi(s, x) > 0$ for all $(s, x) \in D$. We are now prepared to consider the measure $P^c_{(s,x)}$ with creation and killing $c(s, x)$ given by

$$P^c_{(s,x)}[F] = P_{(s,x)}\left[\exp\left\{\int_s^b c(v, X_v)\, dv\right\} F\, 1_{\{b < T_s\}}\right]. \qquad (20)$$

In terms of the measure $P^c_{(s,x)}$, the function $\xi(s, x)$ introduced at (19) is nothing but

$$\xi(s, x) = P^c_{(s,x)}[1].$$

Under the integrability assumption (19), the renormalized measure $\bar{P}$ of $P^c_{(s,x)}$ can be defined as

$$\bar{P}[F] = \int_{D_a} d(dx)\, \frac{1}{\xi(a, x)}\, P^c_{(a,x)}[F] \qquad (21)$$

where $d(dx) = d(x)\, dx$, $d(x) > 0$, is any fixed probability measure on $D_a = \{x : (a, x) \in D\}$. The renormalized measure (21) is a typical example of the reference measure $\bar{P}$ which will be introduced in Section 3.

In order to involve the two observed, i. e. prescribed, marginal distributions, we introduce the set $\mathcal{E}(a, b)$ of all probability measures on $(\mathbb{R}^d \times \mathbb{R}^d, \sigma(\mathbb{R}^d \times \mathbb{R}^d))$ with the given pair (q_a, q_b) of probability measures on $\mathbb{R}^d$ as their marginal distributions. Particle dynamics from initial directly to final time is represented by a probability measure $\bar{p}$ on $\mathbb{R}^d \times \mathbb{R}^d$ given as

$$\bar{p}(A \times B) = \int_{D_a} d(dx)\, \frac{1}{\xi(a, x)}\, 1_A(x) P^c_{(a,x)}[1_B(X_b)]$$

for $A, B \in \sigma(\mathbb{R}^d)$. The assumption

$$\exists\, p \in \mathcal{E}(a, b)\colon H(p|\bar{p}) < \infty \qquad (22)$$

imposes a relation between dynamics and marginal constraints where the relative entropy $H(p|\bar{p})$ of p with respect to $\bar{p}$ is defined as

$$H(p|\bar{p}) = \begin{cases} \int \log\left(\frac{dp}{d\bar{p}}\right) dp, & \text{if } p \ll \bar{p}, \\ \infty, & \text{otherwise.} \end{cases} \qquad (23)$$

Condition (22) implies $p(\cdot \times \mathbb{R}^d) \ll \bar{p}(\cdot \times \mathbb{R}^d)$; hence the distribution q_a must be absolutely continuous with respect to the distribution $d(dx)$. Moreover, $p(\mathbb{R}^d \times \cdot) \ll \bar{p}(\mathbb{R}^d \times \cdot)$ yields $\bar{p}(\mathbb{R}^d \times \operatorname{supp} q_b) > 0$ which means that the process with creation and killing must reach the support of the distribution q_b at the final time b with positive probability. Now we can refer to

Lemma 2.1 ([14], [18], [38]). *Let us assume the conditions* (19) *and* (22). *Then there exists the up to multiplicative constants unique non-negative solution* $(\widehat{\varphi}_a, \varphi_b)$ *of the Schrödinger system*

$$\int_{D_a} \widehat{\varphi}_a(x)\, 1_A(x)\, dx\, P^c_{(a,x)}[\varphi_b(X_b)] = q_a(A)$$

$$\int_{D_a} \widehat{\varphi}_a(x)\, dx\, P^c_{(a,x)}[\varphi_b(X_b)\, 1_B(X_b)] = q_b(B)$$

with $P^c_{(a,x)}$ defined at (20). *It satisfies* $\log \widehat{\varphi}_a \in L^1(q_a)$ *and* $\log \varphi_b \in L^1(q_b)$.

For later reference let us formulate Csiszár's Theorems 2.1 and 2.2 in [14] as Lemma 2.2. A detailed investigation of the relative entropy basing on Csiszár's contributions is given in Chapters X and XI of Nagasawa [39].

Lemma 2.2 ([13]). *Let $M_1(\Omega)$ denote the set of probability measures on a measurable space Ω and let $\bar{P} \in M_1(\Omega)$ be fixed. If a subset A of $M_1(\Omega)$ is convex as well as variation closed and if it contains at least one element P with $H(P|\bar{P}) < \infty$, then there exists the unique I-projection $Q \in A$ of $\bar{P}$ on the set A defined by*

$$\inf_{P \in A} H(P|\bar{P}) = H(Q|\bar{P}) \tag{24}$$

where

$$H(P|\bar{P}) \geq H(P|Q) + H(Q|\bar{P}), \quad \forall P \in A. \tag{25}$$

From now on we will call the *I*-projection *Csiszár's projection* and the relation (25) *Csiszár's inequality.*

Remark 2.2. Csiszár [13], Kemperman [25] and Kullback [29] show that the variational distance is dominated by the relative entropy according to

$$|P - Q|_{\text{var } R} = \int_{\Omega} \left| \frac{dP}{dR} - \frac{dQ}{dR} \right| dR \leq \sqrt{2H(P|Q)} \tag{26}$$

where $P, Q, R \in M_1(\Omega)$ with $P \ll Q \ll R$.

Following Csiszár's Corollary 3.2 in [13], Lemma 2.1 is a consequence of Lemma 2.2 applied in an approximation procedure to marginal distributions from exponential families. In fact, $\mathcal{E}(a, b)$ is convex and variation closed. Let $q(dx\,dz)$ be the unique Csiszár projection of $\bar{p}(dx\,dz)$ on the set $\mathcal{E}(a, b)$ defined at (24) which is

$$H(q|\bar{p}) = \inf_{p \in \mathcal{E}(a,b)} H(p|\bar{p}) \tag{27}$$

where in this particular situation the projection q can be given as

$$\frac{dq}{d\bar{p}} = \frac{\xi(a, x)}{d(x)} \widehat{\varphi}_a(x)\, \varphi_b(z) \tag{28}$$

for $\log \widehat{\varphi}_a \in L^1(q_a)$ and $\log \varphi_b \in L^1(q_b)$. Multiplicative constants $c > 0$ and $1/c$ can obviously be chosen arbitrarily for $\widehat{\varphi}_a$ and φ_b, respectively, and even differently in regions separated by nodal surfaces. However, they will not have any influence in the crucial definition (30) given below. In (28) we observe the essential factorization in products of functions in x and z, respectively.

Proposition 2.1 ([38], [2]). *The Csiszár projection Q of the renormalized measure $\bar{P}$ at (21) on the set*

$$A_{a,b} = \{ P \in M_1(\Omega) : P \circ X_r^{-1} = q_r \ for \ r = a, b \} \subset M_1(\Omega) \qquad (29)$$

is represented in terms of the solution $(\widehat{\varphi}_a, \varphi_b)$ of the Schrödinger system in Lemma 2.2 and the measure $P_{(a,x)}^c$ with creation and killing given at (20) by

$$Q[F] = \int_{D_a} \widehat{\varphi}_a \, dx \, P_{(a,x)}^c [F \varphi_b(X_b)] \qquad (30)$$

for bounded and measurable functions F on Ω. $(X_t, a \le t \le b, Q)$ is called the Schrödinger process for the triplet (c, q_a, q_b).

The proof is an application of Lemma 2.2, where the stated uniqueness of the Csiszár projection is essential. Details are found in Lemma 2 of Nagasawa [38] and in Proposition 4.1 of Aebi–Nagasawa [2].

The Schrödinger process for a given triplet (c, q_a, q_b) is a Markov field by its definition at (30). However, (X_t, Q) is also a Markov process. In fact, following Section 5.4 of Nagasawa [39], we assume that the process $(X_t, P_{(s,x)})$ is right-continuous. It is easily verified that the space-time function

$$\varphi_t(x) = P_{(t,x)} \left[\exp \left\{ \int_t^b c(v, X_v) \, dv \right\} \varphi_b(X_b) 1_{\{b < T_t\}} \right]$$

$$= P_{(t,x)}^- \left[\exp \left\{ \int_t^b c^+(v, X_v) \, dv \right\} \varphi_b(X_b) \right]$$

in terms of $P_{(t,x)}^-$ defined at (18) is a solution of the integral equation

$$\varphi_t(x) = P_{(t,x)}^- [\varphi_b(X_b)] + \int_t^b du \, P_{(t,x)}^- [c^+(u, X_u) \, \varphi_u(X_u)].$$

Consequently, $\varphi_t(x)$ is an excessive function of the killed space-time process and hence $\varphi_t(X_t)$ is right-continuous in t according to Blumenthal–Getoor's Theorem 4.8 in [11]. Applying the Markov property of $P_{(s,x)}$, we obtain

$$Q[F] = \int_{D_a} \widehat{\varphi}_a(x) \, dx \, P_{(a,x)} \left[\exp \left\{ \int_a^t c(v, X_v) \, dv \right\} F \varphi_t(X_t) 1_{\{t < T_a\}} \right] \qquad (31)$$

for $t \in [a, b]$. Denoting the zero set of φ by $N = \{(t, x) : \varphi_t(x) = 0\}$, the right-continuity of $\varphi_t(X_t)$ and (31) imply

$$Q[(t, X_t)_{a \leq t \leq b} \text{ does not hit } N] = 1. \tag{32}$$

Now we are prepared to show the Markov property of Q. Let G be any non-negative $\sigma((v, X_v), a \leq v \leq s)$-measurable function and let f be any non-negative measurable function on $[a, b] \times \mathbb{R}^d$. Then the Markov property of $P_{(s,x)}$ yields

$$Q[G\,f(t, X_t)]$$
$$= \int_{D_a} \widehat{\varphi}_a(x)\,dx\,P_{(a,x)}\left[\exp\left\{\int_a^s c(v, X_v)\,dv\right\} G\varphi_s(X_s)\,q(s, X_s; t, f)\,1_{\{s < T_a\}}\right]$$
$$= Q[Gq(s, X_s; t, f)]$$

for $s < t$ and $D_a = \{x \in \mathbb{R}^d : (a, x) \in D\}$ where the transition probability $q(s, x; t, f)$ is well-defined according to (32) by

$$q(s, x; t, f) = P_{(s,x)}\left[\varphi_s(X_s)^{-1}\exp\left\{\int_s^t c(v, X_v)\,dv\right\}\varphi_t(X_t)f(t, X_t)\,1_{\{t < T_s\}}\right].$$

§3 Microscopical systems

In order to construct microscopical systems, we endow the space $M_1(\Omega)$ of probability measures on Ω with Csiszár's τ_0-topology: For $\varepsilon > 0$ and finite $\sigma(\Omega)$-measurable partitions $\mathcal{P}_k(\Omega) = \{\Omega_1, \ldots, \Omega_k\}$, $k \in \mathbb{N}$, of Ω, the basic neighborhoods of an element $P \in M_1(\Omega)$ are defined according to Csiszár [15] by

$$U_0(P, \varepsilon, \mathcal{P}_k) = \{R \in M_1(\Omega) : |R(\Omega_i) - P(\Omega_i)| < \varepsilon \text{ for } i = 1, \ldots, k$$
$$\text{and } R \ll P \text{ on } \sigma(\mathcal{P}_k)\} \tag{33}$$

where $\sigma(\mathcal{P}_k)$ is the (trivial) σ-field generated by the partition $\mathcal{P}_k$.

For a given pair of probability measures (q_a, q_b) on $\mathbb{R}^d$, the set $A_{a,b}$ defined at (29) represents the class of continuous stochastic processes on $\mathbb{R}^d$ with prescribed marginal distributions q_a and q_b at finite initial and final time a and b, respectively.

Let any fixed probability measure $\bar{P} \in M_1(\Omega)$ play the role of the renormalized measure with creation and killing at (21) as a reference measure. We assume that

$$\exists P \in A_{a,b} : H(P \,|\, \bar{P}) < \infty \tag{34}$$

which is in the situation of Section 2.2 equivalent to (22) as can be shown by means of desintegration with respect to initial and final states (X_a, X_b). In most of the interesting known situations and hence also in our investigations, the measure $\bar{P}$ itself is not an element of the set $A_{a,b}$.

Let $(\Omega^n, \bar{\mathbb{P}})$ be n independent copies of $(\Omega, \bar{P})$, i. e. $\bar{\mathbb{P}}$ is the n-product of the probability measure $\bar{P}$. Denoting the empirical distribution of $\omega = (\omega_1, \dots, \omega_n) \in \Omega^n$ by

$$L_n(\omega) = \frac{1}{n} \sum_{i=1}^{n} \delta_{\omega_i} \tag{35}$$

we have $L_n(\omega) \in M_1(\Omega)$ for each such ω. In order to give a conditioning we would like the empirical distribution L_n to satisfy the marginal conditions in question. However, $\{L_n \in A_{a,b}\}$ might be empty in general, since L_n takes only discrete values. Hence the set $A_{a,b}$ has to be approximated by enlarged sets. Let us take a sequence of finite measurable partitions $\mathcal{P}_k(\mathbb{R}^d) = \{B_1, \dots, B_k\}$ of $\mathbb{R}^d$, $k \in \mathbb{N}$, such that

$$\mathcal{P}_k(\mathbb{R}^d) \subset \mathcal{P}_{k+1}(\mathbb{R}^d) \quad \text{and} \quad \sigma(\mathcal{P}_k(\mathbb{R}^d)) \nearrow \sigma(\mathbb{R}^d) \quad \text{as } k \nearrow \infty. \tag{36}$$

In terms of the partitions $\mathcal{P}_k(\mathbb{R}^d)$, a family of subsets $A(\varepsilon, k)$ of $M_1(\Omega)$ for $\varepsilon > 0$ and $k \in \mathbb{N}$ is introduced as

$$A(\varepsilon, k) = \{\, P \in M_1(\Omega) : |P[X_r \in B_i] - q_r(B_i)| \le \varepsilon 2^{-k} \text{ for all } B_i \in \mathcal{P}_k(\mathbb{R}^d)$$
$$\text{and } P \circ X_r^{-1} \ll \bar{P} \circ X_r^{-1}$$
$$\text{on } \sigma(\mathcal{P}_k(\mathbb{R}^d)) \text{ for } r = a, b\,\}. \tag{37}$$

We can now define a conditioning of $\bar{\mathbb{P}}$ on the set $A(\varepsilon, k)$ by means of the empirical distribution L_n at (35) in terms of

$$\bar{\mathbb{P}}^{(n,k)}[\,\cdot\,] = \bar{\mathbb{P}}[\,\cdot\, | \, L_n \in A(\varepsilon, k)]. \tag{38}$$

The dependence of the set $A(\varepsilon, k)$ on $\varepsilon > 0$ is not substantial but technical. It will be needed in Section 3, in particular for Lemma 3.2 and Lemma 3.5. When the dependence on $\varepsilon > 0$ is not essential, we take $\varepsilon > 0$ arbitrary but fixed and do not indicate this explicitly.

In order to become more familiar with the conditional probability $\bar{\mathbb{P}}^{(n,k)}$, we notice that its marginals on Ω belong to the set $A(k)$. In fact, the convex combination of elements in $A(k)$

$$\int_{\Omega^n} L_n(\omega)(\cdot) \, d\bar{\mathbb{P}}[\omega \, | \, L_n(\omega) \in A(k)] = \int_{\Omega^n} \frac{1}{n} \sum_{i=1}^{n} \delta_{\omega_i}(\cdot) \, d\bar{\mathbb{P}}[\omega \, | \, L_n(\omega) \in A(k)]$$
$$= \frac{1}{n} \sum_{i=1}^{n} \bar{\mathbb{P}}[\omega_i \in \cdot \, | \, L_n \in A(k)] \tag{39}$$
$$= \bar{\mathbb{P}}[\omega_i \in \cdot \, | \, L_n \in A(k)]$$

is just the i-th marginal distribution of $\bar{\mathbb{P}}^{(n,k)}$ on Ω, $i = 1, \dots, d$, which is now denoted by $\bar{\mathbb{P}}_1^{(n,k)}$. In Lemma 3.3 we will see that $A(k)$ is convex, hence the expression (39) belongs to $A(k)$.

We are going to arrange a series of general lemmas which will be applied to the situation considered in Subsection 2.2. They will yield Theorem 3.1 as well as provide essential ingredients for the proof of propagation of chaos given in Section 4.

Lemma 3.1 ([2]). *Let $A(\varepsilon, k)$ be the subset of $M_1(\Omega)$ defined at (37). Then*

    (i) $A(\varepsilon, k)$ *decreases as* $k \nearrow \infty$,
    (ii) $A(\varepsilon, k)$ *decreases as* $\varepsilon \searrow 0$,
    (iii) $A_{a,b} = \bigcap_{k \in \mathbb{N}} A(\varepsilon, k)$ *for all* $\varepsilon > 0$,

where $A_{a,b}$ is the subset of $M_1(\Omega)$ defined at (29) satisfying the assumption (34).

For proofs of this and the two following lemmas we refer to Section 2 in Csiszár [15] as well as to Section 3 in Aebi–Nagasawa [2].

Lemma 3.2 ([2]). *Let $\mathring{A}(\varepsilon, k)$ denote the interior of the set $A(\varepsilon, k)$ with respect to the τ_0-topology given at (33). Then*

$$\mathring{A}(\varepsilon_0, k) \supset \bigcup_{0 < \varepsilon < \varepsilon_0} A(\varepsilon, k) \tag{40}$$

holds for every $\varepsilon_0 > 0$.

Lemma 3.3 ([15], [2]). *The set $A(\varepsilon, k)$ at (37) is completely convex in the sense of Csiszár [15], in particular convex, and variation closed.*

In view of simplicity, let us employ the notation $H(A \,|\, \bar{P}) = \inf_{P \in A} H(P \,|\, \bar{P})$ for subsets $A \subset M_1(\Omega)$.

Lemma 3.4 ("Sanov relation", [15], [2]). *Let $\varepsilon > 0$ and $k \in \mathbb{N}$. Assuming (34), it holds that*

$$-H(\mathring{A}(\varepsilon, k) \,|\, \bar{P}) \leq \liminf_{n \nearrow \infty} \frac{1}{n} \log \bar{\mathbb{P}}[L_n \in A(\varepsilon, k)]$$

$$\leq \limsup_{n \nearrow \infty} \frac{1}{n} \log \bar{\mathbb{P}}[L_n \in A(\varepsilon, k)] \tag{41}$$

$$\leq -H(A(\varepsilon, k) \,|\, \bar{P})$$

where $H(A(\varepsilon, k) \,|\, \bar{P})$ is attained by Csiszár's projection $P_{\varepsilon, k}$ of $\bar{P}$ on $A(\varepsilon, k)$, i.e.

$$H(A(\varepsilon, k) \,|\, \bar{P}) = H(P_{\varepsilon, k} \,|\, \bar{P}). \tag{42}$$

In case of the upper bound in (41), the refined estimation

$$\frac{1}{n} \log \bar{\mathbb{P}}[L_n \in A(\varepsilon, k)] \leq -\frac{1}{n} H(\bar{\mathbb{P}}^{(n,k)} \,|\, P_{\varepsilon,k}^n) - H(P_{\varepsilon, k} \,|\, \bar{P}) \tag{43}$$

holds for $n, k \in \mathbb{N}$. Suppose that $A(\varepsilon, k)$ possesses the Sanov property, i. e. (41) holds with equality, then (43) relates the speed of convergence in the Sanov property to the speed of convergence in entropy of the conditional probability $\bar{\mathbb{P}}_1^{(n,k)}$ to $P_{\varepsilon,k}$.

For a proof we refer to Csiszár's Theorem 1 and Lemma 4.1 in [15] as well as to Aebi–Nagasawa's Lemma 3.4 in [2] from where we also receive that

$$H(\bar{\mathbb{P}}_1^{(n,k)} \,|\, P_{\varepsilon,k}) \leq \frac{1}{n} H(\bar{\mathbb{P}}^{(n,k)} \,|\, P_{\varepsilon,k}^n).$$

Lemma 3.5 ([2]). *Under the assumption* (34) *it holds that:*

(i) *There exists Csiszár's projection Q of $\bar{P}$ on the subset $A_{a,b}$ of $M_1(\Omega)$ defined at* (29), *i. e.*

$$H(A_{a,b} \,|\, \bar{P}) = H(Q \,|\, \bar{P}). \tag{44}$$

(ii) *The sequence $(A(\varepsilon, k))_{k \in \mathbb{N}}$ defined at* (37) *approximating $A_{a,b}$ satisfies*

$$\lim_{k \nearrow \infty} H(\mathring{A}(\varepsilon, k) \,|\, \bar{P}) = \lim_{k \nearrow \infty} H(A(\varepsilon, k) \,|\, \bar{P}) = H(Q \,|\, \bar{P}), \quad \forall \varepsilon > 0. \tag{45}$$

(iii) *Csiszár's projection $P_{\varepsilon,k}$ of $\bar{P}$ on $A(\varepsilon, k)$ converges to Csiszár's projection Q of $\bar{P}$ on $A_{a,b}$ in entropy and hence weakly as k tends to infinity.*

Proof. Lemma 3.5 in Aebi–Nagasawa [2] shows how the previously stated lemmas are combined.

(i) According to (iii) in Lemma 3.1 and Lemma 3.3, the set $A_{a,b}$ is convex and variation closed. Hence there exists Csiszár's projection Q of $\bar{P}$ on $A_{a,b}$ by means of Lemma 2.2 under the entropy condition (34) imposed on the set $A_{a,b}$.

(ii) For $0 < \varepsilon_1 < \varepsilon_2$, Lemma 3.1 and Lemma 3.2 yield

$$H(\mathring{A}(\varepsilon_1, k) \,|\, \bar{P}) \geq H(A(\varepsilon_1, k) \,|\, \bar{P}) \geq H(\mathring{A}(\varepsilon_2, k) \,|\, \bar{P}) \geq H(A(\varepsilon_2, k) \,|\, \bar{P}). \tag{46}$$

By definition we have

$$H(Q \,|\, \bar{P}) = H(A_{a,b} \,|\, \bar{P}) \geq H(A(\varepsilon, k) \,|\, \bar{P}) = H(P_{\varepsilon,k} \,|\, \bar{P}), \quad \forall \varepsilon > 0. \tag{47}$$

According to the lower semi-continuity of the relative entropy, Lemma 3.1 implies

$$\lim_{k \nearrow \infty} H(A(\varepsilon, k) \,|\, \bar{P}) = H(A_{a,b} \,|\, \bar{P}) \quad \forall \varepsilon > 0. \tag{48}$$

As an immediate consequence

$$\lim_{k \nearrow \infty} H(\mathring{A}(\varepsilon, k) \,|\, \bar{P}) = H(A_{a,b} \,|\, \bar{P}), \quad \forall \varepsilon > 0$$

follows because of (46); hence (45) is shown.

(iii) By means of Csiszár's inequality (25)

$$H(Q \,|\, \bar{P}) - H(P_{\varepsilon,k} \,|\, \bar{P}) \geq H(Q \,|\, P_{\varepsilon,k}) \tag{49}$$

holds because of $Q \in A_{a,b} \subset A(\varepsilon, k)$. We notice that the left-hand side of (49) vanishes as k tends to infinity because of (47) and (48). Hence

$$\lim_{k \nearrow \infty} H(Q \,|\, P_{\varepsilon,k}) = 0, \quad \forall \varepsilon > 0 \tag{50}$$

follows and Remark 2.2 completes the proof. $\square$

The lemmas of this section applied to the situation introduced in Subsection 2.2 yield

Theorem 3.1 ([2]). *Let $\bar{P}$ be the renormalized measure of the measure $P_{(a,x)}^c$ with creation and killing $c(s,x)$ at (21) considered under the assumption (19). The Schrödinger process Q defined at (30) uniquely exists if (22) is provided. Then it holds that:*

(i) *The set $A_{a,b}$ at (29) possesses the* approximate Sanov property

$$\lim_{k \nearrow \infty} \lim_{n \nearrow \infty} \frac{1}{n} \log \bar{\mathbb{P}}[L_n \in A(k)] = - \inf_{P \in A_{a,b}} H(P|\bar{P})$$

$$= -H(Q|\bar{P})$$

where $A(k) \searrow A_{a,b}$ as $k \nearrow \infty$. In other words, the probability that the rare event $\{L_n \in A(k)\}$ occurs is given by

$$\bar{\mathbb{P}}[L_n \in A(k)] \approx e^{-nH(Q|\bar{P})} \quad \text{as } n \nearrow \infty \text{ and } k \nearrow \infty.$$

(ii) *The* asymptotical quasi-independence

$$\lim_{n \nearrow \infty} \frac{1}{n} H(\bar{\mathbb{P}}^{(n,k)} | P_{\varepsilon,k}^n) = 0 \qquad (51)$$

holds for the conditional probability $\bar{\mathbb{P}}^{(n,k)}$ defined at (38) with the Csiszár projection $P_{\varepsilon,k}$ as limiting distribution for $k \in \mathbb{N}$.

The conditional process $(\mathbb{X}_t = (X_1(t,\cdot),\dots,X_n(t,\cdot)), \bar{\mathbb{P}})$ on Ω^n in the second assertion of Theorem 3.1 is in general not Markovian. Hence we are now going to quote Lemma 4.2 of Aebi–Nagasawa [2] which will provide the (in a certain sense) unique Markovian modification $\mathbb{Q}^{(n,k)}$ of $\bar{\mathbb{P}}^{(n,k)}$.

Let $\mathbb{P}$ and $\mathbb{Q}$ be probability measures on the space of right-continuous paths with the time parameter t running in $[a,b]$. If $\mathbb{Q}$ is Markovian and the marginal distributions at time t of $\mathbb{P}$ and $\mathbb{Q}$ coincide for all $t \in [a,b]$, then $\mathbb{Q}$ is called a Markovian modification of $\mathbb{P}$.

Lemma 3.6 ([2]). *Let $\mathbb{P}$ and $\mathbb{P}_0$ be probability measures on the space of right-continuous (resp. continuous) paths such that*

$$H(\mathbb{P}|\mathbb{P}_0) < \infty$$

holds. If $\mathbb{P}_0$ is a Markov (resp. diffusion) process, then there exists a unique Markovian (resp. diffusion) modification $\mathbb{Q}$ of $\mathbb{P}$ satisfying

$$H(\mathbb{Q}|\mathbb{P}_0) \leq H(\mathbb{P}|\mathbb{P}_0).$$

§4 PROPAGATION OF CHAOS

Our situation of Schrödinger processes in Proposition 2.1 furnished with micro-scopical systems arising from (38) provides propagation of chaos. In fact, we claim

Theorem 4.1. *Let us assume the conditions* (19) *and* (22). *Then:*

(i) (see Aebi–Nagasawa [2]) *There exists a unique Markovian modification* $\mathbb{Q}^{(n,k)}$ *of* $\bar{\mathbb{P}}^{(n,k)}$ *satisfying*

$$H(\mathbb{Q}^{(n,k)} | P^n_{\varepsilon,k}) \le H(\bar{\mathbb{P}}^{(n,k)} | P^n_{\varepsilon,k}) \tag{53}$$

for $n, k \in \mathbb{N}$ *where* $P_{\varepsilon,k}$ *is the Csiszár projection given at* (42).

(ii) (see Nagasawa [39]) $(\mathbb{X}_t = (X_1(t,\cdot),\dots,X_n(t,\cdot)), \mathbb{Q}^{(n,k)})$ *is a system of in-teracting diffusion processes with the Markovian drift coefficient* $b^{(n,k)}(t,x)$, $x = (x_1,\dots,x_n) \in (\mathbb{R}^d)^n$, *describing the interaction in terms of*

$$b^{(n,k)}(t,x) = (b_i^{(n,k)}(t,x))_{i=1,\dots,n} \tag{54}$$

where $b_i^{(n,k)}$ *is the drift vector of the process* X_i *with respect to* $P^n_{\varepsilon,k}$.

(iii) *Propagation of chaos in entropy defined by* (1) *and* (2) *holds for the Mar-kovian modification* $((X_1,\dots,X_n), \mathbb{Q}^{(n,k)})$ *with the Schrödinger process* Q *as limiting distribution when* n *and* k *tend to infinity.*

In the proof of the third statement of Theorem 4.1, we will employ

Lemma 4.1. ([38]) *For a sequence* $\mathbb{P}^{(n)} \subset M_1(\Omega^n)$, $n \in \mathbb{N}$, *let* $\mathbb{P}_m^{(n)}$, $m \le n$, *denote the marginal distribution of* $\mathbb{P}^{(n)}$ *on* $\Omega^m = \Omega_1 \times \cdots \times \Omega_m$. *We assume that there exists* $P \in M_1(\Omega)$ *such that* $H(\mathbb{P}^{(n)} | P^n) < \infty$ *for* $n \in \mathbb{N}$. *If the marginal distributions of* $\mathbb{P}^{(n)}$ *on* $\Omega_{m\nu+1} \times \cdots \times \Omega_{m(\nu+1)}$ *for* $\nu = 1,2,\dots$ *equal* $\mathbb{P}_m^{(n)}$, *then the inequality*

$$H(\mathbb{P}_m^{(n)} | P^m) \le m \frac{n}{n-r} \frac{1}{n} H(\mathbb{P}^{(n)} | P^n)$$

holds where r *is a non-negative integer smaller than* m.

A proof due to E. Bolthausen can be found in Lemma 11.3 of Nagasawa [39].

Proof of Theorem 4.1. Assertions (i) and (ii) are discussed in Theorem 4.1 of Aebi–Nagasawa [2] and in Theorem 8.3 of Nagasawa [39], respectively. We notice that (i) follows by means of the identification $\mathbb{P} = \bar{\mathbb{P}}^{(n,k)}$ and $\mathbb{P}_0 = P^n_{\varepsilon,k}$ from Lemma 3.6 where its assumptions are satisfied because of (41) and (43).

Let us now prove assertion (iii) on propagation of chaos in the sense of Def-inition 1.1. This obviously requires to deal with an expression of asymptotical quasi-independence as introduced in (ii) of Theorem 3.1. In fact, it is the only limit property we have for use in order to verify the two properties (1) and (2). However, our situation based on equation (38) is particularly symmetrical in ω_i, $i = 1,\dots,n$. Not only since L_n is symmetrical in ω_i, but also because of the conditioning in the definition of $\bar{\mathbb{P}}^{(n,k)}$ at (38). Hence $\mathbb{Q}^{(n,k)}$ inherits the symmetry in ω_i, $i = 1, \ldots, n$, according to Lemma 3.6. As a consequence, property (1) for $m = 1$ implies property (2) of propagation of chaos. In fact, since modifications of stochastic processes have identical one-dimensional marginal distributions, relation (39) remains valid for the Markovian modification $\mathbb{Q}^{(n,k)}$, i. e.

$$\mathbb{Q}^{(n,k)}[L_n] = \frac{1}{n} \sum_{i=1}^{n} \mathbb{Q}^{(n,k)}[\delta_{\omega_i}] = \mathbb{Q}_1^{(n,k)}$$

holds for $n, k \in \mathbb{N}$.

In order to show (1) for arbitrary but fixed $m \in \mathbb{N}$, considerable effort is required. In view of an application of Lemma 4.1, we recall that $\bar{\mathbb{P}}^{(n,k)}$ and $P_{\varepsilon,k}^n$ are asymptotically quasi-independent, i. e. (51) holds according to (41) and (43). Moreover, the marginal distributions of $\bar{\mathbb{P}}^{(n,k)}$ and according to Lemma 3.6 of $\mathbb{Q}^{(n,k)}$ on $\Omega_{m\nu+1} \times \cdots \times \Omega_{m(\nu+1)}$ for $\nu = 0, 1, 2, \ldots$ coincide obviously because of the symmetry of $\bar{\mathbb{P}}^{(n,k)}$ in ω_i, $i = 1, \ldots, n$. By means of the identification $\mathbb{P}^{(n)} = \mathbb{Q}^{(n,k)}$ and $P = P_{\varepsilon,k}$, Lemma 4.1 yields

$$H(\mathbb{Q}_m^{(n,k)} \,|\, P_{\varepsilon,k}^m) \leq m \,\frac{n}{n-r}\, \frac{1}{n}\, H(\mathbb{Q}^{(n,k)} \,|\, P_{\varepsilon,k}^n) \tag{55}$$

for a non-negative integer r smaller than m. According to (53), the right-hand side of (55) vanishes as n tends to infinity because of (51).

By means of

$$\frac{dQ^m}{d\mathbb{Q}_m^{(n,k)}} = \frac{dQ^m}{dP_{\varepsilon,k}^m} \frac{dP_{\varepsilon,k}^m}{d\mathbb{Q}_m^{(n,k)}} \qquad \text{on supp } P_{\varepsilon,k}^m$$

we intend to handle the different positions of arguments in the relative entropy expressions (1) and (51). In fact, it follows

$$\begin{aligned} H(Q^m \,|\, \mathbb{Q}_m^{(n,k)}) &= \int \log\left(\frac{dQ^m}{d\mathbb{Q}_m^{(n,k)}}\right) dQ^m \\ &= \int \log\left(\frac{dQ^m}{dP_{\varepsilon,k}^m}\right) dQ^m + \int \log\left(\frac{dP_{\varepsilon,k}^m}{d\mathbb{Q}_m^{(n,k)}}\right) dQ^m \\ &= m\, H(Q \,|\, P_{\varepsilon,k}) - \int \log\left(\frac{d\mathbb{Q}_m^{(n,k)}}{dP_{\varepsilon,k}^m}\right) dQ^m \end{aligned} \tag{56}$$

which will be justified by verifying that

$$\int \log\left(\frac{d\mathbb{Q}_m^{(n,k)}}{dP_{\varepsilon,k}^m}\right) dQ^m \tag{57}$$

has the same limit as $H(\mathbb{Q}_m^{(n,k)} \,|\, P_{\varepsilon,k}^m)$ for n and k tending to infinity.

Because of $\bar{P} \notin A_{a,b}$, it follows from Lemma 3.1 that there exists $k_0 \in \mathbb{N}$ such that $\bar{P} \notin A(k)$ for $k \geq k_0$. This allows us to assume that $\{L_n \in A(k)\}$ is monotonously decreasing under $\bar{P}^n$ as n tends to infinity for each $k \geq k_0$. Obviously, any $B^{(m)} \subset \Omega^m$ depends exclusively on those first m ω_j's which loose their influence in $(1/n) \sum_{j=1}^n \delta_{\omega_j}$ as n tends to infinity. Hence we can get a kind of dominated asymptotical independence of $B^{(m)}$ and $\{L_n \in A(k)\}$ under $\bar{P}^n$. Actually it holds that

$$\bar{\mathbb{P}}[B^{(m)} \cap \{L_n \in A(k)\}] \leq \bar{\mathbb{P}}\left[B^{(m)} \cap \left\{\frac{1}{n-m} \sum_{j=m+1}^n \delta_{\omega_j} \in A(k)\right\}\right]$$

$$= \bar{\mathbb{P}}[B^{(m)}] \, \bar{\mathbb{P}}\left[\frac{1}{n-m} \sum_{j=m+1}^n \delta_{\omega_j} \in A(k)\right]$$

which yields

$$\bar{\mathbb{P}}[B^{(m)} \mid L_n \in A(k)] \leq p_{m,n} \, \bar{\mathbb{P}}[B^{(m)}] \tag{58}$$

for sufficiently large n with

$$p_{m,n} = \frac{\bar{\mathbb{P}}\left[\frac{1}{n-m} \sum_{j=m+1}^n \delta_{\omega_j} \in A(k)\right]}{\bar{\mathbb{P}}\left[\frac{1}{n} \sum_{j=1}^n \delta_{\omega_j} \in A(k)\right]}$$

where $p_{m,n} \geq 1$ and $p_{m,n} \to 1$ as $n \nearrow \infty$. As a Markovian modification according to Lemma 3.6, $\mathbb{Q}^{(n,k)}$ coincides with $\bar{\mathbb{P}}^{(n,k)} = \bar{\mathbb{P}}[\cdot \mid L_n \in A(k)]$ on events, say $B_{(t)}$, which depend only on finitely many points of time. We recall that these so-called cylinder sets generate the canonical σ-algebra on Ω and we notice that the event $\{L_n \in A(k)\}$ is such a cylinder set. As a consequence of (58),

$$\mathbb{Q}_m^{(n,k)}[B_{(t)}^{(m)}] = \bar{\mathbb{P}}^{(n,k)}[B_{(t)}^{(m)}] \leq p_{m,n} \bar{\mathbb{P}}[B_{(t)}^{(m)}]$$

holds for sufficiently large n, where $B_{(t)}^{(m)} \subset \Omega^m$ is an arbitrary cylinder set. Hence it follows that

$$\frac{d\mathbb{Q}_m^{(n,k)}}{dP_{\varepsilon,k}^m}\bigg|_{\mathrm{supp}\, P_{\varepsilon,k}^m} \leq p_{m,n} \frac{d\bar{P}^m}{dP_{\varepsilon,k}^m}\bigg|_{\mathrm{supp}\, P_{\varepsilon,k}^m} \qquad P_{\varepsilon,k}^m\text{-a. s. on } \Omega^m \tag{59}$$

where in view of its right-hand side,

$$\frac{dP_{\varepsilon,k}}{d\bar{P}}\bigg|_{\mathrm{supp}\, Q} = \frac{dP_{\varepsilon,k}}{dQ}\bigg|_{\mathrm{supp}\, Q} \frac{dQ}{d\bar{P}}\bigg|_{\mathrm{supp}\, Q}, \qquad Q\text{-a. s.}$$

leads to

$$\log\left(\frac{dP_{\varepsilon,k}}{d\bar{P}}\bigg|_{\mathrm{supp}\, Q}\right) = -\log\left(\frac{dQ}{dP_{\varepsilon,k}}\bigg|_{\mathrm{supp}\, Q}\right) + \log\left(\frac{dQ}{d\bar{P}}\bigg|_{\mathrm{supp}\, Q}\right), \qquad Q\text{-a. s.,}$$

which provides

$$\int \log\left(\frac{dP_{\varepsilon,k}}{d\bar{P}}\right) dQ = -H(Q\,|\,P_{\varepsilon,k}) + H(Q\,|\,\bar{P}). \tag{60}$$

We conclude from the definition of the conditional probability $\bar{\mathbb{P}}^{(n,k)}$ at (38) that

$$\frac{d\mathbb{Q}_m^{(n,k)}}{dQ^m} = \frac{d\bar{\mathbb{P}}[\cdot \cap \{L_n \in A(k)\}]}{\bar{\mathbb{P}}[L_n \in A(k)]\,dQ^m}, \qquad Q^m\text{-a. s.}$$

Employing Lemma 3.1, Lemma 2.2, and Theorem 3.1, the strong law of large numbers in case of ω with $\omega_i \in \mathrm{supp}\,Q$, $i \in \mathbb{N}$, provides $n(\omega,k) \in \mathbb{N}$ such that $L_n(\omega) \in A(k)$ for $n \geq n(\omega,k)$ where $k \in \mathbb{N}$. Consequently,

$$\frac{d\mathbb{Q}_m^{(n,k)}}{dQ^m}(\omega) = \begin{cases} \dfrac{1}{\bar{\mathbb{P}}[L_n \in A(k)]} \displaystyle\prod_{i=1}^{m} \frac{d\bar{P}}{dQ}(\omega_i) & \text{for } n \geq n(\omega,k), \\[4mm] 0 & \text{for } n < n(\omega,k), \end{cases} \tag{61}$$

follows. We are merely interested in the logarithm of (61) integrated with respect to the measure Q^m which charges all components $\omega_1,\dots,\omega_m$ equally and independently. The symmetry of L_n at (35) and of $A(k)$ at (37) in the components ω_i, $i = 1,\dots,n$, implies that the set $\{n \geq n(\omega,k)\}$ is closed under permutations of components of ω. Hence the logarithm of (61), investigated in view of an integration with respect to Q^m, may be replaced by

$$-\infty\,1_{\{n<n(\omega,k)\}}(\omega) + \left[m\log\left(\frac{d\bar{P}}{dQ}(\omega_1)\right) - \log\bar{\mathbb{P}}[L_n \in A(k)]\right]1_{\{n\geq n(\omega,k)\}}(\omega)$$

which performs a sequence of increasing functions as n tends to infinity. Hence we conclude by means of Beppo–Levi's theorem that

$$\int \log\left(\frac{d\mathbb{Q}_m^{(n,k)}}{dQ^m}\right) dQ^m \nearrow \quad \text{as } n \nearrow \infty. \tag{62}$$

According to (56), (57) is given by

$$\int \log\left(\frac{d\mathbb{Q}_m^{(n,k)}}{dP_{\varepsilon,k}^m}\right) dQ^m = \int \log\left(\frac{d\mathbb{Q}_m^{(n,k)}}{dQ^m}\right) dQ^m + mH(Q\,|\,P_{\varepsilon,k})$$

where the first term on the right-hand side is monotonously increasing as n increases because of (62) and the second term on the right-hand side is independent of n, it actually vanishes as k tends to infinity according to (50). Consequently,

$$\int \log\left(\frac{d\mathbb{Q}_m^{(n,k)}}{dP_{\varepsilon,k}^m}\right) dQ^m \nearrow \quad \text{as } n \nearrow \infty$$

and it is bounded according to (59) and (60) by

$$\log p_{m,n} + m(H(Q\,|\,P_{\varepsilon,k}) - H(Q\,|\,\bar{P})) \tag{63}$$

which is finite because of (58) and (50) as well as (44). Hence, (57) stays finite as n tends to infinity where k may also tend to infinity.

Seeking now for the limit of (57) as n and k tend to infinity, we refer to Remark 2.2. Two applications of (26) yield

$$|Q^m - P^m_{\varepsilon,k}|_{\mathrm{var}\,P^m_{\varepsilon,k}} \le \sqrt{2H(Q^m\,|\,P^m_{\varepsilon,k})} = \sqrt{2mH(Q\,|\,P_{\varepsilon,k})} \to 0 \tag{64}$$

as $k \nearrow \infty$ because of (50) and

$$|\mathbb{Q}_m^{(n,k)} - P^m_{\varepsilon,k}|_{\mathrm{var}\,P^m_{\varepsilon,k}} \le \sqrt{2H(\mathbb{Q}_m^{(n,k)}\,|\,P^m_{\varepsilon,k})} \to 0 \tag{65}$$

as $n \nearrow \infty$ and $k \nearrow \infty$ because of (51), (53), and (55). We denote

$$f_{k,n}^{(m)} = \log\left(\frac{d\mathbb{Q}_m^{(n,k)}}{dP^m_{\varepsilon,k}}\right)$$

and claim that

$$
\begin{aligned}
&\left|\int f_{k,n}^{(m)}\,dQ^m - \int f_{k,n}^{(m)}\,d\mathbb{Q}_m^{(n,k)}\right| \\
&\le \left|\int f_{k,n}^{(m)}\,dQ^m - \int_{\{|f_{k,n}^{(m)}|\le M\}} f_{k,n}^{(m)}\,dQ^m\right| \\
&\quad + \left|\int_{\{|f_{k,n}^{(m)}|\le M\}} f_{k,n}^{(m)}\,dQ^m - \int_{\{|f_{k,n}^{(m)}|\le M\}} f_{k,n}^{(m)}\,dP^m_{\varepsilon,k}\right| \\
&\quad + \left|\int_{\{|f_{k,n}^{(m)}|\le M\}} f_{k,n}^{(m)}\,dP^m_{\varepsilon,k} - \int_{\{|f_{k,n}^{(m)}|\le M\}} f_{k,n}^{(m)}\,d\mathbb{Q}_m^{(n,k)}\right| \\
&\quad + \left|\int_{\{|f_{k,n}^{(m)}|\le M\}} f_{k,n}^{(m)}\,d\mathbb{Q}_m^{(n,k)} - \int f_{k,n}^{(m)}\,d\mathbb{Q}_m^{(n,k)}\right| \to 0
\end{aligned}
\tag{66}
$$

as $n \nearrow \infty$ and $k \nearrow \infty$. In fact, the first and fourth term on the right-hand side of (66) become arbitrarily small for M large because of (63) and (55), respectively. The second term on the right-hand side of (66) vanishes as k tends to infinity because of (64) and the third term vanishes as n tends to infinity because of (65). In fact, (64) and (65) mean that

$$M \sup_{\substack{f\ \mathrm{measurable} \\ \|f\|_\infty = 1}} \left|\int f\,dQ^m - \int f\,dP^m_{\varepsilon,k}\right| \to 0 \quad \text{as } k \nearrow \infty$$

and

$$M \sup_{\substack{f \text{ measurable} \\ \|f\|_\infty = 1}} \left| \int f \, dP_{\varepsilon,k}^m - \int f \, d\mathbb{Q}_m^{(n,k)} \right| \to 0 \quad \text{as } n \nearrow \infty, \, k \nearrow \infty$$

hold, respectively.

Finally we obtain (1) through (56) by means of

$$|H(Q^m \,|\, \mathbb{Q}_m^{(n,k)})| \leq m|H(Q \,|\, P_{\varepsilon,k})|$$

$$+ \left| \int \log\left(\frac{d\mathbb{Q}_m^{(n,k)}}{dP_{\varepsilon,k}^m} \right) dQ^m - \int \log\left(\frac{d\mathbb{Q}_m^{(n,k)}}{dP_{\varepsilon,k}^m} \right) d\mathbb{Q}_m^{(n,k)} \right|$$

$$+ |H(\mathbb{Q}_m^{(n,k)} \,|\, P_{\varepsilon,k}^m)| \to 0$$

as $n \nearrow \infty$ and $k \nearrow \infty$. In fact, the first term on the right-hand side vanishes as k tends to infinity because of (50), the second term vanishes as n and k tend to infinity because of (66) and the third term vanishes according to (51), (52) and (55). □

Theorem 4.1 has some significance in quantum physics. Following the motivation briefly given in Remark 2.1, Nagasawa [37, 38] and Aebi [5] show that Schrödinger equations can be associated with Schrödinger processes and vice versa, so far as solutions exist and the corresponding Schrödinger processes can be constructed. Hence Theorem 4.1 formulated for propagation of chaos in the sense of Definition 1.1 yields

Corollary 4.1 ([38, 39], [2] modified). *Provided (19) and (22), the distribution of a "typical particle" under the law $\mathbb{Q}^{(n,k)}$ as n and k tend to infinity is determined by the Schrödinger process Q and hence by the corresponding Schrödinger equation. In other words, a Schrödinger equation is a "Boltzmann equation" for a system of interacting particles represented by a system of interacting diffusion processes $((X_1, \ldots, X_n), \mathbb{Q}^{(n,k)})$ as n and k tend to infinity.*

We have assigned "Boltzmann" to Schrödinger equations, in order to place them in the concept of propagation of chaos. Moreover, a Schrödinger process Q can be considered as a Gibbs state of the distribution $\mathbb{Q}^{(n,k)}$ representing a microscopical system as n and k tend to infinity. In fact, Q determines the rate function of the large deviation principle called approximate Sanov property at Theorem 3.1. Hence, Q is the "most probable" diffusion process under the given circumstances and consequently, it is the limiting distribution claimed by the fundamental hypothesis of statistical mechanics, cf. e. g. Lanford [31].

Remark 4.1. We have formulated our results for diffusion processes on $\mathbb{R}^d$, but the continuity of paths is not essential and can be avoided. Let S be a Polish space and let $((t, X_t) \in [a, b] \times S, P_{(s,x)})$ be a (strong) Markov process of right-continuous paths with left limits. Then all Theorems and Lemmas remain valid for this process, except the statements on drift coefficients.

Remark 4.2. We experienced various difficulties caused by the non-symmetrical relative entropy-"distance". However, Theorem 4.1 provides a comparably strong result in terms of convergence in relative entropy. In fact, following Definition 1.1, the crucial relation (26) immediately implies that propagation of chaos in entropy yields generally propagation of chaos in variation and propagation of chaos in variation yields propagation of chaos in weak convergence.

References

1. R. Aebi, *N M-transformed α-Diffusion with Singular Drift*, Doktorarbeit, Universität Zürich, 1989.

2. R. Aebi and M. Nagasawa, *Large deviations and the propagation of chaos for Schrödinger processes*, Probab. Theory Relat. Fields **94** (1992), 53–68.

3. R. Aebi, *Itô's formula for non-smooth functions*, Publ. RIMS, vol. 28, Kyoto Univ., 1992, pp. 595–602.

4. R. Aebi, *A solution to Schrödinger's problem of non-linear integral equations*, Techn. Bericht No. 35 des IMSV, Univ. Bern, 1992.

5. R. Aebi, *Diffusions with singular drift related to wave functions*, Probab. Theory Relat. Fields **96** (1993), 107–121.

6. R. Aebi, *Watching clouds – prediction from past and future*, Habilitationsschrift, Universität Bern, 1993.

7. R. Aebi, *Schrödinger's view of diffusions*, Techn. Bericht No. 36 des IMSV, Univ. Bern, 1993.

8. R. Aebi, *Diffusions in time-symmetrical representation*, Techn. Bericht No. 37 des IMSV, Univ. Bern, 1993.

9. S. Bernstein, *Sur les liaisons entre les grandeurs aléatoires*, Verhandlungen des Internat. Math. Kongresses Zürich, vol. 1, 1932, pp. 288–309.

10. A. Beurling, *An automorphism of product measures*, Ann. Math **72,** no. 1 (1960), 189–200.

11. R. M. Blumenthal and R. K. Getoor, *Markov processes and potential theory*, Academic Press, New York and London, 1968.

12. L. Boltzmann, *Vorlesungen über Gastheorie*, J. A. Barth Verlag, Leipzig, 1896.

13. L. Boltzmann, *Über die sogenannte H-Kurve*, Ges. Abh. III, Nr. 128, Math. Ann. **50** (1898), 325.

14. I. Csiszár, *I-Divergence geometry of probability distributions and minimization problems*, Ann. Probab. **3** (1975), 146–158.

15. I. Csiszár, *Sanov property, generalized I-projection and a conditional limit theorem*, Ann. Probab. **12** (1984), 768–793.

16. D. A. Dawson and J. Gärtner, *Large deviations, free energy functional and quasi-potential for a mean field model of interacting diffusions*, Memoirs of the AMS, vol. 78, no. 398, 1989.

17. D. Dawson, L. Gorostiza, and A. Wakolbinger, *Schrödinger processes and large deviations*, J. Math. Phys. **31, 10** (1990), 2385–2388.

18. H. Föllmer, *Random fields and diffusion processes*, Ecole d'Eté de Saint Flour XV-XVII (1985–87), Lecture Notes Math., vol. 1362, Springer-Verlag, 1988.

19. R. Fortet, *Résolution d'un système d'equation de M. Schrödinger*, J. Math. Pures et Appl. **9** (1940), 83–95.

20. E. Gutkin, and M. Kac, *Propagation of chaos and Burger's equation*, SIAM J. Appl. Math. **43,** no. 4 (1983), 971–980.

21. K. Itô, *On a formula concerning stochastic differentials*, Nagoya Math. J. **3** (1951), 55–65.

22. K. Itô, *Lectures on stochastic processes*, Tata Institute, Bombay, 1961.

23. B. Jamison, *Reciprocal processes*, Z. Wahrsch. verw. Geb. **30** (1974), 65–86.

24. B. Jamison, *The Markov processes of Schrödinger*, Z. Wahrsch. verw. Geb. **32** (1975), 323–331.
25. J. H. B. Kemperman, *On the optimum rate of transmitting information*, Probability and Information Theory, Lecture Notes in Mathematics, Springer-Verlag, Berlin, 1967, pp. 126–169.
26. A. I. Khinchin, *Mathematical foundations of information theory*, Dover Publications, New York, 1957.
27. A. Kolmogorov, *Analytische Methoden in Wahrscheinlichkeitsrechnung*, Math. Ann. **104** (1931), 415–448.
28. A. Kolmogorov, *Zur Umkehrbarkeit der Statistischen Naturgesetze*, Math. Ann. **113** (1937), 766–772.
29. S. Kullback, *A lower bound for discrimination information in terms of variation*, IEEE Trans. Information Theory **IT-13** (1967), 126–127.
30. S. Kusuoka and Y. Tamura, *Gibbs measures for mean field potentials*, J. Fac. Sci. Univ. Tokyo, Sec. IA Math. **31** (1984), 223–245.
31. O. E. Lanford, *Entropy and equilibrium states in classical statistical mechanics*, In: Statistical Mechanics and Mathematical Problems, Lecture Notes in Phys. (A. Lenard, ed.), vol. 20, Springer-Verlag, Berlin, 1973, pp. 1–113.
32. H. P. McKean, *A class of Markov processes associated with non-linear parabolic equations*, Proc. Natl. Acad. Sci. **56** (1966), 1907–1911.
33. H. P. McKean, *Propagation of chaos for a class of non-linear parabolic equations*, Lecture Series in Differential Equations, Catholic Univ., 1967, pp. 41–57.
34. M. Nagasawa and H. Tanaka, *Propagation of chaos for diffusing particles of two types with singular mean field interaction*, Probab. Theory Relat. Fields **71** (1986), 69–83.
35. M. Nagasawa and H. Tanaka, *Diffusion with interactions and collisions between coloured particles and the propagation of chaos*, Probab. Theory Relat. Fields **74** (1987), 161–198.
36. M. Nagasawa and H. Tanaka, *A proof of the propagation of chaos for diffusion processes with drift coefficients not of average form*, Tokyo J. Math. **10** (1987), 403–418.
37. M. Nagasawa, *Transformations of diffusion and Schrödinger processes*, Probab. Theory Relat. Fields **82** (1989), 109–136.
38. M. Nagasawa, *Can the Schrödinger equation be a Boltzmann equation?* Proceedings of the conference on diffusion processes and related problems in analysis, Northwestern Univ. (1989) (M. Pinsky, ed.), Birkhäuser Boston Inc., 1990.
39. M. Nagasawa, *Schrödinger equations and diffusion theory*, Monographs in Mathematics, vol. 86, Birkhäuser Verlag, Basel, 1993.
40. M. Nagasawa, *New Mathematical Foundations of Quantum Mechanics*, Lecture Notes, University of Zürich, 1993.
41. E. Nelson, *Dynamical Theories of Brownian Motion*, Princeton University Press, 1967.
42. K. Ölschläger, *Many-particle Systems and the Continuum Description of Their Dynamics*, Habilitation, Univ. Heidelberg, 1989.
43. E. Schrödinger, *Über die Umkehrung der Naturgesetze*, Sitzungsberichte der Preußischen Akademie der Wissenschaften, physikalisch-mathematische Klasse, 1931, pp. 144–153.
44. E. Schrödinger, *Sur la théorie relativiste de l'électron et l'interprétation de la mécanique quantique*, Ann. Inst. Henri Poincaré, vol. II, Paris, 1932, pp. 269–319.
45. A.-S. Sznitman, *Topics in propagation of chaos*, Ecole d'Eté de Probabilités de Saint Flour, 1989.
46. H. Tanaka, *Limit theorems for certain diffusion processes with interaction*, In: Stochastic Analysis (K. Itô, ed.), Kinokuniya, Tokyo, 1984, pp. 469–488.

ROBERT AEBI, IMSV, UNIVERSITY OF BERNE, CH-3012 BERNE, SWITZERLAND

Progress in Probability, Vol. 36
© 1995 Birkhäuser Verlag Basel/Switzerland

A REMARK ON STOCHASTIC DYNAMICS
ON THE INFINITE-DIMENSIONAL TORUS

S. ALBEVERIO, YU. G. KONDRATIEV, AND M. RÖCKNER

ABSTRACT. We prove the uniqueness of the stochastic dynamics associated with Gibbs measures on the infinite dimensional torus.

§1 INTRODUCTION

In recent years there has been a growing interest in the investigation of lattice systems when the spin space of every particle (single spin space) is a compact Riemannian manifold (see e. g. [1], [12], [20], [21] and references therein). Equilibrium states of such systems are described by Gibbs measures on infinite products of manifolds. The stochastic dynamics corresponding to these states is given by a semigroup. The generator of this semigroup coincides on smooth cylinder functions with the Dirichlet operator associated with a Gibbs measure. An important question is about the uniqueness of this dynamics. In the case of the L^2-dynamics the uniqueness corresponds to essential self-adjointness on a natural minimal domain for the Dirichlet operator of the Gibbs measure considered. We also refer the reader to [8], [9] and [18], [19] for details concerning uniqueness in the sense of martingale problems for the associated diffusions. In the present paper we solve this question in the simplest case when the role of the single spin space is played by the unit circle. This model is known also under the name of "stochastic X-Y-model" ([14], [16]). Our method in the present paper is purely analytic. The corresponding problem for the case where the single spin space is the real line has been treated by a different method, which combines analytic and probabilistic tools, in [2]. For an extension of the method of the present paper to the case of an arbitrary compact Riemannian manifold replacing the circle see [5].

§2 SETTING AND MAIN RESULT

Let us introduce some notations and objects we will be working with.

The lattice used in our models will be the d-dimensional square lattice $\mathbb{Z}^d$ for some $d \in \mathbb{Z}_+$. For $k \in \mathbb{Z}^d$ we shall use the Euclidean norm $|k| = \|k\|_{\mathbb{R}^d}$ generated

1991 *Mathematics Subject Classification.* 32K22; 60H15; 47B25; 60J45.

Key words and phrases. Dirichlet forms, Dirichlet operators, essential self-adjointness, Gibbs measures, stochastic dynamics.

by the natural embedding $\mathbb{Z}^d \subset \mathbb{R}^d$. Given $\Lambda \subset \mathbb{Z}^d$, let $\Lambda^c = \mathbb{Z}^d \setminus \Lambda$ denote the complement of Λ, $|\Lambda|$ denote the cardinality of Λ and $\Lambda + k$ denote the translate $\{j + k \mid j \in \Lambda\}$ of Λ by $k \in \mathbb{Z}^d$.

The single spin space for our models will be the one-dimensional unit circle T in the plane. We identify the circle with $[0, 2\pi]$, where 0 and 2π are considered to be the same point. We use the normalized Lebesgue measure $\frac{1}{2\pi} dx = d\mu_0(x)$ on $[0, 2\pi]$.

The configuration space of the models we consider is the infinite-dimensional torus

$$\mathcal{X} = T^{\mathbb{Z}^d} \ni x = (x_k)_{k \in \mathbb{Z}^d}$$

endowed with the product topology. Given $\Lambda \subset \mathbb{Z}^d$,

$$\mathcal{X} \ni x \mapsto x_\Lambda = (x_k)_{k \in \Lambda} \in T^\Lambda$$

denotes the natural projection taking $\mathcal{X}$ onto T^Λ. For any $p \in \mathbb{Z}_+ \cup \{\infty\}$ the symbol $C_\Lambda^p(\mathcal{X})$ denotes the set of functions on $\mathcal{X}$ of the form

$$\mathcal{X} \ni x \mapsto f(x_\Lambda) \in \mathbb{R},$$

where f runs over the set $C^p(T^\Lambda)$ of continuously differentiable (up to the order p) real-valued functions on T^Λ. For $p = 0$ we write $C_\Lambda^0(\mathcal{X}) = C_\Lambda(\mathcal{X})$. We shall also use the set of finitely based functions

$$\mathcal{F}C^p(\mathcal{X}) = \bigcup_{\substack{\Lambda \subset \mathbb{Z}^d \\ |\Lambda| < \infty}} C_\Lambda^p(\mathcal{X})$$

for $p \in \mathbb{Z}_+ \cup \{\infty\}$.

For the construction of a stochastic dynamics on $\mathcal{X}$ we shall use so-called Gibbs measures on the Borel σ-algebra $\mathcal{B}(\mathcal{X})$. For simplicity of notations below, we restrict ourselves to the case of one- and two-particle interactions. But all our results also hold for any translation invariant interactions of finite range.

Let two fixed real-valued functions

$$V \in C^3(T), \quad W \in C^3(T^2), \quad W(x, y) = W(y, x), \quad (x, y) \in T^2,$$

be given. V and W are called one- and two-particle potential respectively. For a given volume $\Lambda \subset \mathbb{Z}^d$ with $|\Lambda| < \infty$ we introduce the energy of the interaction in Λ with fixed boundary condition $\xi \in \mathcal{X}$ in the following way:

$$U_\Lambda(x_\Lambda|\xi) = \sum_{k \in \Lambda} V(x_k) + \sum_{\langle k,j \rangle \subset \Lambda} W(x_k, x_j) + \sum_{\langle k,j \rangle : k \in \Lambda, j \in \Lambda^c} W(x_k, \xi_j), \quad (1)$$

where $\langle k, j \rangle$ denotes pairs of nearest neighbours $k, j \in \mathbb{Z}^d$, i.e., $|k - j| = 1$. We define the corresponding Gibbs measure in the volume Λ with fixed boundary condition $\xi \in \mathcal{X}$ as a measure on $\mathcal{B}(T^\Lambda)$ of the following form

$$d\mu_\Lambda(x_\Lambda | \xi) = \frac{1}{Z_\Lambda(\xi)} e^{-U_\Lambda(x_\Lambda | \xi)} \, d\mu_0(x_\Lambda), \tag{2}$$

where

$$d\mu_0(x_\Lambda) = \bigotimes_{k \in \Lambda} d\mu_0(x_k)$$

and

$$Z_\Lambda(\xi) = \int_{T^\Lambda} e^{-U_\Lambda(x_\Lambda | \xi)} \, d\mu_0(x_\Lambda).$$

Due to the assumptions about the potentials V and W, these measures are well-defined for any finite volume Λ and all boundary conditions $\xi \in \mathcal{X}$.

A probability measure μ on $\mathcal{B}(\mathcal{X})$ is called a Gibbs measure (for the given potentials V and W) if for any $\Lambda \subset \mathbb{Z}^d$, with $|\Lambda| < \infty$ and all $\xi \in \mathcal{X}$ the conditional measure

$$\mu(\cdot \,|\, x_{\Lambda^c} = \xi_{\Lambda^c})$$

coincides with $\mu_\Lambda(\cdot \,|\, \xi)$ on $\mathcal{B}(T^\Lambda)$. We will denote the set of all such Gibbs measures by $\mathcal{G}$. It is well-known that in the situation considered we have $\mathcal{G} \neq \varnothing$ (see e.g. [17]).

Let $\mu \in \mathcal{G}$ be fixed. The (classical) (real-valued) pre-Dirichlet form which corresponds to μ is defined for $u, v \in \mathcal{F}C^1(\mathcal{X})$ as

$$\mathcal{E}_\mu(u, v) = \int_{\mathcal{X}} \sum_{k \in \mathbb{Z}^d} \frac{\partial u}{\partial x_k} \frac{\partial v}{\partial x_k} \, d\mu. \tag{3}$$

Let us introduce for any $k \in \mathbb{Z}^d$ the operator ∇_k in $L^2(\mathcal{X}, \mu)$ on the domain $\mathcal{F}C^\infty(\mathcal{X})$ by the formula

$$L^2(\mathcal{X}, \mu) \supset \mathcal{F}C^\infty(\mathcal{X}) \ni f \mapsto \nabla_k f = \frac{\partial f}{\partial x_k} \in \mathcal{F}C^\infty(\mathcal{X}).$$

Then the definition of the Gibbs measure μ gives the following representation for the adjoint operator ∇_k^*:

$$(\nabla_k^* f)(x) = -(\nabla_k f)(x) - \beta_k(x) f(x), \quad f \in \mathcal{F}C^\infty(\mathcal{X}), \tag{4}$$

with

$$\beta_k(x) = -\sum_{j:|j-k|=1} \frac{\partial W(x_k, x_j)}{\partial x_k} - V'(x_k). \tag{5}$$

This is easily verified (see e. g. [1]). We introduce the differential operator H_μ on the domain $\mathcal{D}(H_\mu) = \mathcal{F}C^2(\mathcal{X})$ in $L^2(\mathcal{X}, \mu)$ by the formula

$$(H_\mu f)(x) = -\sum_{k\in\mathbb{Z}^d} \frac{\partial^2 f(x)}{\partial x_k^2} - \sum_{k\in\mathbb{Z}^d} \beta_k(x)\frac{\partial f(x)}{\partial x_k}, \quad f \in \mathcal{F}C^2(\mathcal{X}), \; x \in \mathcal{X},$$

or, in short notations,

$$(H_\mu f)(x) = -\Delta f(x) - \langle \beta(x), \nabla f(x)\rangle,$$

where

$$\beta(\cdot) = (\beta_k(\cdot))_{k\in\mathbb{Z}^d} : \mathcal{X} \to \mathbb{R}^{\mathbb{Z}^d},$$

$$\nabla f(\cdot) = (\nabla_k f(\cdot))_{k\in\mathbb{Z}^d} : \mathcal{X} \to \mathbb{R}_0^{\mathbb{Z}^d}$$

and $\langle \cdot, \cdot \rangle$ denotes the pairing between the space $\mathbb{R}^{\mathbb{Z}^d}$ of all real-valued sequences on $\mathbb{Z}^d$ and the space $\mathbb{R}_0^{\mathbb{Z}^d}$ of finite support sequences on $\mathbb{Z}^d$ which is given by the scalar product in the Hilbert space

$$l_2(\mathbb{Z}^d) = \left\{ h \in \mathbb{R}^{\mathbb{Z}^d} \; \middle| \; \langle h, h\rangle = \sum_{k\in\mathbb{Z}^d} h_k^2 < \infty \right\}.$$

By the definition of $\mathcal{E}_\mu$ and (4) we have

$$(H_\mu u, v)_{L^2(\mu)} = \mathcal{E}_\mu(u, v), \quad u, v \in \mathcal{F}C^2(\mathcal{X}),$$

i. e., the operator H_μ is the pre-Dirichlet operator which is associated with the pre-Dirichlet form $\mathcal{E}_\mu$. In particular, the pre-Dirichlet form $\mathcal{E}_\mu$ with domain $\mathcal{F}C^2(\mathcal{X})$ is closable on $L^2(\mathcal{X}, \mu)$. Let us remark that the operator H_μ is a symmetric operator in $L^2(\mathcal{X}, \mu)$ for any $\mu \in \mathcal{G}$ and moreover, such symmetry is a characterization of Gibbs measures, see e. g. [1].

In many applications it is an important question whether one has essential self-adjointness for the Dirichlet operator H_μ, i. e., self-adjointness of the closure $\tilde{H}_\mu$ of H_μ in $L^2(\mathcal{X}, \mu)$. The reason is that in general there are many lower bounded self-adjoint extensions H_μ' of H_μ in $L^2(\mathcal{X}, \mu)$ which therefore define a stochastic dynamics via the symmetric semigroups

$$T_t := e^{-tH_\mu'}, \quad t \geq 0,$$

generated by them. There always exists one such extension called the *Friedrichs extension* which is the operator corresponding to the closure of the pre-Dirichlet form (3) (cf. [6], [13]). If H_μ is essentially self-adjoint there is hence only one such semigroup, consequently only one such dynamics associated with the Gibbs measure μ.

Theorem 1. *For any potentials $V \in C^3(T)$ and $W \in C^3(T^2)$, and for any Gibbs measure $\mu \in \mathcal{G}$, the pre-Dirichlet operator H_μ defined on $\mathcal{F}C^2(\mathcal{X})$ is an essentially self-adjoint operator in $L^2(\mathcal{X}, \mu)$.*

Remark. It is easy to deduce from Theorem 1 that also H_μ defined on $\mathcal{F}C^\infty(\mathcal{X})$ is an essentially self-adjoint operator on $L^2(\mathcal{X}, \mu)$ (see Lemma 6 in [2]).

§3 PROOF OF THEOREM 1

Our proof is a modification of the one which was used in the case of a linear spin space in [2] (see also [3], [4]).

Let us take a sequence $\{\Lambda_n\}_{n\in\mathbb{N}}$ of volumes $\Lambda_n \subset \mathbb{Z}^d$ with $|\Lambda_n| < \infty$ such that $\Lambda_n \subset \Lambda_{n+1}$ for all $n \in \mathbb{N}$ and $\Lambda_n \to \mathbb{Z}^d$ as $n \to \infty$ in van Hove's sense, i. e., for any $\Lambda \subset \mathbb{Z}^d$ with $|\Lambda| < \infty$, there exists $n \in \mathbb{N}$ such that $\Lambda_n \supset \Lambda$. For any $n \in \mathbb{N}$ we define a differential operator H_n on the domain $\mathcal{F}C^2(\mathcal{X}) \subset L^2(\mathcal{X}, \mu)$ by the formula

$$H_n u = -\Delta u - \langle \beta^{\Lambda_n}, \nabla u \rangle, \tag{6}$$

where the vector field $\beta^{\Lambda_n} = \{\beta_k^{\Lambda_n} \mid k \in \mathbb{Z}^d\}$ is given by

$$\beta_k^{\Lambda_n}(x) = \begin{cases} \beta_k(x) & \text{for } k \in \Lambda_n, \\ 0 & \text{for } k \in \Lambda_n^c. \end{cases}$$

We shall use the following general parabolic criterium of essential self-adjointness [10, Ch. 2], [11, Ch. 5].

Let us consider the following Cauchy problems

$$\frac{d}{dt} u_n(t) + H_n u_n(t) = 0, \quad t \in [0, 1],$$
$$u_n(0) = f, \tag{7}$$

where $f \in \mathcal{F}C^\infty(\mathcal{X})$ is arbitrary. If we can prove the existence of strong solutions u_n for (7) such that

   (i) $u_n : [0, 1] \to L^2(\mathcal{X}, \mu)$,
   (ii) $u_n(t) \in \mathcal{D}(H_\mu)$ for all $n \in \mathbb{N}$ and $t \in [0, 1]$,
   (iii) $\int_0^1 \|(H_\mu - H_n)u_n(t)\|_{L^2(\mu)}\, dt \to 0$ as $n \to \infty$,

then the operator H_μ is essentially self-adjoint.

First of all we note that the problem of existence of strong solutions for (7) reduces to a finite-dimensional one. In fact, for any fixed $f \in \mathcal{F}C^\infty(\mathcal{X})$ we have $f \in C_{\Lambda_0}^\infty(\mathcal{X})$ for some $\Lambda_0 \subset \mathbb{Z}^d$ with $|\Lambda_0| < \infty$. Let us take $n_0 \in \mathbb{N}$ such that $\Lambda_0 \subset \Lambda_{n_0}$. Due to (5) the components of β^{Λ_n} are depending only on variables $x_k \in T$ with $k \in \bar{\Lambda}_n := \{k \in \mathbb{Z}^d \mid d(k, \Lambda_n) \leq 1\}$, where $d(k, \Lambda_n) := \min\{|k - j| : j \in \Lambda_n\}$. Then for any $n \geq n_0$ the solution of (7) coincides with the solution of the following Cauchy problem:

$$\frac{\partial}{\partial t} u_n(t, x_{\bar{\Lambda}_n}) = \Delta_{\bar{\Lambda}_n} u_n(t, x_{\bar{\Lambda}_n}) + \langle \beta^{\Lambda_n}(x_{\bar{\Lambda}_n}), \nabla_{\bar{\Lambda}_n} u_n(t, x_{\bar{\Lambda}_n}) \rangle,$$
$$u_n(0, x_{\bar{\Lambda}_n}) = f(x_{\Lambda_0}), \tag{8}$$

where

$$\Delta_{\bar{\Lambda}_n} = \sum_{k\in\bar{\Lambda}_n} \frac{\partial^2}{\partial x_k^2}, \quad \nabla_{\bar{\Lambda}_n}(\cdot) = (\nabla_k(\cdot))_{k\in\bar{\Lambda}_n}.$$

It is well-known that under our conditions on V and W the solution of (8) exists and, moreover, $u_n(t, \cdot) \in C^2_{\bar{\Lambda}_n}(\mathcal{X})$, and it is a C^1-function with respect to t, see e. g. [15] . Hence (ii) is satiesfied.

Let us now check the last condition (iii) of the parabolic criterium. To this end we take $\{ p_k := (1 + |k|)^{-\alpha} \,|\, k \in \mathbb{Z}^d \}$, where $\alpha > 1$ is such that $\sum_{k \in \mathbb{Z}^d} p_k < \infty$. We introduce the Hilbert space $\mathcal{H}_+ \subset l_2(\mathbb{Z}^d)$ as

$$\mathcal{H}_+ = \left\{ h \in l_2(\mathbb{Z}^d) \,\Big|\, |h|^2_+ := \sum_{k \in \mathbb{Z}^d} h_k^2 p_k^{-1} < \infty \right\}$$

and its dual space $\mathcal{H}_- \supset l_2(\mathbb{Z}^d)$

$$\mathcal{H}_- = \left\{ h \in \mathbb{R}^{\mathbb{Z}^d} \,\Big|\, |h|^2_- := \sum_{k \in \mathbb{Z}^d} h_k^2 p_k < \infty \right\}.$$

Using (6) we have

$$(H_\mu - H_n)u_n(t, x) = - \sum_{k \in \Lambda_n^c} \beta_k(x) \nabla_k u_n(t, x).$$

Therefore

$$\begin{aligned}
&\|(H_\mu - H_n)u_n(t)\|^2_{L^2(\mu)} \\
&\qquad \le \left\| \left\{ \sum_{k \in \Lambda_n^c} |\beta_k(\cdot)|^2 p_k \right\}^{1/2} \left\{ \sum_{k \in \Lambda_n^c} |\nabla_k u_n(t)|^2 p_k^{-1} \right\}^{1/2} \right\|^2_{L^2(\mu)} \\
&\qquad \le C_\beta \sup_{x \in \mathcal{X}} \left\{ \sum_{k \in \mathbb{Z}^d} |\nabla_k u_n(t, x)|^2 p_k^{-1} \right\} \sum_{k \in \Lambda_n^c} p_k \\
&\qquad = C_\beta \sup_{x \in \mathcal{X}} |\nabla u_n(t, x)|^2_+ \sum_{k \in \Lambda_n^c} p_k,
\end{aligned} \tag{9}$$

where

$$C_\beta := \sup_{x \in \mathcal{X}} |\beta_k(x)|$$

and the latter expression does not depend on $k \in \mathbb{Z}^d$ (see (5)). Now we need to obtain a uniform estimate for $|\nabla u_n(t, x)|_+$.

Let us set $v_{n,j}(t, x) = \nabla_j u_n(t, x)$ and introduce a vector field v_n by defining $v_n(t, x) = \nabla u_n(t, x) = (v_{n,j}(t, x))_{j \in \mathbb{Z}^d}$. We have $v_{n,j}(t, x) \equiv 0$ for any $j \in \bar{\Lambda}_n^c$, so

$$v_n(t, \cdot) : \mathcal{X} \to \mathbb{R}_0^{\mathbb{Z}^d} \subset \mathcal{H}_+.$$

By using equation (8) we see that for any $j \in \mathbb{Z}^d$

$$\frac{\partial v_{n,j}}{\partial t} = -H_n v_{n,j} + \sum_{k \in \Lambda_n} (\nabla_j \beta_k) v_{n,k} = -H_n v_{n,j} + (R(\cdot) v_n)_j, \tag{10}$$

$$v_{n,j}(0, x) = \nabla_j f(x),$$

where

$$R(x) = \nabla \beta(x) = (r_{kj}(x))_{k,j \in \mathbb{Z}^d}, \quad r_{kj}(x) = \nabla_j \beta_k(x), \quad x \in \mathcal{X},$$

denotes the derivative of the vector field $\beta(\cdot)$. Let us remark that due to (5) $r_{kj}(x) \equiv 0$ if $|k - j| \geq 2$ and $R(x)$, $x \in \mathcal{X}$, is a family of bounded symmetric operators in $l_2(\mathbb{Z}^d)$. Moreover, these operators are uniformly bounded also in the space $\mathcal{H}_+$ and

$$\sup_{x \in \mathcal{X}} \|R(x)\|_{\mathcal{L}(\mathcal{H}_+, \mathcal{H}_+)} = C_R < \infty.$$

To prove this we note that

$$|(Rh)_j|^2 = \left| \sum_{k:|k-j| \leq 1} r_{kj} h_k \right|^2 \leq C_d \sum_{k:|k-j| \leq 1} r_{kj}^2 h_k^2.$$

Then for $\bar{C}_R := \sup_{k,j,x} r_{kj}(x)$, where $\bar{C}_R < \infty$ by (5),

$$|Rh|_+^2 = \sum_{j \in \mathbb{Z}^d} (Rh)_j^2 p_j^{-1}$$

$$\leq C_d \bar{C}_R \sum_{j \in \mathbb{Z}^d} \left\{ \sum_{k:|k-j| \leq 1} h_k^2 p_j^{-1} \right\}$$

$$= C_d \bar{C}_R \sum_{j \in \mathbb{Z}^d} \left\{ \sum_{e \in \mathbb{Z}^d : |e|=1} h_{j+e}^2 (1 + |j+e|)^\alpha \left(\frac{1 + |j|}{1 + |j+e|} \right)^\alpha \right\}$$

$$\leq C_R^2 |h|_+^2,$$

where $C_R^2 := 2^{\alpha+d} C_d \bar{C}_R$.

A standard fact of finite-dimensional diffusion theory is that the differential operator H_n generates a positivity preserving contractive semigroup e^{-tH_n}, $t \geq 0$, in the space $C_{\bar{\Lambda}_n}(\mathcal{X})$. By using this semigroup we can rewrite equation (10) as

$$v_{n,j}(t, x) = (e^{-tH_n} \nabla_j f)(x) + \int_0^t e^{-(t-s)H_n} (R(\cdot) v_n(s, \cdot))_j (x) \, ds.$$

Therefore, for any $h \in \mathcal{H}_-$

$$\langle v_n(t, x), h \rangle = \sum_{j \in \mathbb{Z}^d} v_{n,j}(t, x) h_j$$

$$= (e^{-tH_n} \langle \nabla f(\cdot), h \rangle)(x) + \int_0^t \left(e^{-(t-s)H_n} \langle R(\cdot) v_n(s, \cdot), h \rangle \right)(x) \, ds.$$

Due to the contraction property of the semigroup we have

$$\sup_{x \in \mathcal{X}} |\langle v_n(t,x), h \rangle| \leq \sup_{x \in \mathcal{X}} |\langle \nabla f(x), h \rangle| + \int_0^t \sup_{x \in \mathcal{X}} |\langle R(x) v_n(s,x), h \rangle| \, ds$$

$$\leq \sup_{x \in \mathcal{X}} |\nabla f(x)|_+ \, |h|_- + C_R \int_0^t \sup_{x \in \mathcal{X}} |v_n(s,x)|_+ \, |h|_- \, ds,$$

which implies

$$\sup_{x \in \mathcal{X}} |\nabla u_n(t,x)|_+ \leq \sup_{x \in \mathcal{X}} |\nabla f|_+ + C_R \int_0^t \sup_{x \in \mathcal{X}} |\nabla u_n(s,x)|_+ \, ds.$$

An application of Gronwall's inequality then gives

$$\sup_{x \in \mathcal{X}} |\nabla u_n(t,x)|_+ \leq \sup_{x \in \mathcal{X}} |\nabla f(x)|_+ e^{C_R t}, \quad 0 \leq t \leq 1. \tag{11}$$

Combining (11) with (9) we obtain that

$$\|(H_\mu - H_n) u_n(t)\|_{L^2(\mu)} \leq C_\beta e^{C_R} \|f\|_{C^1_{\Lambda_0}(\mathcal{X})} \sum_{k \in \Lambda_n^c} p_k \to 0 \quad \text{as } n \to \infty,$$

uniformly with respect to $t \in [0,1]$. This proves (iii) and completes the proof of Theorem 1.

REFERENCES

1. Albeverio, S., Antonjuk, A. Val., Antonjuk, A. Vict., Kondratiev, Yu. G., *Stochastic dynamics in some lattice spin systems*, Bochum-preprint Nr. 188, 1993 (to appear in Ukrainean Math. J.).
2. Albeverio, S., Kondratiev, Yu. G., Röckner, M., *Dirichlet operators via stochastic analysis*, BiBoS-preprint Nr. 571, 1993 (to appear in J. Funct. Anal.).
3. Albeverio, S., Kondratiev, Yu. G., Röckner, M., *An approximate criterium of essential self-adjointness of Dirichlet operators*, Potential Anal. **1** (1992), 307–317.
4. Albeverio, S., Kondratiev, Yu. G., Röckner, M., Addendum to the paper *An approximative criterium of essential self-adjointness of Dirichlet operators*, Potential Anal. **2** (1993), 195–198.
5. Albeverio, S., Kondratiev, Yu. G., Röckner, M., *Uniqueness of the stochastic dynamics for continuous spin systems on a lattice*, SFB 256-Preprint (1994).
6. Albeverio, S., Röckner, M., *Classical Dirichlet forms on topological vector spaces – closability and a Cameron–Martin formula*, J. Funct. Anal. **88** (1990), 395–436.
7. Albeverio, S., Röckner, M., *Dirichlet form methods for uniqueness of martingale problems and applications*, SFB 256-Preprint (1993) (to appear in "Proceedings of the Summer Research Institute on Stochastic Analysis 1993").
8. Albeverio, S., Röckner, M., Zhang, T.-S., *Markov uniqueness and its applications to martingale problems, stochastic differential equations and stochastic quantization*, C. R. Math. Rep. Acad. Sci. Canada **XV** (1993), 1–6.

9. Albeverio, S., Röckner, M., Zhang, T.-S., *Markov uniqueness for a class of infinite-dimensional Dirichlet operators*, In: Stochastic Processes and Optimal Control. Stochastic Monographs (H. J. Engelbert et al., eds.), vol. 7, Gordon and Breach, 1993.
10. Berezansky, Yu. M., *Self-adjoint operators in spaces of functions of infinitely many variables*, Translations of Amer. Math. Soc. (1986).
11. Berezansky, Yu. M., Kondratiev, Yu. G., *Spectral methods in infinite-dimensional analysis*, Kluwer Academic Publishers, 1993.
12. Deuschel, J.-D., Stroock, D. W., *Hypercontractivity and spectral gap of symmetric diffusions with applications to the stochastic Ising model*, J. Funct. Anal. **92** (1990), 30–48.
13. Fukushima, M., *Dirichlet forms and Markov processes*, North-Holland, Amsterdam–Oxford–New York, 1980.
14. Holley, R., *The one-dimensional stochastic X-Y-model*, in Random walks, Brownian motion, and interacting particle systems, Progress in Probability, vol. 28, Birkhäuser, 1991, pp. 295–307.
15. Henry, D., *Geometric theory of semilinear parabolic equations*, Lecture Notes in Math. **840** (1981), Springer-Verlag.
16. Holley, R., Stroock, D. W., *Diffusions on the infinite dimensional torus*, J. Funct. Anal. **42** (1981), 29–63.
17. Preston, C., *Random fields*, Springer-Verlag, 1976.
18. Röckner, M., Zhang, T.-S., *On uniqueness of generalized Schrödinger operators and applications*, J. Funct. Anal. **105** (1992), 187–231.
19. Röckner, M., Zhang, T.-S., *Uniqueness of generalized Schrödinger operators – Part II*, J. Funct. Anal. **119** (1994), 455–467.
20. Stroock, D. W., Zegarliński, B., *The equivalence of the logarithmic Sobolev inequality and Dobrushin–Shlosman mixing condition*, Comm. Math. Phys. **144** (1992), 303–323.
21. Stroock, D. W., Zegarliński, B., *The logarithmic Sobolev inequality for continuous spin systems on a lattice*, J. Funct. Anal. **104** (1992), 299–326.

S. ALBEVERIO, RUHR-UNIVERSITÄT, BOCHUM, FRG; BiBoS RESEARCH CENTRE, BIELEFELD, FRG; SFB 237 ESSEN-BOCHUM-DÜSSELDORF; CERFIM, LOCARNO, SWITZERLAND

YU. G. KONDRATIEV, BiBoS RESEARCH CENTRE, BIELEFELD, FRG; INSTITUTE OF MATHEMATICS, KIEV, UKRAINE

M. RÖCKNER, INSTITUT FÜR ANGEWANDTE MATHEMATIK, UNIVERSITÄT BONN, FRG

Progress in Probability, Vol. 36
© 1995 Birkhäuser Verlag Basel/Switzerland

DIFFUSION-APPROXIMATION FOR
THE ADVECTION-DIFFUSION
OF A PASSIVE SCALAR BY A SPACE-TIME GAUSSIAN
VELOCITY FIELD[1]

RENÉ A. CARMONA[2] AND JEAN PIERRE FOUQUE

ABSTRACT. We study the asymptotic behavior, as ϵ goes to zero, of a passive scalar $T^\epsilon(x,t)$ solution of the following advection-diffusion equation:

$$\frac{\partial T^\epsilon}{\partial t} = \frac{\nu}{2}\Delta T^\epsilon + \frac{1}{\epsilon}V\left(x, \frac{t}{\epsilon^2}\right) \cdot \nabla T^\epsilon, \quad t > 0,$$
$$T^\epsilon(x,0) = T_0(x), \quad x \in \mathbb{R}^d,$$

where ν is a strictly positive diffusion constant and $\{V(x,t): x \in \mathbb{R}^d, t \geq 0\}$ is a mean zero homogeneous Gaussian field. We assume that the covariance is of the form

$$\mathbb{E}\{V(x,t)V^*(y,s)\} = \Gamma(x-y)\exp(-a|t-s|),$$

and under some mild regularity assumption on Γ, we prove that $T^\epsilon(x,t)$ converges in distribution to the solution of a stochastic partial differential equation. We derive the effective diffusion coefficient from this result. This work is a generalization of previous works by Bouc–Pardoux [3] and Kushner–Huang [8] where the velocity field is of the form $\frac{1}{\epsilon}V(x, Z_{t/\epsilon^2})$ for some finite-dimensional ergodic noise process Z. Our situation is an example of infinite-dimensional noise.

§1 INTRODUCTION

The advection-diffusion of a passive scalar $T^\epsilon(x,t)$ by a velocity field $V_\epsilon(x,t)$ on a small scale ϵ $(0 < \epsilon \ll 1)$ is given by the equation

$$\frac{\partial T}{\partial t} = \frac{\nu}{2}\Delta T + V_\epsilon \cdot \nabla T, \quad t > 0, \tag{1}$$
$$T(x,0) = T_0(\epsilon x), \quad x \in \mathbb{R}^d.$$

We assume that the diffusion constant ν is strictly positive and that V_ϵ is a d-dimensional Gaussian velocity field. Adjusting the scaling to the nature of the

---

[1]This work is partially supported by a joint NSF-CNRS grant
[2]Partially supported by ONR N00014-91-1010

singularity of the spectral density of this Gaussian field, Avellaneda and Majda
have studied several renormalization procedures for this equation. We refer to
their works [1] and [2] for details. They obtained different limiting regimes: among
them, the mean field and the so-called anomalous regimes. In this paper we study
the diffusion-approximation regime which takes place when $V_\epsilon(x,t)$ is given by
rescaling of a single (non-singular) field $V(x,t)$. More precisely, we assume that
$V_\epsilon(x,t)$ is given by $V(\epsilon x,t)$ with the scaling law $(\frac{x}{\epsilon}, \frac{t}{\epsilon^2})$. Denoting $T(\frac{x}{\epsilon}, \frac{t}{\epsilon^2})$ by
$T^\epsilon(x,t)$ or $T_t^\epsilon(x)$, our problem (1) becomes:

$$\frac{\partial T^\epsilon}{\partial t} = \frac{1}{2}\nu\Delta T^\epsilon + \frac{1}{\epsilon}V(x, \frac{t}{\epsilon^2}) \cdot \nabla T^\epsilon, \qquad (2)$$

$$T^\epsilon(x,0) = T_0(x), \qquad (3)$$

where we suppose the initial condition T_0 in $L^2(\mathbb{R}^d)$ and where we choose, as a
Markovian model, a very simple covariance given by

$$\mathbb{E}\{V(x,s)V^*(y,t)\} = \Gamma(x-y)\exp(-a|t-s|) \qquad (4)$$

with a constant $a > 0$ and a C^∞-space covariance Γ. In the divergence free case
(i.e. in the incompressible case for which $\operatorname{div} V = 0$), the problem (1) can be
used in the two-dimensional case $d = 2$ as a model for the temperature field at
the surface of the ocean. This problem is of great importance. In particular, the
diffusion approximation result which we prove for the law of T^ϵ as $\epsilon \searrow 0$ gives the
justification for a way to close the equations for the moments of T^ϵ. Such closed
equations can be found for example in [9].

   In the next section we show how the solution of (2) can be represented by using
stochastic flows and deduce a priori estimates.

   In the third section we prove the convergence of $(T_t^\epsilon)_{t \geq 0}$ as processes taking
their values in the space of distributions $\mathcal{D}'(\mathbb{R}^d)$. In the last section we study the
particular case of incompressible velocity fields for which we have stronger a priori
estimates. This work is a generalization of previous works by Bouc–Pardoux [3]
and Kushner–Huang [8] where the velocity field is of the form $\frac{1}{\epsilon}V(x, Z_{t/\epsilon^2})$ for
some finite-dimensional ergodic noise process Z. Our situation is an example of
infinite-dimensional noise.

§2 REPRESENTATION OF THE SOLUTION AND A SIMPLE A PRIORI ESTIMATE

   We shall denote the $L^2(\mathbb{R}^d)$-norm by $|\cdot|_2$. Let us multiply both sides of equa-
tion (2) by T_t^ϵ and let us integrate over $\mathbb{R}^d \times [0,t]$. We get:

$$|T_t^\epsilon|_2^2 - |T_0|_2^2 = \nu \int_0^t \int_{\mathbb{R}^d} (\Delta T_s^\epsilon(x))T_s^\epsilon(x)\, dx\, ds$$

$$+ \frac{2}{\epsilon} \int_0^t \int_{\mathbb{R}^d} \left(V\left(x, \frac{s}{\epsilon^2}\right) \cdot \nabla T_s^\epsilon(x)\right)T_s^\epsilon(x)\, dx\, ds.$$

Integration by parts and substitution in the previous equation give:

$$|T_t^\epsilon|_2^2 - |T_0|_2^2 = -\nu \int_0^t |\nabla T_s^\epsilon|_2^2 \, ds - \frac{1}{\epsilon} \int_0^t \int_{\mathbb{R}^d} (T_s^\epsilon(x))^2 \, \text{div} \, V\left(x, \frac{s}{\epsilon^2}\right) dx \, ds. \quad (5)$$

In order to compensate the last term of this equation one may try a perturbation method as in [8]. Unfortunately $\text{div} \, V(x, \frac{s}{\epsilon^2})$, as a homogeneous Gaussian field is unbounded and the method fails. Nevertheless we shall use this approach in the incompressible case ($\text{div} \, V = 0$) to get strong estimates (last section).

We now describe another way to represent the solution $T_t^\epsilon(x)$. Given a d-dimensional standard Brownian motion $\{B_t\}_{t\geq 0}$ independent of V we consider the following stochastic differential equation:

$$d\Phi_s^{t,\epsilon}(x) = -\frac{1}{\epsilon} V\left(\Phi_s^{t,\epsilon}(x), \frac{s}{\epsilon^2}\right) ds + \nu^{1/2} \, dB_s, \quad 0 \leq s \leq t,$$

$$\Phi_t^{t,\epsilon}(x) = x. \quad (6)$$

We shall denote by $\mathbb{E}_B$ the expectation with respect to the Brownian motion B. Since the Gaussian random field V is homogeneous and since its covariance is continuous and tends to zero at infinity, the growth of the sample realizations of this field V is sublinear. This implies that equation (6) defines a unique stochastic flow Φ. If we suppose that the initial data T_0 is smooth and with compact support then we have:

$$T^\epsilon(x, t) = \mathbb{E}_B\{T_0(\Phi_0^{t,\epsilon}(x))\}. \quad (7)$$

We refer to [5] or [6] for details concerning this representation formula. The following lemma shows that T_t^ϵ is almost surely a function in $L^2(\mathbb{R}^d)$:

Lemma 2.1. *For every $t > 0$ and $\epsilon > 0$, $\mathbb{E}\{|T_t^\epsilon|_2^2\} \leq |T_0|_2^2$.*

Proof. Setting $\Psi_s^{t,\epsilon}(x) = \Phi_s^{t,\epsilon}(x) - x$, we get:

$$d\Psi_s^{t,\epsilon}(x) = -\frac{1}{\epsilon} V\left(\Psi_s^{t,\epsilon}(x) + x, \frac{s}{\epsilon^2}\right) ds + \nu^{1/2} \, dB_s,$$

$$\Psi_t^{t,\epsilon}(x) = 0, \quad (8)$$

which implies, by the stationarity of the velocity field V, that for fixed x the law of $(\Phi_s^{t,\epsilon}(x), 0 \leq s \leq t)$ is the same as the law of $(\Phi_s^{t,\epsilon}(0) + x, 0 \leq s \leq t)$. Now:

$$\mathbb{E}\{|T_t^\epsilon|_2^2\} = \mathbb{E}\left\{\int_{\mathbb{R}^d} \mathbb{E}_B\{T_0(\Phi_0^{t,\epsilon}(x))\}^2 \, dx\right\}$$

$$\leq \int_{\mathbb{R}^d} \mathbb{E}_B\{\mathbb{E}\{T_0^2(\Phi_0^{t,\epsilon}(x))\}\} \, dx$$

$$= \int_{\mathbb{R}^d} \mathbb{E}_B\{\mathbb{E}\{T_0^2(\Phi_0^{t,\epsilon}(0) + x)\}\}$$

$$= \mathbb{E}_B\left\{\mathbb{E}\left\{\int_{\mathbb{R}^d} T_0^2(\Phi_0^{t,\epsilon}(0) + x) \, dx\right\}\right\}$$

$$= |T_0|_2^2. \qquad \square$$

A possible approach would be to try to generalize the convergence results of stochastic flows given in Kunita [6] and to use the representation (7) to prove the convergence of $\{T_t^\epsilon\}$. This would require the generalization of some of the results of [6] because of the unboundedness of the velocity field V in the present situation. It would also give convergence of (T_t^ϵ) as continuous processes in spaces like C^k. Such an approach has been used in [10] for a class of velocity fields which do not contain our model.

Instead, we have chosen a more functional analytic approach which relates to the parabolic nature of the problem. It consists in studying the weak form of (2) for a well suited set of test functions.

§3 CONVERGENCE AS DISTRIBUTION-VALUED PROCESSES

Let θ be an element of the Schwartz space $\mathcal{D}(\mathbb{R}^d)$ of real-valued smooth C^∞ test functions with compact supports. We denote by $(\cdot, \cdot)$ the scalar product in $L^2(\mathbb{R}^d)$ as well as the canonical pairing $(\mathcal{D}', \mathcal{D})$ between the space $\mathcal{D}(\mathbb{R}^d)$ and its dual $\mathcal{D}'(\mathbb{R}^d)$. With these notations, the weak form of equation (1) can be written as

$$(T_t^\epsilon, \theta) - (T_0, \theta) = \frac{\nu}{2} \int_0^t (T_s^\epsilon, \Delta\theta)\, ds - \frac{1}{\epsilon} \int_0^t \left(T_s^\epsilon, \mathrm{div}\left(\theta V\left(\cdot, \frac{s}{\epsilon^2}\right)\right) \right) ds. \quad (9)$$

We chose the smallest space of test functions. This is bound to weaken some of the results. Here are the reasons for our choice. On one hand, the regularity of θ is needed to give a meaning to the equation: we have chosen C^∞ test functions in order to consider our $L^2(\mathbb{R}^d)$-valued process (T_t^ϵ) as a process with values in a nuclear space. This will enable us to use the simple tightness criteria of [4] which we recall in the next section. On the other hand the compactness of the support of the test fuctions will be essential when it comes to handle the unboundedness of the Gaussian field V. As a consequence, we shall consider (T_t^ϵ) as a $\mathcal{D}'(\mathbb{R}^d)$-valued process. Note nevertheless that, in the incompressible case, we are able to prove the convergence in a stronger sense. See the last section for details.

3.1 Tightness. According [4], a family of $\mathcal{D}'(\mathbb{R}^d)$-valued continuous (resp. right-continuous with left limits) processes $\{(T^\epsilon), 0 < \epsilon \leq 1\}$ is tight if and only if, for every test function θ in $\mathcal{D}(\mathbb{R}^d)$, the family of real-valued continuous (resp. right-continuous with left limit) processes $\{(T^\epsilon, \theta), 0 < \epsilon \leq 1\}$ is tight. As in [8], we prove the tightness of the right-continuous processes with left limits $\{f((T^\epsilon, \theta)), 0 < \epsilon \leq 1\}$ for all the possible choices of twice differentiable compactly supported real valued function f. This is done by a perturbation method and an adequate tightness criterium developed in [7] (see also [8]). From (9) we deduce an equation for $f(T_t^\epsilon, \theta)$:

$$f(T_t^\epsilon, \theta) - f(T_0, \theta) = \frac{\nu}{2} \int_0^t f'(T_s^\epsilon, \theta)(T_s^\epsilon, \Delta\theta)\, ds$$

$$- \frac{1}{\epsilon} \int_0^t f'(T_s^\epsilon, \theta)\left(T_s^\epsilon, \mathrm{div}\left(\theta V\left(\cdot, \frac{s}{\epsilon^2}\right)\right) \right) ds. \quad (10)$$

Denoting by $\mathcal{F}_t^\epsilon$ the σ-algebra generated by $\{V(x,u),\ x \in \mathbb{R}^d,\ u \le t/\epsilon^2\}$ and by $\mathbb{E}_t^\epsilon$ the conditional expectation with respect to $\mathcal{F}_t^\epsilon$, we define the *first pertubation* of equation (10) by:

Definition 3.1.

$$f_1^\epsilon(t) = \frac{1}{\epsilon} \int_t^{+\infty} \mathbb{E}_t^\epsilon\left\{ f'(T_t^\epsilon, \theta)\left(T_t^\epsilon, \operatorname{div}\left(\theta V\left(\cdot, \frac{s}{\epsilon^2}\right)\right)\right)\right\} ds. \tag{11}$$

Equation (1) implies that T_t^ϵ is $\mathcal{F}_t^\epsilon$-measurable. Moreover, the form of the covariance given in (4) implies that $\mathbb{E}_t^\epsilon V(x, \frac{s}{\epsilon^2}) = V(x, \frac{t}{\epsilon^2})\exp(-a(s-t)/\epsilon^2)$. From these facts one easily gets:

$$f_1^\epsilon(t) = \frac{\epsilon}{a} f'(T_t^\epsilon, \theta)\left(T_t^\epsilon, \operatorname{div}\left(\theta V\left(\cdot, \frac{t}{\epsilon^2}\right)\right)\right). \tag{12}$$

The next lemma tells us how small the perturbation $f_1^\epsilon(t)$ is:

Lemma 3.2. *We have:*

(a) $\sup_t \mathbb{E}\{|f_1^\epsilon(t)|\} = \epsilon \cdot O(1)$.
(b) *For every* $t_0 \ge 0$, $\sup_{t \le t_0} |f_1^\epsilon(t)| \longrightarrow 0$ *in probability as* $\epsilon \searrow 0$.

Proof. (a) From (12) and Lemma 2.1:

$$\mathbb{E}\{|f_1^\epsilon(t)|\} \le \frac{\epsilon}{a}|f'|_\infty \mathbb{E}\left\{\left|\left(T_t^\epsilon, \operatorname{div}\left(\theta V\left(\cdot, \frac{t}{\epsilon^2}\right)\right)\right)\right|\right\}$$

$$\le \frac{\epsilon}{a}|f'|_\infty |T_0|_2 \, \mathbb{E}\left\{\left|\theta \operatorname{div} V\left(\cdot, \frac{t}{\epsilon^2}\right) + V\left(\cdot, \frac{t}{\epsilon^2}\right) \cdot \nabla \theta\right|_2^2\right\}^{1/2}.$$

Using the notation $M = \sup_{x \in \operatorname{supp}\theta} |x|$ we have:

$$\mathbb{E}\left\{\left|\theta \operatorname{div} V\left(\cdot, \frac{t}{\epsilon^2}\right)\right|_2^2\right\} \le |\theta|_2^2 \, \mathbb{E}\left\{\sup_{|x|\le M}\left|\operatorname{div} V\left(x, \frac{t}{\epsilon^2}\right)\right|^2\right\}$$

$$= |\theta|_2^2 \, \mathbb{E}\left\{\sup_{|x|\le M} |\operatorname{div} V(x,0)|^2\right\}.$$

Similarly,

$$\mathbb{E}\left\{\left|V\left(\cdot, \frac{t}{\epsilon^2}\right)\cdot\nabla\theta\right|_2^2\right\} \le |\nabla\theta|_2^2 \, \mathbb{E}\left\{\sup_{|x|\le M} |V(x,0)|^2\right\}.$$

Therefore,

$$\mathbb{E}\{|f_1^\epsilon(t)|\} \le \frac{\epsilon}{a}|f'|_\infty |T_0|_2 (C_1|\theta|_2 + C_2|\nabla\theta|_2)$$

where the constants C_1 and C_2 depend on θ or more precisely on its support.

(b) From (12), (7), and the boundedness assumption on T_0, we get:

$$\sup_{t \le t_0} |f_1^\epsilon(t)| \le C(M) \frac{\epsilon}{a} |f'|_\infty |T_0|_\infty$$

$$\times \left(|\theta|_2 \sup_{|x| \le M, t \le t_0} \left| \operatorname{div} V\left(x, \frac{t}{\epsilon^2} \right) \right| + |\nabla \theta|_2 \sup_{|x| \le M, t \le t_0} \left| V\left(x, \frac{t}{\epsilon^2} \right) \right| \right).$$

Since the homogeneity of the velocity field implies that one can find a (random) constant C such that

$$\sup_{|x| \le M, t \le t_0} \left| V\left(x, \frac{t}{\epsilon^2} \right) \right| \le C \log \left(\frac{(2M)^d t_0}{\epsilon^2} \right)$$

and a similar inequality for $\operatorname{div} V$, we can conclude that $\sup_{t \le t_0} |f_1^\epsilon(t)| \longrightarrow 0$ almost surely as $\epsilon \searrow 0$. $\square$

Following Kushner [7, Chapter 3], we define the *pseudo-generator* A^ϵ acting on any $\mathcal{F}_t^\epsilon$-measurable function $f(t)$ such that $\sup_t |f(t)| < \infty$ as the L^1-limit of $\delta^{-1}(\mathbb{E}_t^\epsilon\{f(t+\delta)\} - f(t))$ whenever this limit exists. The latter will be denoted by $A^\epsilon f$. The main property of this pseudo-generator is that

$$f(t) - \int_0^t A^\epsilon f(s)\, ds$$

is a martingale with respect to the filtration $(\mathcal{F}_t^\epsilon)$ ([7, Chapter 3, Th. 1]). In the present situation we choose $f(t) = f^\epsilon(t) = f(T_t^\epsilon, \theta) - f_1^\epsilon(t)$ in order to cancel the terms in $1/\epsilon$ in the pseudo-generator:

Proposition 3.3. *With $f^\epsilon(t) = f(T_t^\epsilon, \theta) - f_1^\epsilon(t)$ and $\Psi_t^\epsilon(\theta) = \operatorname{div}(\theta V(\cdot, \frac{t}{\epsilon^2}))$, we have:*

$$A^\epsilon f^\epsilon(t) = \frac{\nu}{2} f'(T_t^\epsilon, \theta)(T_t^\epsilon, \Delta\theta) + \frac{1}{a} f''(T_t^\epsilon, \theta)(T_t^\epsilon, \Psi_t^\epsilon(\theta))^2$$

$$+ \frac{1}{a} f'(T_t^\epsilon, \theta) (T_t^\epsilon, \Psi_t^\epsilon(\Psi_t^\epsilon(\theta))) \tag{13}$$

$$- \frac{\epsilon\nu}{2a} \{ f''(T_t^\epsilon, \theta)(T_t^\epsilon, \Delta\theta) + f'(T_t^\epsilon, \theta)(T_t^\epsilon, \Delta\Psi_t^\epsilon(\theta)) \}.$$

Proof. The form (10) implies that $f(T_t^\epsilon, \theta)$ is differentiable in t. Notice that A^ϵ acts on the latter as a differentiation operator:

$$A^\epsilon f(T_t^\epsilon, \theta) = \frac{\nu}{2} f'(T_t^\epsilon, \theta)(T_t^\epsilon, \Delta\theta) - \frac{1}{\epsilon} f'(T_t^\epsilon, \theta)(T_t^\epsilon, \Psi_t^\epsilon(\theta)). \tag{14}$$

The first term on the right hand side of (13) is the first term of the right hand side of (14). Using (12) and $\mathbb{E}_t^\epsilon\{\Psi_{t+\delta}^\epsilon(\theta)\} = \Psi_t^\epsilon(\theta)\exp(-a\delta/\epsilon^2)$, we get:

$$
\begin{aligned}
\mathbb{E}_t^\epsilon\{f_1^\epsilon(t+\delta)\} = \frac{\epsilon}{a}\; \mathbb{E}_t^\epsilon\Big\{ & \left(f'(T_{t+\delta}^\epsilon,\theta) - f'(T_t^\epsilon,\theta)\right)(T_{t+\delta}^\epsilon,\Psi_{t+\delta}^\epsilon(\theta)) \\
& + \frac{\epsilon}{a}f'(T_t^\epsilon,\theta)\mathbb{E}_t^\epsilon(T_{t+\delta}^\epsilon - T_t^\epsilon,\Psi_{t+\delta}^\epsilon(\theta)) \\
& + \frac{\epsilon}{a}f'(T_t^\epsilon,\theta)(T_t^\epsilon,\Psi_t^\epsilon(\theta))\exp\left(-\frac{a\delta}{\epsilon^2}\right)\Big\}.
\end{aligned}
\tag{15}
$$

Substracting $f_1^\epsilon(t)$, dividing by δ and taking a limit as $\delta \searrow 0$ gives:

$$
\begin{aligned}
A^\epsilon f_1^\epsilon(t) = & -f''(T_t^\epsilon,\theta)(T_t^\epsilon,\Psi_t^\epsilon(\theta))^2 - f'(T_t^\epsilon,\theta)\,(T_t^\epsilon,\Psi_t^\epsilon(\Psi_t^\epsilon(\theta))) \\
& + \frac{\epsilon\nu}{2a}\{f''(T_t^\epsilon,\theta)(T_t^\epsilon,\Delta\theta) + f'(T_t^\epsilon,\theta)(T_t^\epsilon,\Psi_t^\epsilon(\theta))\} \\
& - \frac{1}{\epsilon}f'(T_t^\epsilon,\theta)(T_t^\epsilon,\Psi_t^\epsilon(\theta))
\end{aligned}
\tag{16}
$$

where we have used the continuity of $\Psi_t^\epsilon(\theta)$ and $(T_t^\epsilon,\Psi_t^\epsilon(\theta))$. The fact that these limits are in L^1 is justified by the same type of estimates that we have obtained in the proof of Lemma 3.2.
The equality $f^\epsilon(t) = f(T_t^\epsilon,\theta) - f_1^\epsilon(t)$ together with (14) and (16) give (13) because of the cancellation of the $1/\epsilon$-terms. □

Theorem 3.4. *The laws of $(T_\cdot^\epsilon)_{0<\epsilon\leq 1}$ are tight on $D([0,+\infty);\mathcal{D}'(\mathbb{R}^d))$.*

Proof. As explained at the beginning of this section, it is enough to show that the laws of $(f(T_\cdot^\epsilon,\theta))_{0<\epsilon\leq 1}$ are tight on $D([0,+\infty);\mathbb{R})$ for each fixed $\theta \in \mathcal{D}(\mathbb{R}^d)$ and each C^2-function f with compact support. We apply the tightness criterium given in Kushner [7, Chapter 3, Th. 4] based on the martingale property of $f^\epsilon(t) - \int_0^t A^\epsilon f^\epsilon(s)\,ds$ and the fact that the term $f(T_t^\epsilon,\theta) - f^\epsilon(t) = f_1^\epsilon(t)$ is small. Lemma 3.2(b) implies that

$$
\lim_{\epsilon\searrow 0} \mathbb{P}\Big\{\sup_{t\leq t_0}|f^\epsilon(t) - f(T_t^\epsilon,\theta)| \geq \alpha\Big\} = 0
$$

for every $\alpha > 0$ and $t_0 < +\infty$.

It then remains to prove that for every finite t_0, the family $\{A^\epsilon f^\epsilon(t), 0 < \epsilon \leq 1, t \leq t_0\}$ is uniformly integrable. This is done for each term appearing in $A^\epsilon f^\epsilon(t)$ in (13), in a similar way we proved Lemma 3.2(a); the compact support property of θ enables us to take a supremum in x for $|x| \leq M$ and by stationarity of V and its derivatives we remove the dependence in t/ϵ^2. □

3.2 Identification of the limit. Denoting by (T^ϵ) a weakly convergent subsequence and by $(T_\cdot)$ its limit, according to [4], in order to identify the law of the limit, it is enough to identify, for every $\theta_1,\ldots,\theta_n$ in $\mathcal{D}(\mathbb{R}^d)$, the finite-dimensional distributions of $\{(T_\cdot,\theta_1),\ldots,(T_\cdot,\theta_n)\}$.

The martingale property of $f^\epsilon(t) - \int_0^t A^\epsilon f^\epsilon(s)\,ds$ implies that for every bounded continuous function h and for every sequence $0 < s_1 < \cdots < s_n \le s < t$ we have:

$$\mathbb{E}\left\{\left(f^\epsilon(t) - f^\epsilon(s) - \int_s^t A^\epsilon f^\epsilon(u)\,du\right)h((T_{s_1}^\epsilon, \theta), \ldots, (T_{s_n}^\epsilon, \theta))\right\} = 0. \qquad (17)$$

Denoting by $A_i^\epsilon(t)$, $i = 1, 2, 3, 4$ the four terms in $A^\epsilon f^\epsilon(t)$ given by (13) we have that:

- $A_1^\epsilon(t)$ is uniformly bounded in ϵ and converges weakly to

$$\frac{\nu}{2} f'(T_t, \theta)(T_t, \Delta\theta).$$

- $\mathbb{E}\{\int_s^t |A_4^\epsilon f^\epsilon(u)|\,du\}$ converges to *zero* for the same reason as in the proof of Lemma 3.1(a).

The last two terms are given by:

$$A_2^\epsilon(t) = \frac{1}{a} f''(T_t^\epsilon, \theta)(T_t^\epsilon, \Psi_t^\epsilon(\theta))^2$$

$$A_3^\epsilon(t) = \frac{1}{a} f'(T_t^\epsilon, \theta)\,(T_t^\epsilon, \Psi_t^\epsilon(\Psi_t^\epsilon(\theta))).$$

One needs a *second perturbation* in order to center these two terms and get convergence in (17). We introduce special notations for the quantities which appear in the limit:

Definition 3.5. For θ, θ_1, θ_2 in $\mathcal{D}(\mathbb{R}^d)$ and ϕ in $L^2(\mathbb{R}^d)$ define

$$H_{\theta_1, \theta_2}(x, y) = \sum_{i=1}^d \sum_{j=1}^d \partial_{x_i}\partial_{y_j}[\theta_1(x)\theta_2(y)\Gamma_{i,j}(x - y)],$$

$$A_2(\phi) = \int_{\mathbb{R}^d}\int_{\mathbb{R}^d} \phi(x)\phi(y)H_{\theta,\theta}(x, y)\,dx\,dy,$$

$$B = \sum_{i=1}^d \sum_{j=1}^d \Gamma_{i,j}(0)\partial_{x_i}\partial_{x_j},$$

$$b = \left\{b_j = \sum_{i=1}^d (\partial_{x_i}\Gamma_{i,j})(0), j = 1, \ldots, d\right\},$$

$$A_3(\phi) = (\phi, b \cdot \nabla\theta + B\theta).$$

With these notations, an easy computation shows:

Lemma 3.6. $\mathbb{E}\{(\phi, \Psi_t^\epsilon(\theta))^2\} = A_2(\phi)$ *and* $\mathbb{E}\{(\phi, \Psi_t^\epsilon(\Psi_t^\epsilon(\theta)))\} = A_3(\phi)$.

The *second perturbation* involves the following two terms:

Definiton 3.7.

$$f_2^\epsilon(t) = \frac{1}{a} f''(T_t^\epsilon, \theta) \int_t^{+\infty} \mathbb{E}_t^\epsilon \{(T_t^\epsilon, \Psi_s^\epsilon(\theta))^2 - A_2(T_t^\epsilon)\} \, ds,$$

$$f_3^\epsilon(t) = \frac{1}{a} f'(T_t^\epsilon, \theta) \int_t^{+\infty} \mathbb{E}_t^\epsilon \{(T_t^\epsilon, \Psi_s^\epsilon(\Psi_s^\epsilon(\theta))) - A_3(T_t^\epsilon)\} \, ds.$$

Using the conditional covariance:

$$\mathbb{E}_t^\epsilon \left\{ V_i \left(x, \frac{s}{\epsilon^2} \right) V_j \left(y, \frac{s}{\epsilon^2} \right) \right\} =$$
$$\Gamma_{i,j}(x-y) \left(1 - \exp\left(-2a \frac{s-t}{\epsilon^2} \right) \right) + V_i \left(x, \frac{t}{\epsilon^2} \right) V_j \left(y, \frac{t}{\epsilon^2} \right) \exp\left(-2a \frac{s-t}{\epsilon^2} \right)$$

one easily gets:

$$f_2^\epsilon(t) = \frac{\epsilon^2}{2a^2} f''(T_t^\epsilon, \theta) \left((T_t^\epsilon, \Psi_t^\epsilon(\theta))^2 - A_2(T_t^\epsilon) \right), \tag{18}$$

and

$$f_3^\epsilon(t) = \frac{\epsilon^2}{2a^2} f'(T_t^\epsilon, \theta) \left((T_t^\epsilon, \Psi_t^\epsilon(\Psi_t^\epsilon(\theta))) - A_3(T_t^\epsilon) \right). \tag{19}$$

Using the explicit forms (18) and (19) for $f_2^\epsilon(t)$ and $f_3^\epsilon(t)$ and the argument given in Lemma 3.2(a) for $f_1^\epsilon(t)$, we get:

Lemma 3.8.

$$\sup_t \mathbb{E}\{|f_2^\epsilon(t)|\} = \epsilon^2 \, O(1)$$

and

$$\sup_t \mathbb{E}\{|f_3^\epsilon(t)|\} = \epsilon^2 \, O(1).$$

As we computed $A^\epsilon f_1^\epsilon(t)$ in the proof of Proposition 3.3, after a long but straightforward computation, we get:

Proposition 3.9.

$$A^\epsilon f_2^\epsilon(t) = \frac{1}{a} f''(T_t^\epsilon, \theta) \left(A_2(T_t^\epsilon) - (T_t^\epsilon, \Psi_t^\epsilon(\theta))^2 \right) + R_2^\epsilon(t),$$

$$A^\epsilon f_3^\epsilon(t) = \frac{1}{a} f'(T_t^\epsilon, \theta) \left(A_3(T_t^\epsilon) - (T_t^\epsilon, \Psi_t^\epsilon(\Psi_t^\epsilon(\theta))) \right) + R_3^\epsilon(t)$$

with $\sup_t \mathbb{E}\{|R_2^\epsilon(t)|\} = O(\epsilon)$ *and* $\sup_t \mathbb{E}\{|R_3^\epsilon(t)|\} = O(\epsilon)$.

Let us set $R^\epsilon(t) = A_4^\epsilon(t) + R_2^\epsilon(t) + R_3^\epsilon(t)$. Recall that $A_4^\epsilon(t)$ is the fourth term of the decomposition of $A^\epsilon f^\epsilon(t)$ given in (13). Combining Proposition 3.3 and

Proposition 3.9, we obtain that the quantity $M_t^{f,\epsilon}(\theta)$ defined by

$$
\begin{aligned}
M_t^{f,\epsilon}(\theta) = {} & f(T_t^\epsilon, \theta) - f_1^\epsilon(t) + f_2^\epsilon(t) + f_3^\epsilon(t) \\
& - \int_0^t \left[f'(T_s^\epsilon, \theta) \left(\frac{\nu}{2}(T_s^\epsilon, \Delta\theta) + \frac{1}{a} A_3(T_s^\epsilon) \right) \right. \\
& \left. + \frac{1}{a} f''(T_s^\epsilon, \theta) A_2(T_s^\epsilon) + R^\epsilon(s) \right] ds
\end{aligned}
\tag{20}
$$

is a martingale with respect to $(\mathcal{F}_t^\epsilon)$. Introducing the nuclear operator K_ϕ, from $\mathcal{D}$ to $\mathcal{D}'$, defined for ϕ in L^2 by

$$
(\theta_1, K_\phi \theta_2) = \frac{2}{a} \int_{\mathbb{R}^d} \int_{\mathbb{R}^d} \phi(x)\phi(y) H_{\theta_1, \theta_2}(x, y) \, dx \, dy,
\tag{21}
$$

we can use the results of Lemmas 3.2 and 3.8 and Propositions 3.3 and 3.9 to take a limit in the martingale problem (20). We then conclude that the limiting law of the converging subsequence (T^ϵ) is a solution to the martingale problem

$$
\begin{aligned}
M_t^f(\theta) = {} & f(T_t, \theta) - \int_0^t \left[f'(T_s, \theta) \left(\frac{\nu}{2}(T_s, \Delta\theta) + \frac{1}{a} A_3(T_s) \right) \right. \\
& \left. + \frac{1}{2} f''(T_s, \theta)(\theta, K_\phi \theta) \right] ds.
\end{aligned}
\tag{22}
$$

By the representation formula (7) and the boundedness hypothesis on T_0 we get:

$$
|(T_t^\epsilon, \theta)| \leq C(M)|T_0|_\infty |\theta|_2
$$

and therefore $\mathbb{E}\{(T_t^\epsilon, \theta)^2\}$ converges to $\mathbb{E}\{(T_t, \theta)^2\}$. The same argument shows that the martingale property in (22) holds for $f(r) = r$ and $f(r) = r^2$ which enables us to conclude that

$$
M_t(\theta) = (T_t, \theta) - \int_0^t \left(\frac{\nu}{2}(T_s, \Delta\theta) + \frac{1}{a} A_3(T_s) \right) ds
\tag{23}
$$

is a martingale with quadratic variation given by

$$
\langle M(\theta) \rangle_t = \int_0^t (\theta, K_{T_s}\theta) \, ds.
\tag{24}
$$

A similar computation with two test functions θ_1 and θ_2 gives the following quadratic covariation:

$$
\langle M(\theta_1), M(\theta_2) \rangle_t = \int_0^t (\theta_1, K_{T_s}\theta_2) \, ds.
\tag{25}
$$

The martingale problem (23), (25) has been studied by several authors. For instance it is shown in [8] that $M_t(\theta) = (M_t, \theta)$ defines an $L^2(\mathbb{R}^d)$-valued continuous square-integrable martingale; this martingale can be represented as $M_t = \int_0^t K_{T_s}^{1/2} dW_s$, where W denotes a cylindrical Wiener process. The law of (T_t) is then the same as the law of the unique solution of the following stochastic partial differential equation:

$$dT_t = \left(\frac{\nu}{2}\Delta + \frac{1}{a}B - \frac{1}{a}b \cdot \nabla\right)T_t \, dt + K_{T_t}^{1/2} \, dW_t. \tag{26}$$

Applying an infinite-dimensional Itô formula (see for instance [11]) to the square of the $L^2(\mathbb{R}^d)$-norm, one easily gets the following stochastic *a priori* estimates for every $t_0 < +\infty$:

$$\sup_{t \leq t_0} \mathbb{E}\{|T_t|_2^2\} < \infty, \quad \mathbb{E}\int_0^{t_0} \|T_t\|_{H^1}^2 \, dt < \infty, \quad \lim_{\alpha \nearrow +\infty} \mathbb{P}\Big\{\sup_{t \leq t_0} |T_t|_2 \geq \alpha\Big\} = 0.$$

We summarize the results obtained in this section by:

Theorem 3.10. *The laws of the solutions $(T^\epsilon)_{0 \leq \epsilon \leq 1}$ converge weakly on the space $D([0, +\infty); \mathcal{D}'(\mathbb{R}^d))$ to the $L^2(\mathbb{R}^d)$-valued unique solution of the stochastic partial differential equation* (26).

Remark 3.11.

  (1) In the equation (26) the term $\frac{1}{a}(B - b \cdot \nabla)T_t$ is the Stratonovich correction to the stochastic integral.
  (2) The first moment $m(t, x) = \mathbb{E}\{T_t(x)\}$ satisfies:

$$\frac{\partial m}{\partial t} = \left(\frac{\nu}{2}\Delta + \frac{1}{a}(B - b \cdot \nabla)\right)m, \quad m_0 = T_0.$$

  (3) The equation (26) can also be written as follows:

$$dT_t = \frac{\nu}{2}\Delta T_t \, dt + (2/a)^{1/2}\nabla T_t \circ dW_t^\Gamma$$

  where W^Γ is an infinite-dimensional Brownian motion with spatial covariance Γ.
  (4) If the velocity field V is spatially symmetric (i. e. if the spatial covariance satisfies $\Gamma(-x) = \Gamma(x)$), then $b = 0$ in (26). In particular, this is the case for *isotropic* velocity fields.
  (5) If the velocity field V is *divergence free*, then we also have $b = 0$ and much more, as shown in the next section.

§4 THE DIVERGENCE FREE CASE

In this section we assume that

$$\sum_{i=1}^{d}\sum_{j=1}^{d}\partial_{x_i}\partial_{x_j}\Gamma_{i,j} = 0, \tag{27}$$

which ensures the fact that our velocity field V has a null divergence. Going back to equation (5), we deduce the following uniform estimates giving deterministic bounds:

$$|T_t^\epsilon| \le |T_0|, \qquad \int_0^t \|T_s^\epsilon\|_{H^1}^2 \, ds \le \left(t + \frac{1}{\nu}\right)|T_0|^2. \tag{28}$$

The laws of (T^ϵ) are naturally supported by the space of continuous $L^2(\mathbb{R}^d)$-valued trajectories which are $H^1(\mathbb{R}^d)$-locally integable. Considering the weak topologies on these L^2-spaces, we have:

Proposition 4.1. *The laws of (T^ϵ) for $0 < \epsilon \le 1$ are tight on the space*

$$\Omega = D([0,+\infty); L_w^2(\mathbb{R}^d)) \cap L_{w,\mathrm{loc}}^2([0,+\infty); H^1(\mathbb{R}^d))$$

equipped with its Borel σ-field.

Proof. For the tightness on $D([0,+\infty); L_w^2(\mathbb{R}^d))$ we have to prove the tightness of $\{(T^\epsilon,\theta)\}_{0<\epsilon\le1}$ on $D([0,+\infty);\mathbb{R})$ for every θ in $L^2(\mathbb{R}^d)$. According to the first part of (28) it is enough to prove it for a dense subset in $L^2(\mathbb{R}^d)$. This is what we have done in Section 3.1 for θ in $\mathcal{D}(\mathbb{R}^d)$. The second part of (28) implies the boundedness in $L_{\mathrm{loc}}^2([0,+\infty); H^1(\mathbb{R}^d))$ and therefore the relative compactness in $L_{w,\mathrm{loc}}^2([0,+\infty); H^1(\mathbb{R}^d))$. To conclude, we notice, as in [3], that tightness and relative compactness are equivalent on Ω. $\square$

The identification of the limit is done as in Section 3.2 and we obtain:

Theorem 4.2. $\{(T^\epsilon), 0 < \epsilon \le 1\}$ *converges in law on Ω to the continuous process (T_t), which is the unique solution of the stochastic partial differential equation*

$$T_t = T_0 + \int_0^t \left(\frac{\nu}{2}\Delta + \frac{1}{a}B\right)T_s \, ds + (2/a)^{1/2} \int_0^t \nabla T_s \cdot dW_s^\Gamma \tag{29}$$

where $B = \sum_{i=1}^{d}\sum_{j=1}^{d}\Gamma_{i,j}(0)\partial_{x_i}\partial_{x_j}$ and $(W_t^\Gamma(x))_{t\ge0}$ has a covariance given by: $\Gamma(x-y)\delta(t-s)$.

Our final remark is that, in the *divergence free* case, the first moment of the solution satisfies a diffusion equation with an *effective diffusivity* equal to $\frac{\nu}{2}\Delta + \frac{1}{a}B$.

REFERENCES

1. M. Avellaneda and A. Majda, *Mathematical models with exact renormalization for turbulent transport*, Comm. Math. Phys. **131** (1990), 381–429.
2. M. Avellaneda and A. Majda, *An integral representation and bounds on the effective diffusivity in passive advection by laminar and turbulent flows*, Comm. Math. Phys. **138** (1991), 339–391.
3. R. Bouc and E. Pardoux, *Asymptotic analysis of P. D. E. s with wide-band noise disturbances, and expansion of the moments*, Stochastic Analysis and Applications **2** (1984), 369–422.
4. J. P. Fouque, *La convergence en loi pour les processus à valeurs dans un espace nucléaire*, Ann. Inst. Henri Poincaré **20** (1984), 225–245.
5. M. Friedlin, *Functional Integration and Partial Differential Equations*, Annals of Mathematics Series **109** (1985), Princeton University Press, Princeton, N. J.
6. H. Kunita, *Stochastic Flows and Differential Equations*, Cambridge University Press, 1990.
7. H. Kushner, *Approximation and Weak Convergence Methods for Random Processes, with Applications to Stochastic Systems Theory*, MIT Press, Cambridge Mass, 1984.
8. H. Kushner and Huang, *Limits for parabolic partial differential equations with wide band stochastic coefficients and an application to filtering theory*, Stochastics **14** (1985), 115–148.
9. S. A. Molchanov and L. Pitterbarg, *Heat Propagation in a Random Flow*, in press, Russ. J. Math. Phys. (1993).
10. B. Rozovskii, *Some results on a diffusion approximation to the induction equation*, preprint (1992).
11. M. Yor, *Existence et unicité de diffusions à valeurs dans un espace de Hilbert*, Ann. Inst. Henri Poincaré **10** (1974), 55–88.

RENÉ A. CARMONA, DEPARTMENT OF MATHEMATICS, UNIVERSITY OF CALIFORNIA IRVINE, IRVINE, CA. 92717 USA

JEAN PIERRE FOUQUE, CNRS-CMAP, ECOLE POLYTECHNIQUE, 91128 PALAISEAU CEDEX FRANCE

Progress in Probability, Vol. 36

A NEW SPACE OF WHITE NOISE DISTRIBUTIONS
AND APPLICATIONS TO SPDE'S

RENÉ A. CARMONA[1] AND J. A. YAN[2]

ABSTRACT. This paper deals with the so-called white noise calculus. Some of the shortcomings of the existing spaces of generalized functions are discussed and a new space of distributions is introduced. This new space of distributions is shown to be larger than the existing ones. We give a characterization of its elements in terms of a local S-transform. Finally an application to stochastic partial differential equations is given.

§1 INTRODUCTION

This paper is concerned with the use of generalized functionals over white noise space as a model for the noise in stochastic partial differential equations (SPDE's for short). We shall first explain how and why the existing spaces are not satisfactory. Only then will it be natural to consider a new space of test functionals and a new space of generalized functionals. These spaces are defined in Section 2 below where we compare them to the existing ones. A natural characterization is given in Section 3 in terms of a local S-transform. Finally, Section 5 is devoted to an application to a SPDE.

Let $S'(\mathbb{R}^d)$ denote the Schwartz space of tempered distributions and μ denote the white noise measure on $S'(\mathbb{R}^d)$. Recall that μ is the mean zero Gaussian measure on the Borel σ-field of $S'(\mathbb{R}^d)$ given by its characteristic function (Fourier transform)

$$C(\mu, \xi) = \int e^{i\langle x, \xi \rangle} \, d\mu(x) = e^{-\frac{1}{2}|\xi|^2}, \quad \xi \in S(\mathbb{R}^d).$$

We use the notation $\langle \cdot, \cdot \rangle$ for the $L^2 = L^2(\mathbb{R}^d)$-inner product and $|\cdot|$ for the corresponding norm. We shall denote by (L^2) the Hilbert space of μ-equivalence classes of measurable functions which are square integrable with respect to μ, i.e. $(L^2) = L^2(S'(\mathbb{R}^d), d\mu)$. Gelfand triples based on the space (L^2) have been introduced to bear to the space (L^2) the same relation as the Hilbert space $L^2 =$

---

[1]Partially supported by ONR N00014-91-1010
[2]Partially supported by the National Science Foundation of China

$L^2(\mathbb{R}^d, dx)$ to the classical Schwartz space of tempered distributions. The following three Gelfand triples are standard examples:

$$(\mathcal{S}) \subset (L^2) \subset (\mathcal{S})^*, \qquad \mathcal{M} \subset (L^2) \subset \mathcal{M}^*, \qquad (\mathcal{S})^\beta \subset (L^2) \subset (\mathcal{S})^{-\beta}.$$

The first one is the so-called triple of Hida's test functionals and Hida's distributions. See for example [9]. The other ones were introduced more recently in [6] and [4] respectively. They satisfy the following inclusion relationships:

$$\mathcal{M} \subset (\mathcal{S})^\beta \subset (\mathcal{S}) \subset (L^2) \subset (\mathcal{S})^* \subset (\mathcal{S})^{-\beta} \subset \mathcal{M}^*.$$

The reader is refered to [4] for a complete account of the theory of these spaces.

We now explain why they are still not completely satisfactory when it comes to the analysis of natural operations with mathematical models for radom media. The following is an example of a typical shortcoming of the theory. Let us consider for example the random Hamiltonian most used in the localization theory, namely the random Schrödinger operator $H = -\Delta + V$. See [1] for background and recent results. The Feynman–Kac formula is a very convenient tool in the analysis of the properties of such an operator. Among other things it gives the solution of the heat equation

$$\frac{\partial u(t, x)}{\partial t} = -\Delta u(t, x) + V(x)u(t, x) \tag{1}$$

with initial condition $u(0, x) = u_0(x)$ for some given function $u_0(x)$. Indeed, the (typically unique) solution $u(t, x)$ can be written in the form

$$u(t, x) = \mathbb{E}_x\{u_0(X_t)e^{-\int_0^t V(X_s)\,ds}\} \tag{2}$$

where $\{X_s\}_{s \geq 0}$ is a process of Brownian motion in $\mathbb{R}^d$. The potential function $V(x)$ is typically a realization of a homogeneous random field. There are many reasons to extend the analysis to include random potentials whose sample realizations are merely generalized functions. Let us assume for example that V is a random Hida's distribution and let us try to give a meaning to the Feynman–Kac formula. It is relatively easy to show that, for almost every sample path of the Brownian motion, the Hida's distribution V can be evaluated along the path, i. e. that $V(X_s)$ can be given a meaning. Unfortunately, the integral

$$\int_0^t V(X_s)\,ds$$

does not make sense in general because it cannot be interpreted as a Bochner integral, neither in $(\mathcal{S})^*$ nor in $\mathcal{M}^*$ in general! We introduce a new space of white noise distributions in order to overcome this kind of difficulty. One could expand on the example of the heat equation when the potential is a function of white noise, or one could also use the idea behind the use of the Feynman–Kac formula to solve the Dirichlet problem in a random potential of this type, very much in the spirit of [2]. Instead, we decided to revisit some earlier work of these authors.

The last section of the paper is devoted to the discussion of a model for fluid flow in a porous medium first proposed by Lindstrom et al. in [5]. The analysis of their approach was the main driving force behind the present work. These authors realized that the spaces of distributions over white noise were not well suited to many of the standard PDE models of applied mathematics when white noise was used as a model for the source of randomness in the coefficients. We show that our space of distribution can provide a better setting. Indeed, we show in Section 5 that the formula given by Lindstrom et al. in [5] actually defines a distribution over white noise in our space $\widetilde{\mathcal{M}}^*$ and of course that it solves the equation they consider. We prove that this equation can be solved when the random permeability is a homogeneous field constructed from an L^2-function over the white noise space, while it seems that they could only treat the L^1-case. Our approach is not much different from the strategy defined and used in [5] and [2]. The reader is refered to [8] for a reformulation of this strategy in terms of the S-transform. In this approach, one merely checks that the candidate for a solution does make sense once the appropriate framework of spaces has been chosen.

We believe that the new spaces which we introduce in the present paper have an interest of their own. For this reason, we are convinced that they will be useful in the solution of other problems.

§2 A New Space of Test Functionals
and the Corresponding Space of Distributions

As usual, we denote by A the harmonic oscillator operator. It is the self-adjoint operator

$$A = -\frac{d^2}{dx^2} + x^2 + 1$$

on the Hilbert space $L^2 = L^2(\mathbb{R}, dx)$. We shall use the notation A_d for the d-fold tensor product of A with itself. It is the self-adjoint operator

$$A_d = -\Delta + |x|^2 + d$$

on the Hilbert space $L^2(\mathbb{R}^d, dx)$. We shall ignore the dependence upon the dimension d whenever no confusion is possible. For each real number $p \geq 0$ we denote by $\mathcal{S}_p(\mathbb{R}^d) = \mathcal{D}(A^p)$ the domain of the operator A^p and by $\mathcal{S}_{-p}(\mathbb{R}^d)$ the dual of $\mathcal{S}_p(\mathbb{R}^d)$ when the latter is equiped with the norm $|f|_{2,p}$ defined by

$$|f|_{2,p} = |(A_d)^p f|_2$$

where we use the notation $|\cdot|_2$ for the norm of the Hilbert space $L^2(\mathbb{R}^d, dx)$. It is well-known that

$$\mathcal{S}(\mathbb{R}^d) = \bigcap_{p \geq 0} \mathcal{S}_p(\mathbb{R}^d) \quad \text{and} \quad \mathcal{S}'(\mathbb{R}^d) = \bigcup_{p \geq 0} \mathcal{S}_{-p}(\mathbb{R}^d).$$

Each element f of (L^2) has the Wiener -Itô decomposition into the sum of multiple Wiener integrals

$$f = \sum_{n=0}^{\infty} I_n(f_n) \tag{3}$$

for a sequence $(f_n)_n$ of symmetric kernels, $f_n \in L^2(\mathbb{R}^d)^{\hat{\otimes}(s)n}$ where the subscript $_{(s)}$ emphasizes the symmetrization. We have

$$\|f\|_2^2 = \sum_{n=0}^{\infty} n! \, |f_n|_2^2.$$

Let us fix momentarily the real numbers $p \geq 0$ and $q > 0$. We denote by $\mathcal{M}_{p,q}$ the collection of elements $f = (f_n)_n$ of (L^2) for which

$$\|f\|_{p,q}^2 = \sum_{n=0}^{\infty} (n!)^2 2^{-qn} |f_n|_{2,p}^2 < \infty. \tag{4}$$

$\|\cdot\|_{p,q}$ defines a norm on the space $\mathcal{M}_{p,q}$. $\mathcal{M}_{p,q}$ is a Banach space. We denote by $\mathcal{M}_{-p,-q}$ the dual of $\mathcal{M}_{p,q}$. It is also a Banach space. Its norm is given by

$$\|f\|_{-p,-q}^2 = \sum_{n=0}^{\infty} 2^{qn} |f_n|_{2,-p}^2. \tag{5}$$

Notice that the norm $\|\cdot\|_{-p,-q}$ is different from the norm obtained by plugging $-p$ and $-q$ in (4) in lieu of p and q. The meaning of the norm $\|\cdot\|_{p,q}$ depends upon the sign of the parameters p and q. The space $\mathcal{M}$ of test functionals and the space $\mathcal{M}^*$ of distributions can now be defined by

$$\mathcal{M} = \bigcap_{p \geq 0\, q > 0} \bigcup \mathcal{M}_{p,q} = \bigcap_{p \geq 0} \mathcal{M}_p \quad \text{and} \quad \mathcal{M}^* = \bigcup_{p \geq 0\, q > 0} \bigcap \mathcal{M}_{-p,-q} = \bigcup_{p \geq 0} \mathcal{M}_{-p},$$

provided we set

$$\mathcal{M}_p = \bigcup_{q > 0} \mathcal{M}_{p,q} \quad \text{and} \quad \mathcal{M}_{-p} = \bigcap_{q > 0} \mathcal{M}_{-p,-q}.$$

Notice that for each $p \geq 0$, the space $\mathcal{M}_{-p}$ is the dual of the space $\mathcal{M}_p$ and that the space $\mathcal{M}^*$ is the dual of the space $\mathcal{M}$ once the appropriate topologies are chosen. The spaces $\mathcal{M}$ and $\mathcal{M}^*$ were introduced by Meyer and Yan in [6]. They proved that $\mathcal{M}$ contains the Wick exponentials

$$\mathcal{E}(\xi) = \exp\left[\langle \cdot, \xi \rangle - \frac{1}{2}|\xi|_2^2\right], \quad \xi \in \mathcal{S}(\mathbb{R}^d), \tag{6}$$

and that it is stable under Wiener products. This indicates that the space $\mathcal{M}$ is not too small, and consequently that the dual space $\mathcal{M}^*$ is not too large. Nevertheless, it was shown in [4] that the dual $\mathcal{M}^*$ is not too small and that it contains, among other things, the Poisson random measures. These are the main reasons why $\mathcal{M}$ was regarded as a reasonable space of test functionals. We explained in the introduction why the space $\mathcal{M}^*$ is not large enough. It is not stable under as simple an operation as the Bochner integral and this last operation is crucial in the solution of SPDE's. We were led to the following definition.

Definition 2.1. We define the collection $\widetilde{\mathcal{M}}$ of test functionals and the corresponding collection $\widetilde{\mathcal{M}}^*$ of distributions over white noise space by

$$\widetilde{\mathcal{M}} = \bigcap_{p \geq 0, q > 0} \mathcal{M}_{p,q} \quad \text{and} \quad \widetilde{\mathcal{M}}^* = \bigcup_{p \geq 0, q > 0} \mathcal{M}_{-p,-q},$$

and we endow $\widetilde{\mathcal{M}}$ and $\widetilde{\mathcal{M}}^*$ with the projective and the inductive topologies respectively.

Obviously we have the following continuous embeddings:

$$\widetilde{\mathcal{M}} \hookrightarrow \mathcal{M} \hookrightarrow (L)^2 \hookrightarrow \mathcal{M}^* \hookrightarrow \widetilde{\mathcal{M}}^*.$$

The main properties of these new spaces are summarized in the following:

Theorem 2.2. $\widetilde{\mathcal{M}}$ *is a nuclear Fréchet space which is stable under Wick and Wiener products. Moreover* $\widetilde{\mathcal{M}}^*$ *is the topological dual of the locally convex topological vector space* $\widetilde{\mathcal{M}}$.

The duality properties of the new spaces are due to the existence of a set of Hölder like inequalities which we prove in the next two lemmas.

Lemma 2.3. *Let us assume that* $p \geq 0$ *and that* $q > 0$ *and* $\alpha, \beta > 0$ *are such that*

$$2^{-\alpha} + 2^{-\beta} < 1.$$

Then there exists a constant $c = c_{q,\alpha,\beta}$ *satisfying*

$$\|fg\|_{p,q} \leq c_{q,\alpha,\beta} \|f\|_{p+\alpha,q} \|g\|_{p+\beta,q}.$$

Proof. Let $(f_n)_n$ and $(g_n)_n$ be the kernel sequences of f and g. Then we have

$$I_m(f_m)I_n(g_n) = \sum_{k=0}^{m \wedge n} k! \binom{m}{k}\binom{n}{k} I_{m+n-2k}(f_m \otimes_k^{(s)} g_n)$$

where $f_m \otimes_k^{(s)} g_n$ stands for the symmetrization of $f_m \otimes_k g_n$. Recall that the latter is defined by

$$f_m \otimes_k g_n = \int_{\mathbb{R}^{dk}} f_m(u_1, \dots, u_k, \dots, u_m) g_n(u_1, \dots, u_k, \dots, u_n)\, du_1 \dots du_k.$$

Consequently we have

$$\|I_m(f_m)I_n(g_n)\|_{p,q} \leq \sum_{k=0}^{m \wedge n} k! \binom{m}{k}\binom{n}{k} \|I_{m+n-2k}(f_m \otimes_k^{(s)} g_n)\|_{p,q}$$

$$\leq \sum_{k=0}^{m \wedge n} k! \binom{m}{k}\binom{n}{k} (m+n-2k)!\, 2^{q(m+n-2k)/2} |f_m|_{2,p} |g_n|_{2,q}$$

$$= \sum_{k=0}^{m \wedge n} \frac{2^{kq}}{k!} \binom{m+n-2k}{n-k} \|I_m(f_m)\|_{p,q} \|I_n(g_n)\|_{p,q}.$$

It is proven in [11, Lemma 2.3] that, for any $a > 0$ we have

$$\sum_{k=0}^{m \wedge n} \binom{m+n-2k}{n-k} \leq (1+a)^m (1+a^{-1})^n.$$

This implies that

$$\|I_m(f_m)I_n(g_n)\|_{p,q} \leq e^{2^q}(1+a)^m(1+a^{-1})^n\|I_m(f_m)\|_{p,q}\|I_n(g_n)\|_{p,q}. \qquad (7)$$

Because of our assumption $2^{-\alpha} + 2^{-\beta} < 1$, we can find $\epsilon > 0$ such that

$$2^{-(\alpha-\epsilon)} + 2^{-(\beta-\epsilon)} = 1.$$

Setting $a = 2^{\alpha-\epsilon} - 1$, we have $a^{-1} = 2^{\beta-\epsilon} - 1$ and estimate (7) implies

$$\|I_m(f_m)I_n(g_n)\|_{p,q} \leq e^{2^q}2^{-(m+n)\epsilon}\|I_m(f_m)\|_{p+\alpha,q}\|I_n(g_n)\|_{p+\beta,q},$$

which gives the desired result after summation. $\quad\square$

The second technical result concerns the Wick product. It implies, in particular, that the spaces $\widetilde{\mathcal{M}}$ and $\widetilde{\mathcal{M}}^*$ are stable for the Wick product.

Lemma 2.4. *If we have as before $p \geq 0$ and $q > 0$ and if $\alpha, \beta > 0$ are such that $2^{-\alpha} + 2^{-\beta} = 1$ then we have*

$$\|f : g\|_{p,q} \leq \|f\|_{p+\alpha,q}\|g\|_{p+\beta,q} \qquad (8)$$

where as usual $:$ denotes the Wick product. In addition we also have

$$\|f : g\|_{-p,-q} \leq \|f\|_{-p,-q}\|g\|_{-p,-q} \qquad (9)$$

Proof. Recall that, if $f = (f_m)_m$ and $g = (g_n)_n$, then $f : g = (h_k)_k$ with

$$h_k = \sum_{i+j=k} f_i \otimes^{(s)} g_j. \qquad (10)$$

Hence, for $a = 2^\beta - 1$ we have $1 + a^{-1} = 2^\alpha$ and

$$\|I_k(h_k)\|_{p,q}^2 = (k!)^2 2^{-qk}|h_k|_{2,p}^2$$

$$\leq (k!)^2 2^{-qk}\left(\sum_{i+j=k} |f_i|_{2,p}|g_j|_{2,p}\right)^2$$

$$= \left(\sum_{i+j=k} \binom{k}{i}\|I_i(f_i)\|_{p,q}\|I_j(g_j)\|_{p,q}\right)^2$$

$$\leq \left(\sum_{i+j=k} \binom{k}{i}^2 a^{2i}\right)\left(\sum_{i+j=k} \|I_i(f_i)\|_{p,q}^2\|I_j(g_j)\|_{p,q}^2 a^{-2i}\right)$$

$$\leq (1+a)^{2k} \sum_{i+j=k} \|I_i(f_i)\|_{p,q}^2\|I_j(g_j)\|_{p,q}^2 a^{-2i}$$

$$= \sum_{i+j=k} (1+a^{-1})^{2i}(1+a)^{2j}\|I_i(f_i)\|_{p,q}^2\|I_j(g_j)\|_{p,q}^2$$

$$= \sum_{i+j=k} \|I_i(f_i)\|_{p+\alpha,q}^2\|I_j(g_j)\|_{p+\beta,q}^2.$$

which gives (8). The second estimate (9) follows immediately from (10) and (5). $\square$

We can now prove the theorem.

Proof. For each integer $n \geq 1$ we set $\mathcal{H}_n = \mathcal{M}_{n,1/n}$. Then $\mathcal{H}_n$ is a separable Hilbert space and $\mathcal{H}_{n+1} \subset \mathcal{H}_n$. Moreover it is easy to see that $\widetilde{\mathcal{M}} = \bigcap_n \mathcal{H}_n$. Since the second statement of the theorem is an easy consequence of the two lemmas above, we need only the nuclearity, and this will follow if we can prove that the natural embedding, say i_n, of $\mathcal{H}_{n+1}$ into $\mathcal{H}_n$ is a Hilbert–Schmidt operator. We consider only the case $d = 1$ in order to keep reasonably simple notations. Let $\{e_j\}_{j \geq 1}$ be a complete orthonormal system of eigenvectors of the operator A such that $Ae_j = 2je_j$. For each multi-index $\alpha = (\alpha_1, \alpha_2, \cdots)$ we set $e_\alpha = \hat{\otimes}_j e_j^{\otimes \alpha_j}$ and $H_\alpha = I_{|\alpha|}(e_\alpha)$. Then, the collection $\{H_\alpha\}_\alpha$ is a complete orthogonal system in each $\mathcal{H}_n$. Moreover

$$\|H_\alpha\|_{\mathcal{H}_n}^2 = (|\alpha|!)^2 2^{-|\alpha|/n} |e_\alpha|_{2,n}^2 = (|\alpha|!)^2 2^{-|\alpha|/n} \frac{\alpha!}{|\alpha|!} \prod_k (2k)^{2n\alpha_k}$$

$$= \alpha! \, |\alpha|! \, 2^{-|\alpha|/n} \prod_k (2k)^{2n\alpha_k}$$

from which it is easy to compute the Hilbert–Schmidt norm of the inclusion map i_n. We get

$$\|i_n\|_{\mathrm{HS}}^2 = \sum_\alpha \left\| \frac{H_\alpha}{\|H_\alpha\|_{\mathcal{H}_{n+1}}} \right\|_{\mathcal{H}_n}^2 \leq \sum_\alpha \prod_k (2k)^{-2\alpha_k}$$

and this quantity is finite because of a result of [13] (see also [12]). This concludes the proof. $\square$

§3 LOCAL S-TRANSFORM AND CHARACTERIZATIONS

Recall the usual notation

$$\mathcal{E}(\xi) = \exp\left[\langle \cdot, \xi \rangle - \frac{1}{2}|\xi|_2^2\right]$$

for the Wick exponential of the linear random variable $\langle \cdot, \xi \rangle$ corresponding to the test function $\xi \in \mathcal{S}(\mathbb{R}^d)$. This exponential belong to the space $\mathcal{M}$ of test functionals introduced by Meyer and Yan, but it does not necessarily belong to the smaller space $\widetilde{\mathcal{M}}$ which we introduced. We show how to get around this drawback.

Now, for $\varphi \in \mathcal{M}_{p,q}$ and $p \geq 0$ and $q > 0$ we can define the S-transform $S\varphi$ by

$$S\varphi(\xi) = \langle \varphi, \mathcal{E}(\xi) \rangle, \quad \xi \in \mathcal{S}(\mathbb{R}^d).$$

This makes sense because $\mathcal{E}(\xi) \in \mathcal{M}_{-p,-q}$ for all $p \geq 0$ and $q > 0$. In fact, one has

$$\|\mathcal{E}(\xi)\|_{-p,-q} \leq \sum_{k=0}^{\infty} 2^{kq}(k!)^{-2}|\xi|_{2,-p}^{2k} < \infty.$$

From this one can see that, if $\varphi \in \mathcal{M}_{p,q}$ with $p \geq 0$ and $q > 0$, then the S-transform $S\varphi(\xi)$ can be defined on $\mathcal{S}_{-p}(\mathbb{R}^d)$. However, for any non-zero $\xi \in \mathcal{S}(\mathbb{R}^d)$ one has that $\mathcal{E}(\xi) \notin \widetilde{\mathcal{M}}$. In other words, the S transform cannot be defined for a general element of $\widetilde{\mathcal{M}}^*$. Fortunately, if $\xi \in \mathcal{S}_p(\mathbb{R}^d)$ satisfies $|\xi|_{2,p}^2 < 2^q$, then $\mathcal{E}(\xi) \in \mathcal{M}_{p,q}$ and for $\varphi \in \mathcal{M}_{-p,-q}$, the formula giving the definition of the S-transform still makes sense. This is the rationale behind the definition of the local S-transform which we now give.

Definition 3.1. For $p, q \in \mathbb{R}$ we set

$$\mathcal{B}_{p,q} = \{ \xi \in \mathcal{S}(\mathbb{R}^d); \ |\xi|_{2,p}^2 < 2^q \} \tag{11}$$

and for $p \geq 0$ and $q > 0$ and $\varphi \in \mathcal{M}_{-p,-q}$ we define the local $S\varphi$ by

$$[S\varphi](\xi) = \langle \varphi, \mathcal{E}(\xi) \rangle, \quad \xi \in \mathcal{B}_{p,q}. \tag{12}$$

Our goal is now to give characterizations of our space of test functionals and our space of distributions in terms of the local S-transform introduced above. In order to do that we need the notion of ray-entire functions.

Definition 3.2. Let F be a complex-valued function defined on $\mathcal{S}(\mathbb{R}^d)$ (resp. $\mathcal{B}_{p,q}$). F is said to be *ray-entire* if for all $f, g \in \mathcal{S}(\mathbb{R}^d)$ (resp. $f, g \in \mathcal{B}_{p,q}$) the map

$$t \mapsto F(g + tf), \quad t \in \mathbb{R}, \quad (\text{resp. } |t| < |f|_{2,p}^{-1}[2^{q/2} - |g|_{2,p}])$$

has an analytic extension to the entire complex plane $\mathbb{C}$ (resp. to the disk $\mathcal{D} = \{ \lambda \in \mathbb{C}; \ |\lambda| < |f|_{2,p}^{-1}[2^{q/2} - |g|_{2,p}] \}$).

We shall use the same notation $F(g + \lambda f)$ for the analytic extension of the function $F(g + tf)$. The following theorem characterizes the elements of $\widetilde{\mathcal{M}}$. We skip the proof because it is similar to the the proof of Theorem 4.9 in [4].

Theorem 3.3. *For every $\varphi \in \widetilde{\mathcal{M}}$, $S\varphi$ is a ray-entire function on $\mathcal{S}(\mathbb{R}^d)$ and for each $p \geq 0$ and $K > 1/2$ there exists a constant $c > 0$ such that*

$$\sup_{f \in B_p^*} |[S\varphi](\lambda f)| \leq c e^{K|\lambda|}, \quad \lambda \in \mathbb{C},$$

where we used the notation B_p^* for the unit ball in $\mathcal{S}_p(\mathbb{R}^d)$, i. e. $B_p^* = \{ f \in \mathcal{S}(\mathbb{R}^d); \ |f|_{2,-p} \leq 1 \}$. Conversely, for any ray-entire function F which satisfies the condition above, namely such that

$$\sup_{f \in B_p^*} |F(\lambda f)| \leq c e^{K|\lambda|}, \quad \lambda \in \mathbb{C},$$

for some real number K and a constant $c > 0$, there exists a unique element $\varphi \in \widetilde{\mathcal{M}}$ such that $S\varphi = F$.

The following result gives a characterization of the space $\widetilde{\mathcal{M}}^*$ of distributions in terms of the local S-transform introduced earlier. The simplicity of this characterization suggests that this space of distributions may be convenient for applications which could not be handled by the previously known spaces because of the limitations discussed in the introduction. We shall illustrate this fact in the next section on applications to stochastic partial differential equations.

Theorem 3.4. *If $\varphi \in \mathcal{M}_{-p,-q}$ with $p \geq 0$ and $q > 0$, then the local S-transform $S\varphi$ is ray-entire on $\mathcal{B}_{p,q}$. Conversely, if F is a ray-entire function on some $\mathcal{B}_{p,q}$ with $p \geq 0$ and $q > 0$, then there exists a unique $\varphi \in \mathcal{M}_{-p_1,-q_1}$ where $p_1 \geq 0$ and $q_1 > 0$ such that $S\varphi = F$ on $\mathcal{B}_{p,q}$. More precisely, if $p_1 \geq p + 1$ and $q_1 > 0$ are such that*

$$2^{2(p_1 - p - 1) + q} > e^2 \qquad and \qquad 0 < q_1 < [2(p_1 - p - 1) + q] \log 2 - 2, \qquad (13)$$

then $\varphi \in \mathcal{M}_{-p_1, -q_1}$.

Notice that such a couple (p_1, q_1) always exists!

Proof. The first claim is clear so we prove only the second statement. As before we restrict ourselves to the case $d = 1$ for the sake of simplicity. Let us set

$$\widetilde{\mathcal{B}}_{p,q} = \{\, \xi + i\eta; \; \xi, \eta \in \mathcal{S}(\mathbb{R}^d), \; |\xi|_{2,p}^2 + |\eta|_{2,p}^2 \leq 2^q \,\}.$$

Then, F can be extended to $\widetilde{\mathcal{B}}_{p,q}$ as an entire function. Its Taylor expansion reads

$$F(\lambda) = \sum_{n=0}^{\infty} \frac{1}{n!} d^n F(0)(\lambda), \quad \lambda \in \widetilde{\mathcal{B}}_{p,q}.$$

By the kernel theorem we have

$$d^n F(0)(\lambda) = \langle F^{(n)}, \lambda^{\otimes n} \rangle, \quad \lambda \in \widetilde{\mathcal{B}}_{p+1, q+2} \subset \widetilde{\mathcal{B}}_{p,q},$$

for some kernels $F^{(n)} \in \widehat{\mathcal{S}}_{-(p+1)}(\mathbb{R}^n)$. We set

$$f_n = \frac{1}{n!} F^{(n)} \qquad and \qquad \varphi = \sum_{n=0}^{\infty} I_n(f_n)$$

and we prove that $\varphi \in \widetilde{\mathcal{M}}^*$. Let $p_1 > 0$ and $q_1 > 0$ be such that condition (13) is satisfied. Then there exists $r > 0$ such that

$$2^{2(p_1 - p - 1) + q} > r^2 > e^2 \qquad and \qquad 0 < q_1 < 2(\log r - 1).$$

We have

$$|\lambda|_{2,p_1-1}^2 = r^2 \quad \Rightarrow \quad |\lambda|_{2,p}^2 \leq 2^{-2(p_1-p-1)}r^2 < 2^q,$$

and Cauchy's inequality (see [7]) implies that

$$|f_n|_{2,-p_1} \leq \frac{1}{n!}\|d^n F(0)\|_{C(S_{p_1-1,c}^n(\mathbb{R}),\mathbb{C})} \leq \frac{n^n}{n!}\frac{1}{r^n} \sup_{|\lambda|_{2,p_1-1}=r} |F(\lambda)|$$

$$\leq c\left(\frac{e}{r}\right)^n = c\delta^n 2^{-nq_1/2},$$

where $\|\cdot\|_{C(S_{p_1-1,c}^n(\mathbb{R}),\mathbb{C})}$ denotes the norm of n-linear forms, $S_{p_1-1,c}(\mathbb{R})$ denotes the complexification of $S_{p_1-1}(\mathbb{R})$ and $\delta = e2^{q_1/2}/r < 1$. The above estimate implies that $\varphi \in \mathcal{M}_{-p_1,-q_1}$. Moreover, it is clear that $S\varphi = F$ on $\mathcal{B}_{p_1,q_1}$ because $\mathcal{B}_{p_1,q_1} \subset \mathcal{B}_{p,q}$ since $q_1 - 2(p_1 - p) < q$. This completes the proof. $\square$

The above characterization is stable under pointwise product: indeed, the product of the local S-transforms of two distributions in $\widetilde{\mathcal{M}}^*$ is still the local S-transform of a distribution in $\widetilde{\mathcal{M}}^*$. This remark is useful to extend the notion of Wick product to the space $\widetilde{\mathcal{M}}^*$. Indeed, if $\varphi, \psi \in \widetilde{\mathcal{M}}^*$ we can define the Wick product $\varphi : \psi$ as the unique element of $\widetilde{\mathcal{M}}^*$ having the product of $S\varphi$ and $S\psi$ as its local S-transform. In other words, $\varphi : \psi$ is characterized by

$$[S\varphi : \psi](\xi) = [S\varphi](\xi)[S\psi](\xi), \quad \xi \in \mathcal{S}(\mathbb{R}^d).$$

The following theorems will be very useful in the sequel. We omit their proofs because they are very similar to the proofs of the corresponding results in the case of the spaces of Hida's distributions given in [9] and [3].

Theorem 3.5. *A sequence $\{\varphi_n; n \geq 1\}$ in $\widetilde{\mathcal{M}}^*$ converges toward an element φ of $\widetilde{\mathcal{M}}^*$ for the (strong) topology of $\widetilde{\mathcal{M}}^*$ if and only if there exists $p \geq 0$ and $q > 0$ such that $\varphi_n \in \mathcal{M}_{-p,-q}$ for all $n \geq 1$ and the limit $\lim_{n\to\infty}[S\varphi](f)$ exists for every $f \in \mathcal{B}_{p,q}$ in such a way that*

$$\sup_{n\geq 1} \sup_{\lambda\in\mathbb{C},|\lambda|=r} \sup_{f\in\mathcal{B}_{p,q}} |[S\varphi_n](\lambda f)| < \infty$$

for some $0 < r < 1$.

Theorem 3.6. *Let $(\Omega, \mathcal{F}, m)$ be a measure space and let $\varphi : \Omega \ni \omega \mapsto \varphi(\omega) \in \mathcal{M}_{-p,-q}$ be measurable (for some $p \geq 0$ and $q > 0$ as usual). If for some real number $0 < r < 1$ there exists an integrable function $C \in L^1(\Omega, m)$ such that*

$$\sup_{\lambda\in\mathbb{C},|\lambda|=r} \sup_{f\in\mathcal{B}_{p,q}} |[S\varphi(\omega)](\lambda f)| \leq C(\omega)$$

for m-almost all $\omega \in \Omega$, then the Bochner integral $\int_\Omega \varphi(\omega)\,dm(\omega)$ exists in some space $\mathcal{M}_{-p_1,-q_1}$ for some $p_1 \geq p$ and $q_1 \geq q$ and we have

$$\left[S\left(\int_\Omega \varphi(\omega)\,dm(\omega)\right)\right](f) = \int_\Omega [S\varphi(\omega)](f)\,dm(\omega), \quad f \in \mathcal{B}_{p_1,q_1}.$$

§4 HOMOGENEOUS NOISE FIELDS

Let us consider the case of a stationary stochastic process as a motivation for the definitions we are about to introduce. The notion of a stationary stochastic process $\{X_t; t \in \mathbb{R}\}$ is best understood over a dynamical system $(\Omega, \mathcal{F}, (\theta_t)_{t\in\mathbb{R}}, \mathbb{P})$ where $(\theta_t)_{t\in\mathbb{R}}$ is a group of measurable transformations of $(\Omega, \mathcal{F})$ which preserve the measure $\mathbb{P}$. In this case the stationary process $\{X_t; t \in \mathbb{R}\}$ is completely determined by the single random variable X_0 since $X_{s+t} = X_s \circ \theta_t$ for every $s, t \in \mathbb{R}$ (at least $\mathbb{P}$-almost surely). The stationarity of the field is given by the fact that the process is determined by the shifts of one single variable by a group of transformations (of the probability space) which leave the probability measure invariant and which represent the group of translations of the parameter ($s \mapsto s+t$ in the present situation).

We generalize this approach in the following way. For each $a \in \mathbb{R}^d$ we denote by τ_a the translation $\mathbb{R}^d \ni x \mapsto \tau_a(x) = x + a \in \mathbb{R}^d$. This translation is lifted to the functions over $\mathbb{R}^d$ in a covariant way. If f is a function over $\mathbb{R}^d$, the translated function $\tau_a f$ is defined by

$$[\tau_a f](x) = f(x - a), \quad x \in \mathbb{R}^d. \tag{14}$$

It is clear that τ_a so defined is an isometry of L^2. Obviously, it is also unitary since $\tau_a^* = (\tau_a)^{-1} = \tau_{-a}$. Moreover, τ_a is also a continuous operator from $\mathcal{S}(\mathbb{R}^d)$ into itself. More precisely we have

Lemma 4.1. *For every $p \geq 0$, τ_a is a continuous operator from $\mathcal{S}_p(\mathbb{R}^d)$ into itself. In fact,*

$$|\tau_a f|_{2,p}^2 \leq 2^{4p-1}(1 + |a|^{4p}2^{-2dp})|f|_{2,p}^2, \quad f \in \mathcal{S}_p(\mathbb{R}^d). \tag{15}$$

Proof. We have

$$\tau_a^* A \tau_a = -\Delta + d + \tau_a^*|x|^2\tau_a = -\Delta + d + |x + a|^2 \leq 2(A + |a|^2)$$

where the inequality has to be understood in the sense of quadratic forms on L^2. $\square$

Translation can be defined for tempered distributions by duality:

$$\langle \tau_a^* f, g \rangle = \langle f, \tau_a g \rangle, \quad f \in \mathcal{S}'(\mathbb{R}^d), g \in \mathcal{S}(\mathbb{R}^d).$$

We set $\theta_a = \tau_a^*$. $\{\theta_a; a \in \mathbb{R}^d\}$ is a group of measurable transformations of the space $(\mathcal{S}'(\mathbb{R}^d), \mathcal{B})$ which leave the white noise measure μ invariant. Indeed, for every $\xi \in \mathcal{S}(\mathbb{R}^d)$ we have

$$\int e^{i\langle x,\xi \rangle}\, d[\theta_a \mu](x) = \int e^{i\langle \theta_a x,\xi \rangle}\, d\mu(x) = \int e^{i\langle x,\tau_{-a}\xi \rangle}\, d\mu(x)$$

$$= \exp\left[-\frac{1}{2}|\tau_{-a}\xi|_2^2\right] = \exp\left[-\frac{1}{2}|\xi|_2^2\right].$$

We can now mimic the classical definition recalled at the beginning of this section. If φ is a random variable over the white noise space $(S'(\mathbb{R}^d), \mathcal{B}, \mu)$, then for each $a \in \mathbb{R}^d$ we define the random variable $\theta_a \varphi$ by

$$\theta_a \varphi(x) = \varphi(\theta_a x), \quad x \in S'(\mathbb{R}^d).$$

As in the classical case, one checks that θ_a defines a bounded linear transformation of the spaces of test functionals and by duality, of the spaces of distributions over the white noise space. Moreover, θ_a is a unitary transformation of the Hilbert space (L^2).

Definition 4.2. We shall say that $\{V(x); x \in \mathbb{R}^d\}$ is a *homogeneous field* over white noise if

$$V(x) = \theta_x \varphi, \quad x \in \mathbb{R}^d,$$

for some $\varphi \in \widetilde{\mathcal{M}}^*$.

We now come to the analysis of specific examples.

Example 4.3. Let $\varphi = \mathrm{Exp}(\langle \cdot, \delta_0 \rangle)$, δ_0 being the Dirac "delta function" at the origin. Recall that Exp stands for the Wick exponential. If we set $V(x) = \theta_x \varphi$ as in the above definition, then one has

$$[SV(x)](\xi) = [S(\theta_x \varphi)](\xi) = [S\varphi](\tau_{-x} \xi)$$
$$= \exp\langle \tau_{-x} \xi, \delta_0 \rangle = \exp\langle \xi, \tau_x \delta_0 \rangle = \exp\langle \xi, \delta_x \rangle = e^{\xi(x)}$$

which shows that $V(x) = \mathrm{Exp}(\langle \cdot, \delta_x \rangle) = \mathrm{Exp}(W_x)$ where W_x denotes the d-parameter white noise. It is also clear that this noise is positive. A smoothed out version of this noise model was introduced in [5].

Example 4.4. More generally, let $f \in S'(\mathbb{R}^d)$ and let us set $\varphi = \mathcal{E}(f)$ (i.e. $\varphi = \mathrm{Exp}\langle \cdot, f \rangle$). As before we define the homogeneous field $\{V(x); x \in \mathbb{R}^d\}$ by $V(x) = \theta_x \varphi$. Now we have

$$[SV(x)](\xi) = [S(\theta_x \varphi)](\xi) = [S\varphi](\tau_{-x} \xi) = \exp\langle \tau_{-x} \xi, f \rangle = \exp\langle \xi, \tau_x f \rangle,$$

in other words, $V(x) = \mathcal{E}(\tau_x f)$. This is also a positive homogeneous noise.

§5 APPLICATIONS TO SPDE'S

This last section is devoted to the discussion of the model for fluid flow in a porous medium first proposed by Lindström et al. in [5].

Let D be a bounded domain in $\mathbb{R}^d$ and let f be a bounded measurable function on D. We consider the stationary (i. e. time independent) boundary value problem

$$\mathrm{div}\,(K(x)\nabla p(x)) = -f \quad \text{in } D$$
$$p(x) = 0 \qquad \text{on } \partial D, \tag{16}$$

where $K(x)$ is a random term modelling the permeability of the medium. This term has to be non-negative. Moreover, physical measurements suggest that $K(x)$ be taken as a function of white noise. Lindström et al. proposed to take $K(x) = \text{Exp}(W_x)$. Recall Examples 4.3 and 4.4. It turns out that equation (16) cannot be solved in the space $(\mathcal{S})^*$ of Hida's distributions. This problem was one of the reasons we had to introduce a new space of distributions. Indeed, if one takes the S-transform, at least formally, of both sides of equation (16) we get

$$\text{div}\,([SK(x)](\xi)\nabla[Sp(x)](\xi)) = -f \tag{17}$$

in D with the boundary condition

$$[Sp(x)](\xi) = 0$$

for $x \in \partial D$. Notice that $Sf = f$ because f is deterministic. As explained in the introduction, solving this transformed problem is the way to look for a solution of the problem (16). Indeed, for each fixed $\xi \in \mathcal{S}(\mathbb{R}^d)$ equation (17) is a deterministic boundary value problem which can be solved by classical methods. For convenience we set $k_\xi(x) = [SK(x)](\xi)$ and $p_\xi(x) = [Sp(x)](\xi)$ for the S-transforms. With these notations, equation (17) can be rewritten as

$$\text{div}(k_\xi(x)\nabla p_\xi(x)) = -f \tag{18}$$

in D with the boundary condition

$$p_\xi(x) = 0$$

on the boundary ∂D. Because of our background (and taste) we follow [5] in using the probabilistic representation of the solution of (17). Is is given by the formula

$$p_\xi(x) = \frac{1}{\sqrt{k_\xi(x)}}\widehat{\mathbb{E}}_x\left\{\int_0^{T_D} f(X_t)\sqrt{k_\xi(X_t)}e^{\int_0^t V_\xi(X_s)\,ds}\,dt\right\} \tag{19}$$

where we used the notation

$$V_\xi(x) = \frac{1}{2}\Delta(\log k_\xi)(x) + |\nabla(\log k_\xi)(x)|^2$$

and where $(X_t, \widehat{\mathbb{P}}_x)$ is an auxiliary (independent) process of Brownian motion in $\mathbb{R}^d$, the variance of X_t being $2t$ and the first exit time from D being denoted by T_D. This formula can be understood in a simple way if one develops the divergence. The operator in the left hand side becomes

$$k_\xi(x)\Delta p_\xi(x) + \nabla k_\xi(x) \cdot \nabla p_\xi(x).$$

In the probabilistic representation of the solution of second order elliptic problems, the multiplication by k_ξ corresponds to a time change while the first order gives a Cameron–Martin type change of measure. Using the fact that the coefficient of the first order term is a gradient and Itô's formula one can transform the stochastic integral appearing in the Cameron–Martin formula into a standard additive functional. This gives a Feynman–Kac type exponential term involving the potential $V_\xi(x)$ as defined above.

What remains to be done is now clear: one merely has to invert the S-transform, replacing the usual products with Wick products and the exponentials with Wick exponentials. With a modicum of care necessary to check that the product and integral operations do not take the quantities of interest outside our space $\widetilde{\mathcal{M}}^*$, the final part of the route to the solution should be an easy ride.

Theorem 5.1. *Let us assume that the random permeability is of the form* $K(x) = \mathrm{Exp}(\langle \cdot, \tau_x g \rangle) = \mathcal{E}(\tau_x g)$ *for some* $g \in L^2(\mathbb{R}^d)$. *Then equation* (16) *has a unique weak solution in* $\widetilde{\mathcal{M}}^*$. *This solution is given by*

$$p(x) = \mathrm{Exp}\left[-\frac{1}{2}\langle \cdot, \tau_x g \rangle\right] \widehat{\mathbb{E}}_x \left\{ \int_0^{T_D} f(X_t) \, \mathrm{Exp}\left[\frac{1}{2}\langle \cdot, \tau_{X_t} g \rangle \right. \right.$$
$$\left. \left. - \int_0^t \left(\frac{1}{4}\langle \cdot, \tau_{X_s} \nabla g \rangle \diamond \langle \cdot, \tau_{X_s} \nabla g \rangle + \frac{1}{2}\langle \cdot, \tau_{X_s} \Delta g \rangle\right) ds\right] dt \right\} \quad (20)$$

where we used the notation

$$\langle \cdot, \tau_{X_s} \nabla g \rangle \diamond \langle \cdot, \tau_{X_s} \nabla g \rangle = \sum_{j=1}^d \left\langle \cdot, \tau_{X_s} \frac{\partial g}{\partial x_j} \right\rangle \left\langle \cdot, \tau_{X_s} \frac{\partial g}{\partial x_j} \right\rangle.$$

The reader has to be aware that the existence part of the statement does not need any justification since the notion of a weak solution in white noise calculus is based on the S-transform: a solution is nothing but a generalized function (over the white noise space) whose S-transform is, for each ξ for which the local S-transform is defined, a solution of the corresponding deterministic problem.

Proof of Theorem 5.1. Let us fix momentarily a Brownian path $0 < s < t$ and $\xi \in \mathcal{S}(\mathbb{R}^d)$. We have

$$[S(\langle \cdot, \tau_{X_s} \nabla g \rangle \diamond \langle \cdot, \tau_{X_s} \nabla g \rangle)](\xi) = \langle \xi, \tau_{X_s} \nabla g \rangle^t \langle \xi, \tau_{X_s} \nabla g \rangle = \langle \tau_{-X_s} \nabla \xi, g \rangle^t \langle \tau_{-X_s} \nabla \xi, g \rangle$$

where we used the notation ^{t} for the transpose of a d-dimensional vector. In other words, $a^t a = \sum_{j=1}^d a_j^2$. Consequently, if $\lambda \in \mathbb{C}$ we have

$$\left| [S(\langle \cdot, \tau_{X_s} \nabla g \rangle \diamond \langle \cdot, \tau_{X_s} \nabla g \rangle)](\lambda \xi) \right| \leq |\lambda|^2 \, |\langle \tau_{-X_s} \nabla \xi, g \rangle|^2$$

$$\leq |\lambda|^2 \sum_{j=1}^d \left|\frac{\partial \xi}{\partial x_j}\right|_2^2 |g|_2^2 \leq |\lambda|^2 \, |g|_2^2 \, |\xi|_{2,1}^2.$$

Similarly we have

$$|[S\langle\cdot,\tau_{X_s}\Delta g\rangle](\lambda\xi)| \le |\lambda|\,|g|_2\,|\xi|_{2,1}.$$

This implies that, for each Brownian path and for each $t > 0$, the integral

$$I_t = \int_0^t V_\xi(X_s)\,ds = \int_0^t \left(\frac{1}{4}\langle\cdot,\tau_{X_s}\nabla g\rangle \diamond \langle\cdot,\tau_{X_s}\nabla g\rangle + \frac{1}{2}\langle\cdot,\tau_{X_s}\Delta g\rangle\right)ds$$

exists in $(\mathcal{S})^*$ in the sense of Bochner. Now, for each complex number λ, we have:

$$\left|\left[S\operatorname{Exp}\left[\frac{1}{2}\langle\cdot,\tau_{X_t}g\rangle - I_t\right]\right](\lambda\xi)\right|$$
$$\le \exp\left[\frac{|\lambda|}{2}|g|_2\,|\xi|_2\right]\exp\left[t\left(\frac{d}{8}|\lambda|^2\,|g|_2^2\,|\xi|_{2,1}^2 + |\lambda|\,|g|_2\,|\xi|_{2,1}\right)\right].$$

This implies that, for each path $\widehat{\omega}$ of the Brownian motion the integral

$$\varphi(\widehat{\omega}) = \int_0^{T_D(\widehat{\omega})} f(X_t(\widehat{\omega}))\operatorname{Exp}\left[\frac{1}{2}\langle\cdot,\tau_{X_t(\widehat{\omega})}g\rangle - I_t\right]dt$$

exists in $(\mathcal{S})^*$ in the sense of Bochner and for each complex number λ we have

$$|[S\varphi(\omega)](\lambda\xi)| \le c\exp\left[\frac{|\lambda|}{2}|g|_2\,|\xi|_2\right]\exp\left[T_D(\omega)\left(\frac{d}{8}|\lambda|^2\,|g|_2^2\,|\xi|_{2,1}^2 + |\lambda|\,|g|_2\,|\xi|_{2,1}\right)\right].$$

We used the fact that $xe^x \le e^{2x}$ for $x \ge 0$. Since we assume that the domain D is bounded, there exists an $\epsilon > 0$ such that

$$\widehat{\mathbb{E}}_x\{e^{\epsilon T_D}\} < \infty$$

for each $x \in D$. Let us choose $p \ge 1$ and $q > 0$ in such a way that

$$\frac{d}{4}2^{-2(p-1)+q}|g|_2^2 + 2e^{-(p-1)+q/2}|g|_2 < \epsilon.$$

Then, for any $\xi \in \mathcal{B}_{p,q}$ and each complex number λ satisfying $|\lambda| \le 1$, we have

$$|[S\varphi(\omega)](\lambda\xi)| \le c'e^{\epsilon T_D(\omega)}$$

and consequently, $\varphi(\omega)$ is integrable with respect to $\widehat{\mathbb{P}}_x$ in the sense of Bochner in some Hilbert space $\mathcal{M}_{-p_1,-q_1}$ and for $\xi \in \mathcal{B}_{p,q}$ we have:

$$[S\widehat{\mathbb{E}}_x\{\varphi\}](\xi) = \widehat{\mathbb{E}}_x\left\{\int_0^{T_D} f(X_t)\exp\left[\frac{1}{2}\langle\xi,\tau_{X_t}g\rangle\right.\right.$$
$$\left.\left. - \int_0^t \left(\frac{1}{2}|\langle\xi,\tau_{X_s}\nabla g\rangle|^2 + \langle\xi,\tau_{X_s}\Delta g\rangle\right)ds\right]dt\right\}.$$

From this we can conclude that, when $\xi \in \mathcal{B}_{p,q}$, the S-transform of $p(x)$ defined by formula (20) does exist and it is given by formula (19) above. Hence, $p(x)$ is the unique solution of (16) which is the unique solution of equation (17). This concludes the proof of the theorem. □

Acknowledgements: This work was done while the second named author (J. A. Yan) was visiting the Department of Mathematics of the University of California at Irvine. The latter should be thanked for its warm hospitality and for financial support.

REFERENCES

1. R. Carmona and J. Lacroix, *Spectral Theory of Random Schrödinger Operators*, Birkhäuser, Boston, 1990.
2. H. Holden, T. Lindström, B. Øksendal, J. Uboe, and T. S. Zhang, *Stochastic boundary value problem: a white noise approach*, Probab. Theory Relat. Fields **95** (1993), 391–419.
3. D. C. Khandeker and L. Streit, *Constructing the Feynman integrand*, preprint, 1992.
4. Ju. G. Kondratiev and L. Streit, *Spaces of white noise distributions: constructions, descriptions, applications. I*, preprint, 1992.
5. T. Lindström, B. Øksendal and J. Uboe, *Stochastic differential equations involving positive noise*, Stochastic Analysis (M. Barlow and N. Bingham, eds.), Univ. Press, Cambridge, 1991.
6. P. A. Meyer and J. A. Yan, *Les fonctions "characteristiques" des distributions sur l'espace de Wiener*, in Semmin. Proba. XXV, Lect. Notes in Math. **1485** (1991), 61–78.
7. L. Nachbin, *Topologies on Spaces of Holomorphic Mappings*, Springer-Verlag, New York, N. Y., 1969.
8. J. Potthoff and L. Streit, *A characterization of Hida's distributions*, Stochastic Partial Differential Equations and Their Applications, Lect. Notes in Control and Info. Sci. (B. L. Rozovskii and R. B. Sowers, eds.), vol. 171, 1991, pp. 238–251.
9. J. Potthoff and L. Streit, *A characterization of Hida's distributions*, J. Functional Anal. **101** (1991), 212–229.
10. L. Streit and W. Westerkamp, *A generalization of the characterization theorem for generalized functionals of white noise*, preprint, 1991.
11. J. A. Yan, *Inequalitites for products of white noise functionals*, Stochastic Processes. A Festschrift in Honor of G. Kallianpur (S. Cambanis et al., eds.), Springer-Verlag, New York, N. Y., 1993, pp. 349–358.
12. J. A. Yan, *A Cameron–Martin expansion approach to white noise analysis*, preprint (1993).
13. T. S. Zhang, *Characterizations of the white noise functionals and Hida's distributions*, Stochastics and Stochastic Reports **41** (1992), 71–87.

RENÉ A. CARMONA, DEPARTMENT OF MATHEMATICS, UNIVERSITY OF CALIFORNIA IRVINE, IRVINE, CA 92717, USA

J. A. YAN, INSTITUTE OF APPLIED MATHEMATICS, ACADEMIA SINICA, BEIJING 100080, CHINA

Progress in Probability, Vol. 36
© 1995 Birkhäuser Verlag Basel/Switzerland

DISSIPATIVITY OF THREE-DIMENSIONAL
STOCHASTIC NAVIER–STOKES EQUATION

HANS CRAUEL AND FRANCO FLANDOLI

ABSTRACT. A three-dimensional stochastic Navier–Stokes equation is considered. The aim is to prove that the associated stochastic flow has a compact random absorbing set.

§1 INTRODUCTION

The paper is concerned with the three-dimensional Navier–Stokes equation for an incompressible fluid in a bounded domain, with a random perturbation of the body forces in the form of a Gaussian space–time random field that is white noise in time. Preliminary results on strong solutions are proved. Strong solutions exist locally in time, and are unique, but existence of global strong solutions is an open problem, as in the deterministic case. *Assuming that such solutions are global,* i. e. that singularities do not develop in finite time, we prove that the system is dissipative, in the sense that there exists a compact random absorbing set for the associated stochastic flow. This is the basic property in order to investigate existence of a compact random attractor and invariant measures, see [3]. The analysis developed here is strongly based on the ideas of [2] and [9].

§2 FORMULATION

Let D be a regular bounded open domain of $\mathbb{R}^3$. We consider the three-dimensional stochastic Navier–Stokes equation in D

$$\frac{\partial u(t, x)}{\partial t} = \Delta u(t, x) - (u(t, x) \cdot \nabla)u(t, x) - \nabla p(t, x) + f(x) + n(t, x)$$

with the incompressiblity condition

$$\operatorname{div} u(t, x) = 0$$

and the boundary condition

$$u(t, x) = 0, \quad x \in \partial D.$$

1991 *Mathematics Subject Classification.* primary: 58F11, 58F12; secondary: 34D45, 35Q30, 60H15, 76D05.

Key words and phrases. Stochastic Navier–Stokes equation, attractors, invariant measures.

Here $n(t,x)$ is a Gaussian random field, white noise in time, subject to the restrictions imposed below. Definitions and assumptions concerning this equation will be given at the level of the classical abstract formulation that we are going to introduce.

Let $\mathcal{V}$ be the space of infinitely differentiable three-dimensional vector fields $u(x)$ on D with compact support strictly contained in D, satisfying $\operatorname{div} u(x) = 0$. We denote by V_α the closure of $\mathcal{V}$ in $[H^\alpha(D)]^3$, for all real $\alpha \geq 0$, and we set in particular

$$H = V_0, \quad V = V_1.$$

We denote by $|\cdot|$ and $\langle \cdot, \cdot \rangle$ the norm and inner product in H.

Moreover, we set $D(A) = [H^2(D)]^3 \cap V$, and define the linear operator $A : D(A) \subset H \to H$ as $Au = -\Delta u$. Since V coincides with $D(A^{1/2})$, we can endow V with the norm $\|u\| = |A^{1/2}u|$. The operator A is positive self-adjoint with compact resolvent; we denote by $0 < \lambda_1 \leq \lambda_2 \leq \cdots$ the eigenvalues of A, and by $e_1, e_2, \ldots$ a corresponding complete orthonormal system of eigenvectors. We remark that $\|u\|^2 \geq \lambda_1 |x|^2$.

We define the bilinear operator $B(u,v) : V \times V \to V'$ as

$$\langle B(u,v), z \rangle = \int_D z(x) \cdot (u(x) \cdot \nabla) v(x)\, dx$$

for all $z \in V$. By the incompressibility condition we have

$$\langle B(u,v), v \rangle = 0, \quad \langle B(u,v), z \rangle = -\langle B(u,z), v \rangle.$$

By a classical projection procedure we formally arrive to the abstract equation

$$du(t) + Au(t)\, dt + B(u(t), u(t))\, dt = f\, dt + dw(t). \tag{1}$$

Here we assume that $w(t)$ is a finite-dimensional Brownian motion of the form

$$w(t) = \sum_{j=1}^n \sigma_j \beta_j(t) e_j$$

where $\beta_1, \ldots, \beta_n$ are independent standard Brownian motions on a complete probability space $(\Omega, \mathcal{F}, P)$ (with expectation denoted by E), and σ_j are real coefficients. It is possible to consider an infinite-dimensional Brownian motion, but not white noise in space; to restrict considerably the length of the computations we shall only deal with the finite dimensional case.

We also assume that, for each j, $\beta_j(t)$ is a two-sided Brownian motion, i. e. it is defined for all real t. We shall denote by $\mathcal{F}_s$ the σ-algebra generated by $w(\tau)$ for $\tau < s$.

§3 Main Results

3.1 Existence, uniqueness and regularity.

Definition 3.1. Let t_0 be a given initial time, and $T = T(\omega)$ be a random variable with $T > t_0$ *P*-a. s. We say that a stochastic process $u(t, \omega)$ is a *weak solution* of equation (1) over the (random) time interval $[t_0, T]$ if

$$u(\cdot, \omega) \in L^\infty(t_0, T(\omega); H) \cap L^2(t_0, T(\omega); V)$$

for *P*-a. e. $\omega \in \Omega$, the extension of u equal to 0 after $T(\omega)$ is progressively measurable in these topologies, and *P*-a. s. equation (1) is satisfied in the integral sense

$$\langle u(t), \phi \rangle + \int_{t_0}^t \langle u(s), A\phi \rangle \, ds - \int_{t_0}^t \langle B(u(s), \phi), u(s) \rangle \, ds$$

$$= \langle u(t_0), \phi \rangle + \int_{t_0}^t \langle f, \phi \rangle \, ds + \langle w(t) - w(t_0), \phi \rangle$$

for all $t \in [t_0, T]$ and all $\phi \in D(A)$. If in addition

$$u(\cdot, \omega) \in C(t_0, T(\omega); V) \cap L^2(t_0, T(\omega); D(A))$$

for *P*-a. e. $\omega \in \Omega$ and the previous extension of u is progressively measurable in these topologies, than we say that u is a *strong solution*.

Theorem 3.2.

(i) *For each (deterministic) time interval $[t_0, T]$, each initial condition $u_{t_0} \in H$ and each forcing term $f \in L^2(t_0, T; V')$, there exists at least one weak solution u of equation (1) over $[t_0, T]$ with the initial condition $u(t_0) = u_{t_0}$.*

(ii) *For each t_0, each initial condition $u_{t_0} \in V$ and each forcing term $f \in L^2(t_0, T; H)$, we can find a random interval $[t_0, T(\omega)]$ such that there exists one and only one strong solution u of equation (1) over $[t_0, T(\omega)]$ with the initial condition $u(t_0) = u_{t_0}$.*

(iii) *Given a (deterministic) time interval $[t_0, T]$ and a forcing term f in the space $L^2(t_0, T; H)$, assume that for all $u_{t_0} \in V$ there exists at least one global strong solution on $[t_0, T]$; then:*

(a) *it is unique, also in the class of weak solutions;*

(b) *P-a. s. the mapping $u_0 \mapsto u(\cdot, u_0)$ is bounded from V to $C(t_0, T; V)$: for every $\rho > 0$ there exists a random variable $C(\rho, \omega)$ such that*

$$\sup_{0 \le t \le T; \|u_0\| \le \rho} \|u(t, u_0)\| \le C(\rho, \omega);$$

(c) *P-a. s., for each $t \in [t_0, T]$ the mapping $u_0 \mapsto u(t, u_0)$ is continuous in V.*

The proof of this theorem is given in Section 4.

3.2 Dissipativity. The main result of this paper is:

Theorem 3.3. *Let $f(t) = f \in H$ be given. Assume that the stochastic Navier–Stokes equation (1) does not develop singularities in finite time, in the sense that for all $t_0 < T$ and all $u_{t_0} \in V$ there exists at least one global strong solution on $[t_0, T]$. Denote by $u(t, \omega; t_0, u_{t_0})$ the unique strong solution of equation (1) over $[t_0, \infty)$, starting from the point $u(t_0) = u_{t_0}$. Then there exists a real random variable $\rho(\omega)$ such that for P-a. e. $\omega \in \Omega$ the following property holds true: for all bounded sets $B \subset H$ there exists $\tau(B, \omega) < 0$ such that for all $t_0 < \tau(B, \omega)$ and all $u_{t_0} \in B$ we have*

$$\|u(0, \omega; t_0, u_{t_0})\| \le \rho(\omega).$$

Because of this property, the ball $B_V(0, \rho(\omega))$ in V can be called a compact (in H) random absorbing set. The previous property is a quite strong form of stochastic dissipation.

The proof of the theorem is given in the next section.

§4 PROOFS OF THEOREMS 3.2 AND 3.3

Step 1: Consider the auxiliary Ornstein–Uhlenbeck equation

$$dz_\alpha(t) + A z_\alpha(t) \, dt + \alpha z_\alpha(t) \, dt = dw(t)$$

where α is a positive real number. This is in fact a finite-dimensional equation, since the noise is finite-dimensional and commutes with A. Therefore it is clear that the process

$$z_\alpha(t) = \int_{-\infty}^{t} e^{(t-s)(-A-\alpha)} \, dw(s)$$

is a stationary ergodic solution with continuous trajectories, taking values for instance in $D(A)$ (but in fact in any power $D(A^k)$, because the eigenvectors e_j belong to such spaces). Of course in the case of infinite-dimensional noise more care and lengthy details would be necessary.

Let $\alpha \ge 0$ be given. Denote for simplicity $z_\alpha(t)$ by $z(t)$. By the classical change of variable $v(t) = u(t) - z(t)$ we obtain the differential equation

$$\frac{dv(t)}{dt} + Av(t) + B(v(t) + z(t), v(t) + z(t)) = \alpha z(t) + f. \tag{2}$$

For almost all given paths of the process $z(t)$ we can study this equation as a deterministic evolution equation and then deduce results for $u(t)$.

Step 2: We prove Theorem 3.2(i). Let P_n be the projection operator in H onto the space spanned by $e_1, \ldots, e_n$. Consider the ordinary differential equation

$$\frac{dv_n(t)}{dt} + Av_n(t) + P_n B(v_n(t) + z(t), v_n(t) + z(t)) = \alpha P_n z(t) + P_n f \tag{3}$$

with the initial condition

$$v_n(t_0) = P_n(u_{t_0} - z(t_0)).$$

Its maximal (unique) solution satisfies

$$\frac{1}{2}\frac{d}{dt}|v_n|^2 + \|v_n\|^2 = -\langle B(v_n + z, v_n + z), v_n\rangle + \alpha\langle z, v_n\rangle + \langle f, v_n\rangle$$

$$\leq \varepsilon\|v_n\|^2 + C(\varepsilon)|z|_{L^4}^{8/3}|v_n|^2 + C(\varepsilon)|z|_{L^4}^4$$

$$+ C(\varepsilon)\alpha^2|z|^2 + \varepsilon|v_n|^2 + C(\varepsilon)|f|_{V'}^2 + \varepsilon\|v_n\|^2.$$

Here we have used the first part of Lemma 4.1 below; moreover, ε is an arbitrary positive number and $C(\varepsilon)$ is a positive constant. Taking ε sufficiently small (and denoting the corresponding constant by C) we have

$$\frac{1}{2}\frac{d}{dt}|v_n|^2 \leq \left(-\frac{\lambda_1}{2} + C|z|_{L^4}^{8/3}\right)|v_n|^2 + C|z|_{L^4}^4 + C\alpha^2|z|^2 + C|f|_{V'}^2.$$

Therefore, for all $t \in [t_0, T]$ (renaming C)

$$|v_n(t)|^2 \leq e^{\int_{t_0}^t (-\lambda_1 + C|z(s)|_{L^4}^{8/3})\,ds}|v_n(t_0)|^2$$

$$+ \int_{t_0}^t e^{\int_\sigma^t (-\lambda_1 + C|z(s)|_{L^4}^{8/3})\,ds} C(|z(\sigma)|_{L^4}^4 + \alpha^2|z(\sigma)|^2 + |f|_{V'}^2)\,d\sigma. \quad (4)$$

With possibly a different choice of ε above, we also obtain

$$\frac{1}{2}\frac{d}{dt}|v_n|^2 + \frac{\lambda_1}{2}\|v_n\|^2 \leq C|z|_{L^4}^{8/3}|v_n|^2 + C|z|_{L^4}^4 + C\alpha^2|z|^2 + C|f|_{V'}^2$$

which gives us, over a generic interval $[r, s] \subset [t_0, T]$ (renaming C)

$$\int_r^s \|v_n(\sigma)\|^2\,d\sigma \leq |v_n(r)|^2 + \int_r^s C(|z|_{L^4}^{8/3}|v_n|^2 + |z|_{L^4}^4 + \alpha^2|z|^2 + |f|_{V'}^2)\,d\sigma. \quad (5)$$

First, (4) gives us an a priori bound on the maximal solution which implies that v_n is a global solution, over $[t_0, T]$. Therefore, from (4) and (5) we obtain that the sequence v_n is bounded in $L^\infty(t_0, T; H)$ and in $L^2(t_0, T; V)$. By classical arguments (see [6], [9]), that at this point do not require any novelty, one can prove further bounds on the time derivative of v_n, and reasoning on weakly and strongly convergent subsequences one gets the existence of a global weak solution v, for a. e. ω. The processes $v_n(t, \omega)$ are progressively measurable, being obtained by limiting procedures over successive approximations which are progressively measurable at each step (by their explicit definition). By difference, the processes $u_n(t, \omega)$ are also progressively measurable. It is not difficult to deduce now the same property for v and u, analyzing in detail the various limiting procedures for v_n and u_n.

Finally, the solution v just obtained satisfies the two inequalities

$$|v(t)|^2 \le e^{\int_{t_0}^t (-\lambda_1 + C|z(s)|_{L^4}^{8/3}) \, ds} |v(t_0)|^2$$
$$+ \int_{t_0}^t e^{\int_\sigma^t (-\lambda_1 + C|z(s)|_{L^4}^{8/3}) \, ds} C\big(|z(\sigma)|_{L^4}^4 + \alpha^2 |z(\sigma)|^2 + |f|_{V'}^2\big) \, d\sigma. \quad (6)$$

for all $t \in [t_0, T]$, and

$$\int_r^s \|v(\sigma)\|^2 \, d\sigma \le |v(r)|^2 + \int_r^s C\big(|z|_{L^4}^{8/3}|v|^2 + |z|_{L^4}^4 + \alpha^2|z|^2 + |f|_{V'}^2\big) \, d\sigma \quad (7)$$

over a generic interval $[r, s] \subset [t_0, T]$.

We conclude this step by the lemma announced above.

Lemma 4.1. *For all $\varepsilon > 0$ there exists a constant $C(\varepsilon) > 0$ such that: for all $v \in V$ and $z \in [L^4(D)]^3$*

$$|\langle B(v+z, v+z), v \rangle| \le \varepsilon \|v\|^2 + C(\varepsilon)|z|_{L^4}^{8/3}|v|^2 + C(\varepsilon)|z|_{L^4}^4; \quad (8)$$

for all $v, z \in D(A)$,

$$\langle Av, B(v+z, v+z) \rangle \le \varepsilon |Av|^2 + C(\varepsilon)\{\|v\|^6 + |Az|^6 + 1\}. \quad (9)$$

Proof. By the Hölder inequality, for all $v \in V$, $u, z \in [L^4(D)]^3$ we have (C denotes a generic positive constant)

$$\langle B(u, v), z \rangle \le C\|v\| \, |u|_{L^4}|z|_{L^4}.$$

Therefore

$$\begin{aligned}
\langle B(v+z, v+z), v \rangle &= -\langle B(v+z, v+z), z \rangle \\
&= -\langle B(v+z, v), z \rangle \\
&= -\langle B(v, v), z \rangle - \langle B(z, v), z \rangle \\
&\le C\|v\| \, |v|_{L^4}|z|_{L^4} + C\|v\| \, |z|_{L^4}^2 \\
&\le C\|v\| \, |v|_{H^{3/4}}|z|_{L^4} + C\|v\| \, |z|_{L^4}^2 \\
&\le C\|v\| \, \|v\|^{1/4}|v|^{3/4}|z|_{L^4} + C\|v\| \, |z|_{L^4}^2 \\
&\qquad \text{(by an interpolation inequality)} \\
&= C\|v\|^{5/4}|v|^{3/4}|z|_{L^4} + C\|v\| \, |z|_{L^4}^2.
\end{aligned}$$

This bound easily implies (8), recalling the Young inequality: for all $a, b, \varepsilon > 0$, $r > 1$ and $r' = \frac{r}{r-1}$,

$$ab \leq \frac{\varepsilon}{r}a^r + \frac{1}{r'\varepsilon^{r'/r}}b^{r'}.$$

To prove (9), we use an interpolation inequality of Agmon (cf. [9]):

$$\begin{aligned}
\langle Av, B(v + z, v + z)\rangle &\leq C|Av|\,\|v + z\|\,|v + z|_{L^\infty} \\
&\leq C|Av|\,\|v + z\|^{3/2}|A(v + z)|^{1/2} \\
&\leq C\|v\|^{3/2}|Av|^{3/2} + C\|v\|^{3/2}|Az|^{1/2}|Av| \\
&\quad + C\|z\|^{3/2}|Av|^{3/2} + C\|z\|^{3/2}|Az|^{1/2}|Av| \\
&\leq \varepsilon|Av|^2 + C(\varepsilon)\{\|v\|^6 + \|v\|^3|Az| + \|z\|^6 + \|z\|^3|Az|\} \\
&\leq \varepsilon|Av|^2 + C(\varepsilon)\{\|v\|^6 + |Az|^6 + 1\}.
\end{aligned}$$

The proof is complete. $\square$

Step 3: We prove Theorem 3.2(ii). Let v_n be the approximating sequence defined at Step 2. We have

$$\begin{aligned}
\frac{1}{2}\frac{d}{dt}\|v_n\|^2 + |Av_n|^2 &= -\langle B(v_n + z, v_n + z), Av_n\rangle + \alpha\langle z, Av_n\rangle + \langle f, Av_n\rangle \\
&\leq \varepsilon|Av_n|^2 + C(\varepsilon)\{\|v_n\|^6 + |Az|^6 + 1\} \\
&\quad + \varepsilon|Av_n|^2 + C(\varepsilon)|z|^2 + \varepsilon|Av_n|^2 + C(\varepsilon)|f|^2.
\end{aligned}$$

For a suitable ε we have (estimating some terms in a crude way)

$$\frac{1}{2}\frac{d}{dt}\|v_n\|^2 + \frac{1}{2}|Av_n|^2 \leq C\{\|v_n\|^6 + |Az|^6 + 1 + |f|^2\}.$$

The conclusion is identical to [8], pp. 20–21.

Step 4: We outline the proof of Theorem 3.2(iii). In view of part (a), let v_1 be a weak solution (ω may be considered as given) and v_2 a strong solution. Then,

$$\frac{d(v_1 - v_2)}{dt} + A(v_1 - v_2) = B(v_1 + z, v_1 - v_2) + B(v_1 - v_2, v_2 + z).$$

Therefore

$$\begin{aligned}
\frac{1}{2}\frac{d|v_1 - v_2|^2}{dt} + \|v_1 - v_2\|^2 &\leq |\langle B(v_1 - v_2, v_2 + z), v_1 - v_2\rangle| \\
&= |\langle B(v_1 - v_2, v_1 - v_2), v_2 + z\rangle| \\
&\leq C\|v_1 - v_2\|\,|v_1 - v_2|\,|v_2 + z|_{L^\infty} \\
&\leq \varepsilon\|v_1 - v_2\|^2 + C(\varepsilon)|v_1 - v_2|^2|v_2 + z|_{L^\infty}^2
\end{aligned}$$

and by the Gronwall Lemma (and a suitable choice of ε)

$$|v_1(t) - v_2(t)|^2 \leq \exp\left(\int_{t_0}^t C|v_2(s) + z(s)|_{L^\infty}^2 \, ds\right)|v_1(t_0) - v_2(t_0)|^2$$

where we recall that $\int_{t_0}^T |v_2(s) + z(s)|_{L^\infty}^2 \, ds < \infty$ by the properties of z and the assumption that v_2 is a strong solution. It follows that $v_1 = v_2$ if the initial conditions coincide. From this inequality one can also deduce statements analogous to parts (b) and (c) of Theorem 3.2(iii), but relative to the H-topology instead of the V-topology.

The proofs of (b) and (c) are a little lengthy and similar in all the details to the deterministic case (cf. [9], pp. 382–384), so we omit them.

Step 5: We finally prove Theorem 3.3. We begin with a dissipativity property in H. To this end recall inequality (6).

We have assumed that $w(t)$ is finite-dimensional, so that the analysis of $z_\alpha(t)$ reduces completely to the analysis of its components

$$z_\alpha^j(t) = \int_{-\infty}^t e^{(t-s)(-\lambda_j - \alpha)}\sigma_j \, d\beta_j(s).$$

Thus we may readily use the following simple facts from ergodic theory (see [3] for more detailes). For α sufficiently large, $E|z(0)|_{L^4}^{8/3}$ is arbitrarily small (it is sufficient to compute explicitly $E|z(0)|_{L^4}^4$). Therefore we may choose α such that

$$-\lambda_1 + CE|z(0)|_{L^4}^{8/3} \leq -\frac{\lambda_1}{2}$$

where C is the constant appearing in (6).

With this choice of α the Ergodic Theorem gives us

$$\lim_{t_0 \to -\infty} \frac{1}{-t_0}\int_{t_0}^0 \left(-\lambda_1 + C|z(s)|_{L^4}^{8/3}\right) ds = -\lambda_1 + CE|z(0)|_{L^4}^{8/3} \leq -\frac{\lambda_1}{2}.$$

Therefore, given $\omega \in \Omega$, there exists $\tau(\omega) < 0$ such that

$$\int_{t_0}^0 \left(-\lambda_1 + C|z(s,\omega)|_{L^4}^{8/3}\right) ds \leq -\frac{\lambda_1}{4}(-t_0) \quad \forall \, t_0 < \tau(\omega).$$

Moreover, by the continuity of the trajectories of z,

$$\sup_{\tau(\omega) \leq t_0 \leq 0} \int_{t_0}^0 \left(-\lambda_1 + C|z(s,\omega)|_{L^4}^{8/3}\right) ds \leq C(\omega) < \infty.$$

for some constant $C(\omega)$. Introducing the process $\lambda(t_0, \omega)$ defined by

$$\lambda(t_0, \omega) = -\frac{\lambda_1}{4}(-t_0) \quad \forall t_0 < \tau(\omega)$$

and

$$\lambda(t_0, \omega) = C(\omega) \quad \forall \tau(\omega) \le t_0 \le 0,$$

we finally have

$$\int_{t_0}^0 \left(-\lambda_1 + C|z(s, \omega)|_{L^4}^{8/3}\right) ds \le \lambda(t_0, \omega)$$

for all t_0. Assuming without restriction that $\tau \le -1$, we also have, for all $t \in [-1, 0]$,

$$\int_{t_0}^t \left(-\lambda_1 + C|z(s, \omega)|_{L^4}^{8/3}\right) ds \le \lambda(t_0, \omega) + C(\omega).$$

Consider now a strong solution $u(t) = u(t, \omega; t_0, u_{t_0})$, and let

$$v(t) = v(t, \omega; t_0, v_{t_0}) := u(t) - z(t).$$

Going back to inequality (6), we now have, for all $t \in [-1, 0]$ and $t_0 \le t$,

$$|v(t, \omega)|^2 \le e^{\lambda(t_0, \omega) + C(\omega)}|v(t_0, \omega)|^2$$
$$+ \int_{t_0}^t e^{\lambda(\sigma, \omega) + C(\omega)} C\left(|z(\sigma, \omega)|_{L^4}^4 + \alpha^2|z(\sigma, \omega)|^2 + |f|_{V'}^2\right) d\sigma.$$

We also have

$$\lim_{t \to -\infty} \frac{|z(t, \omega)|_{L^4}^4}{t^4} = 0 \tag{10}$$

for P-a.e. ω, and also, as a by-product,

$$\lim_{t \to -\infty} \frac{|z(t)|^2}{t^2} = 0 \quad P\text{-a.s.}$$

Recalling the definition of $\lambda(\sigma, \omega)$, it is now clear that there exists a random variable $r_1(\omega)$, a.s. finite, such that

$$|u(t, \omega; t_0, u_{t_0})|^2 \le r_1(\omega) \tag{11}$$

for all $t \in [-1, 0]$, $t_0 \le t$ and $u_{t_0} \in V$. This is the dissipativity property in H.

Using this fact in the inequality (7), we can find a random variable $r_2(\omega)$, a.s. finite, such that

$$\int_{-1}^0 \|u(t, \omega; t_0, u_{t_0})\|^2 dt \le r_2(\omega) \tag{12}$$

for all $t_0 \le -1$ and $u_{t_0} \in V$.

The final argument of the proof is now similar to the deterministic one (cf. 9, p. 385). Let ω be fixed. By (12), for all $t_0 \leq -1$ and $u_{t_0} \in V$ there exists $t_{(t_0, u_{t_0})}$ such that

$$\|u(t_{(t_0, u_{t_0})}, \omega; t_0, u_{t_0})\|^2 \leq r_2(\omega). \tag{13}$$

We say that from (11), (12), (13) it follows that there exists $r_3(\omega)$ such that

$$\|u(0, \omega; t_0, u_{t_0})\|^2 \leq r_3(\omega) \tag{14}$$

for all $t_0 \leq -1$ and $u_{t_0} \in V$, which is precisely the claim of Theorem 3.3. By contradiction, if (14) is not true, we can find a sequence of strong solutions $u_n(t)$ of equation (1) over $[-1, 0]$ (ω is fixed, and the meaning of strong solution is the same as that defined in Definition 3.1) that are equibounded in $L^\infty(-1, 0; H)$ and $L^2(-1, 0; V)$, weak* convergent to some u^* in $L^\infty(-1, 0; H)$ and weakly convergent to u^* in $L^2(-1, 0; V)$, and there exists a sequence $t_n \in [-1, 0]$ convergent to some $t^* \in [-1, 0]$, with $\|u_n(t_n)\|$ bounded and $u_n(t_n)$ V-weakly convergent to some $x \in V$, but such that $\|u_n(0)\| \to \infty$. Note also that $u_n(t^*)$ converges weakly to x in V. If we consider the strong solution $u(t)$ starting from x at time t^*, it is possible to show that $u^* = u$ and reach a contradiction, as in [9], pp. 382–383. The proof is complete. $\square$

References

1. A. Bensoussan, R. Temam, *Equations stochastiques du type Navier–Stokes*, J. Func. Anal. **13** (1973), 195–222.
2. P. Constantin, C. Foias, R. Temam, *Attractors representing turbolent flows*, Memoirs of AMS **53, n. 314**, n. 314 (1985).
3. H. Crauel, F. Flandoli, *Attractors for random dynamical systems*, preprint n. 148, Scuola Normale Superiore, Pisa, (1992) (to appear on Prob. Th. Rel. Fields).
4. G. Da Prato, J. Zabczyk, *Stochastic Equations in Infinite Dimensions*, Cambridge, 1992.
5. H. Fujita Yashima, *Equations de Navier–Stokes Stochastiques non Homogenes et Applications*, Scuola Normale Superiore, Pisa, 1992.
6. J. L. Lions, *Quelques Méthodes de Résolution des Problèmes aux Limites non Linéaires*, Dunod, Paris, 1969.
7. A. Pazy, *Semigroups of Linear Operators and Applications to Partial Differential Equations*, Springer-Verlag, New York, 1983.
8. R. Temam, *Navier–Stokes Equations and Nonlinear Functional Analysis*, SIAM, Philadelphia, 1983.
9. R. Temam, *Navier–Stokes Equations*, North-Holland, Amsterdam, 1984.
10. R. Temam, *Infinite Dimensional Dynamical Systems in Mechanics and Physics*, Springer-Verlag, New York, 1988.

HANS CRAUEL, FACHBEREICH MATHEMATIK, UNIVERSITÄT DES SAARLANDES, 6600 SAARBRÜKKEN 11, FEDERAL REPUBLIC OF GERMANY

FRANCO FLANDOLI, SCUOLA NORMALE SUPERIORE, PIAZZA DEI CAVALIERI 7, 56100 PISA, ITALY

Progress in Probability, Vol. 36
© 1995 Birkhäuser Verlag Basel/Switzerland

BERNSTEIN DIFFUSIONS
AND EUCLIDEAN QUANTUM FIELD THEORY

A. B. Cruzeiro[1], Z. Haba[2], and J. C. Zambrini[1]

ABSTRACT. We extend to an infinite-dimensional context the construction as well as the calculus of variations associated with Bernstein processes. We focus on the aspects which are relevant for the physical applications to two-dimensional quantum fields.

§0 Introduction

Euclidean quantum field theory refers usually to the various approaches originated in the works of J. Schwinger and K. Symanzik, dating back to the sixties [1, 2, 3]. The first mathematical realization of these ideas is due to E. Nelson and can be formulated as the problem to give a mathematical sense via the theory of Markov processes to the formal functional integrals manipulated by theoretical physicists. Other approaches have been focused on the relations between Minkowski and Euclidean Green functions or on the connection between Euclidean quantum field theory and (classical) statistical mechanics [4, 5]. The Euclidean approach to quantum physics we are referring to here is quite distinct from the above mentioned ones. It is founded on a purely probabilistic analogue to non-relativistic quantum mechanics proposed by E. Schrödinger in 1931–1932 [6]. The original remark of Schrödinger has been completed some years ago and shown to be, indeed, a new possible starting point for Euclidean physics, in other words a new "Euclidean Quantum Mechanics" [7]. In the simplest non-relativistic situation, i.e., the dynamics of a system of particles in potentials, the relevant stochastic processes are Markovian diffusions. They have been introduced under the name of "Bernstein diffusions" because S. Bernstein stressed that Schrödinger's idea appeared to be related with some kind of probabilistic extension of the classical calculus of variations [7, 8, 9, 10]. Various realizations of Bernstein processes are known today; they are sometimes refered to as "Schrödinger processes" in the literature (cf. [11,

---

1991 *Mathematics Subject Classification.* 60H07, 60G20, 81T99.
Key words and phrases. Euclidean quantum mechanics, Bernstein processes, quantum fields.
[1]Partly supported by JNICT
[2]Supported by Fundação Calouste Gulbenkian

12] and references therein) although some of their general properties are studied for their own mathematical sake, without relation with quantum mechanics.

Malliavin's stochastic calculus of variations [13, 14] has proved to be a very natural tool for the study of the functionals of Bernstein diffusions associated with our version of Euclidean quantum mechanics. Indeed, those "path integrals" require a functional calculus which should be the Euclidean (i.e. well-defined) counterpart of the formal one designed originally by R. Feynman [17]. Such calculus has been shown to be a simple consequence of Malliavin's framework in the case where the state space of the diffusions is finite-dimensional, i.e., for non-relativistic quantum mechanics [15, 16]. It is the purpose of the present article to extend some of these results when the state space is a Banach space of continuous functions or a distribution space. This is what is required for quantum field theory.

The main results, in this infinite-dimensional context, have no direct connection with Euclidean quantum field theory, and they are the object of a longer publication [18]. Here, we shall focus on these aspects relevant for the physical applications to two-dimensional quantum fields (where the dimension of a field theory refers here to the dimension of the domain of the field).

More precisely, our construction will start from a certain regularized two-parameter Ornstein–Uhlenbeck process, one parameter for the time and another for space; when indexed by time, the process takes its values in a space of functions.

This starting process, a regular random field, is only a regularization of the physically relevant one (which is a generalized random field) but it has continuous sample paths. We can, then, use some known results of Malliavin calculus in infinite dimensions. The results relevant for Euclidean quantum field theory follow by removing the regularization.

The present paper is a contribution to the above mentioned program of Euclidean quantum mechanics inspired by Schrödinger and therefore deals with time reversible measures generally distinct from the ones used in conventional Euclidean quantum field theory.

§1 PRELIMINARIES

A_ε will denote the pseudo-differential operator $(-\Delta + m^2)^{1/2} + \frac{\varepsilon}{m}(-\Delta)$, where $m > 0$, acting on functions defined on the (space) interval $\Lambda = [-\ell, \ell]$ and with Dirichlet boundary conditions. Consider, for each $\varepsilon > 0$, the real scalar product

$$(h, g)_\varepsilon = \int_{-\ell}^{\ell} (A_\varepsilon h)(x) g(x) \, dx. \tag{1.1}$$

Denote by $\mathcal{H}^\varepsilon$ the corresponding (real) Hilbert space. According to the expression of A_ε,

$$\|h\|_\varepsilon^2 = \int_{-\ell}^{\ell} (-\Delta + m^2)^{1/2} h \cdot h \, dx + \frac{\varepsilon}{m} \int_{-\ell}^{\ell} (-\Delta) h \cdot h \, dx.$$

This norm is, for each $\varepsilon > 0$, equivalent to the Sobolev norm of H^1 over $[-\ell, \ell]$.

Let us consider the centered Gaussian measure μ_ε defined by

$$E_{\mu_\varepsilon}[\exp i(h, \varphi)_\varepsilon] = \exp\left(-\frac{1}{4}\|h\|_\varepsilon^2\right). \tag{1.2}$$

The triplet $(\Omega, \mathcal{H}^\varepsilon, \mu_\varepsilon)$ where $\Omega = \{\varphi \in C([-\ell, \ell]) : \varphi(-\ell) = \varphi(\ell) = 0\}$ is an abstract Wiener space.

On this space we shall consider differentiation in the sense of Malliavin calculus. Namely, for a functional F defined on Ω, the derivative of F along a direction h in the Cameron–Martin space $\mathcal{H}_\varepsilon$ is defined by the μ_ε-a. s. limit

$$D_h^\varepsilon F(\varphi) = \lim_{\lambda \to 0} \frac{1}{\lambda}[F(\varphi + \lambda h) - F(\varphi)]. \tag{1.3}$$

$D_h^\varepsilon F(\varphi)$ defines a linear operator on $\mathcal{H}^\varepsilon$, denoted afterwards by $\nabla^\varepsilon F(\varphi)$. Therefore,

$$(\nabla^\varepsilon F(\varphi), h)_\varepsilon = D_h^\varepsilon F(\varphi). \tag{1.4}$$

Remark 1. $\nabla = A^\varepsilon \nabla^\varepsilon$ is the L^2-derivative usually considered in two-dimensional quantum field theory (cf. for example [20] and [21]).

Let us define the Malliavin–Sobolev spaces of Wiener functionals $W_r^p(\Omega, \mu_\varepsilon)$ for $1 \leq p, r < +\infty$ by

$$W_r^p(\Omega, \mu_\varepsilon) = \{F \in L^p(\Omega, \mu_\varepsilon) : E_{\mu_\varepsilon}[\|(\nabla^\varepsilon)^i F\|_{\mathrm{HS}}^p] < +\infty, \forall 1 \leq i \leq r\},$$

where $\| \cdot \|_{\mathrm{HS}}$ stands for the Hilbert–Schmidt norm of the corresponding i-linear operator

$$\|(\nabla^\varepsilon)^i F(\varphi)\|_{\mathrm{HS}}^2 = \sum_{k_1, \ldots, k_i = 1}^{+\infty} |D_{e_{k_1}}^\varepsilon \ldots D_{e_{k_i}}^\varepsilon F(\varphi)|^2$$

and $\{e_n\}$ is an orthonormal basis of the Hilbert space $\mathcal{H}^\varepsilon$.

The measure μ_ε is an invariant measure for the Ornstein–Uhlenbeck process X_t^ε, $X_t^\varepsilon(\cdot) \in \Omega$. This process solves the (forward) stochastic differential equation

$$\begin{aligned} dX_t^\varepsilon &= A_\varepsilon^{1/2} dW_t^\varepsilon - A_\varepsilon X_t^\varepsilon dt, \quad t \geq 0, \\ X_0^\varepsilon &= \varphi \in \Omega, \end{aligned} \tag{1.5}$$

associated with the increasing filtration representing the past history of the Ornstein–Uhlenbeck process.

With these notations, $X_t^\varepsilon(x) \in \mathbb{R}$ and W_t^ε is a Wiener process on the abstract Wiener space, with covariance $E[W_t^\varepsilon(x) W_{t'}^\varepsilon(y)] = \min(t, t') A_\varepsilon^{-1}(x, y)$. The generator of the Ornstein–Uhlenbeck process is the Ornstein–Uhlenbeck operator, namely

$$(\mathcal{L}^\varepsilon F)(\varphi) = \frac{1}{2} \mathrm{Tr}(A_\varepsilon^2 (\nabla^\varepsilon)^2 F)(\varphi) - (A_\varepsilon \varphi, \nabla^\varepsilon F(\varphi))_\varepsilon \tag{1.6}$$

with domain $\mathcal{D}(\mathcal{L}^\varepsilon)$ contained in the Malliavin–Sobolev space of Wiener functionals $W_2^p(\Omega, \mu_\varepsilon)$.

Remark 2. In terms of the L^2-derivative ∇, the generator of the Ornstein–Uhlenbeck process reduces to

$$(\mathcal{L}^\varepsilon F)(\varphi) = \frac{1}{2} \operatorname{Tr} \nabla^2 F(\varphi) - (A_\varepsilon \varphi, \nabla F(\varphi))$$

where $(\cdot, \cdot)$ refers to the L^2-scalar product.

We shall need the Krée–Meyer inequalities (cf. [14]):

$$C_1 \|\mathcal{L}^\varepsilon F\|_{L^p_{\mu_\varepsilon}} \le \|(\nabla^\varepsilon)^2 F\|_{L^p_{\mu_\varepsilon}} \le C_2 \|\mathcal{L}^\varepsilon F\|_{L^p_{\mu_\varepsilon}} \tag{1.7}$$

for $1 < p < +\infty$ and positive constants C_1, C_2. We shall also use the notion of capacity on the Wiener space associated to $\mathcal{L}^\varepsilon$. It is defined, for an open set $O \subset \Omega$ by

$$\operatorname{Cap}^\varepsilon(O) = \inf\{ \|u\|_{W_1^2(\Omega, \mu_\varepsilon)} : u > 0, u \ge 1 \text{ a. e. in } O \}$$

and, for general sets $\Gamma \subset \Omega$ by

$$\operatorname{Cap}^\varepsilon(\Gamma) = \inf_{\substack{O \supset \Gamma \\ O \text{ open}}} \operatorname{Cap}^\varepsilon(O).$$

Theorem 1. *X_t^ε is a stationary Gaussian process with an invariant measure μ_ε. Let ν_ε be the measure corresponding to the process X_t^ε; then the weak limits $\nu_\varepsilon \to \nu_0$ and $\mu_\varepsilon \to \mu_0$ exist and μ_0 is the Gaussian measure defined by*

$$\int d\mu_0 \exp i(\varphi, h)_0 = \exp\left(-\frac{1}{4}\|h\|_0^2\right). \tag{1.8}$$

Moreover ν_0 is the Gaussian measure defined by

$$\int d\nu_0 \exp i \int (X_\tau, h_\tau)\, d\tau$$

$$= \exp \int (h_\tau, \exp[-\tau A_0]\varphi)\, d\tau \tag{1.9}$$

$$= \exp\left\{ -1/4 \iint \left(h_{\tau'}, [\exp(-A_0|\tau - \tau'|) - \exp(-A_0|\tau + \tau'|)] A_0^{-1} h_\tau\right) d\tau'\, d\tau \right\}$$

where A_0 denotes the pseudo-differential operator $(-\Delta + m^2)^{1/2}$.

Proof. X_t^ε is the Gaussian process with the characteristic function defined by eq. (1.9), with A_ε replacing A_0. The weak convergence of ν_ε to ν_0 follows from the strong convergence in L^2 of $A_\varepsilon^{-1} \to A_0^{-1}$ and $\exp -tA_\varepsilon \to \exp -tA_0$. It is easy to see that the unique invariant measure for X_t^ε is equal to the measure defined by the characteristic function

$$\int d\mu_\varepsilon \exp i(\varphi, A_\varepsilon h) = \exp\left(-\frac{1}{4}(h, A_\varepsilon^{-1}h)\right).$$

The weak convergence of μ_ε follows from the strong convergence in L^2 of $A_\varepsilon^{-1} \to A_0^{-1}$.

In order to characterize the measure corresponding to the Ornstein–Uhlenbeck process (1.5) we note that X_t^ε is a Gaussian process. So, it is sufficient to compute the mean and the covariance resulting from eq. (1.5). We get the result expressed in eq. (1.9) (with $A_0 \to A_\varepsilon$) . By the previous argument the limit $A_\varepsilon \to A_0$ exists, leading to formula (1.9). □

Remark 3. The stochastic differential equation (1.5) is expressed with respect to the Wiener process W^ε. We can make the ε-dependence explicit: $W^\varepsilon = A_\varepsilon^{-1/2}W$, where $E[W_t(x)W_{t'}(y)] = \min(t, t')\,\delta(x - y)$. So, we can take the paths of the ε-independent Wiener process W as a model of the probability space defining the expectation values $E[\,\cdot\,]$.

§2 Feynman–Kac formulae

We proved in [18] a general Feynman–Kac formula on the classical abstract Wiener space. In the present framework, and for each $\varepsilon > 0$, the statement is the following:

Theorem 2. *Let $V \geq -K_\varepsilon$, K_ε a positive constant, be a functional belonging to the Sobolev space $W_4^p(\Omega, \mu_\varepsilon)$, for some $1 < p < +\infty$. Then the operator $H_\varepsilon = -\mathcal{L}^\varepsilon + V$ generates a strongly continuous contraction semigroup $e^{-t(H_\varepsilon + K_\varepsilon)}$ such that the following representation holds, for $\Theta_\varepsilon \in W_2^p(\Omega, \mu_\varepsilon)$ and $t > 0$,*

$$e^{-tH_\varepsilon}\Theta_\varepsilon(\varphi) = E_\varphi\left[\Theta_\varepsilon(X_t^\varepsilon)\exp\left(-\int_0^t V(X_s^\varepsilon)\,ds\right)\right] \quad \mu_\varepsilon\text{-a. e.} \qquad (2.1)$$

Sketch of the proof. We follow a scheme which is close to the one in [22]. Define

$$\Psi_\lambda(\varphi) = E_\varphi \int_0^{+\infty} V(X_s^\varepsilon)e^{-\lambda s}\,ds.$$

Then $(-\mathcal{L}^\varepsilon + \lambda)\Psi_\lambda = V$ and, since we have assumed $V \in W_4^p$, the functional Ψ_λ belongs to W_2^p for every λ (cf. the inequalities (1.7)). Therefore $\Psi_\lambda(\varphi) < +\infty$ except on a set of ε-capacity zero. This implies in particular that, for every $t > 0$, and every λ,

$$P_\varphi\left(\int_0^t V(X_s^\varepsilon)e^{-\lambda s}\,ds = \infty\right) = 0$$

and

$$P_\varphi\left(\lim_{t\to 0+}\int_0^t V(X_s^\varepsilon)\,ds = 0\right) = 1.$$

Since the Ornstein–Uhlenbeck process X_t^ε never hits sets of ε-capacity zero, we also have that $P_{X_\tau^\varepsilon}(\lim_{t\to 0+}\int_0^t V(X_s^\varepsilon)\,ds = 0) = 1$ for every $\tau > 0$.

Since V is bounded below, and using the Markov property of the Ornstein–Uhlenbeck process, the right-hand side of (2.1) defines a strongly continuous semigroup bounded by $\exp tK_\varepsilon$. The equality, i. e., the fact that H_ε is the generator of such a semigroup, follows from the Sobolev assumptions on V and Θ_ε. □

Remark 4. This Feynman–Kac formula breaks the time symmetry underlying our whole construction. However, it is just apparent. We could as well start from the formula valid for $t \le 0$

$$e^{tH}\Theta_\varepsilon^*(\varphi) = E_\varphi\left[\Theta_\varepsilon^*(X_t^\varepsilon)\exp\left(-\int_t^0 V(X_s^\varepsilon)\,ds\right)\right] \qquad (2.1^*)$$

where X_t^ε solves now the backward stochastic differential equation

$$\begin{aligned}
d_*X_t^\varepsilon &= A_\varepsilon^{1/2}d_*W_{*t}^\varepsilon + A_\varepsilon X_t^\varepsilon\,dt, \quad t \ge 0, \\
X_0^\varepsilon &= \varphi \in \Omega,
\end{aligned} \qquad (1.5^*)$$

with W_{*t}^ε denoting a backward Wiener process on the abstract Wiener space, and eq. (1.5^*) is associated with the decreasing filtration representing the future of the Ornstein–Uhlenbeck process. In this sense (cf. the statement of the Theorem 7), the apparently time-asymmetrical steps of the construction which follows can always be made time symmetric.

In order to introduce interactions in the limiting case $\varepsilon \to 0$, we define first Wick powers (cf. [20] and [21]). Let X_t^ε be the Ornstein–Uhlenbeck process solving eq. (1.5). We remark that we can make it start from a Gaussian random variable $\varphi \in \operatorname{supp}\mu_0$. We shall define a (forward) stochastic process Y_t by the relation

$$X_t^\varepsilon = Z_t^\varepsilon + Y_t^\varepsilon \equiv e^{-tA_\varepsilon}\varphi + Y_t^\varepsilon. \qquad (2.2)$$

Let ψ be either Z or Y. We define the Wick power of ψ by the formal power series

$$:\exp(\psi, f): = \exp\left(-\frac{1}{2}E[(\psi, f)^2]\right)\exp(\psi, f) \equiv \sum_{n=0}^\infty \frac{1}{n!}:(\psi, f)^n: \qquad (2.3)$$

Now, define as well

$$:X^n: = :(Z+Y)^n: = \sum_{m=0}^\infty \binom{n}{m}:Z^{n-m}::Y^m: \qquad (2.4)$$

In particular, as long as $\varepsilon > 0$, $E[(\psi^\varepsilon(x))^2] < \infty$. Hence, we may consider $:(\psi^\varepsilon(x))^n:$ in eq. (2.3) (which corresponds to $f_x(y) = \delta(x - y)$). $:\psi^\varepsilon(x)^n:$ is a random field with a. s. continuous paths. So, we can define the "polynomial interaction"

$$V_n^\varepsilon(t, \Lambda; X_t^\varepsilon) = \int_\Lambda :(X_t^\varepsilon(x))^n: dx \tag{2.5}$$

where $\Lambda \subset \mathbb{R}$ is a finite interval.

We shall denote the left-hand side of eq. (2.5) by $V_n^\varepsilon(t, \Lambda)$ for brevity. We shall also consider the formal power series defining the "exponential interaction"

$$V_E^\varepsilon(t, \Lambda) = \sum_{n=0}^\infty \frac{\alpha^n}{n!} V_n^\varepsilon(t, \Lambda). \tag{2.6}$$

(E for "Exponential", and the dependence in the parameter α is omitted in the left-hand side.) V_E^ε is just the integral over Λ of $:\exp \alpha X_t^\varepsilon:$. For this reason

$$V_E^\varepsilon(t, \Lambda) > 0. \tag{2.7}$$

Coming back to eq. (2.5),

$$V_{2n}^\varepsilon(t, \Lambda) \geq -|\Lambda| K_{2n}(\varepsilon) \tag{2.8}$$

where $|\Lambda|$ denotes the length of the interval Λ and $K_{2n}(\varepsilon)$ is a constant, which tends to infinity when $\varepsilon \to 0$.

The restrictions on V in Theorem 2 are fulfilled for our Wick ordered potentials (2.5) as long as $\varepsilon > 0$. This defines therefore a semigroup $\exp(-tH_\varepsilon)$. We will be interested in the limit $\varepsilon \to 0$.

Theorem 3. *Assuming that $\varphi \in \operatorname{supp} \mu_0$ we have:*

$$\int d\mu_0 \, E[(V_n^\varepsilon(t, \Lambda) - V_n^{\varepsilon'}(t, \Lambda))^2] \to 0 \tag{2.9}$$

when $\varepsilon, \varepsilon' \to 0$. Moreover, the L^2-convergence (2.9) holds true for V_E^ε defined in (2.6) if $\alpha^2 < 2\pi$.

Sketch of the proof. The idea of the proof is easy to see for V_E^ε (see ref. [29]). Indeed,

$$V_E^\varepsilon - V_E^{\varepsilon'} = \int_\Lambda dx \left(:\exp \alpha Z_\varepsilon: \exp(\alpha Y_\varepsilon) \exp\left(-\frac{\alpha^2}{2} E[Y_\varepsilon^2]\right) \right.$$

$$\left. - :\exp \alpha Z_{\varepsilon'}: \exp(\alpha Y_{\varepsilon'}) \exp\left(-\frac{\alpha^2}{2} E[Y_{\varepsilon'}^2]\right) \right). \tag{2.10}$$

Using eq. (2.10), we can easily compute the Gaussian integral of (2.9) explicitly. It is expressed by exponentials of the covariances. The condition $\alpha^2 < 2\pi$ follows from the requirement

$$\int_\Lambda dx\, dx'\, \exp\big(\alpha^2\, \mathrm{E}[Y_\varepsilon(x)Y_\varepsilon(x')]\big) < \infty \tag{2.11}$$

for any $\varepsilon \geq 0$.

Although $\mathrm{E}[Y_\varepsilon^2(x)] \to \infty$ as $\varepsilon \to 0$, the L^2-convergence holds true, because the terms $\mathrm{E}[Y_\varepsilon^2(x)]$ cancel each other. The integral depends only on the covariances $\mathrm{E}[Y_\varepsilon(x)Y_{\varepsilon'}(x')]$, where $x \neq x'$. These covariances do converge as $\varepsilon, \varepsilon' \to 0$. The same argument applies to polynomials V_n^ε. The L^2-integrals are expressed by the covariances, which have indeed a limit, when $\varepsilon \to 0$. □

From Theorem 3 it follows that there exist $V(t, \Lambda) \in L_{\mu_0}^2$. Moreover, we could check that the Kolmogorov criterion for continuity of the sample paths of $V(t, \Lambda)$ is fulfilled. Hence, we shall define

$$V([s, t], \Lambda) = \int_s^t V(\tau, \Lambda)\, d\tau. \tag{2.12}$$

Theorem 4. *Let V be either V_{2n} or V_{E}. Then, the strong limit $\varepsilon \to 0$ of $\exp(-tH^\varepsilon)$ exists in $L_{\mu_0}^2$. The limit semigroup $\exp(-tH)$ maps $L_{\mu_0}^2$ onto $L_{\mu_0}^2$. It can be expressed by the Feynman–Kac formula*

$$\big(\exp(-tH)\Theta\big)(\varphi) = \mathrm{E}_\varphi\big[\Theta(X_t)\exp\big(-V([0, t], \Lambda)\big)\big] \tag{2.13}$$

where X_t is the Ornstein–Uhlenbeck process defined by eq. (1.9). Moreover, there is a constant K such that $\exp[-t(H + K)] \leq 1$.

Sketch of the proof. Let us begin with the exponential interaction V_{E} (see ref. [29]). Then, the proof is elementary. We use the estimate

$$|\exp(-V_{\mathrm{E}}^\varepsilon) - \exp(-V_{\mathrm{E}}^{\varepsilon'})| \leq |V_{\mathrm{E}}^\varepsilon - V_{\mathrm{E}}^{\varepsilon'}|\big(\exp(-V_{\mathrm{E}}^\varepsilon) + \exp(-V_{\mathrm{E}}^{\varepsilon'})\big). \tag{2.14}$$

We have proved in Theorem 3 that $V_{\mathrm{E}}^\varepsilon - V_{\mathrm{E}}^{\varepsilon'} \to 0$ in L^2. Owing to $\exp(-V_{\mathrm{E}}) \leq 1$ we get from eq. (2.14) the convergence of $\exp(-V_{\mathrm{E}}^\varepsilon)$ in L^2. In the exponential case the constant $K = 0$. The proof for polynomial interactions is more involved, because of the difficulty to estimate $\exp(-V_{2n}^\varepsilon)$. The bound (2.8) tends to $-\infty$ when $\varepsilon \to 0$. So, V_{2n} fails to be bounded from below. Nevertheless, the set where $-V_{2n}$ is large has small measure, so that $\exp(-V_{2n})$ is integrable (see Nelson [20], and [21]). Using the additivity of V, i.e.

$$V([0, t], \Lambda) = V([0, s], \Lambda) + V([s, t], \Lambda),$$

and the Markov property of the Ornstein–Uhlenbeck process, we can show that $\exp(-tH)$ admits the representation (2.13). The proof that $\exp(-tH) \leq \exp tK$ for a certain constant K is due to Glimm and Jaffe [19]. □

Remark 5. It would also be possible to treat trigonometric interactions (see [30]) with somewhat different methods.

§3 THE INFINITE-DIMENSIONAL BERNSTEIN PROCESSES

The existence of Wiener space valued Bernstein processes is proved in [18]. We shall sketch here the proof for each $\varepsilon > 0$ and obtain, by a limiting procedure, the Bernstein processes whose state space is a subspace of the Schwartz space of distributions $\mathcal{S}'(\mathbb{R})$.

Let us consider a variable time t in the time interval $[0, T]$ and let us introduce the notations $\eta_t^\varepsilon = e^{-(T-t)H^\varepsilon}\Theta_\varepsilon$ and $\eta_t = e^{-(T-t)H}\Theta$, where $H^\varepsilon = -\mathcal{L}^\varepsilon + V^\varepsilon$, $H = -\mathcal{L}^0 + V$ are the associated Hamiltonians.

Lemma 5. *Assume that V_ε fulfills the conditions of Theorem 2 and, moreover, that $\Theta_\varepsilon > 0$ almost surely, $\Theta_\varepsilon \in L^2_{\mu_\varepsilon}$, and for any $\varepsilon \geq 0$, $\log\Theta_\varepsilon \in L^2_{\mu_\varepsilon}$. Then, for the same ε and $t \geq 0$, $\log\eta_t^\varepsilon \in L^2_{\mu_\varepsilon}$.*

Proof. Note that $\int d\mu_\varepsilon(\varphi)\, \mathrm{E}_\varphi[\log^2\Theta_\varepsilon(X_t^\varepsilon)] = \int\log^2\Theta_\varepsilon\, d\mu_\varepsilon$. Then, from the Jensen inequality and Feynman–Kac formula (2.1) and (2.13),

$$\eta_t^\varepsilon(\varphi) \geq \exp\big(-\mathrm{E}_\varphi[V^\varepsilon([0,t],\Lambda)] + \mathrm{E}_\varphi[\log\Theta(X_t^\varepsilon)]\big).$$

Hence,

$$
\begin{aligned}
\int (\log\eta_t^\varepsilon)^2\, d\mu_\varepsilon &\\
= \int_{\{\eta_t^\varepsilon \geq 1\}} (\log\eta_t^\varepsilon)^2\, d\mu_\varepsilon &+ \int_{\{\eta_t^\varepsilon < 1\}} (\log\eta_t^\varepsilon)^2\, d\mu_\varepsilon \\
\leq \int (\eta_t^\varepsilon)^2\, d\mu_\varepsilon &+ \int d\mu_\varepsilon(\varphi)\,(\mathrm{E}_\varphi^\varepsilon[V([0,t],\Lambda) - \log\Theta_\varepsilon(X_t^\varepsilon)])^2 \\
< \infty. &
\end{aligned}
\tag{3.1}
$$

The first term on the right-hand side of eq. (3.1) is finite because of Theorem 4, and the second one from our assumption and the Theorem 3. □

The following estimate will be fundamental to prove the existence of Bernstein processes:

Lemma 6. *Under the assumptions of Lemma 5, we have for any $\varepsilon \geq 0$ and $T \geq 0$*

$$\mathrm{E}_{\mu_\varepsilon}\int_0^T \|A_\varepsilon^{1/2}\nabla^\varepsilon\log\eta_t^\varepsilon\|_\varepsilon^2\, dt < +\infty.$$

Proof. Let us start with $\varepsilon > 0$. By definition of η_t^ε we have

$$\frac{\partial}{\partial t}\log\eta_t^\varepsilon = -\frac{\mathcal{L}^\varepsilon\eta_t^\varepsilon}{\eta_t^\varepsilon} + V^\varepsilon.$$

On the other hand,

$$\mathcal{L}^\varepsilon \log \eta_t^\varepsilon = \frac{\mathcal{L}^\varepsilon \eta_t^\varepsilon}{\eta_t^\varepsilon} - \frac{1}{2}\|A_\varepsilon^{1/2}\nabla^\varepsilon \log \eta_t^\varepsilon\|_\varepsilon^2.$$

Therefore,

$$\frac{1}{2}\,\mathrm{E}_{\mu_\varepsilon}\left[\|A_\varepsilon^{1/2}\nabla^\varepsilon \log \eta_t^\varepsilon\|_\varepsilon^2\right] = \mathrm{E}_{\mu_\varepsilon}\,V^\varepsilon - \mathrm{E}_{\mu_\varepsilon}\left[\frac{\partial}{\partial t}\log \eta_t^\varepsilon\right]$$

and

$$\mathrm{E}_{\mu_\varepsilon}\left[\int_0^T \|A_\varepsilon^{1/2}\nabla^\varepsilon \log \eta_t^\varepsilon\|_\varepsilon^2\,dt\right] = 2\,\mathrm{E}_{\mu_\varepsilon}[\log \eta_0^\varepsilon - \log \eta_T^\varepsilon] + 2\,\mathrm{E}_{\mu_\varepsilon}\int_0^T V^\varepsilon\,dt. \quad (3.2)$$

Now, on the basis of Lemma 5, both terms on the right-hand side of eq. (3.2) are separately finite. □

Theorem 7. *Let* $V_\varepsilon \geq -K_\varepsilon$, $V_\varepsilon \in W_4^p(\Omega, \mu_\varepsilon)$ *with* $1 < p < +\infty$, *and let* Θ_ε, $\Theta_\varepsilon^* > 0$ *be bounded functionals in* $W_2^p(\Omega, \mu_\varepsilon)$ *such that* $\mathrm{E}_{\mu_\varepsilon} \log^2 \Theta_\varepsilon < +\infty$. *Assume the normalization condition* $\mathrm{E}_{\mu_\varepsilon}(\Theta_\varepsilon^*(\varphi)e^{-2TH_\varepsilon}\Theta_\varepsilon(\varphi)) = 1$. *Then there exists a Bernstein process* z_t^ε, $t \in [-T,T]$, *with state space* Ω, *which is a weak solution of the forward stochastic differential equation*

$$dz_t^\varepsilon = A_\varepsilon^{1/2}\,dW_t^\varepsilon - \left(A_\varepsilon z_t^\varepsilon - A_\varepsilon \nabla^\varepsilon \log \eta_t^\varepsilon(z_t^\varepsilon)\right) dt \quad (3.3)$$

with initial probability

$$P(z_0^\varepsilon \in \Gamma) = \int_\Gamma \Theta_\varepsilon^*(\varphi)e^{-2TH_\varepsilon}\Theta_\varepsilon(\varphi)\,d\mu_\varepsilon(\varphi), \quad (3.4)$$

where $(\eta_t^{*\varepsilon})(\varphi) = e^{-(t+T)H_\varepsilon}\Theta_\varepsilon^*(\varphi)$. *Furthermore, for each* $-T \leq t \leq T$, *we have*

$$P(z_t^\varepsilon \in \Gamma) = \int_\Gamma (\eta_t^{*\varepsilon})\eta_t^\varepsilon\,d\mu_\varepsilon. \quad (3.5)$$

Sketch of the proof. Denote by P^ε the probability measure induced by the Ornstein–Uhlenbeck process X_t^ε with initial density given by (3.4) on the space of paths, $B = C([0,T],\Omega)$. Let us consider the following functional on this space:

$$\rho_t^\varepsilon(z^\varepsilon) = \exp\left\{ \int_0^t \left(A_\varepsilon^{1/2}\nabla^\varepsilon \log \eta_s^\varepsilon(z_s^\varepsilon),\,dW_s^\varepsilon\right)_\varepsilon \right. $$
$$\left. - \frac{1}{2}\int_0^t \|A_\varepsilon^{1/2}\nabla^\varepsilon \log \eta_s^\varepsilon(z_s^\varepsilon)\|_\varepsilon^2\,ds \right\} \quad (3.6)$$

with $t \in [0,T]$, and where $(\cdot,\cdot)_\varepsilon$ also denotes the extension to $\mathcal{H}_\varepsilon \times \Omega$ of the scalar product of $\mathcal{H}_\varepsilon$. Lemmas 5 and 6 insure that the functional ρ_t^ε is well defined.

Itô's formula gives:

$$\log \eta_t^\varepsilon(z_t^\varepsilon) - \log \eta_0^\varepsilon(z_0^\varepsilon)$$
$$= \int_0^t \left(A_\varepsilon^{1/2} \nabla^\varepsilon \log \eta_s^\varepsilon(z_s^\varepsilon), \, dW_s^\varepsilon \right)_\varepsilon + \int_0^t \left(\frac{\partial}{\partial s} + \mathcal{L}^\varepsilon \right) \log \eta_s^\varepsilon(z_s^\varepsilon) \, ds. \tag{3.7}$$

The proof that Itô's formula is true in this framework can be found in [18]. Using the identity

$$\left(\frac{\partial}{\partial t} + \mathcal{L}^\varepsilon \right) \log \eta_t^\varepsilon = V^\varepsilon - \frac{1}{2} \| A_\varepsilon^{1/2} \nabla^\varepsilon \log \eta_t^\varepsilon \|_\varepsilon^2$$

and after exponentiating (3.7), we obtain

$$\rho_t^\varepsilon(z^\varepsilon) = \frac{\eta_t^\varepsilon(z_t^\varepsilon)}{\eta_0^\varepsilon(\varphi)} \exp \left(- \int_0^t V^\varepsilon(z_s^\varepsilon) \, ds \right). \tag{3.8}$$

Now, by Feynman–Kac's formula,

$$\mathrm{E}^{P^\varepsilon} \rho_t^\varepsilon(z^\varepsilon) = \int_0^t (\eta_0^*)^\varepsilon(\varphi) \, \mathrm{E}_\varphi \left[\eta_t^\varepsilon(z_t^\varepsilon) \exp \left(- \int_0^t V^\varepsilon(z_s^\varepsilon) \, ds \right) \right] d\mu_\varepsilon(\varphi)$$
$$= \int \Theta_\varepsilon^*(\varphi) e^{-2TH_\varepsilon} \Theta_\varepsilon(\varphi) \, d\mu_\varepsilon(\varphi), \tag{3.9}$$

which is equal to one by our normalization assumption and the symmetry of the Hamiltonian. One also deduces that this expression is equal to $\int \eta_t \eta_t^* \, d\mu_\varepsilon$, since $\eta_t, \eta_t^* \in L^1_{\mu_\varepsilon}$.

Girsanov's theorem therefore provides the existence of a weak solution of the equations (3.3) and (3.4), defined for $t \in [0, T]$. $\square$

Remark 6. The above mentioned argument involves, actually, only z_t^ε for $t \in [0, T]$. We can see, however, that (3.3) can be extended to $[-T, T]$, as well as the backward stochastic differential equation

$$d_* z_t^\varepsilon = A_\varepsilon^{1/2} \, d_* W_{*t}^\varepsilon + \left(A_\varepsilon z_t^\varepsilon - A_\varepsilon \nabla^\varepsilon \log \eta_t^{*\varepsilon}(z_t^\varepsilon) \right) dt \tag{3.3*}$$

so that relation (3.5) holds for every $t \in [-T, T]$. The reasoning is the same as in finite dimension (cf. [7, 16, and 23]).

Remark 7. Property (3.5) is the one which makes Bernstein processes special and constitutes the key of our new strategy for Euclidean quantum physics. For the elementary quantum mechanics of a system in $\mathbb{R}^3$ with self-adjoint Hamiltonian H, and with respect to the Lebesgue measure, it expresses Schrödinger's idea [6] that

$$\int_{\Gamma \subset \mathbb{R}^3} \eta_t^* \eta_t(x) \, dx,$$

where η_t^* is a positive solution of the heat equation $-\frac{\partial \eta^*}{\partial t} = H\eta^*$ and η_t a positive solution of the adjoint heat equation $\frac{\partial \eta}{\partial t} = H\eta$, should be the Euclidean version of Born's probabilistic interpretation of the quantum mechanical wave function ψ. Indeed it is axiomatic, in this context, that

$$\int_\Gamma \Psi_t \bar{\Psi}_t(x)\, dx = P\{\text{system is in } \Gamma \subset \mathbb{R}^3 \text{ at time } t\}$$

where P denotes a probability whose nature, or origin, are not specified mathematically. Here Ψ_t solves the initial value problem for the Schrödinger equation $i\frac{\partial \psi}{\partial t} = H\psi$ and $\bar{\Psi}_t$ the complex conjugate problem. The associated Bernstein diffusions (indexed by the time) have been introduced in [7], [8], and [9]. Notice that the relation between η and η_* can be interpreted, under proper restrictions on the probabilities at the boundary of a given time interval [23], as a time reversal. It follows, therefore, that all the resulting probability measures are invariant under time reversal by construction although, in general, not stationary.

Eq. (3.9) determines the law ν_Θ^ε of the process z_t^ε. We choose $\Theta \in L_{\mu_0}^2$ independent of ε; then the initial distribution of z_t^ε reads:

$$d\mu_\varepsilon^\Theta(\varphi) = d\mu_\varepsilon(\varphi)\, \Theta(\varphi) e^{-2TH_\varepsilon} \Theta^*(\varphi)$$

(an equivalent expression for eq. (3.4)). The law of z^ε is determined by the Ornstein–Uhlenbeck measure ν^ε (see Theorem 1)

$$
\begin{aligned}
d\nu_\Theta^\varepsilon &= d\mu_\varepsilon^\Theta(\varphi)\, d\nu_\varepsilon(X^\varepsilon) \frac{\eta_T^\varepsilon(\varphi)}{\Theta(\varphi)} \exp -V^\varepsilon([0,T],\Lambda) \\
&= d\mu_\varepsilon(\varphi)\, d\nu_\varepsilon(X^\varepsilon)\, \Theta^*(\varphi) \eta_T^\varepsilon(\varphi) \exp -V^\varepsilon([0,T],\Lambda)
\end{aligned}
\tag{3.11}
$$

where V^ε is defined in eqs. (2.5) and (2.12). We will introduce a functional $\chi_\varepsilon(\varphi, X)$ so that (3.11) is equal to

$$d\mu_\varepsilon(\varphi)\, d\nu_\varepsilon(X^\varepsilon)\, \chi_\varepsilon(\varphi, X^\varepsilon).$$

We will need the following:

Lemma 8. ν_Θ^ε *has a weak limit as* $\varepsilon \to 0$. *The limit measure* ν_Θ^0 *is continuous with respect to* $\mu_0 \otimes \nu_0$, *explicitely:*

$$d\nu_\Theta^0 = d\mu_0(\varphi)\, d\nu_0(X^\varepsilon)\, \eta_T^*(\varphi) \eta_T(\varphi) \exp -V([0,T],\Lambda), \tag{3.12}$$

where η_T *is defined by the left-hand side of eq. (2.13), for* $t = T$.

Sketch of the proof. Let us write in eq. (3.11)

$$d\nu_\Theta^\varepsilon = (d\mu_\varepsilon(\varphi) - d\mu_0(\varphi))\, d\nu_\varepsilon(X^\varepsilon)\, \chi_\varepsilon(\varphi, X^\varepsilon) + d\mu_0(\varphi)\, d\nu_\varepsilon(X^\varepsilon)\, \chi_\varepsilon(\varphi, X^\varepsilon). \tag{3.13}$$

Let us consider first the second term in eq. (3.13). We have proved in Theorem 4 that $\eta_T^\varepsilon(\varphi)$, as well as $\exp -V^\varepsilon([0,t], \Lambda)$, have an L^2-limit as $\varepsilon \to 0$. Hence, the second term in eq. (3.13) has a weak limit. There remains to show that the first term in eq. (3.13) tends weakly to zero. For this purpose, we may use the following formula, valid for any integrable F,

$$\int d\mu_\varepsilon(\varphi)\, F(\varphi) = \int d\mu_0(\varphi)\, F(A_0^{1/2} A_\varepsilon^{-1/2}\varphi).$$

Then, for arbitrary integrable $\mathcal{F}$,

$$\int (d\mu_\varepsilon(\varphi) - d\mu_0(\varphi)) \int d\nu_\varepsilon(X^\varepsilon)\, \chi_\varepsilon(\varphi, X^\varepsilon)\mathcal{F}(\varphi, X^\varepsilon)$$
$$= \int d\mu_0(\varphi)\, d\nu_\varepsilon(X^\varepsilon) \big[\chi_\varepsilon(A_0^{1/2} A_\varepsilon^{-1/2}\varphi, X^\varepsilon)\mathcal{F}(A_0^{1/2} A_\varepsilon^{-1/2}\varphi, X^\varepsilon) \tag{3.14}$$
$$- \chi_\varepsilon(\varphi, X^\varepsilon)\mathcal{F}(\varphi, X^\varepsilon) \big].$$

By means of the methods used in the proof of Theorem 4, we prove that the right-hand side of eq. (3.14) tends to zero when $A_0^{1/2} A_\varepsilon^{-1/2} \to 1$ (strongly in $L^2(dx)$). It is sufficient to restrict $\mathcal{F}$ to trigonometric polynomials in eq. (3.14). $\square$

Using the notation introduced in (1.4), we have the

Lemma 9.
$$\left\| \int_0^T A_0^{1/2}\nabla^0 \log \eta_\tau^0(X_\tau^0(\varphi))\, d\tau \right\|_0 \in L^2(\nu_\Theta^0).$$

Proof. By Cauchy–Schwarz,

$$\int \left\| \int_0^T A_0^{1/2}\nabla^0 \log \eta_\tau^0 \right\|_0^2 d\nu_\Theta^0 \le T \int d\nu_\Theta^0 \int_0^T \|A_0^{1/2}\nabla^0 \log \eta_\tau^0\|_0^2\, d\tau$$

and the right-hand side is finite owing to Lemma 6. $\square$

Theorem 10. ν_Θ^0 *is the measure corresponding to the solution of the stochastic differential equation*

$$dz_t^0 = A_0^{1/2}\, dW_t^0 - (A_0 z_t^0 - A_0\nabla^0 \log \eta_t^0)\, dt \quad on\ \mathcal{S}'(\mathbb{R}), \tag{3.15}$$

where z_0^0 has the probability distribution $d\mu_0^\Theta(\varphi) = d\mu_0(\varphi)\, \Theta^*(\varphi)e^{-2TH}\Theta(\varphi)$. Moreover, ν_Θ^0 determines a Bernstein process in the sense that

$$\int d\nu_\Theta^0\, F(z_t) = \int d\mu_0(\varphi)\, e^{-(t+T)H}\Theta^*(\varphi)e^{-(T-t)H}\Theta(\varphi)F(\varphi) \tag{3.16}$$

for any integrable functional F.

Proof. We show first that

$$\xi_t = z_t^0 + \int_0^t (A_0 z_\tau^0 - A_0 \nabla^0 \log \eta_\tau^0) \, d\tau \tag{3.17}$$

is the Wiener process $A_0^{1/2} W_t^0$. Let $h \in \mathcal{H}_0$ and consider $(\xi_t, h)_0$.

According to Lemma 9, $(z_t^0, h)_0$ and $((\xi_t - z_t^0), h)_0$ are elements of $L^2(\nu_\Theta^0)$. Moreover, from Theorem 4 and Lemma 6 it follows that each term in $(\xi_t, h)_0$ is a $L^2_{\nu_\Theta^0}$-limit of $(\xi_t^\varepsilon, h)_0$ when $\varepsilon \to 0$, where

$$\xi_t^\varepsilon = z_t^\varepsilon + \int_0^t (A_\varepsilon z_\tau^\varepsilon - A_\varepsilon \nabla^\varepsilon \log \eta_\tau^\varepsilon) \, d\tau.$$

From Theorem 7, $\xi_t^\varepsilon = A_\varepsilon^{1/2} W_t^\varepsilon$, hence $\xi_t = \lim_{\varepsilon \to 0} A_\varepsilon^{1/2} W_t^\varepsilon = A_0^{1/2} W_t^0$. Formula (3.16) is a consequence of eq. (3.12) and Theorem 4.

§4 FUNCTIONAL CALCULUS

Let us consider, as before, the space of continuous paths $B = C([0; T]; \Omega) \simeq C([0, T] \times [-\ell, \ell]; \mathbb{R})$ endowed with the measure P^ε (the law of the Ornstein–Uhlenbeck process X_t^ε with initial distribution given by (3.4); P^ε is equal to ν_Θ^ε (3.11) when $V = 0$). This space, denoted by B, can be realized as an abstract Wiener space where the corresponding Cameron–Martin Hilbert space is the subspace of B consisting of functions $\psi(t, x)$ satisfying:

$$\int_0^T \left\| A_\varepsilon^{1/2} \psi(t, \cdot) + A_\varepsilon^{-1/2} \frac{\partial}{\partial t} \psi(t, \cdot) \right\|_\varepsilon^2 dt < +\infty. \tag{4.1}$$

From the stochastic differential equation (1.5) solved by X_t^ε, we deduce that

$$d(X_t^\varepsilon + \lambda \psi) = A_\varepsilon^{1/2} \, dW_t^\varepsilon - A_\varepsilon (X_t^\varepsilon + \lambda \psi) \, dt + \left(\lambda A_\varepsilon \psi + \lambda \frac{\partial \psi}{\partial t} \right) dt.$$

By Cameron–Martin–Girsanov's theorem, the law $P^{\varepsilon, \lambda}$ of the process $X_t^\varepsilon + \lambda \psi$, is absolutely continuous with respect to P^ε and its Radon–Nikodym density is given by the following expression:

$$\frac{dP^{\varepsilon, \lambda}}{dP^\varepsilon} = \exp \left[\int_0^T \lambda \left(\left(A_\varepsilon^{1/2} \psi + A_\varepsilon^{-1/2} \frac{\partial \psi}{\partial t} \right), (A_\varepsilon^{-1/2} \, dX_t^\varepsilon + A_\varepsilon^{1/2} X_t^\varepsilon \, dt) \right) \right.$$
$$\left. - \frac{\lambda^2}{2} \int_0^T \left\| A_\varepsilon^{1/2} \psi + A_\varepsilon^{-1/2} \frac{\partial \psi}{\partial t} \right\|_\varepsilon^2 dt \right]. \tag{4.2}$$

A differential calculus can be built for functionals defined on two-parameter function spaces, like B. This has been done in various frameworks, like for example

[24]. In our framework, the derivative of a functional $F \in L^2(B, P^\varepsilon)$ along a "direction" ψ satisfying (4.1) is defined by the P^ε-a. s. limit:

$$\mathbb{D}_\psi^\varepsilon F(z^\varepsilon) = \lim_{\lambda \to 0} \frac{1}{\lambda}[F(z^\varepsilon + \lambda\psi) - F(z^\varepsilon)]. \tag{4.3}$$

According to this definition, and using (4.2), one obtains the following integration-by-parts formula, when $\mathbb{D}_\psi F \in L^2(B, P^\varepsilon)$ and the right-hand side is integrable:

$$E_s^\varepsilon(\mathbb{D}_\psi F(z^\varepsilon))$$

$$= E_s^\varepsilon \left[F(z^\varepsilon) \int_0^T \left(\left(A_\varepsilon^{1/2}\psi + A_\varepsilon^{-1/2}\frac{\partial\psi}{\partial t} \right), (A_\varepsilon^{-1/2} dz_t^\varepsilon + A_\varepsilon^{1/2} z_t^\varepsilon \, dt) \right)_\varepsilon dt \right] \tag{4.4}$$

P^ε-a. e., where E_s^ε stands for the conditional expectation with respect to the past filtration $\mathcal{P}_s$ at time $s \in [0, T]$.

Combining the integration-by-parts formula obtained for the Ornstein–Uhlenbeck process X_t^ε with the Radon–Nikodym density (3.8) for the law of the Bernstein processes z_t^ε, we obtain an integration-by-parts formula for these processes. Namely, the following theorem holds (cf. [18] for the proof):

Theorem 11. *Assume that the conditions of Theorem 7 hold. Let $F \in L^2(B, P^\varepsilon)$ for some $1 \le p < +\infty$ and let $\psi : B \to B$ be a $\mathcal{P}_t$-adapted functional, $\psi_t \in L_{P^\varepsilon}^2$ for each $t \in [s, T]$, $\psi_t = 0$ for $t \in [0, s]$, and such that*

$$E^{P^\varepsilon} \left[\int_0^T \left\| A_\varepsilon^{1/2}\psi + A_\varepsilon^{-1/2}\frac{\partial\psi}{\partial t} \right\|_\varepsilon^2 dt \right] < +\infty$$

and $\mathbb{D}_\psi F \in L_{P^\varepsilon}^2$. Then the following integration-by-parts formula holds:

$$E_s^\varepsilon(\mathbb{D}_\psi F(z^\varepsilon))$$

$$= E_s \left(F(z^\varepsilon) \left[\int_s^T \left(\left(A_\varepsilon^{1/2}\psi_t + A_\varepsilon^{-1/2}\frac{\partial\psi}{\partial t} \right), (A_\varepsilon^{-1/2} dz_t^\varepsilon + A_\varepsilon^{1/2} z_t^\varepsilon \, dt) \right)_\varepsilon \right. \right.$$

$$\left. + \int_s^T \left(A_\varepsilon^{1/2}\psi_t, A_\varepsilon^{-1/2}\nabla^\varepsilon V^\varepsilon(z_t^\varepsilon) \right)_\varepsilon dt \right] \right)$$

$$+ E_s \left(F(z^\varepsilon) \left[\left(A_\varepsilon^{1/2}\psi_s(z_s^\varepsilon), A_\varepsilon^{-1/2}\nabla^\varepsilon \log \eta_s^\varepsilon(z_s^\varepsilon) \right)_\varepsilon \right. \right. \tag{4.5}$$

$$\left. \left. - \left(A_\varepsilon^{1/2}\psi_t(z_T^\varepsilon), A_\varepsilon^{-1/2}\nabla^\varepsilon \log \eta_T^\varepsilon(z_T^\varepsilon) \right)_\varepsilon \right] \right).$$

A simple but nevertheless important consequence of this theorem is the derivation of the laws of motion (regularized Euler–Lagrange equations) for Bernstein processes. For this we shall use Nelson's notions (cf. [25]) of forward and backward derivatives defined, for a functional F on (B, P^ε) respectively by the following a. s. limits:

$$D^\varepsilon F(z_t^\varepsilon) = \lim_{\lambda \to 0} \frac{1}{\lambda} E_t^\varepsilon \left(F(z_{t+\lambda}^\varepsilon) - F(z_t^\varepsilon) \right)$$

$$D_*^\varepsilon F(z_t^\varepsilon) = \lim_{\lambda \to 0} \frac{1}{\lambda} E_t^\varepsilon \left(F(z_t^\varepsilon) - F(z_{t-\lambda}^\varepsilon) \right)$$
(4.6)

where the last expectation is the conditional expectation with respect to the decreasing filtration $\mathcal{F}_t$ representing the future of z_t^ε.

Applying the integration-by-parts formula (4.5) to the functional $F \equiv 1$, as suggested by Feynman [17], we obtain

$$0 = E_s \left[\int_s^T \left(\left(A_\varepsilon^{1/2} \psi_t + A_\varepsilon^{-1/2} \frac{\partial \psi}{\partial t} \right), \left(A_\varepsilon^{-1/2} Dz_t^\varepsilon + A_\varepsilon^{1/2} z_t^\varepsilon \, dt \right) \right)_\varepsilon \right]$$

$$+ E_s \left[\int_s^T \left(A_\varepsilon^{1/2} \psi_t, A_\varepsilon^{-1/2} \nabla^\varepsilon V^\varepsilon(z_t^\varepsilon) \right)_\varepsilon dt \right]$$

$$+ E_s \left[\left(A_\varepsilon^{1/2} \psi_s(z_s^\varepsilon), A_\varepsilon^{-1/2} \nabla^\varepsilon \log \eta_s^\varepsilon(z_s^\varepsilon) \right)_\varepsilon \right.$$

$$\left. - \left(A_\varepsilon^{1/2} \psi_T(z_T^\varepsilon), A_\varepsilon^{-1/2} \nabla^\varepsilon \log \eta_T^\varepsilon(z_T^\varepsilon) \right)_\varepsilon \right].$$

According to Itô's calculus,

$$E_s \int_s^T \left(A_\varepsilon^{1/2} \frac{\partial \psi}{\partial t}, A_\varepsilon^{-1/2} D^\varepsilon z_t^\varepsilon \right)_\varepsilon dt$$

$$= -E_s \int_s^T \left(A_\varepsilon^{1/2} \psi_t, A_\varepsilon^{-3/2} D^\varepsilon D^\varepsilon z_t^\varepsilon \right)_\varepsilon dt$$

$$+ E_T \left(A_\varepsilon^{1/2} \psi_T(z_T^\varepsilon), D^\varepsilon z_T^\varepsilon \right)_\varepsilon - E_s \left(A_\varepsilon^{1/2} \psi_s(z_s^\varepsilon), D^\varepsilon z_s^\varepsilon \right)_\varepsilon.$$

From the definition of the processes z_t^ε and since ψ_t is an "arbitrary" adapted direction of derivation, we finally obtain (cf. [18] for details):

$$E_s \left(D^\varepsilon D^\varepsilon z_t^\varepsilon - A_\varepsilon^2 z_t^\varepsilon - A_\varepsilon \nabla^\varepsilon V^\varepsilon(z_t^\varepsilon) \right) = 0.$$

Since the expression inside the parenthesis is $\mathcal{P}_t$-adapted, the following law of motion is valid:

Corollary 12. *Under the assumptions of Theorem 7, the Bernstein processes z_t^ε are solutions of the following regularized Euler–Lagrange equations:*

$$a. s. \qquad D^\varepsilon D^\varepsilon z_t^\varepsilon = A_\varepsilon^2 z_t^\varepsilon + A_\varepsilon \nabla^\varepsilon V^\varepsilon(z_t^\varepsilon), \quad 0 \le t \le T, \qquad (4.7)$$

with the final boundary condition $D^\varepsilon z_T^\varepsilon = -A_\varepsilon z_T^\varepsilon + \nabla^\varepsilon \log \eta_T^\varepsilon(z_t^\varepsilon)$ a. s.

Remark 8. Eq. (4.7) is a relativistic (but Euclidean) counterpart of Feynman's "probabilistic" interpretation of elementary quantum dynamics (eq. (7.42) in [17]).

The relevant Euclidean (then rigorous) version of Feynman's result has been given in [15] and [16]. Cf. also the next paragraph for the relation with Feynman's path integral approach.

When $\varepsilon \to 0$ in eq. (4.7) then the sample paths of z_t^ε become singular. In order to define, nevertheless, a forward regularized derivative for the generalized process z_t^0, we apply a method natural in the context of distributions theory.

Lemma 13. *Let $\delta z(t)$ be a $\mathcal{P}_t$-adapted Wiener space valued process of bounded variation such that $\delta z(s) = 0$ and let $F(z_t^\varepsilon, t)$ be a $\mathcal{P}_t$-adapted process such that $D^\varepsilon F(z_t^\varepsilon, t)$ exists and $t \mapsto D^\varepsilon F(z^\varepsilon(t), t)$ is continuous. Then, for $s < u$,*

$$\mathrm{E}_s\left[\int_s^u \left(\delta z(t), D^\varepsilon F(z_t^\varepsilon, t)\right) dt\right]$$

$$= \mathrm{E}_s\left[\left(\delta z(u), F(z_u^\varepsilon, u)\right)\right] - \mathrm{E}_s\left[\int_s^u \left(\frac{d}{dt}\delta z(t), F(z_t^\varepsilon, t)\right) dt\right]. \tag{4.8}$$

Proof. The proof is the same as in finite dimensions (see [15], Lemma 4.4). □

Lemma 14. *For any test function h in $L^2(dx)$,*

$$D^0(z_t^0, h) = -(A^0 z_t^0, h) + (A_0 \nabla^0 \log \eta_t^0, h) \tag{4.9}$$

and $D^0(z_t^0, h)$ fulfills the identity (4.8).

Proof. We can compute the conditional expectation value (4.6) using the stochastic differential equation (3.15)

$$D(z_t^0, h) = \lim_{\Delta t \to 0} \mathrm{E}_t\left[\frac{1}{\Delta t}\int_t^{t+\Delta t} d(z_t^0, h)\right]. \tag{4.10}$$

The a. s. limit (4.10) exists and is equal to the drift of eq. (3.15). We can show that eq. (4.8) is fulfilled for $D(z_t^0, h)$ either by copying the derivation of eq. (4.8) in ref. [15] and [16] or by a direct proof that

$$\left(\delta z(s), z_s^\varepsilon\right) - \mathrm{E}_s\left[\int_s^u \left(\frac{d}{dt}\delta z(t), z_t^\varepsilon\right) dt\right]$$

has a limit when $\varepsilon \to 0$ (this follows from the weak convergence of z_t^ε, cf. Lemma 8 and Theorem 10). □

Definition. We define the second order derivative $D^0 D^0 z_t^0 \equiv (D^0)^2 z_t^0$ via the integration-by-parts formula (4.8), when it holds:

$$\mathrm{E}_s\left[\int_s^u \left(\delta z(t), (D^0)^2 z_t^0\right) dt\right]$$

$$\equiv \left(\delta z(s), D^0 z_s^0\right) - \mathrm{E}_s\left[\int_s^u \left(\frac{d}{dt}\delta z(t), D z_t^0\right)\right] dt. \tag{4.11}$$

We define as well $\nabla F(z_t^0)$ by the limit in $L_{\nu_\Theta^0}^2$:

$$(\nabla F, h) = \lim_{\lambda \to 0} \frac{1}{\lambda}\left(F(z_t^0 + \lambda h) - F(z_t^0)\right).$$

Theorem 15. *The Bernstein process z_t^0 satisfies the forward equations of motion*

$$(D^0)^2 z_t^0 = -A_0^2 z_t^0 - \nabla V(z_t^0) \quad on \ t \in [0, T], \tag{4.12}$$

where V is either the polynomial or the exponential interaction (eqs. (2.5)–(2.6)).

Sketch of the proof. We have proved in Corollary 12 that eq. (4.12) is true with the ε-regularization. The assumptions of Lemma 13 are fulfilled for $F = (Dz_t^\varepsilon, h)$. So, from eq. (4.8) and Corollary 12 we get

$$-\mathrm{E}_s \left[\int_s^u \left(\delta z(t), \nabla V(z_t^\varepsilon) + A_\varepsilon z_t^\varepsilon \right) dt \right]$$
$$= \left(\delta z(u), D^\varepsilon z_u^\varepsilon \right) - \mathrm{E}_s \left[\int_s^u \left(\frac{d}{dt} \delta z(t), D^\varepsilon z_t^\varepsilon \right) dt \right]. \tag{4.13}$$

It follows from the argument of Lemma 6 that $D^\varepsilon z_u^\varepsilon \to D^0 z_u^0$. Now, according to eq. (4.11), the limit of the right-hand side of eq. (4.13) gives the definition of $(D^0)^2 z_t^0$. Hence, the right-hand side of eq. (4.13) has a limit equal to

$$\mathrm{E}_s \left[\int_s^u \left(\delta z(t), (D^0)^2 z_t^0 \right) dt \right]. \tag{4.14}$$

On the left-hand side of eq. (4.13), $z_t^\varepsilon \to z_t^0$ weakly (Lemma 8 and Theorem 10). Moreover, using the formula (3.11) we can prove that

$$\int d\mu_0(\varphi) \, \mathrm{E}_\varphi \left[\left((\nabla V(z_t^\varepsilon), h) - (\nabla V(z_t^0), h) \right)^2 \right]$$
$$\leq c \int d\mu_0(\varphi) \, \mathrm{E}_\varphi \left[\left((\nabla V(X_t^\varepsilon), h) - (\nabla V(X_t^0), h) \right)^2 \right] \tag{4.15}$$

where X_t^ε is the Ornstein–Uhlenbeck process (see Theorem 1). It follows in the same way as in Theorem 3 that $(\nabla V(z_t^\varepsilon), h) \to (\nabla V(z_t^0), h)$. Therefore, the left-hand side of eq. (4.13) has a limit as $\varepsilon \to 0$. This limit is equal to

$$-\mathrm{E}_s \left[\int_s^u \left(\delta z(t), \nabla V(z_t^0) + A_0 z_t^0 \right) dt \right]. \tag{4.16}$$

Hence, (4.14) is equal to (4.16) as limits of eq. (4.13). □

Remark 9. All the arguments of the present section can be repeated with respect to the backward filtration $\mathcal{F}_t$ so that the time symmetry of the theory is preserved. We shall, however, not do it here.

§5 PATH INTEGRALS AND VARIATIONAL PRINCIPLES

We define, for each $\varepsilon > 0$, the (Euclidean) Lagrangian associated to the Hamiltonian H^ε by

$$L^\varepsilon(z,^\varepsilon, D^\varepsilon z^\varepsilon) = \frac{1}{2}\|A_\varepsilon^{-1/2}(D^\varepsilon z_t^\varepsilon + A_\varepsilon z_t^\varepsilon)\|_\varepsilon^2 + V^\varepsilon(z_t^\varepsilon). \tag{5.1}$$

The corresponding (forward) regularized action functional is then defined by

$$J^\varepsilon(z^\varepsilon) = \mathrm{E}_s \int_s^T L^\varepsilon(z^\varepsilon, D^\varepsilon z^\varepsilon)\, dt - \mathrm{E}_s[\log \eta_T^\varepsilon(z_T^\varepsilon)] \tag{5.2}$$

with $0 < s < T$.

Then the laws of motion (4.7) can be derived as the Euler–Lagrange equation associated to this regularized action functional. More precisely (cf. [18] for proof and complete statements), a necessary and sufficient condition for a Bernstein process to be an extremal of the action functional J^ε is that it solves the equations (4.7). The domain of the functional J^ε consists of processes y_t^ε which are weak solutions of Ω-valued stochastic differential equations of the form

$$dy_t^\varepsilon = A_\varepsilon^{1/2}\, dW_t^\varepsilon - A_\varepsilon y_t^\varepsilon\, dt + b(y_t^\varepsilon)\, dt,$$

where b is any vector field such that $A_\varepsilon^{-1/2} b \in \mathcal{H}^\varepsilon$, the law of y_t^ε is absolutely continuous with respect to P^ε. By definition, here, a process y^ε is extremal of a functional J^ε if $\mathbb{D}_\psi^\varepsilon J^\varepsilon(y^\varepsilon) = 0$ a. s. along every direction ψ, in the sense of eq. (4.3).

By computing the derivative D of $\log \eta_t^\varepsilon(z_t^\varepsilon)$ and after time integration on $[s,t]$, one obtains the following forward path integral representation of η_t^ε in terms of the Bernstein processes z_t^ε:

$$\eta_t^\varepsilon(z_t^\varepsilon) = \exp\left\{-\mathrm{E}_s \int_s^t L^\varepsilon(z^\varepsilon, D^\varepsilon z^\varepsilon)\, d\tau + \mathrm{E}_s[\log \eta_t^\varepsilon(z_t^\varepsilon)]\right\} \tag{5.3}$$

as well as an analogous formula for the time reversed functional $\eta_s^{*\varepsilon}$. This last path integral requires, of course, to use the characteristics of z_t^ε with respect to the backward filtration.

When $\varepsilon \to 0$ in eq. (5.1), we obtain formally the Lagrangian

$$\begin{aligned} L^0(z^0, Dz^0) &= \frac{1}{2}\|A_0^{-1/2}(Dz^0 + A_0 z^0)\|_0^2 + V(z^0) \\ &= \frac{1}{2}\|Dz^0 + A_0 z^0\|_{L^2}^2 + V(z^0). \end{aligned} \tag{5.4}$$

This is a well-defined object: the quadratic part is well-defined and $V(z^0)$ is defined like in the proof of Theorem 15 as the L^2-limit of $V(z^\varepsilon)$. The Lagrangian gives rise to the same law of motion as another one, namely

$$L(z^0, Dz^0) = \frac{1}{2}\|Dz^0\|_{L^2}^2 + \frac{1}{2}\|A_0 z^0\|_{L^2}^2 + V(z^0). \tag{5.5}$$

Since $A_0 = (-\Delta + m^2)^{1/2}$, this is a regularized version of the classical Lagrangian used in two-dimensional Euclidean quantum field theory, namely

$$L = \int_\Lambda \left(\frac{1}{2} \left| \frac{\partial}{\partial t} z \right|^2 + \frac{1}{2} \left| \frac{\partial}{\partial x} z \right|^2 + \frac{1}{2} m^2 z^2 + v(z) \right) dx. \qquad (5.6)$$

The Euler–Lagrange equation associated to eq. (5.5) is the regularized Euclidean Klein–Gordon equation:

$$DDz_t^0 = (-\Delta + m^2) z_t^0 + \nabla V(z_t^0) \qquad \text{a. s.} \qquad (5.7)$$

Remark 10. Although $V(z^0)$ and $\nabla V(z^0)$ can be defined as L^2-limits, there is some difficulty in expressing them as powers of z_t^0. These problems are discussed in references [27] and [28].

If the sample paths $t \mapsto z_t^0$ were smooth, the left-hand side of eq. (5.7) would re-duce to $\frac{\partial^2}{\partial t^2} z_t^0$ and therefore this equation would reduce to the familiar (Euclidean) Klein–Gordon equation associated, in a finite time interval, to the Lagrangian (5.6). This is in this sense that the whole procedure for our approach to Euclidean quantum field theory can also be regarded as a new kind of regularization for quantum dynamics and path integrals. In the particular case considered here, we have overcome some of the difficulties associated with the singularity of the path integral mentioned in [26], §10.

REFERENCES

1. J. Schwinger, Proc. Natl. Aca. Sci. USA **44** (1958), 956.
2. K. Symanzik, in Local Quantum Theory (R. Jost, eds.), Acad. Press, Varenna, 1968.
3. E. Nelson, in Constructive Quantum Field Theory, Lect. Notes in Physics (G. Velo and A. Wightman, eds.), vol. 25, Springer-Verlag, 1973.
4. K. Osterwalder and R. Schrader, Com. in Math. Physics **31** (1973), 83; and **42** (1975), 281.
5. F. Guerra, L. Rosen, and B. Simon, Ann. Math. **101, 111.**
6. E. Schrödinger, Ann. Inst. H. Poincaré **2** (1932), 269.
7. J. C. Zambrini, J. Math. Phys. **27** (1986), 2307.
8. _____, *New probabilistic approach to the classical heat equation*, in Proceed. of Swansea Meeting, Lect. Notes in Math. (A. Truman and I. M. Davies, eds.), vol. 1325, Springer-Verlag, 1988.
9. _____, *Euclidean quantum mechanics: an alternative starting point for Euclidean field theory*, in Proceed. of IX IAMP, Congress, Swansea (B. Simon, A. Truman, and I. M. Davies, eds.), Hilger, Bristol, 1989, p. 260.
10. S. Bernstein, *Sur les liaisons entre les grandeurs aléatoires*, Verh. Int. Math. Zürich, Band 1, 1932.
11. H. Föllmer, *Random fields and diffusion processes*, in Proc. of Summer School XV-XVII, St. Flour, Lect. Notes in Math. (P. L. Hennequin, ed.), Springer-Verlag, 1988, p. 1362.
12. A. Wakolbinger, J. Math. Phys. **30** (1989), 2943.
13. P. Malliavin, *Stochastic Calculus of variations and hypoelliptic operators*, Proc. Int. Symp. S. D. E., Kyoto 1976 (K. Itô, ed.), Kinokuniya–Wiley, 1978.
14. _____, *Implicit functions in finite corank on the Wiener space*, Tanniguchi Symp. 82, Kinokuniya, Tokyo, 1984.

15. A. B. Cruzeiro and J. C. Zambrini, J. of Funct. Anal. **96, 1** (1991), 62.
16. _____, *Feynman's functional calculus and stochastic calculus of variations*, in: Stochastic analysis and applications: Prog. in Prob. (A. B. Cruzeiro and J. C. Zambrini, eds.), Series, vol. 26, Birkhäuser, Boston, 1991.
17. R. Feynman and A. Hibbs, *Quantum Mechanics and Path Integrals*, Mc Graw–Hill, N. York, 1965, §7.2.
18. A. B. Cruzeiro and J. C. Zambrini, *Malliavin calculus and Euclidean quantum mechanics II* (to appear in J. Funct. Anal.).
19. J. Glimm and A. Jaffe, Comm. Math. Physics **22** (1971), 253.
20. E. Nelson, *Probability theory and Euclidean field theory*, in Constructive Quantum field theory (G. Velo and A. Wightman, eds.), Springer-Verlag, N. Y., 1973.
21. B. Simon, *The $P(\phi)_2$ Euclidean (Quantum) Field Theory*, Princeton Univ. Press, 1974.
22. H. P. McKean, J. of Math. Phys. **18, 6** (1977), 1277.
23. S. Albeverio, K. Yasue, and J. C. Zambrini, Ann. Inst. M. Poincaré, Phys. Theor. **49, 3** (1989), 259.
24. D. Nualart and M. Sanz, Z. Wahrscheinl. verw. Gebiete **70** (1985), 573.
25. E. Nelson, *Dynamical Theories of Brownian Motion*, Princeton Univ. Press, Princeton N. J., 1967.
26. T. Kolsrud and J. C. Zambrini, *The general mathematical framework of Euclidean quantum mechanics: an outline*, in: Stochastic analysis and Applications (A. B. Cruzeiro and J. C. Zambrini, eds.), vol. 26, Birkhäuser, Boston, 1991.
27. R. Schrader, Fortschr. Phys. **22** (1974), 611.
28. J. Feldman and R. Raczka, Ann. Phys. **10** (1977), 212.
29. S. Albeverio and R. Hoegh-Krohn, J. Funct. Anal. **16** (1974), 39.
30. _____, Comm. Math. Phys. **30** (1973), 171.
31. J. Fröhlich, Comm. Math. Phys. **47** (1976), 233.

A. B. Cruzeiro, Complexo II, Universidade de Lisboa, Av. Prof. Gama Pinto, 2 – 1699 Lisboa Codex, Portugal

Z. Haba, Institute of Theoretical Physics, 50-204 Wroclaw, Pl Max Borna 9, University of Wroclaw, Wroclaw, Poland

J. C. Zambrini, Complexo II, Universidade de Lisboa, Av. Prof. Gama Pinto, 2 – 1699 Lisboa Codex, Portugal

Progress in Probability, Vol. 36
© 1995 Birkhäuser Verlag Basel/Switzerland

A FUBINI THEOREM
FOR GENERALIZED STRATONOVICH INTEGRALS

ROSARIO DELGADO*AND MARTA SANZ-SOLÉ*

§0 INTRODUCTION

The purpose of this paper is to analyze under which conditions the multiple generalized (that means, non necessarily adapted) Stratonovich integral with respect to the Brownian sheet can be iterated. The motivation of this problem comes from a previous work by the authors. Indeed, in [2] an iteration of a double Stratonovich integral with respect to a non adapted "semimartingale" in the plane is needed in order to prove a Green formula. In comparison with that article the situation considered here is more simple, since our integrator is the Brownian sheet, but also more general, because we are considering multiple integrals of any order $k \geq 2$. The basic ingredients which are needed are the Hu-Meyer formula established in [1], the Fubini's theorem for the multiple Skorohod integral (see [3]) and some results on the iteration of traces.

First, in Section 1, we introduce the suitable spaces of processes where the results will hold. Section 2 deals with properties on iteration of the traces. It should be mentioned that the integral representation of the traces (see Proposition 2.2 in [2]) provides a useful tool to study this problem. In the last section we prove the Fubini theorem (see Theorem 3.2). The idea is to translate this problem in terms of iteration of multiple Skorohod integrals. This can be done using the above mentioned version of the Hu-Meyer formula and some additional formulas for the traces of multiple Skorohod integrals of processes, which are also trace terms. These results are based on the iteration of traces established before.

§1 NOTATION AND BASIC DEFINITIONS

The parameter space of our processes is $T = [0, 1]^2$ endowed with the order relation defined coordinatewise. Given two points $z = (s, t)$ and $z' = (s', t')$, $z, z' \in T$, we

---

1991 *Mathematics Subject Classification.* 60H05.

Key words and phrases. Multiple Skorohod integral, multiple Stratonovich integral, iteration of traces.

*Partially supported by the DGICYT grant no. PB 90-0452

write $z \perp z'$ if $s \leq s'$ and $t \geq t'$. The following order relations will also be considered:

$$z\,R^1\,z' \iff z \leq z', \qquad z\,R^2\,z' \iff z \perp z',$$
$$z\,R^3\,z' \iff z' \leq z, \qquad z\,R^4\,z' \iff z' \perp z.$$

We denote by π any partition of T, $\pi = \{z_i, i = 1, \ldots, r_\pi\}$, which is determined by the product of two grids of $[0,1]$, $\pi^1 = \{0 = s_1 < \cdots < s_{r_1+1} = 1\}$, $\pi^2 = \{0 = t_1 < \cdots < t_{r_2+1} = 1\}$. The mesh of π is defined as

$$|\pi| = \max_{1 \leq i \leq r_1} |s_{i+1} - s_i| + \max_{1 \leq j \leq r_2} |t_{j+1} - t_j|.$$

Let $W = \{W_z, z \in T\}$ be a Brownian sheet defined on some probability space and $(\Omega, \mathcal{F}, P)$ the canonical space associated with W. In the sequel we will deal with the following operators: D^k, the k-th Malliavin derivative and its adjoint δ^k the k-th Skorohod integral. We recall that, for any $k \in \mathbb{N}$ and $p \in [2, \infty)$, $\mathbb{D}^{k,p}$ denotes the space of Wiener functionals $F \in L^p(\Omega)$ with $\sum_{j=1}^{k} \left\| \|D^j F\|_{L^2(T^j)} \right\|_p < \infty$.

We want now to give a description of the main spaces of processes that will be used along this paper. First, we set

$$\mathbb{L}_k^{1,2} = L^2(T^k; \mathbb{D}^{k,2}).$$

For $k = 0$, $\mathbb{L}_k^{1,2} = L^2(\Omega)$, by convention. This space is a subset of $\mathrm{Dom}\,\delta^k$, well suited for the purpose of developing a stochastic calculus concerning the k-th Skorohod integral. However, the Stratonovich anticipating calculus needs more restrictive spaces, in order to ensure the existence of "trace terms" in a sense to be specified later.

In the sequel k will be an integer greater than 1, j an element of the set $\{0, 1, \ldots, [\frac{k}{2}]\}$ and r a number in the set $\{0, 1, \ldots, k - 2j\}$.

Definition 1.1. We denote by $\mathbb{L}_{k,c(j,r)}^{1,2}$ the set of symmetric processes in $\mathbb{L}_k^{1,2}$ such that
(1.1.1)

$$\operatorname*{ess\,sup}_{(t_1,\ldots,t_{2j},t_{2j+1},\ldots,t_{2j+r},s_1,\ldots,s_r)} \left\| D_{s_1,\ldots,s_r}^r X_{(t_1,\ldots,t_{2j},t_{2j+1},\ldots,t_{2j+r},\cdot)} \right\|_{\mathbb{L}_{k-2j-r}^{1,2}} < \infty.$$

(1.1.2) There exists a neighbourhood $V_{j,r}$ of the set

$$\Big\{ (t_1, \ldots, t_{2j}, t_{2j+1}, \ldots, t_{2j+r}, s_1, \ldots, s_r) \in T^{2(j+r)} :$$
$$t_1 = t_{j+1}, \ldots, t_j = t_{2j}, t_{2j+1} = s_1, \ldots, t_{2j+r} = s_r \Big\}$$

such that, for any $(i_1, \ldots, i_r, h_1, \ldots, h_j) \in \{1,2,3,4\}^{r+j}$, there exists a version of $D^r X$ such that the mapping

$$(t_1, \ldots, t_{2j}, t_{2j+1}, \ldots, t_{2j+r}, s_1, \ldots, s_r)$$
$$\longmapsto D_{t_{2j+1},\ldots,t_{2j+r}}^r X_{(t_1,\ldots,t_{2j},s_1,\ldots,s_r,\cdot)},$$

defined on

$$V_{j,r} \cap \{(t_1, \ldots, t_{2j}, t_{2j+1}, \ldots, t_{2j+r}, s_1, \ldots, s_r) \in T^{2(j+r)} :$$
$$s_1 R^{i_1} t_{2j+1}, \ldots, s_r R^{i_r} t_{2j+r}, t_1 R^{h_1} t_{j+1}, \ldots, t_j R^{h_j} t_{2j}\},$$

and taking its values on $\mathbb{L}^{1,2}_{k-2j-r}$, is continuous in the variables $(t_1, \ldots, t_j, s_1, \ldots, s_r)$ uniformly in $(t_{j+1}, \ldots, t_{2j}, t_{2j+1}, \ldots, t_{2j+r})$.

For $X \in \mathbb{L}^{1,2}_{k,c(j,r)}$ we can define

$$\left(D^{(i_1,\ldots,i_r)}_{(h_1,\ldots,h_j)} X \right)_{(t_{j+1},\ldots,t_{2j+r},\cdot)} \tag{1.1}$$
$$= \mathbb{L}^{1,2}_{k-2j-r} - \lim D^r_{t_{2j+1},\ldots,t_{2j+r}} X_{(t_1,\ldots,t_{2j},s_1,\ldots,s_r,\cdot)}$$

as $(s_1, \ldots, s_r) \to (t_{2j+1}, \ldots, t_{2j+r})$ with $s_1 R^{i_1} t_{2j+1}, \ldots, s_r R^{i_r} t_{2j+r}$, and $(t_1, \ldots, t_j)$ $\to (t_{j+1}, \ldots, t_{2j})$ with $t_1 R^{h_1} t_{j+1}, \ldots, t_j R^{h_j} t_{2j}$. Then, we will set

$$(\nabla^r_j X)_{(t_{j+1},\ldots,t_{2j+r},\cdot)} = \frac{1}{4^{j+r}} \sum \left(D^{(i_1,\ldots,i_r)}_{(h_1,\ldots,h_j)} X \right)_{(t_{j+1},\ldots,t_{2j+r},\cdot)}, \tag{1.2}$$

where the sum extends to $i_1, \ldots, i_r, h_1, \ldots, h_j \in \{1, 2, 3, 4\}$. Notice that $(\nabla^r_j X)_{(t_{j+1},\ldots,t_{2j+r},\cdot)} \in \mathbb{L}^{1,2}_{k-2j-r}$. We will set

$$\mathbb{L}^{1,2}_{k,c} = \bigcap_{j=0}^{[\frac{k}{2}]} \bigcap_{r=0}^{k-2j} \mathbb{L}^{1,2}_{k,c(j,r)}, \tag{1.3}$$

with the convention $\mathbb{L}^{1,2}_{k,c(0,0)} = \mathbb{L}^{1,2}_k$.

Remark 1.2. Let $X \in \mathbb{L}^{1,2}_{k,c(j,r)}$ and $m \in \{1, \ldots, k-2j-r\}$. Then, for any $\underline{s} \in T^m$, the process $X_{(\underline{s},\cdot)}$ belongs to $\mathbb{L}^{1,2}_{k-m,c(j,r)}$.

The proof of Fubini's theorem needs some uniformity on the spaces defined previously. In the next two definitions, $D^r X$ is the version specified in (1.1.2).

In the next two definitions we assume $k - 2j - r \geq 1$.

Definition 1.3. We define the space $\mathbb{L}^{1,2,u}_{k,c(j,r)}$ as the subset of $\mathbb{L}^{1,2}_{k,c(j,r)}$ of processes such that

$$(1.3.1) \qquad \operatorname*{ess\,sup}_{(t_1,\ldots,t_{2j},t_{2j+1},\ldots,t_{2j+r+1},s_1,\ldots,s_r)} \left\| D^r_{s_1,\ldots,s_r} X_{(t_1,\ldots,t_{2j+r+1},\cdot)} \right\|_{\mathbb{L}^{1,2}_{k-2j-r-1}} < \infty.$$

(1.3.2) The following property holds: For any $i_1, \ldots, i_r, h_1, \ldots, h_j \in \{1, 2, 3, 4\}$, and any $s \in T$ a. e. we have

$$\left(D^{(i_1,\ldots,i_r)}_{(h_1,\ldots,h_j)} X \right)_{(t_{j+1},\ldots,t_{2j+r},s,\cdot)}$$
$$= \mathbb{L}^{1,2}_{k-2j-r-1} - \lim D^r_{t_{2j+1},\ldots,t_{2j+r}} X_{(t_1,\ldots,t_{2j},s_1,\ldots,s_r,s,\cdot)}$$

as $(s_1, \ldots, s_r) \longmapsto (t_{2j+1}, \ldots, t_{2j+r})$ with $s_1 R^{i_1} t_{2j+1}, \ldots, s_r \ R^{i_r} t_{2j+r}$, and $(t_1, \ldots, t_j) \longmapsto (t_{j+1}, \ldots, t_{2j})$ with $t_1 R^{h_1} t_{j+1}, \ldots, t_j \ R^{h_j} t_{2j}$ uniformly in $(t_{j+1}, \ldots, t_{2j+r}, s)$.

Finally, we need to introduce some subset of the space given in Definition 1.3.

Definition 1.4. Denote by $\overline{\mathbb{L}}_{k,c(j,r)}^{1,2,u}$ the subset of $\mathbb{L}_{k,c(j,r)}^{1,2,u}$ such that for any t a. e.,
(1.4.1)
$$\operatorname*{ess\,sup}_{(t_1, \ldots, t_{2j}, t_{2j+1}, \ldots, t_{2j+r+1}, s_1, \ldots, s_r)} \left\| D_t D_{s_1, \ldots, s_r}^r X_{(t_1, \ldots, t_{2j+r+1}, \cdot)} \right\|_{\mathbb{L}_{k-2j-r-1}^{1,2}} < \infty.$$

(1.4.2) The following property is satisfied: For any $s \in T$ a. e. and for any $i_1, \ldots, i_r, h_1, \ldots, h_j \in \{1, 2, 3, 4\}$, it holds that

$$D_t \left(D_{(h_1, \ldots, h_j)}^{(i_1, \ldots, i_r)} X \right)_{(t_{j+1}, \ldots, t_{2j+r}, s, \cdot)}$$
$$= \mathbb{L}_{k-2j-r-1}^{1,2} - \lim D_t \left(D_{t_{2j+1}, \ldots, t_{2j+r}}^r X_{(t_1, \ldots, t_{2j}, s_1, \ldots, s_r, s, \cdot)} \right)$$

as $(s_1, \ldots, s_r) \longmapsto (t_{2j+1}, \ldots, t_{2j+r})$ with $s_1 R^{i_1} t_{2j+1}, \ldots, s_r \ R^{i_r} t_{2j+r}$, and $(t_1, \ldots, t_j) \longmapsto (t_{j+1}, \ldots, t_{2j})$ with $t_1 R^{h_1} t_{j+1}, \ldots, t_j \ R^{h_j} t_{2j}$, uniformly in $(t_{j+1}, \ldots, t_{2j+r}, s, t)$.

§2 ITERATION OF TRACES

The aim of this section is to study some properties of the correction terms between the multiple Skorohod and Stratonovich integrals. We start by giving the notion of *traces*.

Definition 2.1. Let $X = \{X_{\underline{t}}, \ \underline{t} \in T^k\}$ be a symmetric process belonging to $\mathbb{L}_k^{1,2}$. Fix $j \in \{0, 1, \ldots, [\frac{k}{2}]\}$ and $r \in \{0, 1, \ldots, k-2j\}$. The *trace* $T_{j,r}(X)$ is the process of $\mathbb{L}_{k-2j-r}^{1,2}$ given by the $\mathbb{L}_{k-2j-r}^{1,2}$-limit of the sequence of processes

$$\left\{ \sum_{i_1, \ldots, i_{j+r}} \frac{1}{|\Delta_{i_1}| \ldots |\Delta_{i_{j+r}}|} \int_{\Delta_{i_1}^2 \times \cdots \times \Delta_{i_{j+r}}^2} D_{\underline{s}}^r X_{\underline{t}} \, dt_1 \ldots dt_{2j} \, ds_1 \, dt_{2j+1} \ldots ds_r \, dt_{2j+r} \right\}$$

corresponding to a sequence of grids $\{\pi_n, n \geq 1\}$ of T whose mesh tends to zero as $n \to \infty$, whenever this limit exists.

We also recall the definition of the k-th Stratonovich integral.

Definition 2.2. Let $X = \{X_{\underline{t}}, \ \underline{t} \in T^k\}$ be a symmetric process defined on $(\Omega, \mathcal{F}, P)$, such that $E[\int_{T^k} X_{\underline{t}}^2 d\underline{t}] < \infty$. For any partition π, $\pi = \{\Delta_1, \ldots, \Delta_{r_\pi}\}$, of T we set

$$S_\pi(X) = \sum_{i_1, \ldots, i_k = 1}^{r_\pi} \frac{1}{|\Delta_{i_1}| \ldots |\Delta_{i_k}|} \left(\int_{\Delta_{i_1} \times \cdots \times \Delta_{i_k}} X_{\underline{t}} \, d\underline{t} \right) W(\Delta_{i_1}) \ldots W(\Delta_{i_k}).$$

Then, the process X is said to be Stratonovich integrable if the family $\{S_\pi(X), \pi$ partition of $T\}$ converges in $L^2(\Omega)$ as $|\pi| \to 0$. We will call this limit the k-Stratonovich integral of the process X and it will be denoted by $I_k^s(X)$.

If X is a symmetric process belonging to $\mathbb{L}^{1,2}_{k,c(j,r)}$, for any $j \in \{0, 1, \ldots, [\frac{k}{2}]\}$, and any $r \in \{0, 1, \ldots, k - 2j\}$, then all traces $T_{j,r}(X)$ exist. In [1] we prove that in this situation, $I_k^s(X)$ exists and the following formula holds

$$I_k^s(X) = \sum_{j=0}^{[\frac{k}{2}]} \sum_{r=0}^{k-2j} \binom{k-2j}{r} \frac{k!}{(k-2j)!\, j!\, 2^j} \delta^{k-2j-r}(T_{j,r}(X)). \tag{2.1}$$

This is called the Hu-Meyer formula. In addition (see Proposition 2.2 [2])

$$T_{j,r}(X) = \int_{T^{j+r}} (\nabla_j^r X)_{(t_{j+1}, \ldots, t_{2j+r}, \cdot)}\, dt_{j+1} \ldots dt_{2j+r}, \tag{2.2}$$

where $\nabla_j^r X$ has been defined in (1.2). We quote a first result on iteration of traces proved in [2] that will often be used in the sequel (see Proposition 2.8 in [2]).

(a) Let X be a process belonging to $\mathbb{L}^{1,2}_{3,c(1,0)}$. Then

$$T_{1,0}(\delta(X)) = \delta(T_{1,0}(X)), \tag{2.3}$$

in the sense that both terms in (2.3) exist and coincide.

(b) Let X be a process in $\mathbb{L}^{1,2}_{2,c(0,1)} \cap \mathbb{L}^{1,2}_{2,c(1,0)}$. Then

$$T_{0,1}(\delta(X)) = \delta(T_{0,1}(X)) + T_{1,0}(X), \tag{2.4}$$

that means, all terms appearing in (2.4) exist and the relation (2.4) holds.

The main result of this section is a property on the iteration of the traces. It is a generalization of Proposition 2.5 in [2] for $k > 2$ and it is one of the ingredients in the proof of the Fubini-Stratonovich result.

Proposition 2.3. *Fix $k \geq 2$, and let $X = \{X_{\underline{t}}, \underline{t} \in T^k\}$ be a symmetric process.*

(a) *Let $j \in \{0, 1, \ldots, [\frac{k}{2}] - 1\}$, $r \in \{0, 1, \ldots, k - 2j - 2\}$. Assume that*

$$X \in \mathbb{L}^{1,2,u}_{k,c(j,r)} \cap \mathbb{L}^{1,2}_{k,c(j+1,r)} \quad and \quad T_{j,r}(X) \in \mathbb{L}^{1,2}_{k-2j-r,c(1,0)}.$$

Then, $T_{j+1,r}(X)$ and $T_{1,0}(T_{j,r}(X))$ exist and

$$T_{j+1,r}(X) = T_{1,0}(T_{j,r}(X)). \tag{2.5}$$

(b) *Let $j \in \{0, 1, \ldots, [\frac{k-1}{2}]\}$, $r \in \{0, 1, \ldots, k - 2j - 1\}$. Suppose that*

$$X \in \overline{\mathbb{L}}^{1,2,u}_{k,c(j,r)} \cap \mathbb{L}^{1,2}_{k,c(j,r+1)} \quad and \quad T_{j,r}(X) \in \mathbb{L}^{1,2}_{k-2j-r,c(0,1)}.$$

Then, $T_{j,r+1}(X)$ and $T_{0,1}(T_{j,r}(X))$ exist and

$$T_{j,r+1}(X) = T_{0,1}(T_{j,r}(X)). \tag{2.6}$$

Proof. Proposition 2.2 of [2] yields

$$T_{j+1,r}(X) = \int_{T^{j+r+1}} \left(\nabla^r_{j+1}X\right)_{(t_{j+2},\ldots,t_{2j+r+2},\cdot)} dt_{j+2}\ldots dt_{2j+r+2}$$

and

$$T_{1,0}\big(T_{j,r}(X)\big) = \int_T \nabla^0_1\Big(\int_{T^{j+r}} \left(\nabla^r_j X\right)_{(t_{j+1},\ldots,t_{2j+r},\cdot)} dt_{j+1}\ldots dt_{2j+r}\Big)_{(t_{2j+r+2},\cdot)} dt_{2j+r+2}.$$

The identity (2.5) will be a consequence of the following fact: For any $i_1,\ldots,i_r$, $h_1,\ldots,h_{j+1}\in\{1,2,3,4\}$,

$$\int_{T^{j+r+1}} \left(D^{(i_1,\ldots,i_r)}_{(h_1,\ldots,h_{j+1})}X\right)_{(t_{j+2},\ldots,t_{2j+r+2},\cdot)} dt_{j+2}\ldots dt_{2j+r+2}$$

$$= \int_T D_{(h_{j+1})}\Big(\int_{T^{j+r}} \left(D^{(i_1,\ldots,i_r)}_{(h_1,\ldots,h_j)}X\right)_{(t_{j+1},\ldots,t_{2j+r},\cdot)} dt_{j+1}\ldots dt_{2j+r}\Big)_{(t_{2j+r+2},\cdot)} dt_{2j+r+2}. \tag{2.7}$$

By (1.1) the right-hand side of (2.7) is

$$\int_T \Big\{ \lim_{t_{2j+r+1}\to t_{2j+r+2}} \Big(\int_{T^{j+r}} \left(D^{(i_1,\ldots,i_r)}_{(h_1,\ldots,h_j)}X\right)_{(t_{j+1},\ldots,t_{2j+r},t_{2j+r+1},t_{2j+r+2},\cdot)} dt_{j+1}\ldots dt_{2j+r}\Big)\Big\} dt_{2j+r+2}, \tag{2.8}$$

where the limit is in the $\mathbb{L}^{1,2}_{k-2j-r-2}$ convergence and $t_{2j+r+1}R^{h_{j+1}}t_{2j+r+2}$. For $s,t\in T$, set

$$C(s,t) = \Big\| \int_{T^{j+r}} \Big\{ \left(D^{(i_1,\ldots,i_r)}_{(h_1,\ldots,h_j)}X\right)_{(t_{j+1},\ldots,t_{2j+r},s,t,\cdot)} - \left(D^{(i_1,\ldots,i_r)}_{(h_1,\ldots,h_{j+1})}X\right)_{(t_{j+1},\ldots,t_{2j+r},t,\cdot)} \Big\} dt_{j+1}\ldots dt_{2j+r} \Big\|_{\mathbb{L}^{1,2}_{k-2j-r-2}}$$

Then, it suffices to prove that for each t a. e.,

$$\lim_{\substack{s\to t \\ sR^{h_{j+1}}t}} C(s,t) = 0. \tag{2.9}$$

Let

$$C_1(s,t;t_1,\ldots,t_j,\underline{u}) = \operatorname*{ess\,sup}_{(t_{j+1},\ldots,t_{2j+r})} \Big\| \left(D^{(i_1,\ldots,i_r)}_{(h_1,\ldots,h_j)}X\right)_{(t_{j+1},\ldots,t_{2j+r},s,t,\cdot)} - D^r_{t_{2j+1},\ldots,t_{2j+r}}X_{(t_1,\ldots,t_{2j},\underline{u},s,t,\cdot)} \Big\|_{\mathbb{L}^{1,2}_{k-2j-r-2}}$$

and

$$C_2(s,t;t_1,\ldots,t_j,\underline{u}) = \operatorname*{ess\,sup}_{(t_{j+1},\ldots,t_{2j+r})} \left\| D^r_{t_{2j+1},\ldots,t_{2j+r}} X_{(t_1,\ldots,t_{2j},\underline{u},s,t,\cdot)} \right.$$
$$\left. - \left(D^{(i_1,\ldots,i_r)}_{(h_1,\ldots,h_{j+1})} X \right)_{(t_{j+1},\ldots,t_{2j+r},t,\cdot)} \right\|_{\mathbb{L}^{1,2}_{k-2j-r-2}},$$

with $\underline{u} \in T^r$ and $t_1,\ldots,t_j \in T$. It is obvious that

$$C(s,t) \le C_1(s,t;t_1,\ldots,t_j,\underline{u}) + C_2(s,t;t_1,\ldots,t_j,\underline{u}).$$

Fix $\varepsilon > 0$ and $t \in T$ a. e. Since X belongs to $\mathbb{L}^{1,2,u}_{k,c(j,r)}$ (see Definition 1.3), there exists a neighbourhood of $(t_{j+1},\ldots,t_{2j+r})$ such that for any $(\underline{t},\underline{u})$ in this set, $\underline{t} = (t_1,\ldots,t_j)$, $\underline{u} = (u_1,\ldots,u_r)$, with $t_1 R^{h_1} t_{j+1},\ldots, t_j R^{h_j} t_{2j}, u_1 R^{i_1} t_{2j+1},\ldots, u_r R^{i_r} t_{2j+1}$,

$$\operatorname*{ess\,sup}_s C_1(s,t;t_1,\ldots,t_j,\underline{u}) < \frac{\varepsilon}{2}.$$

Furthermore, X belongs to $\mathbb{L}^{1,2}_{k,c(j+1,r)}$. Thus, there exist $\delta > 0$ such that if $|s-t| < \delta$, $s R^{h_{j+1}} t$,

$$\operatorname*{ess\,sup}_t C_2(s,t;t_1,\ldots,t_j,\underline{u}) < \frac{\varepsilon}{2}.$$

Consequently (2.9) holds and the proof of the statement (a) is complete.

In order to prove part (b) we will check that, if $i_{r+i} \in \{1,2,3,4\}$

$$\int_{T^{j+r+1}} \left(D^{(i_1,\ldots,i_{r+1})}_{(h_1,\ldots,h_j)} X \right)_{(t_{j+1},\ldots,t_{2j+r+1},\cdot)} dt_{j+1}\ldots dt_{2j+r+1}$$
$$= \int_T D^{(i_{r+1})} \left(\int_{T^{j+r}} \left(D^{(i_1,\ldots,i_r)}_{(h_1,\ldots,h_j)} X \right)_{(t_{j+1},\ldots,t_{2j+r},\cdot)} \right. \tag{2.10}$$
$$\left. dt_{j+1}\ldots dt_{2j+r} \right)_{(t_{2j+r+1},\cdot)} dt_{2j+r+1}.$$

The right-hand side of this expression is, by (1.1),

$$\int_T \left\{ \lim_{s_{r+1} \to t_{2j+r+1}} D_{t_{2j+r+1}} \left(\int_{T^{j+r}} \left(D^{(i_1,\ldots,i_r)}_{(h_1,\ldots,h_j)} X \right)_{(t_{j+1},\ldots,t_{2j+r},s_{r+1},\cdot)} \right. \right.$$
$$\left. \left. dt_{j+1}\ldots dt_{2j+r} \right) \right\} dt_{2j+r+1},$$

with the limit in the $\mathbb{L}^{1,2}_{k-2j-r-1}$ topology, and $s_{r+1} R^{i_{r+1}} t_{2j+r+1}$.

Set

$$K(s,t) = \left\| \int_{T^{j+r}} \left\{ D_t \left(\left(D^{(i_1,\ldots,i_r)}_{(h_1,\ldots,h_j)} X \right)_{(t_{j+1},\ldots,t_{2j+r},s,\cdot)} \right) \right. \right.$$
$$\left. \left. - \left(D^{(i_1,\ldots,i_{r+1})}_{(h_1,\ldots,h_j)} X \right)_{(t_{j+1},\ldots,t_{2j+r},t,\cdot)} \right\} dt_{j+1}\ldots dt_{2j+r} \right\|_{\mathbb{L}^{1,2}_{k-2j-r-1}},$$

and

$$K_1(s,t;t_1,\dots,t_j,\underline{u}) = \operatorname*{ess\,sup}_{(t_{j+1},\dots,t_{2j+r})} \left\| D_t \left\{ \left(D_{(h_1,\dots,h_j)}^{(i_1,\dots,i_r)} X \right)_{(t_{j+1},\dots,t_{2j+r},s,\cdot)} \right. \right.$$
$$\left. \left. - D_{t_{2j+1},\dots,t_{2j+r}}^r X_{(t_1,\dots,t_{2j},\underline{u},s,\cdot)} \right\} \right\|_{\mathbb{L}_{k-2j-r-1}^{1,2}},$$

as well as

$$K_2(s,t;t_1,\dots,t_j,\underline{u}) = \operatorname*{ess\,sup}_{(t_{j+1},\dots,t_{2j+r})} \left\| D_{t_{2j+1},\dots,t_{2j+r},t}^{r+1} X_{(t_1,\dots,t_{2j},\underline{u},s,\cdot)} \right.$$
$$\left. - \left(D_{(h_1,\dots,h_j)}^{(i_1,\dots,i_{r+1})} X \right)_{(t_{j+1},\dots,t_{2j+r},t,\cdot)} \right\|_{\mathbb{L}_{k-2j-r-1}^{1,2}},$$

with $\underline{u} \in T^r$ and $s,t,t_1,\dots,t_j \in T$. Then, the hypotheses on X ensure that, given $\varepsilon > 0$ and $t \in T$ a.e., we can choose $\underline{u} \in T^r, t_1,\dots,t_j \in T$, and there exist $\delta > 0$ such that if $s \in T$, $|s-t| < \delta$, $sR^{i_{r+1}}t$, we have

$$\max \left\{ \operatorname*{ess\,sup}_{s} K_1(s,t;t_1,\dots,t_j,\underline{u}), \operatorname*{ess\,sup}_{t} K_2(s,t;t_1,\dots,t_j,\underline{u}) \right\} < \frac{\varepsilon}{2}.$$

Therefore, for each t a.e.,

$$\lim_{\substack{s \to t \\ sR^{i_{r+1}}t}} K(s,t) = 0,$$

proving (2.10). This ends the proof of the proposition. □

§3 THE FUBINI-STRATONOVICH THEOREM

In this section we prove the main result of this paper: A Fubini theorem for the Stratonovich integral. The basic ingredients are the Hu-Meyer formula (2.1), the Fubini theorem for the multiple Skorohod integral (see Proposition 2.6 of [3]), the formulas for the trace of a Skorohod integral (see (2.3) and (2.4)) and the result on iteration of the traces given in Proposition 2.3. The first result generalizes (2.3) and (2.4) to multiple Skorohod integrals.

Lemma 3.1. Let $X = \{X_{\underline{t}}, \; \underline{t} \in T^k\}$ be a symmetric process, $k \geq 2$, such that $X \in \mathbb{L}_{k,c(j,r+1)}^{1,2} \cap \mathbb{L}_{k,c(j+1,r)}^{1,2} \cap \overline{\mathbb{L}}_{k,c(j,r)}^{1,2,u}$ for some $j \in \{0,1,\dots,[\frac{k}{2}]-1\}$ and some $r \in \{0,1,\dots,k-2j-2\}$. Consider the following set of hypotheses:

(i) $\delta^\alpha(T_{j,r}(X)) \in \mathbb{L}_{k-2j-r-\alpha,c(1,0)}^{1,2}$ for any $\alpha \in \{0,\dots,k-2j-r-2\}$.

(ii) $\delta^\alpha(T_{j,r}(X)) \in \mathbb{L}_{k-2j-r-\alpha,c(0,1)}^{1,2}$ for any $\alpha \in \{0,\dots,k-2j-r-1\}$.

Then,

(a) Under (i), it holds that for any $\alpha \in \{0,1,\dots,k-2j-r-2\}$,

$$T_{1,0}\left(\delta^\alpha(T_{j,r}(X))\right) = \delta^\alpha(T_{j+1,r}(X)). \tag{3.1}$$

(b) Under (i) and (ii), it holds that for any $\alpha \in \{1,\dots,k-2j-r-1\}$,

$$T_{0,1}\left(\delta^\alpha(T_{j,r}(X))\right) = \delta^\alpha(T_{j,r+1}(X)) + \alpha\delta^{\alpha-1}(T_{j+1,r}(X)). \tag{3.2}$$

Proof. The assumption (i) for $\alpha = 0$ yields

$$T_{1,0}(T_{j,r}(X)) = T_{j+1,r}(X). \tag{3.3}$$

Similarly, condition (ii) with $\alpha = 0$ ensures

$$T_{0,1}(T_{j,r}(X)) = T_{j,r+1}(X), \tag{3.4}$$

Indeed, it suffices to apply Proposition 2.3.

We will prove (3.1) and (3.2) by induction on α. For $\alpha = 0$, (3.1) is nothing but (3.3). For $\alpha = 1$, assuming $k - 2j - r \geq 3$, it follows from (2.3) and (3.3) that

$$T_{1,0}\big(\delta(T_{j,r}(X))\big) = \delta(T_{j+1,r}(X)).$$

Suppose now that (3.1) holds for $\alpha \in \{0, 1, \ldots, k - 2j - r - 3\}$. Then, Fubini's theorem for the Skorohod integral, the identities (2.3) and (3.3) yield

$$T_{1,0}\big(\delta^{\alpha+1}(T_{j,r}(X))\big) = T_{1,0}\Big(\delta\big(\delta^{\alpha}(T_{j,r}(X))\big)\Big)$$
$$= \delta\Big(T_{1,0}\big(\delta^{\alpha}(T_{j,r}(X))\big)\Big) = \delta^{\alpha+1}(T_{j+1,r}(X)).$$

This finishes the proof of (3.1).

The proof of (3.2) is based on similar ideas. For $\alpha = 1$ the statement follows from (2.4), (3.3) and (3.4). Assume that (3.2) holds for $\alpha \in \{1, \ldots, k - 2j - r - 2\}$. Then, using the Fubini theorem for the Skorohod integral, the identity (2.4) applied to $\delta^{\alpha}(T_{j,r}(X))$ and (3.1), we conclude the induction argument. $\square$

We can now present the main result of this note.

Theorem 3.2. *Let* $X = \{X_{\underline{t}},\ \underline{t} \in T^k\}$, $k \geq 2$, *be a symmetric process. We assume that*

(i) *For any* $j \in \{0, 1, \ldots, [\frac{k}{2}] - 1\}$ *and any* $r \in \{0, 1, \ldots,\ k - 2j - 2\}$ *the hypotheses of Lemma 3.1 are satisfied.*

(ii) $X \in \mathbb{L}^{1,2}_{k,c(j,k-2j)} \cap \overline{\mathbb{L}}^{1,2,u}_{k,c(j,k-2j-1)}$ *and* $T_{j,k-2j-1}(X) \in \mathbb{L}^{1,2}_{1,c(0,1)}$, *for any* $j \in \{0, 1, \ldots, [\frac{k-1}{2}]\}$.

(iii) $\delta^{k-2j-r-1}(T_{j,r}(X)) \in \mathbb{L}^{1,2}_{1,c(0,1)}$, *for any* $j \in \{0, 1, \ldots, [\frac{k-1}{2}]\}$ *and any* $r \in \{0, 1, \ldots, k - 1 - 2j\}$.

(iv) $X \in \mathbb{L}^{1,2}_{k,c([\frac{k}{2}],1)} \cap \overline{\mathbb{L}}^{1,2,u}_{k,c([\frac{k}{2}],0)}$ *and* $T_{[\frac{k}{2}],0}(X) \in \mathbb{L}^{1,2}_{1,c(0,1)}$, *if* k *is odd.*

Then, $I^s_k(X)$ *exists and*

$$I^s_k(X) = I^s\big(I^s_{k-1}(X)\big). \tag{3.5}$$

Proof. It is clear that $I^s_{k-1}(X)$ and $I^s_k(X)$ exist because $X \in \mathbb{L}^{1,2}_{k,c}$ (see (1.3)). Indeed, condition (i) yields that $X \in \mathbb{L}^{1,2}_{k,c(j,r)}$, for any $j \in \{1, \ldots, [\frac{k}{2}]\}$, $r \in$

$\{0, 1, \ldots, k - 2j\}$ and for $j = 0$, $r \in \{0, 1, \ldots, k - 1\}$, and condition (ii) yields that $X \in \mathbb{L}^{1,2}_{k,c(0,k)}$. Moreover, (2.1) yields that, for every $t \in T$ a. e.,

$$I^s_{k-1}(X_{(t,\cdot)})$$

$$= \sum_{j=0}^{[\frac{k-1}{2}]} \sum_{r=0}^{k-1-2j} \binom{k - 1 - 2j}{r} \frac{(k - 1)!}{(k - 1 - 2j)! \, j! \, 2^j} \delta^{k-1-2j-r}\left(T_{j,r}(X_{(t,\cdot)})\right). \quad (3.6)$$

The assumption (iii) ensures that $I^s_{k-1}(X_{(t,\cdot)})$ belongs to $\mathbb{L}^{1,2}_{1,c(0,1)}$. Consequently, by (2.1) again

$$I^s\left(I^s_{k-1}(X_{(t,\cdot)})\right) = \delta\left(I^s_{k-1}(X_{(t,\cdot)})\right) + T_{0,1}\left(I^s_{k-1}(X_{(t,\cdot)})\right). \quad (3.7)$$

We want to show that the right-hand side of (3.7) coincides with the right-hand side of (2.1). Indeed, the identity (3.6) and Fubini's theorem for the multiple Skorohod integral yields

$$\delta\left(I^s_{k-1}(X_{(t,\cdot)})\right)$$

$$= \sum_{j=0}^{[\frac{k-1}{2}]} \sum_{r=0}^{k-1-2j} \binom{k - 1 - 2j}{r} \frac{(k - 1)!}{(k - 1 - 2j)! \, j! \, 2^j} \delta^{k-2j-r}\left(T_{j,r}(X)\right). \quad (3.8)$$

Set

$$A = \sum_{j=0}^{[\frac{k}{2}]-1} \sum_{r=0}^{k-1-2j} \binom{k - 1 - 2j}{r} \frac{(k - 1)!}{(k - 1 - 2j)! \, j! \, 2^j}$$

$$\left[\delta^{k-1-2j-r}\left(T_{j,r+1}(X_{(t,\cdot)})\right) + (k - 1 - 2j - r)\delta^{k-2-2j-r}\left(T_{j+1,r}(X_{(t,\cdot)})\right)\right],$$

where $(k - 1 - 2j - r)\delta^{k-2-2j-r}\left(T_{j+1,r}(X_{(t,\cdot)})\right)$ equals to zero, by convention, if $r = k - 2j - 1$.

Assume that k is *even*, that means $\left[\frac{k-1}{2}\right] = \left[\frac{k}{2}\right] - 1$. In this case,

$$T_{0,1}\left(I^s_{k-1}(X_{(t,\cdot)})\right) = A_1 + A_2, \quad (3.9)$$

where

$$A_1 = \sum_{j=0}^{[\frac{k}{2}]-1} \sum_{r=0}^{k-2-2j} \binom{k - 1 - 2j}{r} \frac{(k - 1)!}{(k - 1 - 2j)! \, j! \, 2^j} T_{0,1}\left(\delta^{k-1-2j-r}\left(T_{j,r}(X_{(t,\cdot)})\right)\right)$$

and

$$A_2 = \sum_{j=0}^{[\frac{k}{2}]-1} \frac{(k - 1)!}{(k - 1 - 2j)! \, j! \, 2^j} T_{0,1}\left(T_{j,k-1-2j}(X_{(t,\cdot)})\right).$$

Hypothesis (i) and (3.2) for $\alpha = k - 2j - r - 1$ yield that

$$
\begin{aligned}
A_1 = \sum_{j=0}^{[\frac{k}{2}]-1} \sum_{r=0}^{k-2-2j} & \binom{k-1-2j}{r} \frac{(k-1)!}{(k-1-2j)!\, j!\, 2^j} \\
& \left[\delta^{k-1-2j-r}\big(T_{j,r+1}(X_{(t,\cdot)})\big) \right. \\
& \left. + (k-1-2j-r)\delta^{k-2-2j-r}\big(T_{j+1,r}(X_{(t,\cdot)})\big) \right],
\end{aligned}
\tag{3.10}
$$

and for $r = k - 2j - 1$, (2.6) ensures that

$$
A_2 = \sum_{j=0}^{[\frac{k}{2}]-1} \frac{(k-1)!}{(k-1-2j)!\, j!\, 2^j} T_{j,k-2j}(X_{(t,\cdot)}).
\tag{3.11}
$$

Here we have used hypothesis (ii). By replacing (3.10) and (3.11) in (3.9) we have that

$$
T_{0,1}\big(I_{k-1}^s(X_{(t,\cdot)})\big) = A.
\tag{3.12}
$$

On the other hand, if k is *odd*, $\left[\frac{k-1}{2}\right] = \left[\frac{k}{2}\right]$ and

$$
T_{0,1}\big(I_{k-1}^s(X_{(t,\cdot)})\big) = A_1 + A_2 + A_3,
\tag{3.13}
$$

where

$$
A_3 = \frac{(k-1)!}{\left[\frac{k}{2}\right]!\, 2^{[\frac{k}{2}]}} T_{0,1}\big(T_{[\frac{k}{2}],0}(X)\big).
$$

Hypothesis (iv) ensures that we can apply (2.6). Then,

$$
A_3 = \frac{(k-1)!}{\left[\frac{k}{2}\right]!\, 2^{[\frac{k}{2}]}} T_{[\frac{k}{2}],1}(X).
\tag{3.14}
$$

Replacing (3.10), (3.11) and (3.14) in (3.13) we have that

$$
T_{0,1}\big(I_{k-1}^s(X_{(t,\cdot)})\big) = A + \frac{(k-1)!}{\left[\frac{k}{2}\right]!\, 2^{[\frac{k}{2}]}} T_{[\frac{k}{2}],1}(X).
\tag{3.15}
$$

Finally, we replace (3.8) and (3.12) (respectively, (3.15)) if k is *even* (respectively, *odd*) in the right-hand side of (3.7). Easy combinatorial arguments yield

$$
I^s\big(I_{k-1}^s(X_{(t,\cdot)})\big) = \sum_{j=0}^{[\frac{k}{2}]} \sum_{r=0}^{k-2j} \binom{k-2j}{r} \frac{k!}{(k-2j)!\, j!\, 2^j} \delta^{k-2j-r}\big(T_{j,r}(X)\big).
$$

Then, (3.5) follows. $\square$

References

1. Delgado, R. and Sanz-Solé, M., *The Hu–Meyer formula for nondeterministic kernels*, Stochastics and Stochastics Reports **38** (1992), 149–158.
2. Delgado, R. and Sanz-Solé, M., *Green formulas in anticipating calculus*, vol. 134, Mathematical Preprint Series, Universitat de Barcelona, 1993.
3. Nualart, D. and Zakai, M., *Generalized multiple stochastic integrals and the representation of Wiener functionals*, Stochastics and Stochastics Reports **23** (1988), 311–330.

ROSARIO DELGADO, DEPARTAMENT DE MATEMÀTIQUES, UNIVERSITAT AUTÒNOMA DE BARCELONA, 08193 BELLATERRA, SPAIN

MARTA SANZ-SOLÉ, FACULTAT DE MATEMÀTIQUES, UNIVERSITAT DE BARCELONA, GRAN VIA 585, 08007 BARCELONA, SPAIN

Progress in Probability, Vol. 36
© 1995 Birkhäuser Verlag Basel/Switzerland

LARGE DEVIATIONS
VIA PARAMETER DEPENDENT CHANGE OF MEASURE,
AND AN APPLICATION TO THE LOWER TAIL
OF GAUSSIAN PROCESSES

AMIR DEMBO[1] AND OFER ZEITOUNI[2]

ABSTRACT. A refinement of the technique of measure tilting in large deviations is presented. This refinement allows for the handling of situations which are not covered by standard fixed exponential tilting. An application to the study of the law of Brownian motion conditioned on staying in a small L^2 ball follows.

§1 INTRODUCTION

Let $\{\mu_\epsilon\}_{\epsilon>0}$ be a family of probability measures on $\mathbb{R}^d$. We say that μ_ϵ satisfies the large deviations principle (LDP) if there exists a lower semicontinuous nonnegative function $I(\cdot)$ (called a rate function) such that, for all Borel measurable $\Gamma \subset \mathbb{R}^d$,

$$- \inf_{x\in\Gamma^\circ} I(x) \le \liminf_{\epsilon\to0} \epsilon \log \mu_\epsilon(\Gamma) \le \limsup_{\epsilon\to0} \epsilon \log \mu_\epsilon(\Gamma) \le - \inf_{x\in\bar\Gamma} I(x), \qquad (1.1)$$

where $\bar\Gamma$ denotes the closure of Γ, Γ° the interior of Γ, and the infimum of a function over an empty set is interpreted as infinity. We say that μ_ϵ satisfies the weak LDP if the upper bound in (1.1) holds only for compact Γ.

When μ_ϵ is the law of Z_ϵ, the empirical mean of $\lfloor 1/\epsilon \rfloor$ i. i. d. random variables, the celebrated theorem of Cramér yields the LDP. More generally, Gärtner [6] and later Ellis [5] have considered the case where the following limit, called the logarithmic moment generating function, exists as an extended real number:

$$\Lambda(\theta) \stackrel{\triangle}{=} \lim_{\epsilon\to0} \epsilon\Lambda_{\mu_\epsilon}\left(\frac{\theta}{\epsilon}\right), \qquad (1.2)$$

1991 *Mathematics Subject Classification.* Primary 60G15, 60F10. Secondary 60J65.
Key words and phrases. Large Deviations. Change of measure.
[1] Partially supported by a US-Israel BSF grant and by NSF DMS92-09712 grant.
[2] Partially supported by a US-Israel BSF grant and by the fund for promotion of research at the Technion.

where

$$\Lambda_{\mu_\epsilon}(\theta) = \log \mathrm{E}\left[e^{\langle \theta, Z_\epsilon \rangle}\right] = \log \int_{\mathbb{R}^d} e^{\langle \theta, x \rangle} \mu_\epsilon(dx), \quad \theta \in \mathbb{R}^d.$$

Under the assumption that $\Lambda(\theta)$ is finite in a neighborhood of the origin and is essentially smooth, Ellis [5] proved that $\{\mu_\epsilon\}_{\epsilon>0}$ satisfies the LDP with the rate function Λ^* that is the Fenchel–Legendre transform of Λ, i. e.,

$$\Lambda^*(x) \overset{\triangle}{=} \sup_{\lambda \in \mathbb{R}^d} \{\langle \lambda, x \rangle - \Lambda(\lambda)\}$$

(see [4, Section 2.3] and the references therein for earlier works, see also [10, Theorem 5.1] where the origin can be a boundary point of a ball in which Λ is finite). The method of proof, which goes in the i. i. d. case back to Cramér, is to make a change of measure of the form

$$\frac{d\eta_\epsilon}{d\mu_\epsilon} = \frac{e^{\langle \theta, Z_\epsilon \rangle / \epsilon}}{e^{\Lambda_{\mu_\epsilon}(\theta/\epsilon)}}. \tag{1.3}$$

By a judicious choice of the parameter θ (independently of ϵ) and Chebycheff's inequalities coupled with convex analysis arguments, the LDP follows.

The purpose of this paper is to explore situations in which the conditions of Gärtner–Ellis are not fulfilled. The additional ingredient here is to make a change of measure of the form (1.3), in which the parameter θ depends on ϵ. It turns out that by doing that, one may avoid some of the difficulties associated with non-essentially smooth logarithmic moment generating functions.

The idea of using parameter dependent change of measure is not new. In particular, such change of measures were used in the context of obtaining refinements and full asymptotic expansions of the LDP. For a representative example, see [2]. It seems however that following the work of Gärtner–Ellis and the infinite-dimensional abstractizations, most modern treatments of the LDP choose to consider the limiting moment generating function even when dealing with non i. i. d. schemes. We first became aware of the possibilities in abandoning this route in [9], where the LDP for lower tails of Gaussian norms was derived. For some other applications, see the references in the beginning of Section 3.

This paper is organized as follows: in the next section, we provide some general theorems combining parameter dependent change of measure and other tools from the theory of large deviations. Section 3 is concerned with the evaluation of the "density" of the path of Brownian motion (or any other Gaussian process) under the conditioning that its L^2 norm is small. We show that the conditioning has a "spreading" effect, in that the measure ceases to concentrate, on the large deviations scale, around the path which is identically zero.

§2 GENERAL RESULTS

We present below bounds of the LDP type under suitable assumptions on the logarithmic moment generating functions involved. Due to their different nature, it will be convenient to separate the discussion of lower and upper bounds. Towards the end of the section, we bring theorems which combine both bounds to a full LDP. Throughout this paper, for any function $f : \mathbb{R}^d \to [-\infty, \infty]$, we denote by $\mathcal{D}_f$ the effective domain of f, i.e. $\mathcal{D}_f \stackrel{\triangle}{=} \{x : f(x) < \infty\}$. The relative interior of $\mathcal{D}_f$ is denoted by ri $\mathcal{D}_f$.

We begin by considering the lower bound. Let $\Lambda_\epsilon(\theta) = \epsilon \Lambda_{\mu_\epsilon}(\theta/\epsilon)$ and $\Lambda_\epsilon^*(x) \stackrel{\triangle}{=} \sup_{\lambda \in \mathbb{R}^d} \{\langle \lambda, x \rangle - \Lambda_\epsilon(\lambda)\}$. We say that a point $x \in \mathbb{R}^d$ is *exponentially exposed* if there exists a $\theta_\epsilon = \theta_\epsilon(x)$ such that

$$\epsilon \max\{1, |\theta_\epsilon|^2\} \sum_{i=1}^{d} \frac{\partial^2}{\partial \theta_i^2} \Lambda_\epsilon(\theta)\big|_{\theta = \theta_\epsilon} \to_{\epsilon \to 0} 0 \tag{2.1}$$

and

$$\max\{1, |\theta_\epsilon|\}|x - \nabla \Lambda_\epsilon(\theta_\epsilon)| \to_{\epsilon \to 0} 0. \tag{2.2}$$

The main reason for considering exponentially exposed points lies in the following:

Theorem 2.3. *Let $x \in \mathbb{R}^d$ be an exponentially exposed point. Then the lower bound is satisfied locally at x, i.e. for each neighborhood G of x,*

$$\liminf_{\epsilon \to 0} \epsilon \log \mu_\epsilon(Z_\epsilon \in G) \geq -I_2(x),$$

where $I_2(x) = \limsup_{\epsilon \to 0} \Lambda_\epsilon^*(x)$.

Proof. Define a new measure by

$$\frac{d\eta_\epsilon}{d\mu_\epsilon} = \frac{e^{\langle \theta_\epsilon, Z_\epsilon \rangle / \epsilon}}{e^{\Lambda_\epsilon(\theta_\epsilon)/\epsilon}}. \tag{2.4}$$

Then, for some positive δ,

$$\mu_\epsilon(Z_\epsilon \in G) \geq \mu_\epsilon(|Z_\epsilon - x| < \delta) \geq \mu_\epsilon(|Z_\epsilon - x| \leq \delta_n) \geq \frac{\eta_\epsilon(|Z_\epsilon - x| \leq \delta_n)}{\sup_{|Z_\epsilon - x| \leq \delta_n}\{\frac{d\eta_\epsilon}{d\mu_\epsilon}\}}$$

where $\delta_n = \delta/2n(\max\{1, |\theta_\epsilon|\})$. Thus,

$$\epsilon \log \mu_\epsilon(Z_\epsilon \in G) \geq \epsilon \log \eta_\epsilon(|Z_\epsilon - x| \max\{1, |\theta_\epsilon|\} \leq \delta/2n) - \Lambda_\epsilon^*(x) - \delta/2n.$$

By Chebycheff's inequality,

$$\eta_\epsilon(|Z_\epsilon - x| \max\{1, |\theta_\epsilon|\} > \delta/2n) \leq \frac{\mathrm{Var}_{\eta_\epsilon}(Z_\epsilon)}{\left(\frac{\delta}{2n \max\{1, |\theta_\epsilon|\}} - |x - E_{\eta_\epsilon}(Z_\epsilon)|\right)^2}$$

$$= \frac{\epsilon \max\{1, |\theta_\epsilon|^2\} \sum_{i=1}^{d} \frac{\partial^2 \Lambda_\epsilon(\theta)}{\partial \theta_i^2}\big|_{\theta = \theta_\epsilon}}{\left(\delta/2n - \max\{1, |\theta_\epsilon|\}|x - \nabla \Lambda_\epsilon(\theta_\epsilon)|\right)^2}.$$

Taking first ϵ to zero and then n to infinity, and using the assumption, yields the result. $\square$

We turn next to the computation of complementary upper bounds. Recall first that if (1.2) is satisfied for each λ, then the large deviations upper bound holds for compact sets, with the rate function Λ^* (see [4, Theorem 4.5.3]). Coupling this with Theorem 2.3, one obtains immediately the following:

Corollary 2.5. *Assume that* (1.2) *holds and every* $x \in \text{ri} \, \mathcal{D}_{\Lambda^*}$ *is exponentially exposed with* $\Lambda^*(x) \geq I_2(x)$. *Then the weak LDP holds with the rate function* Λ^*.

Proof. The only thing which is not yet obvious is that it is enough to consider only $\text{ri} \, \mathcal{D}_{\Lambda^*}$ in the lower bound. This however follows from the inequality

$$\inf_{x \in G \cap \text{ri} \, \mathcal{D}_{\Lambda^*}} \Lambda^*(x) \leq \inf_{x \in G} \Lambda^*(x),$$

which holds true for any open set G (see [4, page 51] for a proof). $\square$

Remarks.

(1) Note that always, under (1.2), $\Lambda^*(x) \leq \liminf_{\epsilon \to 0} \Lambda_\epsilon^*(x)$. However, unlike epigraph convergence, pointwise convergence to a convex lower semicontinuous function does not imply the pointwise convergence of the Fenchel–Legendre transforms, as the following example suggested to us by J. Sadowsky shows. Let $f_\epsilon(\lambda) = (1 - \epsilon)|\lambda|$ for $|\lambda| < 1/\epsilon$ and ∞ otherwise. Then $f_\epsilon^*(x) = 0$ for $|x| \leq (1 - \epsilon)$ and $f_\epsilon^*(x) = (1 - 1/\epsilon) + |x|/\epsilon$ otherwise. On the other hand, $f_\epsilon(\lambda) \to_{\epsilon \to 0} |\lambda| \triangleq f(\lambda)$, with $f^*(1) = 0$, whereas $f_\epsilon^*(1) = 1$ for all ϵ. This example may be modified, by smoothing f_ϵ near the origin and modifying it to be a steep function at $|\lambda| = 1/\epsilon$, to involve essentially smooth, lower semicontinuous f_ϵ and f, with the same conclusion.

(2) Obviously, when the assumptions of Corollary 2.5 are satisfied and in addition μ_ϵ are exponentially tight, then the full LDP, and not just the weak LDP, holds. This is the case, for example, if $0 \in \mathcal{D}_\Lambda^\circ$ (see [4, page 49]).

A drawback of Corollary 2.5 is the fact that one has to compare Λ_ϵ^* to Λ^*: we are particularly interested in situations where some information is lost in the passage to the limit from Λ_ϵ to Λ. The following is one way to avoid assumptions involving Λ^*. Define $I_1(x) = \liminf_{\epsilon \to 0} \Lambda_\epsilon^*(x)$.

Proposition 2.6. *For every compact set* K

$$\limsup_{\epsilon \to 0} \epsilon \log \mu_\epsilon(Z_\epsilon \in K) \leq - \inf_{\beta > 0} \liminf_{\epsilon \to 0} \inf_{\{y : \exists x \in K, |y - x| \leq e^{-\beta/\epsilon}\}} \Lambda_\epsilon^*(y). \qquad (2.7)$$

For every $\delta > 0$ and all $\epsilon > 0$ small enough, let $\theta_\epsilon(x, \delta)$ be of minimal norm such that

$$\langle \theta_\epsilon, x \rangle - \Lambda_\epsilon(\theta_\epsilon) \geq I_1^\delta(x) = \min\left\{I_1(x) - \delta, \frac{1}{\delta}\right\}.$$

Then the LDP upper bound with $I_1(x)$ holds for each compact set K for which for all $\delta > 0$

$$\epsilon \log\left(\sup_{x \in K} |\theta_\epsilon(x, \delta)|\right) \to_{\epsilon \to 0} 0. \qquad (2.8)$$

Proof. Fix K compact and $\beta > 0$. Cover K by a finite number of balls of radius $\rho_\epsilon = e^{-\beta/\epsilon}$ with centers $x_i \in K$, $i = 1, 2, \ldots, N_{\epsilon, K}$, where $N_{\epsilon, K}$ is the (finite) covering number of K with respect to balls of radius ρ_ϵ. Since $K \subset \mathbb{R}^d$, and $N_{\epsilon,K}$ is bounded by the maximal number of rigid balls of radius $\rho_\epsilon/2$ we can place with centers in K, by simple volume considerations it follows that $N_{\epsilon,K} \leq c(d,K)\rho_\epsilon^{-d}$ with $c(d,K) < \infty$ independent of ϵ. By [4, inequality (4.5.6)], for each $x \in K$,

$$\epsilon \log \mu_\epsilon(|Z_\epsilon - x| \leq \rho_\epsilon) \leq - \inf_{|y-x| \leq \rho_\epsilon} \Lambda_\epsilon^*(y) ,$$

and (2.7) follows by the union bound over $i = 1, \ldots, N_{\epsilon,K}$, taking first $\epsilon \to 0$ and then $\beta \to 0$.

Let $\phi_\epsilon = \epsilon \log(\sup_{x \in K} |\theta_\epsilon(x,\delta)|)$ and suppose that (2.8) holds, i.e. $\phi_\epsilon \to 0$ for every fixed $\delta > 0$. Then, for each $x \in K$, every $\beta > 0$, and all $|y - x| \leq e^{-\beta/\epsilon}$,

$$\Lambda_\epsilon^*(y) \geq \langle \theta_\epsilon, x \rangle - \Lambda_\epsilon(\theta_\epsilon) - |\theta_\epsilon| e^{-\beta/\epsilon} \geq I_1^\delta(x) - e^{(\phi_\epsilon - \beta)/\epsilon}.$$

Taking the infimum over $x \in K$, followed by $\epsilon \to 0$ we have by (2.7) that

$$\limsup_{\epsilon \to 0} \epsilon \log \mu_\epsilon(Z_\epsilon \in K) \leq - \inf_{x \in K} I_1^\delta(x),$$

and the announced LDP upper bound follows by taking $\delta \to 0$. $\square$

Remark. By [4, inequality (4.5.6)], Proposition 2.6 holds also for every *convex* closed K (in which case (2.7) applies with $\beta = \infty$).

Recall that the closure of a convex, proper function g, denoted cl g, is its lower semicontinuous hull. One has the following.

Corollary 2.9. *Assume that the limit of $\Lambda_\epsilon^*(x)$ exists as an extended real number for every $x \in \mathbb{R}^d$ (i.e. $I_1(x) = I_2(x)$). Let $I_0(\cdot) = \mathrm{cl}\, I_1(\cdot)$. Further assume that every $x \in \mathrm{ri}\, \mathcal{D}_{I_0}$ is exponentially exposed and that (2.8) holds for every compact K. Then the weak LDP holds with the rate function $I_0(\cdot)$.*

Proof. Note that $I_1(x) \geq I_0(x)$ while $I_0(x) = I_1(x)$ for every $x \in \mathrm{ri}\, \mathcal{D}_{I_0}$. By Proposition 2.6, the upper bound holds for compact sets with the function $I_1(x)$ and hence with the function $I_0(x)$, whereas for the lower bound it is enough, by the same reasoning as in the proof of Corollary 2.5, to get the bound locally at points $x \in \mathrm{ri}\, \mathcal{D}_{I_0}$, where $I_1(x) = I_0(x)$. The latter however holds true in view of Theorem 2.3. Finally, by its definition, $I_0(\cdot)$ is lower semicontinuous. $\square$

When (1.2) holds, an often occurring condition for the existence of the LDP upper bound is that $0 \in \mathcal{D}_\Lambda^\circ$. In that case, not only can the weak LDP be strengthened to a full LDP, but also the conditions for the existence of the LDP may be weakened. To this end, recall that a point $y \in \mathcal{D}_{\Lambda^*}$ is called *exposed* if for some $\lambda \in \mathbb{R}^d$, and all $x \neq y$,

$$\langle \lambda, y \rangle - \Lambda^*(y) > \langle \lambda, x \rangle - \Lambda^*(x). \tag{2.10}$$

λ in (2.10) is called an exposing hyperplane. Denote the set of exposed points whose exposing hyperplane belongs to $\mathcal{D}_\Lambda^\circ$ by $\mathcal{F}$.

Corollary 2.11. *Assume* (1.2). *Further assume that* $0 \in \mathcal{D}^{\circ}_{\Lambda}$ *and that every* $x \in \mathrm{ri}\,\mathcal{D}_{\Lambda^*} \setminus \mathcal{F}$ *is exponentially exposed with* $\Lambda^*(x) \geq I_2(x)$. *Then the LDP holds with the good rate function* Λ^*.

Proof. The only part which does not follow immediately from Theorems 2.3 and the exponential tightness imposed by the fact that $0 \in \mathcal{D}^{\circ}_{\Lambda}$ is the lower bound at points which belong to $\mathcal{F}$. But this is exactly the content of the Gärtner–Ellis theorem as presented in [4, part (b) of Theorem 2.3.6]. $\square$

Remark. Under the assumptions of Corollary 2.11, it holds that every point in $\mathcal{F}$ is also exponentially exposed. The reason for distinguishing between the two lies in the fact that while it is easy to check that a point is exposed, it is typically harder to directly check that it is exponentially exposed.

Corollaries 2.5 and 2.11 are not completely satisfactory for two reasons: first, one needs to assume (1.2) for all λ, and then, one has to compare Λ^*_ϵ to Λ^*. For an example where (1.2) might fail to hold on the boundary of the effective domain of the logarithmic moment generating function, and yet the LDP holds, see [1, proof of Theorem 2.1] which is based on the forthcoming Proposition 2.14. To overcome both difficulties mentioned above, a natural tool is the upper bound introduced by Zabell [13], based on the Mosco (epigraph) convergence. Define

$$\underline{\Lambda}(\theta) = \inf_{\theta_\epsilon \to \theta} \limsup_{\epsilon \to 0} \Lambda_\epsilon(\theta_\epsilon) = \sup_{\delta > 0} \limsup_{\epsilon \to 0} \inf_{|\lambda - \theta| < \delta} \Lambda_\epsilon(\lambda).$$

The convex lower semicontinuous function $\underline{\Lambda}$ is related to [13, Condition MC1]. By [13, Theorem 1.2], the upper bound holds for compact sets with the convex rate function

$$(\underline{\Lambda})^* \stackrel{\triangle}{=} \sup_{\lambda \in \mathbb{R}^d} \{\langle \lambda, x \rangle - \underline{\Lambda}(\lambda)\}.$$

(It is easy to check that $\underline{\Lambda}(0) \leq 0$ and hence $(\underline{\Lambda})^*(\cdot)$ is non negative). In particular, we have the following dual of Corollary 2.5.

Corollary 2.12. *Assume that all* $x \in \mathrm{ri}\,\mathcal{D}_{(\underline{\Lambda})^*}$ *are exponentially exposed with* $(\underline{\Lambda})^*(x) \geq I_2(x)$. *Then the weak LDP holds with the rate function* $(\underline{\Lambda})^*$. *If*

$$\underline{\Lambda}(\theta) < \infty, \quad \text{for all } |\theta| < \delta \text{ and some fixed } \delta > 0, \tag{2.13}$$

then the full LDP holds with the good convex rate function $(\underline{\Lambda})^*$.

Proof. The LDP lower bound is obtained as in Corollary 2.5. Either $(\underline{\Lambda})^*(x) = \infty$ everywhere or else $\underline{\Lambda}(\theta) > -\infty$ everywhere. In the latter case, by the argument in [4, top of page 47], $(\underline{\Lambda})^*$ is a good rate function. Replacing the $2d$ half-spaces used in [4, proof of Theorem 2.3.6 part (a)] by θ_j, $j = 1, \ldots, N$, which form a $\delta/8$-cover of the sphere $\{\theta : |\theta| = \delta/2\}$, it is not hard to check that (2.13) also results with $\{\mu_\epsilon\}$ exponentially tight. The LDP upper bound with $(\underline{\Lambda})^*$ then extends to all closed sets. $\square$

As before, checking that all points are exponentially exposed can be tedious, and it is desirable to reduce the check to a (hopefully small) subset of ri $\mathcal{D}_{(\underline{\Lambda})^*}$. To this end, let $\bar{\mathcal{F}}$ denote the set of exposed points with exposing hyperplane in $\mathcal{D}_{\underline{\Lambda}}^\circ$, where in the definition of exposing points, Λ^* is replaced by $(\underline{\Lambda})^*$. The analog of Corollary 2.11 is now:

Proposition 2.14. *Assume* (2.13) *and that*

$$\underline{\Lambda}(\theta) \leq \sup_{\theta_\epsilon \to \theta} \liminf_{\epsilon \to 0} \Lambda_\epsilon(\theta_\epsilon), \quad \forall \theta \in \mathcal{D}_{\underline{\Lambda}}^\circ. \tag{2.15}$$

If all $x \in \text{ri}\, \mathcal{D}_{(\underline{\Lambda})^*} \setminus \bar{\mathcal{F}}$ are exponentially exposed with $(\underline{\Lambda})^*(x) \geq I_2(x)$, then the LDP holds with the good rate function $(\underline{\Lambda})^*$.

Proof. Suffices to prove that (2.15) results with the LDP lower bound holding at each $x \in \bar{\mathcal{F}}$. Fix such x with exposing hyperplane $\theta \in \mathcal{D}_{\underline{\Lambda}}^\circ$, and follow [4, proof of Theorem 2.3.6, part (b)], now with ϵ-dependent change of measures determined as in (2.4) by θ_ϵ approaching the supremum in (2.15). Since $\tilde{\Lambda}(\cdot)$ corresponding to the tilted measures η_ϵ is bounded above by $\underline{\Lambda}(\cdot + \theta) - \underline{\Lambda}(\theta)$ and $\theta \in \mathcal{D}_{\underline{\Lambda}}^\circ$, it follows that $\{\eta_\epsilon\}$ is exponentially tight (see the proof of Corollary 2.12), and the application of the upper bound of [13, Theorem 1.2] completes the proof. □

We emphasize that (2.15) is not required to hold true on the boundary of $\mathcal{D}_{\underline{\Lambda}}$, and in particular (2.15) is implied by (1.2) holding in $\mathcal{D}_{\underline{\Lambda}}^\circ$. The next lemma orders the various functions we encountered.

Lemma 2.16. *For any family of measures $\{\mu_\epsilon\}$ and at any $x \in \mathbb{R}^d$*

$$\Lambda^*(x) \leq (\underline{\Lambda})^*(x) \leq I_0(x) \leq I_1(x) \leq I_2(x).$$

Proof. Plainly, $I_0 \leq I_1 \leq I_2$ and (when (1.2) holds) $\Lambda^* \leq (\underline{\Lambda})^*$ since $\underline{\Lambda} \leq \Lambda$. By [14, Theorem 1.1]

$$(\underline{\Lambda})^*(x) \leq \sup_{\delta > 0} \liminf_{\epsilon \to 0} \inf_{|x-y|<\delta} \Lambda_\epsilon^*(y) \leq I_1(x),$$

and with $(\underline{\Lambda})^*$ convex lower semicontinuous, also $(\underline{\Lambda})^* \leq I_0$. □

§3 ASYMPTOTIC BEHAVIOUR
OF GAUSSIAN PROCESSES UNDER NORM CONDITIONING

In this section, we provide a representative, two-dimensional example for the applicability of the method described above. Appropriate parameter dependent change of measures are useful in a variety of situations. In particular, it is useful in the analysis of intersymbol interference in communication systems (see [11]), in avoiding steepness constraint in the analysis of the LDP for quadratic forms of Gaussian processes (see [1, proof of Theorem 2.1]), and in the analysis of the probability of error in some communication systems (see [4, Exercise 2.3.24]). It should be mentioned that in the last two examples, condition (2.1) is not satisfied directly, and one has to resort to a suitable change of variables in order to complete the proof of the lower bound.

We turn now to the situation which concerns us here. Let x_t be a Gaussian process. Suppose one is interested in the probability that $\|x\|_2^2 < \eta$, where $\|\cdot\|_2$ denotes the L^2-norm of the process x_t on $[0, T]$. A partial answer to this question was provided by [7], and the large deviations behavior of the random variable $Z_\epsilon = \|x\|_2^2/\eta$ when $\eta(\epsilon) \to 0$ was derived in [9]. The latter is actually a particular case of Corollary 2.5. Specifically, let a_i^2 be the eigenvalues of the Karhunen–Loéve expansion of $x_.$, then $\|x\|_2^2 = \sum_{i=1}^\infty x_i^2/a_i^2$ where x_i are i.i.d. standard Normal random variables. Considering for simplicity the case $a_i^2 \sim i^\beta$, $\beta > 1$, and letting $\epsilon = \eta^{\frac{1}{\beta-1}}$, it is shown there by an appropriate choice of the parameter θ_ϵ that Z_ϵ satisfies the LDP with the rate function $I(x) = c_\beta x^{-1/(\beta-1)}$ for positive x and ∞ otherwise, where

$$c_\beta = \frac{\beta-1}{2}\left(\int_0^\infty \frac{dy}{1+y^\beta}\right)^{\beta/(\beta-1)}$$

Note that this rate function is not good and hence the methods based on the Gärtner–Ellis theorem fail. Refinements of this result, based on appropriate asymptotic expansions, are presented in [3]. Note that analytical approaches to this problem, based on saddle point methods, may be found in [8, 12, 15].

Here, we wish to explore this set up further. Since the conditioning on small L^2-norm appears naturally in the evaluation of Onsager–Machlup functionals, it is of interest to check what is the effect of conditioning on the behavior of the path $z_t \stackrel{\triangle}{=} x_t/\sqrt{\eta}$. One possible way to gauge this effect is to compute the behavior of $B_{\delta,h} \stackrel{\triangle}{=} P(\|z_. - h\|_2 < \delta \mid \|z_.\|_2 < 1)$ as a function of h and δ, where h is a deterministic L^2-function. For the sake of simplicity and concreteness, we consider the case of x_t a Brownian motion on $[0, \pi]$, in which case $a_i/i \to 1$ as $i \to \infty$, and $\epsilon = \eta$. The analysis extends to other T and other a_i.

Define now $A_{\delta,h} \stackrel{\triangle}{=} P(\|z_. - h\|_2 < \delta$ and $\|z_.\|_2 < 1)$, then $B_{\delta,h} = A_{\delta,h}/A_{1,0}$. Since the large deviations behavior of $B_{\delta,h}$ may be read of the one of $A_{\delta,h}$, we concentrate first on the latter.

Theorem 3.1. Let $a = \delta^2$ if $\|h\|_2^2 + \delta^2 \leq 1$, $a = 1$ if $\|h\|_2^2 - \delta^2 \leq -1$ and $[1 - (1 - \delta^2 + \|h\|_2^2)^2/4\|h\|_2^2] \vee 0$ otherwise. Let $I(\delta, \|h\|_2) = c_2/a$. Then

$$\epsilon \log A_{\delta,h} \to_{\epsilon \to 0} -I(\delta, \|h\|_2).$$

Proof. Let $Z_\epsilon = (Z_{\epsilon,1}, Z_{\epsilon,2})$ denote the two-dimensional vector consisting of $\|z_.\|_2^2$ and $2\int_0^\pi x_t h_t\, dt/\sqrt{\epsilon}$. Note that $Z_{\epsilon,2} = -\|z_. - h\|_2^2 + \|z_.\|_2^2 + \|h\|_2^2$ and that $Z_{\epsilon,2}^2 \leq 4Z_{\epsilon,1}\|h\|_2^2$. Let e_i be the eigenfunctions of the Karhunen–Loéve expansion associated with $x_.$, and let $b_i = \int_0^\pi h_t e_i(t)\, dt$. Then $\|h\|_2^2 = \sum_{i=1}^\infty b_i^2$ and

$Z_{\epsilon,1} = \sum_{i=1}^{\infty} x_i^2/\epsilon a_i^2$, $Z_{\epsilon,2} = 2\sum_{i=1}^{\infty} x_i b_i/\sqrt{\epsilon}a_i$. Letting $\theta_\epsilon = (\theta_{\epsilon,1}, \theta_{\epsilon,2})$, one observes that

$$\Lambda_\epsilon(\theta_\epsilon) = -\frac{\epsilon}{2}\sum_{i=1}^{\infty}\log\left(1 - \frac{2\theta_{\epsilon,1}}{\epsilon^2 a_i^2}\right) - \theta_{\epsilon,2}^2\sum_{i=1}^{\infty}\frac{b_i^2}{(\theta_{\epsilon,1} - \epsilon^2 a_i^2/2)}.$$

Let (x, y) be given numbers, with $x > y^2/4\|h\|_2^2$. Then,

$$\frac{\partial}{\partial\theta_{\epsilon,1}}\Lambda_\epsilon(\theta_\epsilon) = \theta_{\epsilon,2}^2\sum_{i=1}^{\infty}\frac{b_i^2}{(\epsilon^2 a_i^2/2 - \theta_{\epsilon,1})^2} + \sum_{i=1}^{\infty}\frac{\epsilon}{2(\epsilon^2 a_i^2/2 - \theta_{\epsilon,1})},$$

$$\frac{\partial}{\partial\theta_{\epsilon,2}}\Lambda_\epsilon(\theta_\epsilon) = 2\theta_{\epsilon,2}\sum_{i=1}^{\infty}\frac{b_i^2}{(\epsilon^2 a_i^2/2 - \theta_{\epsilon,1})}$$

and

$$\frac{\partial^2}{\partial\theta_{\epsilon,1}^2}\Lambda_\epsilon(\theta_\epsilon) = 2\theta_{\epsilon,2}^2\sum_{i=1}^{\infty}\frac{b_i^2}{(\epsilon^2 a_i^2/2 - \theta_{\epsilon,1})^3} + \sum_{i=1}^{\infty}\frac{\epsilon}{2(\epsilon^2 a_i^2/2 - \theta_{\epsilon,1})^2},$$

$$\frac{\partial^2}{\partial\theta_{\epsilon,2}^2}\Lambda_\epsilon(\theta_\epsilon) = 2\sum_{i=1}^{\infty}\frac{b_i^2}{(\epsilon^2 a_i^2/2 - \theta_{\epsilon,1})}.$$

Under the assumption on (x, y), one may choose θ_ϵ such that $x = \frac{\partial}{\partial\theta_{\epsilon,1}}\Lambda_\epsilon(\theta_\epsilon)$ and $y = \frac{\partial}{\partial\theta_{\epsilon,2}}\Lambda_\epsilon(\theta_\epsilon)$. Moreover, one checks that $\theta_{\epsilon,1} \to_{\epsilon\to 0} -\theta_1$ and $\theta_{\epsilon,2} \to_{\epsilon\to 0} \theta_2$, where (θ_1, θ_2) satisfy the equations

$$x = \frac{y^2}{4\|h\|_2^2} + g(\theta_1), \qquad y = \frac{2\theta_2}{\theta_1}\|h\|_2^2,$$

and $g(\theta_1) = \frac{1}{2}\int_0^\infty dz/(\theta_1 + z^2/2) = (\int_0^\infty dz/(1 + z^2))/\sqrt{2\theta_1}$. It follows immediately that (2.1) and (2.2) are satisfied, and hence one concludes that all points in the interior of the support of the law of $\{(Z_{\epsilon,1}, Z_{\epsilon,2})\}$ are exponentially exposed. One also easily checks that (1.2) holds, and that $\Lambda_\epsilon^* \to_{\epsilon\to 0} \Lambda^*$. Applying now Corollary 2.5, one concludes that Z_ϵ satisfies the weak LDP with the rate function

$$F(\theta_1) \overset{\Delta}{=} \frac{1}{2}\int_0^\infty \log(1 + 2\theta_1/z^2)\, dz - \frac{\theta_1}{2}\int_0^\infty dz/(\theta_1 + z^2/2)$$

$$= \frac{\sqrt{2\theta_1}}{2}\int_0^\infty \frac{1}{z^2 + 1}\, dz$$

where θ_1 satisfies the equation $g(\theta_1) = x - y^2/(4\|h\|_2^2)$. Since the two events $\{\|z_\cdot - h\|_2 < \delta, \|z_\cdot\|_2 < 1\}$ and $\{Z_{\epsilon,1} < 1, Z_{\epsilon,2} > Z_{\epsilon,1} + \|h\|_2^2 - \delta^2\}$ are equivalent, it follows that

$$\epsilon\log A_{\delta,h} \to_{\epsilon\to 0} -\inf\{H(a) : 0 \le a = x - y^2/4\|h\|_2^2, x \le 1, x - y + \|h\|_2^2 \le \delta^2\},$$

where $H(a) = F(\theta_1)|_{g(\theta_1)=a} = c_2/a$. The theorem follows from the last relation by observing that $H(\cdot)$ decreases in a, and optimizing over a. $\square$

The following corollary follows from Theorem 3.1 in view of the continuity of $I(\delta, \|h\|_2)$ in δ.

Corollary 3.2.

$$\epsilon \log B_{\delta,h} \to_{\epsilon \to 0} -[I(\delta, \|h\|_2) - I(1,0)] \overset{\triangle}{=} - J(\delta, \|h\|_2) = -c_2[1/a - 1].$$

In particular, Corollary 3.2 implies that $J(\delta, \|h\|_2)$, for a given δ, is an increasing function of $\|h\|_2$ and, moreover, is fixed for all $\|h\|_2 \le \sqrt{1 - \delta^2}$. This reflects the fact that under the conditioning, the law of $z_.$ tends to spread and not concentrate around particular paths, in contrast with the unconditional law of the Brownian motion which concentrates (in an Onsager-Machlup sense) around $z_. \equiv 0$.

Remarks.

(1) One could conjecture, in view of Corollary 3.2, that $B_{\delta,h}$ exhibits exponential decay with rate which depends only on the norm of h, regardless of the type of norm used. That this is not the case may be seen for Brownian motion in the interval $[0, 1]$ by considering the supremum norm. Using the Markov property of Brownian motion, and solving the associated Dirichlet problem, one may show that for smooth h with $|h(0)| < 1$ and $|h(t)| < 2$, $t \in (0, 1]$,

$$\epsilon \log P(\|z_. - h\|_\infty < 1 \mid \|z_.\|_\infty < 1) \to_{\epsilon \to 0} \pi^2 - \int_0^1 \left(\frac{2\pi}{2 - |h(t)|}\right)^2 dt,$$

which depends on h not only through its supremum norm.

(2) In the case $a_i \sim i^{\beta/2}$ with $\beta > 1$, Theorem 3.1 remains valid, with the expression for $I(\delta, \|h\|_2)$ replaced by

$$I(\delta, \|h\|_2) = c_\beta / a^{1/(\beta-1)}.$$

(3) One could replace the Brownian motion $w_.$ by the solution to the stochastic differential equation

$$dx_t = f(x_t) \, dt + dw_t \tag{3.3}$$

where $f(\cdot)$ is a C_b^1 function. Noting that

$$P(x_. \in B) = E(1_{\{x_. \in B\}})$$
$$= E\left(1_{\{w_. \in B\}} \exp\left(\int_0^\pi f(w_t) \, dw_t - \frac{1}{2} \int_0^\pi f^2(w_t) \, dt\right)\right)$$
$$= E\left(1_{\{w_. \in B\}} \exp\left(F(w_T) - \frac{1}{2} \int_0^\pi (f'(w_t) + f^2(w_t)) \, dt\right)\right),$$

where $F(x) = \int_0^x f(\theta) \, d\theta$, one obtains that the drift $f(\cdot)$ does not affect Corollary 3.2, since an additional term of $-\frac{\pi}{2}(f'(0) + f^2(0))$ appears in both $A_{\delta,h}$ and $A_{1,0}$.

REFERENCES

1. W. Bryc and A. Dembo, *Large deviations for quadratic functionals of Gaussian processes*, IMA preprint #1179, 1993.
2. N. R. Chaganty and J. Sethuraman, *Limit theorems in the area of large deviations for some dependent random variables*, Ann. Probab. **15** (1987), 628–645.
3. A. Dembo, E. Mayer-Wolf, and O. Zeitouni, *Exact behavior of Gaussian seminorms* (to appear in: Prob. Statis. Letters).
4. A. Dembo and O. Zeitouni, *Large deviations techniques and applications*, Jones and Bartlett, Boston, 1993.
5. R. S. Ellis, *Large deviations for a general class of random vectors*, Ann. Probab. **12** (1984), 1–12.
6. J. Gärtner, *On large deviations from the invariant measure*, Theory Probab. Appl. **22** (1977), 24–39.
7. J. Hoffman-Jorgensen, L. A. Shepp, and R. M. Dudley, *On the lower tail of Gaussian seminorms*, Annals of Probability **7** (1979), 319–342.
8. I. A. Ibragimov, *Hitting probability of a Gaussian vector with values in a Hilbert space in a sphere of small radius*, J. Sov. Math **20** (1982), 2164–2174.
9. E. Mayer-Wolf and O. Zeitouni, *The probability of small Gaussian ellipsoids and associated conditional moments*, Annals of Probability **21** (1993), 14–24.
10. G. L. O'Brien and W. Vervaat, *Compactness in the theory of large deviations*, Preprint, 1993.
11. E. Sinai, *Asymptotic bounds for intersymbol interference*, Master's thesis, Dept. of EE, Technion, Haifa, Israel, 1993.
12. G. N. Sytaya, *On some asymptotic representations of the Gaussian measure in a Hilbert space*, In: Theory of Stochastic processes, Publication 2, Ukrainan Academy of Science, 1974, pp. 93–104.
13. S. Zabell, *Mosco convergence and large deviations*, In: Probability in Banach Spaces (M. G. Hahn, R. M. Dudley, and J. Kuelbs, eds.), vol. 8, Birkhäuser, 1992, pp. 245–252.
14. S. Zabell, *Mosco convergence in locally convex spaces*, Jour. Funct. Anal. **110** (1992), 226–246.
15. V. M. Zolotarev, *Asymptotic behavior of the Gaussian measure in l_2*, J. Sov. Math **24** (1986), 2330–2334.

AMIR DEMBO, DEPARTMENT OF MATHEMATICS AND DEPARTMENT OF STATISTICS, STANFORD UNIVERSITY, STANFORD, CA 94305, USA

OFER ZEITOUNI, DEPARTMENT OF ELECTRICAL ENGINEERING, TECHNION-ISRAEL INSTITUTE OF TECHNOLOGY, HAIFA 32000, ISRAEL

Progress in Probability, Vol. 36

AN EQUATION MODELLING TRANSPORT OF A SUBSTANCE IN A STOCHASTIC MEDIUM

JON GJERDE, HELGE HOLDEN, BERNT ØKSENDAL,
JAN UBØE, AND TUSHENG ZHANG

ABSTRACT. We find an explicit expression for the (unique) solution $u = u(t, x, \omega)$ of the stochastic partial differential equation

$$\frac{\partial u}{\partial t} = \frac{1}{2}\nu^2 \Delta u + \overrightarrow{W}_x \diamond \nabla u; \quad (t, x) \in \mathbb{R}^+ \times \mathbb{R}^d,$$
$$u(0, \cdot) = f(\cdot),$$

where Δ and ∇ are the Laplacian and gradient operators respectively, with respect to $x = (x_1, \ldots, x_d)$ and $\overrightarrow{W}_x$ is d-dimensional white noise in the d parameters $(x_1, x_2, \ldots, x_d)$. The symbol $\diamond$ denotes the (vector) Wick product, the use of which corresponds to an Itô/Skorohod interpretation of the equation. This equation occurs in many situations. For example, it models the transport of a substance in a turbulent (stochastic) medium.

§1 INTRODUCTION

A substance dissolved in a moving fluid/medium in $\mathbb{R}^d$ is exposed to both a molecular diffusion and to a drift coming from the movement of the fluid. If the fluid is turbulent, a natural model for its velocity at the time t and the point x is d-dimensional white noise $\overrightarrow{W}_x$ in the d parameters $(x_1, \ldots, x_d)$ and $\omega \in \Omega$ is a random parameter. The concentration $u(t, x, \omega)$ of the substance at (t, x) will then satisfy the stochastic partial differential equation

$$\frac{\partial u}{\partial t} = \frac{1}{2}\nu^2 \Delta u + \overrightarrow{W}_x \cdot \nabla u, \quad (t, x) \in \mathbb{R}^+ \times \mathbb{R}^d, \tag{1.1}$$

where $\nu > 0$ is the molecular viscosity of the medium. We assume that the concentration at time $t = 0$ is a known, random function $f(x, \omega)$:

$$u(0, x, \omega) = f(x, \omega), \quad x \in \mathbb{R}^d. \tag{1.2}$$

1991 *Mathematics Subject Classification.* primary 60H15, secondary 35R60, 76Fxx.

Key words and phrases. Transport equation, stochastic medium, white noise analysis, Wick products, generalized white noise distributions (Hida distributions).

Equations of the type (1.1)–(1.2) have been studied by several authors. See e. g. [2], [11] and [12]. However, there are several differences from these papers and ours (see §2 for the precise definitions):

(a) We will adopt the functional process interpretation of the equation (see e. g. [6] and [7]). This means that we smoothen the singular white noise $\overrightarrow{W}_x$ by an x-shifted test function ϕ and consider the solution $u = u^\phi(t, x, \omega)$ of the corresponding equation

$$\frac{\partial u}{\partial t} = \frac{1}{2}\nu^2 \Delta u + \overrightarrow{W}_{\phi_x} \cdot \nabla u \qquad (1.3)$$

where $\phi_x(y) = \phi(y - x)$ and $\overrightarrow{W}_{\phi_x}$ is the white noise smoothed by ϕ. This smoothing is not just a technical convenience; physically it represents taking a macroscopical average so as to obtain a more realistic model. For example, the support of ϕ will determine the maximal distance within which there is a correlation between the fluid velocities.

(b) We interpret the product $\cdot$ in (1.3) as a *Wick product* $\diamond$. This corresponds to considering the equation (1.3) in the Itô/Skorohod sense. Previous authors have all adopted the Stratonovich interpretation, corresponding to using the usual, pointwise product in (1.3). (See e. g. [6, p. 398] for further explanation). Our interpretation leads to the equation

$$\frac{\partial u}{\partial t} = \frac{1}{2}\nu^2 \Delta u + \overrightarrow{W}_{\phi_x} \diamond \nabla u, \qquad (1.4)$$

$$u(0, x, \omega) = f(x, \omega). \qquad (1.5)$$

(c) We allow the initial value $f(x, \omega)$ to be stochastic and anticipating.

The purpose of this paper is to give a rather explicit formula for the solution $u(t, x, \omega)$ of the equation (1.4)–(1.5). Moreover, we show that the solution is unique as an element of the space $(\mathcal{S})^*$ of generalized white noise distributions (Hida distributions).

§2 SOME PRELIMINARIES ON WHITE NOISE

Here we briefly recall some of the basic definitions and results from white noise calculus. For more information the reader is referred to [5]. In the following we fix the parameter dimension d and let $\mathcal{S} = \mathcal{S}(\mathbb{R}^d)$ denote the Schwartz space of rapidly decreasing smooth functions on $\mathbb{R}^d$. The dual space $\mathcal{S}^* = \mathcal{S}^*(\mathbb{R}^d)$ is the space of tempered distributions. By the Bochner–Minlos theorem there exists a probability measure μ on the Borel subsets $\mathcal{B}$ of $\mathcal{S}^*$ with the property that

$$\int_{\mathcal{S}^*} e^{i\langle \omega, \phi \rangle} \, d\mu(\omega) = \exp\left(-\frac{1}{2}\|\phi\|^2\right), \quad \phi \in \mathcal{S}(\mathbb{R}^d), \qquad (2.1)$$

where $\langle \omega, \phi \rangle$ denotes the action of $\omega \in \mathcal{S}^*$ on $\phi \in \mathcal{S}$ and $\|\phi\|^2 = \int_{\mathbb{R}^d} |\phi(x)|^2 \, dx$. The triple $(\mathcal{S}^*, \mathcal{B}, \mu)$ is called the *white noise space*. The (one-dimensional) *white noise process* is the map $W : \mathcal{S} \times \mathcal{S}^* \to \mathbb{R}$ defined by

$$W(\phi, \omega) = W_\phi(\omega) = \langle \omega, \phi \rangle, \quad \omega \in \mathcal{S}^*, \phi \in \mathcal{S}. \qquad (2.2)$$

Expressed in terms of Itô integrals with respect to d-parameter Brownian motion we have

$$W_\phi(\omega) = \int_{\mathbb{R}^d} \phi(x)\, dB_x(\omega), \quad \phi \in \mathcal{S}. \tag{2.3}$$

By the Wiener–Itô chaos theorem, every element $X \in (L^2) = L^2(\mathcal{S}^*(\mathbb{R}^d), \mu)$ admits a chaos decomposition

$$X = \sum_{n=0}^{\infty} I_n(f^{(n)}). \tag{2.4}$$

Here I_n denotes a multiple Itô integral of order n and the integrands $f^{(n)}$ belong to $\hat{L}^2((\mathbb{R}^d)^{\otimes n})$, i.e. the symmetric L^2-space, see e.g. [4]. The (L^2)-norm of X with this expansion is given by

$$\|X\|^2_{(L^2)} = \sum_{n=0}^{\infty} n! \|f^{(n)}\|^2_{L^2((\mathbb{R}^d)^{\otimes n})}. \tag{2.5}$$

Consider the (densely defined) self-adjoint operator $A\colon L^2(\mathbb{R}) \to L^2(\mathbb{R})$ given by

$$Af(x) = -f''(x) + (1 + x^2)f(x). \tag{2.6}$$

Let $\mathcal{P} \subset (L^2)$ denote the algebra generated by functionals of the form $\langle \cdot, \phi \rangle$ with $\phi \in \mathcal{S}(\mathbb{R}^d)$, and for every integer p let $(\mathcal{S})_p$ denote the completion of $\mathcal{P}$ with respect to the norm

$$\|X\|^2_{2,p} := \sum_{n=0}^{\infty} n! \|(A^{\otimes nd})^p f^{(n)}\|^2_{L^2(\mathbb{R}^{nd})}. \tag{2.7}$$

The Hida test function space $(\mathcal{S})$ is then the projective limit of the spaces $(\mathcal{S})_p$. For $\xi \in \mathcal{S}(\mathbb{R}^d)$ the Wick exponential $\mathrm{Exp}[W_\xi]$ is defined by the expression

$$\mathrm{Exp}[W_\xi] = \sum_{n=0}^{\infty} I_n(\xi^{\otimes n}). \tag{2.8}$$

It is easy to see that $\mathrm{Exp}[W_\xi] \in (\mathcal{S})$ and it turns out that

$$\mathrm{Exp}[W_\xi] = \exp\left(W_\xi - \frac{1}{2}\|\xi\|^2\right). \tag{2.9}$$

The space $(\mathcal{S})^*$ of Hida distributions is the dual space of $(\mathcal{S})$. The $\mathcal{S}$-transform of an element $X \in (\mathcal{S})^*$ is a functional on $\mathcal{S}(\mathbb{R}^d)$ defined by

$$\mathcal{S}X(\xi) = \langle X, \mathrm{Exp}[W_\xi] \rangle, \quad \xi \in \mathcal{S}(\mathbb{R}^d), \tag{2.10}$$

where $\langle \cdot, \cdot \rangle$ denotes the dual pairing between $(\mathcal{S})^*$ and $(\mathcal{S})$. The $\mathcal{S}$-transform plays an important role in the study of white noise analysis. The generalized functionals are completely determined by their $\mathcal{S}$-transforms. For our purpose we recall the following definition and two propositions from [13]:

Definition 2.1. A complex valued function F on $\mathcal{S}(\mathbb{R}^d)$ is called a U-functional if for every $\xi, \eta \in \mathcal{S}(\mathbb{R}^d)$, the mapping $\lambda \mapsto F(\eta + \lambda\xi)$, $\lambda \in \mathbb{R}$, has an entire analytic extension, denoted by $F(\eta + z\xi)$, $z \in \mathbb{C}$, and moreover there exist constants K and $p \geq 0$ such that

$$|F(z\xi)| \leq K \exp(K|z|^2\|\xi\|_{2,p}^2), \quad z \in \mathbb{C}, \tag{2.11}$$

where $\|\xi\|_{2,p} := \|(A^{\otimes d})^p \xi\|_{L^2(\mathbb{R}^d)}$.

Proposition 2.1. *A complex valued function F on $\mathcal{S}(\mathbb{R}^d)$ is the $\mathcal{S}$-transform of an element in $(\mathcal{S})^*$ if and only if F is a U-functional.*

Proposition 2.2. *Let $X_n, X \in (\mathcal{S})^*$. Then $X_n \to X$ in $(\mathcal{S})^*$ if and only if the following conditions are satisfied:*

$$\mathcal{S}X_n(\xi) \to \mathcal{S}X(\xi) \tag{2.12}$$

and

$$|\mathcal{S}X_n(z\xi)| \leq K \exp(K|z|^2\|\xi\|_{2,p}^2) \tag{2.13}$$

where the constants K and p do not depend on n.

Before we close this section, let us recall the definition of the *Wick product* [4], [5]. Let F and G be generalized functionals with the $\mathcal{S}$-transforms $\mathcal{S}F(\xi)$ and $\mathcal{S}G(\xi)$. The Wick product of F and G, denoted by $F \diamond G$, is defined to be the unique element in $(\mathcal{S})^*$ with the $\mathcal{S}$-transform $\mathcal{S}F(\xi)\mathcal{S}G(\xi)$. Interpreting the products in a stochastic differential equation as Wick products corresponds to interpreting the equation in the sense of Itô/Skorohod: One can express the Itô/Skorohod integral in a striking way by using the Wick product. See e. g. [1], [8], [9].

2.1 Multidimensional white noise. Finally we mention that to get the m-dimensional white noise, we need to work on the product probability space:

$$\left(\Omega = \prod_{i=1}^{m} \mathcal{S}^*(\mathbb{R}^d), \, \mathcal{B} = \bigotimes_{i=1}^{m} \mathcal{B}(\mathcal{S}^*(\mathbb{R}^d)), \, \mu_m = \prod_{i=1}^{m} \mu \right).$$

If $\omega = (\omega_1, \omega_2, \cdots, \omega_m) \in \Omega$ and $\phi = \phi_1 \otimes \phi_2 \otimes \cdots \otimes \phi_m$ with $\phi_i \in \mathcal{S}(\mathbb{R}^d)$ we define m-*dimensional white noise* $\overrightarrow{W}_\phi = (W_{\phi_1}^1, W_{\phi_2}^2, \ldots, W_{\phi_m}^m)$ by

$$\overrightarrow{W}_\phi(\omega) = (\langle \omega_1, \phi_1 \rangle, \langle \omega_2, \phi_2 \rangle, \ldots, \langle \omega_m, \phi_m \rangle) \in \mathbb{R}^m. \tag{2.14}$$

All the machinery we used on white noise space $(\mathcal{S}^*(\mathbb{R}^d), \mathcal{B}(\mathcal{S}^*(\mathbb{R}^d)), \mu)$ carries over to the above product space $(\Omega, \mathcal{B}, \mu_m)$. The details can be found in [3].

§3 AN EQUATION FROM TURBULENT TRANSPORT

We now return to the stochastic partial differential equation which arises in modelling the transport of a substance in a turbulent medium:

$$\frac{\partial u}{\partial t} = \frac{1}{2}\nu^2 \Delta u + \overrightarrow{W}_{\phi_x} \diamond \nabla u,$$

$$u(0, x, \omega) = f(x, \omega) \tag{3.1}$$

where $u = u(t, x, \omega)$ with $(t, x, \omega) \in \mathbb{R}_+ \times \mathbb{R}^d \times \Omega$ and

$$\overrightarrow{W}_{\phi_x} = \left(W^1_{\phi_{1,x_1}}, W^2_{\phi_{2,x_2}}, \dots, W^d_{\phi_{d,x_d}} \right)$$

is the d-dimensional white noise. Here $\phi \in \mathcal{S} := \bigotimes_{i=1}^d \mathcal{S}(\mathbb{R}^d)$ and $\phi_{j,x_j}(\cdot)$ is the x-shift of ϕ defined by

$$\phi_{j,x_j}(y_j) = \phi_j(y_j - x_j), \quad 1 \le j \le d. \tag{3.2}$$

Note the use of the vector Wick product

$$\overrightarrow{W}_{\phi_x} \diamond \nabla u := W^1_{\phi_{1,x_1}} \diamond \frac{\partial u}{\partial x_1} + \dots + W^d_{\phi_{d,x_d}} \diamond \frac{\partial u}{\partial x_d}$$

in (3.1). As mentioned above, this corresponds to interpreting the equation in the Itô/Skorohod sense.

Let $(b_t(\hat{\omega}), P)$ be a standard Brownian motion in $\mathbb{R}^d$ which is independent of the white noise $\overrightarrow{W}_\phi$ and let E denote the expectation with respect to P. Let $C_b^2(\mathbb{R}^d \to (L^2))$ denote the space of functions $f : \mathbb{R}^d \to (L^2)$ such that f is twice Fréchet differerentiable and bounded. Our main result can be stated as follows:

Theorem 3.1. *Assume the constant $\nu > 0$ and $f \in C_b^2(\mathbb{R}^d \to (L^2))$. Then for all $\phi \in \mathcal{S}$ there is a unique solution $u(t, x, \cdot) \in (\mathcal{S})^*$ of the equation (3.1) given by*

$$u(t, x, \omega) = E\left[f(x + \nu b_t) \diamond \operatorname{Exp}\left[\frac{1}{\nu} \sum_{i=1}^d \int_0^t [W_{\phi_y}(\omega)]_{y=x+\nu b_s} \, db_s^i \right.\right.$$

$$\left.\left. - \frac{1}{2\nu^2} \sum_{i=1}^d \nu^{-2} \int_0^t [W_{\phi_y}^{\diamond 2}(\omega)]_{y=x+\nu b_s} \, ds \right]\right]. \tag{3.3}$$

3.1 Proof. Existence. We will split the proof of this theorem into several lemmas. For simplicity, we assume that the dimension $d = 1$ and that f is deterministic. It is easily seen that the following proof carries over to the general case.

Lemma 3.1. *The functional $u(t, x, \cdot)$ defined by (3.3) is indeed in $(\mathcal{S})^*$.*

Proof. Taking the $\mathcal{S}$-transform inside the expectation in (3.3) we obtain the following functional on $\mathcal{S}(\mathbb{R}^d)$:

$$
F(t, x, \xi) = E\left[f(x + \nu b_t) \exp\left\{\frac{1}{\nu}\int_0^t (\xi, \phi_{x+\nu b_s})\, db_s \right.\right.
$$
$$
\left.\left. - \frac{1}{2\nu^2}\int_0^t (\xi, \phi_{x+\nu b_s})^2\, ds\right\}\right]. \tag{3.4}
$$

According to Proposition 2.1 we need to show that $F(t, x, \xi)$ is a U-functional. For $\eta, \xi \in \mathcal{S}(\mathbb{R}^d)$ and $z \in \mathbb{C}$, define

$$
F(t, x, \eta + z\xi) = E\left[f(x + \nu b_t) \exp\left\{\frac{1}{\nu}\int_0^t (\eta + z\xi, \phi_{x+\nu b_s})\, db_s \right.\right.
$$
$$
\left.\left. - \frac{1}{2\nu^2}\int_0^t (\eta + z\xi, \phi_{x+\nu b_s})^2\, ds\right\}\right]. \tag{3.5}
$$

We want to show that $z \mapsto F(t, x, \eta + z\xi)$ is analytic on the complex plane. Note that $f(x)$ is bounded and $|(\xi, \phi_x)| \leq \|\xi\|_{2,0}\|\phi\|_{2,0}$. For any $M > 0$, it holds that

$$
\sup_{|z|\leq M} E\left[|f(x + \nu b_t)|^2 \left|\exp\left\{\frac{1}{\nu}\int_0^t (\eta + z\xi, \phi_{x+\nu b_s})\, db_s\right.\right.\right.
$$
$$
\left.\left.\left. - \frac{1}{2\nu^2}\int_0^t (\eta + z\xi, \phi_{x+\nu b_s})^2\, ds\right\}\right|^2\right] \tag{3.6}
$$
$$
\leq K\exp(KM^2) E\left[\exp\left\{\frac{2}{\nu}\int_0^t (\eta + \mathrm{Re}(z)\xi, \phi_{x+\nu b_s})\, db_s\right\}\right]
$$

where K is an appropriate constant depending on f, ξ, η, ϕ, and t. This is further equal to

$$
K\exp(KM^2) E\left[\exp\left\{\frac{2}{\nu}\int_0^t (\eta + \mathrm{Re}(z)\xi, \phi_{x+\nu b_s})\, db_s\right.\right.
$$
$$
\left. - \frac{2}{\nu^2}\int_0^t (\eta + \mathrm{Re}(z)\xi, \phi_{x+\nu b_s})^2\, ds\right\}
$$
$$
\left.\times \exp\left\{\frac{2}{\nu^2}\int_0^t (\eta + \mathrm{Re}(z)\xi, \phi_{x+\nu b_s})^2\, ds\right\}\right]
$$
$$
\leq K_1 \exp(K_1 M^2) E\left[\exp\left\{\frac{2}{\nu}\int_0^t (\eta + \mathrm{Re}(z)\xi, \phi_{x+\nu b_s})\, db_s\right.\right. \tag{3.7}
$$
$$
\left.\left. - \frac{2}{\nu^2}\int_0^t (\eta + \mathrm{Re}(z)\xi, \phi_{x+\nu b_s})^2\, ds\right\}\right]
$$
$$
= K_1 \exp(K_1 M^2).
$$

Since the the integrand in the right-hand side of (3.5) is an analytic function, (3.7) implies that $F(t, x, \eta + z\xi)$ is continuous in z. Morever, for any closed curve D in the complex plane we have

$$
\int_D F(t, x, \eta + z\xi)\, dz
$$

$$
= E\left[\int_D f(x + \nu b_t) \exp\left\{\frac{1}{\nu}\int_0^t (\eta + z\xi, \phi_{x+\nu b_s})\, db_s \right.\right.
$$

$$
\left.\left. - \frac{1}{2\nu^2}\int_0^t (\eta + z\xi, \phi_{x+\nu b_s})^2\, ds\right\} dz\right] \tag{3.8}
$$

$$
= 0.
$$

Thus it follows from Morera's Theorem that $F(t, x, \eta + z\xi)$ is analytic. For $\xi \in \mathcal{S}(\mathbb{R}^d)$ and $z \in \mathbb{C}$, we have

$$
|F(t, x, z\xi)| \leq K \exp(K|z|^2 \|\xi\|_{2,0}^2) E\left[\exp\left\{\frac{\mathrm{Re}(z)}{\nu}\int_0^t (\xi, \phi_{x+\nu b_s})\, db_s\right\}\right] \tag{3.9}
$$

where K is a constant. The above is equal to

$$
K \exp(K|z|^2 \|\xi\|_{2,0}^2) E\left[\exp\left\{\frac{\mathrm{Re}(z)}{\nu}\int_0^t (\xi, \phi_{x+\nu b_s})\, db_s\right.\right.
$$

$$
\left. - \frac{(\mathrm{Re}(z))^2}{2\nu^2}\int_0^t (\xi, \phi_{x+\nu b_s})^2\, ds\right\}
$$

$$
\left.\times \exp\left\{\frac{(\mathrm{Re}(z))^2}{2\nu^2}\int_0^t (\xi, \phi_{x+\nu b_s})^2\, ds\right\}\right] \tag{3.10}
$$

$$
\leq \exp(K_1|z|^2 \|\xi\|_{2,0}^2).
$$

This together with the analytic property of F shows that $F(t, x, \xi)$ is indeed a U-functional, which ends the proof of Lemma 3.1. $\square$

Lemma 3.2. $\frac{\partial u(t,x)}{\partial x}$ and $\frac{\partial^2 u(t,x)}{\partial x^2}$ exist in $(S)^*$ and

$$
S\frac{\partial u(t, x)}{\partial x}(\xi) = \frac{\partial Su(t, x)(\xi)}{\partial x}, \tag{3.11}
$$

$$
S\frac{\partial^2 u(t, x)}{\partial x^2}(\xi) = \frac{\partial^2 Su(t, x)(\xi)}{\partial x^2}. \tag{3.12}
$$

Proof. As in Lemma 3.1, we continue to denote the S-transform of $u(t, x)$ by $F(t, x, \xi)$. By saying that $\frac{\partial u(t,x)}{\partial x}$ exists in $(S)^*$ and is equal to $g(t, x) \in (S)^*$, we mean that the following limit holds in $(S)^*$:

$$
\lim_{\Delta x \to 0} \frac{u(t, x + \Delta x) - u(t, x)}{\Delta x} = g(t, x). \tag{3.13}
$$

Since $(\xi, \phi_{x+\nu b_s})$ is bounded and smooth in x, it is not difficult to show that

$$
\frac{\partial F(t,x,\xi)}{\partial x}
$$

$$
= E\left[f'(x+\nu b_t)\exp\left\{\frac{1}{\nu}\int_0^t (\xi,\phi_{x+\nu b_s})\,db_s - \frac{1}{2\nu^2}\int_0^t (\xi,\phi_{x+\nu b_s})^2\,ds\right\}\right]
$$

$$
+ E\left[f(x+\nu b_t)\exp\left\{\frac{1}{\nu}\int_0^t (\xi,\phi_{x+\nu b_s})\,db_s\right.\right.
$$

$$
\left.\left. - \frac{1}{2\nu^2}\int_0^t (\xi,\phi_{x+\nu b_s})^2\,ds\right\}\right. \tag{3.14}
$$

$$
\left.\times\left(-\frac{1}{\nu}\int_0^t (\xi,\phi'_{x+\nu b_s})\,db_s + \frac{1}{\nu^2}\int_0^t (\xi,\phi_{x+\nu b_s})(\xi,\phi'_{x+\nu b_s})\,ds\right)\right]
$$

$$
= I_1(t,x,\xi) + I_2(t,x,\xi).
$$

From the proof of Lemma 3.1, one can see that $I_1(t,x,\xi)$ is a U-functional. As for $I_2(t,x,\xi)$, we note that for $\eta, \xi \in \mathcal{S}(\mathbb{R}^d)$ and $M > 0$

$$
\sup_{|z|\le M} E\left[\left|f(x+\nu b_t)\exp\left\{\frac{1}{\nu}\int_0^t (\eta + z\xi,\phi_{x+\nu b})\,db_s\right.\right.\right.
$$

$$
\left.\left.\left. - \frac{1}{2\nu^2}\int_0^t (\eta + z\xi,\phi_{x+\nu b_s})^2\,ds\right\}\right|^2 \cdot \left|-\frac{1}{\nu}\int_0^t (\eta + z\xi,\phi'_{x+\nu b_s})\,db_s\right.\right.
$$

$$
\left.\left. + \frac{1}{\nu^2}\int_0^t (\eta + z\xi,\phi_{x+\nu b_s})(\eta + z\xi,\phi'_{x+\nu b_s})\,ds\right|^2\right]
$$

$$
\le \sup_{|z|\le M}\left\{E\left[\left|f(x+\nu b_t)\exp\left\{\frac{1}{\nu}\int_0^t (\eta + z\xi,\phi_{x+\nu b_s})\,db_s\right.\right.\right.\right.
$$

$$
\left.\left.\left.\left. - \frac{1}{2\nu^2}\int_0^t (\eta + z\xi,\phi_{x+\nu b_s})^2\,ds\right\}\right|^4\right]^{1/2} \right. \tag{3.15}
$$

$$
\times E\left[\left|-\frac{1}{\nu}\int_0^t (\eta + z\xi,\phi'_{x+\nu b_s})\,db_s\right.\right.
$$

$$
\left.\left. + \frac{1}{\nu^2}\int_0^t (\eta + z\xi,\phi_{x+\nu b_s})(\eta + z\xi,\phi'_{x+\nu b_s})\,ds\right|^4\right]^{1/2}\right\}
$$

$$
\le K\exp(K^2 M^2)[M + M^2]
$$

where K is an appropriate constant. Using this inequality and the method in Lemma 3.1 we conclude that $I_2(t,x,\eta + z\xi)$ is analytic on the complex plane. Moreover, employing similar estimates as in the proof of Lemma 3.1 and the Schwartz inequality, one can show that

$$
\left|\frac{\partial F(t,x,z\xi)}{\partial x}\right| \le K_1\exp(K_1|z|^2\|\xi\|_{2,0}^2) \tag{3.16}
$$

where K_1 is a constant which is independent of x. Thus according to Proposition 2.1, we have proved that $\frac{\partial F(t,x,\xi)}{\partial x}$ is the $\mathcal{S}$-transform of an element in $(\mathcal{S})^*$, which is denoted by $g(t,x)$.

Next we prove that $\frac{\partial u(t,x)}{\partial x} = g(t,x)$: Set

$$\bar{u}(t,x,\triangle x) = \frac{u(t,x+\triangle x) - u(t,x)}{\triangle x}. \tag{3.17}$$

We need to show that

$$\bar{u}(t,x,\triangle x) \to g(t,x) \quad \text{as } \triangle x \to 0. \tag{3.18}$$

The above limit is taken in $(\mathcal{S})^*$. From the definition it follows that

$$S\bar{u}(t,x,\triangle x)(\xi) \to Sg(t,x)(\xi) \quad \text{as } \triangle x \to 0. \tag{3.19}$$

On the other hand, by (3.16) we have

$$|S\bar{u}(t,x,\triangle x)(z\xi)| = \left| \frac{1}{\triangle x} \int_x^{x+\triangle x} \frac{\partial F(t,y,z\xi)}{\partial y}\, dy \right|$$

$$\leq K_1 \exp(K_1 |z|^2 \|\xi\|_{2,0}^2). \tag{3.20}$$

Here K_1 does not depend on $\triangle x$. By proposition 2.2, (3.19) and (3.20) imply that (3.18) indeed holds. The proof for $\frac{\partial u(t,x)}{\partial x}$ is completed. The remaining claim (3.12) can be proved similarly. We omit the details. $\square$

Lemma 3.3. $\frac{\partial u(t,x)}{\partial t}$ *exists in* $(\mathcal{S})^*$ *and*

$$S\frac{\partial u(t,x)}{\partial t}(\xi) = \frac{\partial Su(t,x)(\xi)}{\partial t}. \tag{3.21}$$

Proof. Since

$$f(x+\nu b_t) = f(x) + \nu \int_0^t f'(x+\nu b_s)\, db_s + \frac{1}{2}\nu^2 \int_0^t f''(x+\nu b_s)\, ds, \tag{3.22}$$

we can, using the properties of exponential martingales, rewrite the $\mathcal{S}$-transform of $u(t, x)$ as

$$F(t, x, \xi)$$

$$= f(x) + \nu E\left[\left(\int_0^t f'(x + \nu b_s)\, db_s\right) \exp\left\{\frac{1}{\nu} \int_0^t (\xi, \phi_{x+\nu b_s})\, db_s\right.\right.$$

$$\left.\left. - \frac{1}{2\nu^2} \int_0^t (\xi, \phi_{x+\nu b_s})^2\, ds\right\}\right]$$

$$+ \frac{1}{2}\nu^2 E\left[\left(\int_0^t f''(x + \nu b_s)\, ds\right) \exp\left\{\frac{1}{\nu} \int_0^t (\xi, \phi_{x+\nu b_s})\, db_s\right.\right.$$

$$\left.\left. - \frac{1}{2\nu^2} \int_0^t (\xi, \phi_{x+\nu b_s})^2\, ds\right\}\right]$$

$$= f(x) + E\left[\int_0^t f'(x + \nu b_s)(\xi, \phi_{x+\nu b_s}) \exp\left\{\frac{1}{\nu} \int_0^s (\xi, \phi_{x+\nu b_v})\, db_v\right.\right. \tag{3.23}$$

$$\left.\left. - \frac{1}{2\nu^2} \int_0^s (\xi, \phi_{x+\nu b_v})^2\, dv\right\} ds\right]$$

$$+ \frac{1}{2}\nu^2 E\left[\int_0^t f''(x + \nu b_s) \exp\left\{\frac{1}{\nu} \int_0^s (\xi, \phi_{x+\nu b_v})\, db_v\right.\right.$$

$$\left.\left. - \frac{1}{2\nu^2} \int_0^s (\xi, \phi_{x+\nu b_v})^2\, dv\right\} ds\right]$$

This gives

$$\frac{\partial F(t, x, \xi)}{\partial t} = E\left[f'(x + \nu b_t)(\xi, \phi_{x+\nu b_t}) \exp\left\{\frac{1}{\nu} \int_0^t (\xi, \phi_{x+\nu b_s})\, db_s\right.\right.$$

$$\left.\left. - \frac{1}{2\nu^2} \int_0^t (\xi, \phi_{x+\nu b_s})^2\, ds\right\}\right]$$

$$+ \frac{1}{2}\nu^2 E\left[f''(x + \nu b_t) \exp\left\{\frac{1}{\nu} \int_0^t (\xi, \phi_{x+\nu b_s})\, db_s\right.\right. \tag{3.24}$$

$$\left.\left. - \frac{1}{2\nu^2} \int_0^t (\xi, \phi_{x+\nu b_s})^2\, ds\right\}\right].$$

Using this expression, as in the proof of Lemma 3.1, we can show that $\frac{\partial F(t,x,\xi)}{\partial t}$ is an $\mathcal{S}$-transform and for any $T > 0$, there exists a constant K such that the following holds uniformly on $[0, T] \times \mathbb{R}^d$:

$$\left|\frac{\partial F(t, z, \xi)}{\partial t}\right| \leq K \exp(K|z|^2 \|\xi\|_{2,0}^2). \tag{3.25}$$

This is enough to conclude that $\frac{\partial u(t,x)}{\partial t}$ exists in $(S)^*$ and (3.21) holds. See the proof of Lemma 3.2. □

Now we are ready to complete the proof of the main theorem. From the expression of the S-transform $F(t,x,\xi)$ of $u(t,x)$ one knows that it solves the deterministic equation

$$\frac{\partial F(t,x,\xi)}{\partial t} = \frac{1}{2}\nu^2 \Delta F(t,x,\xi) + (\xi,\phi_x) \cdot \nabla F(t,x,\xi),$$
$$F(0,x,\xi) = f(x). \tag{3.26}$$

This together with the above three lemmas shows that $u(t,x,\omega)$ is indeed a solution.

Uniqueness. If a solution $u(t,x)$ exists, the S-transform of u must satisfy the equation (3.26). From the standard probabilistic representation of the solution of the parabolic equation, we know that the S-transform $Su(t,x)(\xi)$ is given by (3.4). The uniqueness follows.

Acknowledgements. This work is supported by VISTA, a research cooperation between The Norwegian Academy of Science and Letters and Den Norske Stats Oljeselskap (Statoil).

REFERENCES

1. F. E. Benth, *Integrals in the Hida distribution space* $(S)^*$, in: Stochastic Analysis and Related Topics (T. Lindström et al., eds.), Gordon and Breach, 1993, pp. 89–99.
2. P. L. Chow, *Generalized solution of some parabolic equation with a random drift*, J. Appl. Math. Optimization **20** (1989), 81–96.
3. J. Gjerde, *Multidimensional noise*, Cand. Scient. Thesis, Univ. of Oslo, 1993.
4. H. Gjessing, H. Holden, T. Lindström, J. Ubøe, and T.-S. Zhang, *The Wick product*, in: Frontiers in Pure and Applied Probability (H. Niemi et al., eds.), vol. 1, TVP Publishers, Moscow, 1993, pp. 29–67.
5. T. Hida, H.-H. Kuo, J. Potthoff, and L. Streit, *White Noise Analysis*, Kluwer, 1993.
6. H. Holden, T. Lindström, B. Øksendal, J. Ubøe, and T.-S. Zhang, *Stochastic boundary value problems: A white noise functional approach*, Probab. Th. Rel. Fields **95** (1993), 391–419.
7. H. Holden, T. Lindström, B. Øksendal, J. Ubøe, and T.-S. Zhang, *The Burgers equation with a noisy force*, Communications PDE **19** (1994), 119–141.
8. T. Lindström, B. Øksendal, and J. Ubøe, *Stochastic differential equations involving positive noise*, in: Stochastic Analysis (M. Barlow and N. Bingham, eds.), Cambridge Univ. Press, 1991, pp. 261–303.
9. T. Lindström, B. Øksendal, and J. Ubøe, *Wick multiplication and Itô–Skorohod stochastic differential equations*, in: Ideas and Methods in Mathematical Analysis, Stochastics, and Applications (S. Albeverio et al., eds.), Cambridge Univ. Press, 1992, pp. 183–206.
10. T. Lindström, B. Øksendal, and J. Ubøe, *Stochastic modelling of fluid flow in porous media*, in: Control Theory, Stochastic Analysis and Applications (S. Chen and J. Yong, eds.), World Scientific, 1991, pp. 156–172.
11. D. Nualart and M. Zakai, *Generalized Brownian functionals and the solution to a stochastic partial differential equation*, J. Functional Anal. **84** (1989), 279–296.

12. J. Potthoff, *White noise methods for stochastic partial differential equations*, Stochastic Partial Differential Equations and Their Applications (B. L. Rozovskii and R. B. Sowers, eds.), Springer-Verlag, 1992.
13. J. Potthoff and L. Streit, *A characterization of Hida distributions.*, J. Func. Anal. **101** (1991), 212–229.

JON GJERDE, DEPARTMENT OF MATHEMATICS, UNIVERSITY OF OSLO, BOX 1053, BLINDERN, N-0316 OSLO 3, NORWAY

HELGE HOLDEN, DEPARTMENT OF MATHEMATICAL SCIENCES, NORWEGIAN INSTITUTE OF TECHNOLOGY, N-7034 TRONDHEIM, NORWAY

BERNT ØKSENDAL, DEPARTMENT OF MATHEMATICS, UNIVERSITY OF OSLO, BOX 1053, BLINDERN, N-0316 OSLO 3, NORWAY

JAN UBØE, DEPARTMENT OF MATHEMATICS, NATIONAL COLLEGE OF SAFETY ENGINEERING, SKÅREGATEN 103, N-5500 HAUGESUND, NORWAY

TUSHENG ZHANG, DEPARTMENT OF MATHEMATICS, NATIONAL COLLEGE OF SAFETY ENGINEERING, SKÅREGATEN 103, N-5500 HAUGESUND, NORWAY

Progress in Probability, Vol. 36
© 1995 Birkhäuser Verlag Basel/Switzerland

STOCHASTIC REPRESENTATION OF
UNITARY QUANTUM EVOLUTION

Z. HABA

ABSTRACT. We show that unitary Schrödinger evolution can be expressed by a diffusion process. Quantum mechanics is realized as stochastic mechanics on an (extended) complex configuration space. Such a formulation of quantum mechanics may be considered as a rigorous version of the Feynman path integral. We discuss several explicitly soluble models.

§1 INTRODUCTION

We discuss here a probabilistic description of quantum mechanics. We mean by this a representation of mathematical problems of quantum mechanics as problems in the theory of random dynamical systems. For example, we view some random perturbations of chaotic systems as a representation of conventional quantum mechanics. Such a representation may be useful when we are able to apply our experience and intuition in dynamical systems to describe their behaviour after a random perturbation. It can be valuable for computer simulations of quantum mechanical problems. The relation can be useful also in the other direction to get some information about random dynamical systems from quantum mechanics. We should emphasize that a realization of a probabilistic representation of quantum mechanics does not mean that quantum mechanics is a classical mechanics in a random medium (like in old models of de Broglie and Bohm). We find such attempts futile. We know from experiments that a quantum particle behaves in a different way than a classical one. We know that the probabilistic rules of the behaviour of the quantum particle are quite different from the probabilistic rules of classical statistical mechanics [1]. We get the description of quantum mechanics as a classical random dynamical system by an extension (complexification) of the configuration space and particular translation rules from a random system to quantum mechanics. So, we may say that quantum mechanics is a random mechanics but not of this dynamical system, which we usually associate with the classical particle.

1991 *Mathematics Subject Classification.* 81P20.
Key words and phrases. Quantum mechanics, stochastic mechanics, Feynman path integral.

First, we would like to indicate that classical mechanics and quantum mechanics are not so far away. Let us consider the Hamiltonian

$$H = B(x, t)p. \tag{1}$$

Then, the classical equations of motion read

$$\frac{dx}{dt} = B(x, t) \tag{2}$$

and

$$\frac{dp}{dt} = -\nabla B(x, t)p. \tag{3}$$

Equation (3) can just be considered as a definition of the momentum. Equation (2) describes a general dynamical system. The simple relation between a classical and a quantum system is preserved by quantization. Then

$$H = -i\hbar B(x, t)\nabla \tag{4}$$

becomes an operator in $L^2(dx)$, which is symmetric if div $B = 0$. If equation (2) has the unique solution for arbitrary time, then

$$(U_t \psi)(x) = \psi(q_t(x)), \tag{5}$$

where $q_t(x)$ is the solution of equation (2) with the initial condition x, can be considered as a unitary Schrödinger evolution in L^2.

In the conventional quantum mechanics we have Hamiltonians which are quadratic in momenta what makes the quantum mechanics basically different from the classical one. In such a case we get after quantization the Schrödinger equation

$$i\hbar \partial_t \psi = \frac{1}{2m}(i\hbar \partial_\mu + eA_\mu)^2 \psi + V\psi \tag{6}$$

where A is a magnetic vector potential, V is a scalar potential, and e is an electric charge.

Let us make a similarity transformation

$$\psi_t = \chi_t \phi_t \tag{7}$$

where

$$\chi_t(x) = \exp\left(\frac{i}{\hbar} W_t(x)\right). \tag{8}$$

Then, ψ_t fulfills equation (6) if and only if ϕ_t fulfills the equation

$$\partial_t \phi_t(x) = \left(\frac{i\hbar}{2m}\Delta - \frac{1}{m}(\nabla W - eA)\nabla\right)\phi_t(x) - \frac{1}{2m}\operatorname{div}(\nabla W - eA)\phi_t(x)$$
$$- \frac{i}{\hbar}\left(\partial_t W + \frac{1}{2m}(\nabla W - eA)^2 + V\right)\phi_t(x). \tag{9}$$

We get a particularly simple form of equation (9) if χ_t is any solution of the Schrö-dinger equation (6) (it does not need to be square integrable). Then, equation (9) reads

$$\partial_t \phi = \left(\frac{i\hbar}{2m} \Delta - \frac{1}{m} (\nabla W - eA)\nabla \right)\phi. \tag{10}$$

Note that in the formal limit $\hbar \to 0$ we get in equation (10) the flow generated by

$$\frac{1}{m}(\nabla W - eA)\nabla$$

which is determined by the solution of the equation

$$\frac{dx}{dt} = -\frac{1}{m}(\nabla W_t - eA). \tag{11}$$

This is the classical flow (2).

§2 QUANTUM SYSTEM AS A STOCHASTIC DYNAMICAL SYSTEM

In order to get a classical (stochastic) representation of the solution of the second order equation (9) we introduce a complex diffusion process. First, let us consider the elementary case with no potential. Assume that the initial condition ψ is a boundary value of an analytic function. Then, it can easily be seen that the solution of the free Schrödinger equation can be expressed in the form

$$\psi_t(x) = \mathrm{E}[\psi(x + \sigma b_t)] \tag{12}$$

when $t \geq 0$ with

$$\sigma = (1 + i)\left(\frac{\hbar}{2m} \right)^{1/2}, \tag{13}$$

for negative time $s = -t \leq 0$ we express the solution in the form

$$\psi_s(x) = (U_{-t}\psi)(x) = \mathrm{E}[\psi(x + \bar{\sigma} b_t^*)] \tag{14}$$

where b^* is an independent Brownian motion. Note that

$$\overline{(U_{-t}\psi)}(x) = (U_t\bar{\psi})(x). \tag{15}$$

The semigroup property of U_t is a consequence of the Markov property of b_t. The extension (14) to negative time defines a unitary group in the Hilbert space $L^2(\mathbb{R}^n)$. The formula (12) as written does not apply to all $\psi \in L^2(\mathbb{R}^n)$. However, its weak form (as a linear functional) makes sense on $L^2(\mathbb{R}^n)$.

Another way to extend formula (12) to a larger class of functions is to use a representation [2] of a function (or even a distribution) as a sum of a boundary value of a holomorphic function analytic in the upper half-plane and a boundary value of a function holomorphic in the lower half plane. Using this representation for the δ-function

$$-i\,2\pi\delta(x) = (x+i0)^{-1} - (x-i0)^{-1}, \tag{16}$$

we derive the well-known formula for the Feynman propagator

$$\left(\exp\frac{it\hbar}{2}\Delta\right)(x,y) = \mathrm{E}[\delta(x-y-\sigma b_t)]. \tag{17}$$

We wish to obtain a probabilistic representation of the solution of equation (9), where a part or the whole dynamics is contained in the stochastic process. We start with the latter case, i.e. equation (10). We introduce a complex Markov process $q_t(x)$ as a solution of the stochastic differential equation

$$dq_t = -\frac{1}{m}(\nabla W - eA)\,dt + \sigma\,db_t \tag{18}$$

with the initial condition $q|_{t=0} = x$.

Assume that there exists the unique solution $q_t(x)$ of equation (18) till the explosion time $\tau(q)$. Then

$$\psi_t(x) = \chi_t(x)\,\mathrm{E}[\phi(q_t(x))] \tag{19}$$

is the unique solution of the Schrödinger equation (6) with the initial condition $\psi_0 = \chi_0\phi$.

We would like to express the solution of the Schrödinger equation for negative time also in terms of a stochastic process. For this purpose we may use the identity

$$\psi_s(x) = (U_s(e)\psi)(x) = \overline{(U_{-s}(-e)\bar\psi)(x)} \tag{20}$$

where $U_s(e)$ denotes the evolution operator corresponding to the Hamiltonian with charge e. For $-s = t \geq 0$, we express the right-hand side of equation (20) by means of equation (19). Then, the complex conjugations can be moved into a definition of a new Markov process q_t^*. This way we can express ψ_s for negative s as

$$\psi_s(x) = \chi_s(x)\,\mathrm{E}[\phi(q_t^*(x))], \tag{21}$$

where the stochastic process $q_t^*(x)$ is defined as the solution of the stochastic equation

$$dq_t^* = \frac{1}{m}\nabla W_t(q_t^*)\,dt - \frac{e}{m}A_t(q_t^*)\,dt + \bar\sigma\,db_t^*. \tag{22}$$

We wish to express quantum (time-dependent) Heisenberg operators (denoted here by capital script letters) in terms of a stochastic process. The evolution of quantum mechanical operators is defined by

$$\mathcal{F}_t = U_t^+\mathcal{F}U_t. \tag{23}$$

We consider here the case when $\mathcal{F}$ depends only on $\mathcal{Q}$ (for general observables see references [3] and [4]). Then, treating $\mathcal{F}\chi_t{}^{-1}\chi_0$ as ϕ in equation (19), we get for $t \geq 0$

$$(\Phi(\mathcal{Q}_t)\chi_0)(x) = \mathrm{E}\left[\exp\left\{\frac{i}{\hbar}S(t, q^*)\right\}\Phi(q_t^*(x))\right]\chi_0(x) \tag{24}$$

where

$$S(t, q^*) = W_t(q_t^*(x)) - W_0(q_t^*(x)) + W_{-t}(x) - W_0(x). \tag{25}$$

If $s = -t \leq 0$, then the formula (24) reads

$$(\Phi(q_s)\chi_0)(x) = \mathrm{E}\left[\exp\left\{\frac{i}{\hbar}S(-t, q)\right\}\Phi(q_t(x))\right]\chi_0(x) \tag{26}$$

and

$$S(-t, q) = W_{-t}(q_t(x)) - W_0(q_t(x)) + W_{-t}(x) - W_0(x). \tag{27}$$

We can represent in a similar way products of operators at different times, e. g. we have

$$\begin{aligned} &\big(\Phi_1(\mathcal{Q}_\tau)\Phi_2(\mathcal{Q}_t)\chi_0\big)(x) \\ &= \mathrm{E}\left[\exp\left\{\frac{i}{\hbar}S(t, \tau; q^*)\right\}\Phi_1\big(q_\tau^*(x)\big)\Phi_2\big(q_{t,\tau}^*(q_\tau^*(x))\big)\right]\chi_0(x) \end{aligned} \tag{28}$$

where $t \geq \tau \geq 0$. The process $q_{t,\tau}^*$ is the solution of the stochastic equation

$$dq_{t,\tau}^* = \frac{1}{m}\nabla W_{t,\tau}(q_{t,\tau}^*)\,dt - \frac{e}{m}A_t(q_{t,\tau}^*)\,dt + \bar{\sigma}\,db_t \tag{29}$$

and

$$\begin{aligned} S(t, \tau; q^*) &= W_{t,\tau}\big(q_{t,\tau}^*(q_\tau^*(x))\big) - W_0\big(q_{t,\tau}^*(q_\tau^*(x))\big) \\ &\quad + W_{t,\tau}\big(q_\tau^*(x)\big) - W_0\big(q_\tau^*(x)\big) + W_{-\tau}(x) - W_0(x) \end{aligned} \tag{30}$$

where $\exp\frac{i}{\hbar}W_{t,\tau}$ is the solution of the Schrödinger equation with the initial condition $\exp\frac{i}{\hbar}W_\tau$ at $t = \tau$, b and b^* are two independent Brownian motions. Note that non-commutativity of operators is expressed in the stochastic formula by the irreversibility of the composition of the stochastic processes.

We are going to apply the probabilistic representation of quantum dynamics to derive a version of the ergodic theorem in quantum mechanics. It is known that classical systems are in general not ergodic, but an addition of noise can make them ergodic (see e. g. reference [6]). We consider the "quantum noise" which arises from our representation of quantum mechanics as a justification of "experimental ergodicity" and formal ergodicity in quantum mechanics noticed by von Neumann [7].

Let us restrict ourselves here to time-independent Hamiltonians. So, assume the spectral decomposition $\mathcal{E}$ of the Hamiltonian

$$U_t = \int \exp(it\omega)\, d\mathcal{E}(\omega). \tag{31}$$

The Hilbert space $\mathcal{H}$ can be decomposed into a direct sum

$$\mathcal{H} = \mathcal{H}_{ac} \oplus \mathcal{H}_s \oplus \mathcal{H}_{pp} \equiv \mathcal{H}_c \oplus \mathcal{H}_{pp} \tag{32}$$

where $\mathcal{H}_{ac}$ belongs to the absolutely continuous spectrum, $\mathcal{H}_s$ to the singular continuous spectrum and $\mathcal{H}_{pp}$ to the pure point spectrum. If $\mathcal{H} = \mathcal{H}_{pp}$ then the integral (31) becomes a sum

$$U_t = \sum_n \exp(it\omega_n)\mathcal{E}_n. \tag{33}$$

Let us compute now the time average of an observable

$$\lim_{T \to \infty} \frac{1}{T} \int_0^T \mathcal{F}_s\, ds = \lim_{T \to \infty} \frac{1}{T} \int_0^T \sum_{n,m} \exp(-is\omega_n)\mathcal{E}_n\mathcal{F}\mathcal{E}_m \exp(is\omega_m)\, ds. \tag{34}$$

Hence,

$$\mathcal{F} \to \sum_n \mathcal{E}_n\mathcal{F}\mathcal{E}_n. \tag{35}$$

We are going to derive some statements about the large time behaviour on the basis of spectral properties of U_t without a detailed study of the process q_t. Let us define a probability measure

$$d\rho(x) = dx\, |\chi(x)|^2. \tag{36}$$

Then we have the following limit theorem in the sense of the weak convergence in $L^2(dx)$ (see reference [5] for the proof): Assume that χ_0 is an eigenvector of H with the eigenvalue ω_r. If $\psi = \phi\chi_0 \in \mathcal{H} \ominus \mathcal{H}_{pp}$ then

$$\frac{1}{t} \int_0^t \mathrm{E}[\phi(q_\tau(x))]\, d\tau \to 0. \tag{37}$$

If $\psi \in \mathcal{H}_{pp}$ then

$$\lim_{t \to \infty} \frac{1}{t} \int_0^t \mathrm{E}[\phi(q_\tau(x))]\, d\tau = \chi_0(x)^{-1}(\mathcal{E}_r\psi)(x) \tag{38}$$

where $\mathcal{E}_r$ is the projector onto the eigenspace of ω_r. In particular, if ω_r is non-degenerate, then

$$\lim_{t \to \infty} \frac{1}{t} \int_0^t \mathrm{E}[\phi(q_\tau(x))]\, d\tau = \int d\rho(x)\, \phi(x). \tag{39}$$

The right-hand sides of equations (37)–(39) have a Hilbert space meaning and are an expression of the ergodic theorem in quantum mechanics in the formulation of von Neumann [7]. We can see that the limit in the degenerate case (38) depends on the point x, where the process starts. Such a process cannot be ergodic (in an agreement with the intuition that there are additional constants of motion if the energy eigenvalue is degenerate).

§3 MODELS

We discuss now some examples explicitly. We start with some simple models which can be solved exactly.

Let us begin with the well-known free motion of the wave packet

$$\chi(x) = \exp\left(\frac{i}{\hbar}kx - \frac{1}{2\delta^2}x^2\right).$$ (40)

Then,

$$q_t(x) = x + \frac{k}{m}t + \xi_t$$ (41)

where

$$d\xi = -\frac{i\hbar}{m\delta^2}\left(1 + \frac{i\hbar t}{m\delta}\right)^{-1}\xi\, dt + \sigma\, db.$$ (42)

The solution of this equation is

$$\xi_t = \sigma\int_0^t \left(1 + \frac{i\hbar\tau}{m\delta}\right)^{1/\delta}\left(1 + \frac{i\hbar t}{m\delta}\right)^{-1/\delta} db_\tau.$$ (43)

Equations (41) and (43) describe fluctuations around the straight line trajectory of the free particle. The variance of the process ξ_t increases linearly in t (spreading of the wave packet). If we allow δ to be complex, i. e.

$$\chi(x) = \exp\left(\frac{i}{\hbar}kx - \frac{\alpha + i\beta}{2\hbar}x^2\right),$$ (44)

then from the Hamilton–Jacobi interpretation we could suspect an instability, because the momentum is linearly rising with the distance. However, the spreading of the wave packet precludes any global instability. Nevertheless, if $\beta > 0$, then the Markov process will spread faster than in the case of $\beta = 0$. In detail, we get the stochastic equation

$$dq_t = \frac{k}{m}R_t^{-1}(1 + \beta t + i\alpha t)\, dt - i\alpha R_t^{-1}q_t\, dt$$
$$+ R_t^{-1}\left(\beta(1 + \beta t) + \alpha^2 t\right)q_t\, dt + \sigma\, db_t$$ (45)

where

$$R_t = (1 + \beta t)^2 + \alpha^2 t^2.$$ (46)

Equation (45) is a linear equation of the type

$$dq = -i\Omega(t)q\, dt - \Gamma(t)q\, dt + F(t)\, dt + \sigma\, db$$ (47)

with the solution

$$q_t(x) = \exp\left(-i\int_0^t \Omega(\tau)\, d\tau - \int_0^t \Gamma(\tau)\, d\tau\right)x$$
$$+ \int_0^t \exp\left(-i\int_\tau^t \Omega(s)\, ds - \int_\tau^t \Gamma(s)\, ds\right)(F(\tau)\, d\tau + \sigma\, db_\tau).$$ (48)

It should be pointed out that the linearity of equation (47) results from two assumptions. First, the Lagrangian is quadratic and then the initial state is an exponential of a quadratic form.

Let us consider another linear case: an oscillator in the ground state

$$\chi(x) = \exp\left(-\frac{m\omega}{2\hbar}x^2\right). \tag{49}$$

Then

$$dq = -i\omega q\, dt + \sigma\, db. \tag{50}$$

The solution of this equation with the initial condition x is

$$q_t(x) = \exp(-i\omega t)x + \sigma \int_0^t \exp(-i\omega(t-\tau))\, db_\tau. \tag{51}$$

The simplicity of equations (50) and (51) is misleading. It resulted from the choice of the minimal uncertainty state as the initial state. If we take

$$\chi(x) = \exp\left(-(\alpha+i\beta)\frac{x^2}{2\hbar} + \frac{ikx}{\hbar}\right) \tag{52}$$

where α, β, and k are real, then

$$\chi_t(x) = (\alpha+i\beta - i\omega ctg(\omega t))^{-1/2}\exp\left(\frac{i}{2}\omega t\right)\exp\left(\frac{i\omega}{2\hbar}x^2 ctg\omega t\right)$$
$$\times \exp\left(-\frac{1}{2\hbar}\left\{\left(k-\frac{\omega x}{\sin\omega t}\right)(\alpha+i\beta - i\omega ctg(\omega t))^{-1}\left(k-\frac{\omega x}{\sin\omega t}\right)\right\}\right). \tag{53}$$

Now, the stochastic equation can be expressed in the form (47) where

$$\Gamma(t) = \omega ctg(\omega t) + \omega^2(\beta - \omega ctg(\omega t))(\sin(\omega t))^{-2}(\alpha^2 + (\beta - \omega ctg(\omega t))^2)^{-1},$$
$$\Omega(t) = \alpha\omega^2(\sin(\omega t))^{-2}(\alpha^2 + (\beta - \omega ctg(\omega t))^2)^{-1}, \tag{54}$$
$$F(t) = k\omega(\beta + i\alpha)(\sin(\omega t))^{-1}(\alpha^2 + (\beta - \omega ctg(\omega t))^2)^{-1}.$$

This equation describes an oscillator with a periodic enhancement (or damping) of its amplitude and moving in space under an influence of a periodic force. Nevertheless, the trajectory remains in a bounded region with probability 1.

It is easy to see that we get a linear equation of the form (47) as long as the Schrödinger propagator has the form of an exponential of a quadratic form and the initial state is also an exponential of a quadratic form (a wave packet). These examples include a general quadratic time-dependent potential V and a linear time-dependent electromagnetic potential. The quantum mechanical effect on the classical motion comes from the noise b_t as well as from the departure of the state χ_t from its semi-classical or minimal uncertainty form.

We shall discuss next some linear models in higher dimensions. Let

$$W_t = \frac{m}{2}(ax^2 + by^2 + cz^2) + \frac{i\hbar}{2}(a + b + c)t, \tag{55}$$

$$eA = (m\alpha y, -m\beta x, 0), \tag{56}$$

$$V = -\frac{m}{2}[(ax - \alpha y)^2 + (by + \beta x)^2 + c^2 z^2]. \tag{57}$$

In equations (55)–(57) the parameters (α, β) are real, whereas (a, b, c) may be either real or purely imaginary. In the latter case the condition

$$a\alpha = b\beta \tag{58}$$

should be fulfilled.

The stochastic equations read

$$\begin{aligned}
dx &= (-ax + \alpha y)\, dt + \sigma\, db_1, \\
dy &= (-by - \beta x)\, dt + \sigma\, db_2, \\
dz &= -cz\, dt + \sigma\, db_3.
\end{aligned} \tag{59}$$

A diagonalization gives three eigenvalues for the frequencies, namely

$$\lambda_{\mp} = -\frac{1}{2}(a + b) \mp \sqrt{\frac{1}{4}(a - b)^2 - \alpha\beta} \quad \text{and} \quad \lambda_3 = -c. \tag{60}$$

In general, we get a spiraling solution of the equations in (59). Only if (a, b, c) are purely imaginary and

$$\frac{1}{4}(a - b)^2 - \alpha\beta \leq 0, \tag{61}$$

the solution will be oscillatory. If $|\frac{\alpha}{\beta}| < 1$ and (a, b, c) are purely imaginary, then the equations in (59) describe a damped oscillator in an electromagnetic field of a constant strength. In the remaining cases we may have pure damping without any oscillations.

It is interesting to write down explicitly the solution in the case of pure dissipation $((\alpha, \beta) = 0,\ (a, b, c) \in \mathbb{R}^3)$. The solution is

$$\begin{aligned}
x_t &= \exp(-at)x + \sigma \int_0^t \exp(-at + a\tau)\, db_1(\tau), \\
y_t &= \exp(-bt)y + \sigma \int_0^t \exp(-bt + b\tau)\, db_2(\tau), \\
z_t &= \exp(-ct)z + \sigma \int_0^t \exp(-ct + c\tau)\, db_3(\tau).
\end{aligned} \tag{62}$$

The solution (19) of the Schrödinger equation with the inital condition

$$\psi_0 = \exp \frac{i}{\hbar} (ax^2 + by^2 + cz^2) \, \phi(x, y, z)$$

reads (for $t \geq 0$)

$$\psi_t = \exp\left(\frac{i}{\hbar} W_t\right) \mathrm{E}\big[\phi(x_t(x), y_t(y), z_t(z))\big]. \tag{63}$$

The expectation value in equation (63) can explicitly be computed if we use the Fourier transform

$$\phi(x) = \int \Phi(k) \exp(ik_a x^a). \tag{64}$$

Then

$$\mathrm{E}[\exp(ik_a x_t^a)] = \exp i\big(k_1 \exp(-at)x + k_2 \exp(-bt)y + k_3 \exp(-ct)z\big)$$
$$\times \exp\left[-\frac{\sigma^2}{4}\left(\frac{k_1^2}{a}(1 - \exp(-2at)) + \frac{k_2^2}{b}(1 - \exp(-2bt))\right.\right.$$
$$\left.\left. + \frac{k_3^2}{c}(1 - \exp(-2ct))\right)\right].$$

The solution for negative time can be obtained by means of the complex conjugation

$$\psi_{-t}(x) = \overline{(U_t\bar{\psi})(x)}.$$

If

$$\int |\Phi(k)| < \infty$$

then we get from equation (63) that for any $t \in R$

$$|\psi_t(x)| \leq \mathrm{const}\, \exp\left[-\frac{1}{2}(a + b + c)t\right]. \tag{65}$$

In spite of the pointwise decay (for $t > 0$), the unitarity holds true (as can be checked by direct calculations)

$$\int |\psi_t(x)|^2 = \int |\psi_0(x)|^2.$$

It is interesting to note that we would get $|t|$ on the right-hand side of equation (65) (instead of t) if we inserted $\epsilon(t)$ in front of $(ax^2 + by^2 + cz^2)$ in equation (55).

Equation (59) is a linearized (around 0) form of the Lorenz model. Let us consider another simplified (but less trivial) version of this model

$$dx = -ax\,dt + \sigma\,db_1,$$
$$dy = (rx + \gamma xz - by)\,dt + \sigma\,db_2, \qquad (66)$$
$$dz = (-\gamma xy - cz)\,dt + \sigma\,db_3.$$

From the first equation we get

$$x_t = \exp(-at)x_0 + \sigma \int_0^t \exp(-at + a\tau)\,db_1(\tau). \qquad (67)$$

Then, the remaining system (y, z) is linear. It is still not easy to obtain an explicit solution of the equations in (66). However, if $b = c$, then we have an elementary solution (here b is either real or purely imaginary). Set $w = y + iz$, then

$$w_t = r\int_0^t \exp\left(-i\gamma \int_\tau^t x_s\,ds - c(t - \tau)\right)x_\tau\,d\tau + \exp\left(-i\gamma \int_0^t x_s\,ds - ct\right)w_0$$
$$+ \sigma \int_0^t \exp\left(-i\gamma \int_\tau^t x_s\,ds - c(t - \tau)\right)\left(db_2(\tau) + i\,db_3(\tau)\right). \qquad (68)$$

In particular, we can easily calculate the Gaussian integral

$$E[w_t] = E\left[r\int_0^t \exp\left(-i\gamma \int_\tau^t x_s\,ds - c(t - \tau) - a\tau\right)x_0\,d\tau\right.$$
$$\left. + \exp\left(-i\gamma \int_0^t x_s\,ds - ct\right)w_0\right]. \qquad (69)$$

If $a = c = 0$, then we get a periodic solution with the frequency

$$\omega = \gamma x_0 + \frac{\gamma^2 \hbar}{2m}. \qquad (70)$$

If c is purely imaginary then a computation of the expectation value (69) shows that $E[w_t]$ is quasiperiodic with two frequencies $\omega = |c|$ and γx_0.

Equation (66) corresponds to the quantum mechanics with

$$W = \frac{m}{2}(ax^2 + by^2 + cz^2) + \frac{i\hbar}{2}(a + b + c)t,$$
$$eA = (0, m\gamma xz + mrx, -m\gamma xy), \qquad (71)$$
$$V = -\frac{m}{2}[(-\gamma xz + by - rx)^2 + (\gamma xy + cz)^2].$$

§4 STOCHASTIC PERTURBATION OF THE CLASSICAL DYNAMICS

In Sections 2 and 3 we assumed that χ_t in (8) fulfills the Schrödinger equation (6). In such a case we have got a simple formula (19) for the solution of the Schrödinger equation with an arbitrary initial condition $\chi_0\phi$. We assume now that $W \equiv S$ is a solution of the Hamilton–Jacobi equation

$$\partial_t S_t + \frac{1}{2m}(\nabla S_t)^2 + V = 0. \tag{72}$$

Let us define the complex Markov process

$$dq_\tau = -\frac{1}{m}(\nabla S_\tau - eA)\, d\tau + \sigma\, db. \tag{73}$$

Let us write the initial state ψ in the form

$$\psi = \exp\left(\frac{i}{\hbar}S_0\right)\phi.$$

Then, the solution of the Schrödinger equation (6) can be expressed in the form

$$\psi_t(x) = \exp\left(\frac{i}{\hbar}S_t(x)\right)\mathrm{E}\left[\exp\left\{-\frac{1}{2m}\int_0^t \mathrm{div}(\nabla S_\tau - eA)(q_\tau(x))\, d\tau\right\}\phi(q_0(x))\right] \tag{74}$$

where q_τ is the solution of the stochastic equation (73) with the boundary condition $q_\tau(x)|_{\tau=t} = x$.

Note that $\hbar$ enters equation (74) only through q. The formal classical limit ($\hbar \to 0$) of equations (73)–(74) is

$$\psi_t(x) = \exp\left(\frac{i}{\hbar}S_t(x)\right)\exp\left(-\frac{1}{2m}\int_0^t \mathrm{div}(\nabla S_\tau - eA)(x_\tau(x))\, d\tau\right)\phi(x_0(x)). \tag{75}$$

$$\frac{dx_\tau}{d\tau} = -\frac{1}{m}(\nabla S_\tau(x) - eA(\tau, x)) \tag{76}$$

where $x_\tau|_{\tau=t} = x$. The representation (74) is appealing because of its close connection to the conventional classical mechanics. Its usefulness depends on whether we can control the exponential factor in the expectation value (74). We studied a physical model where the expression (74) is well under control. We consider the model with $A = 0$ and

$$V = \frac{m}{4}(\alpha^2 - \beta^2)\cos(2x + 2\omega t) - \frac{m\alpha\beta}{2}\sin(2x + 2\omega t)$$
$$+ m\alpha\omega\sin(x + \omega t) + m\beta\omega\cos(x + \omega t). \tag{77}$$

The potential (77) describes a particle in a plane electromagnetic wave. We can find a particular solution of the classical Hamilton–Jacobi equation. The solution is

$$S_t(x) = m\alpha\cos(x + \omega t) - m\beta\sin(x + \omega t) + \frac{m}{2}(\alpha^2 + \beta^2)t. \tag{78}$$

The classical equation of motion of the model (77)–(78) reads

$$\frac{dx}{dt} = \alpha \sin(x + \omega t) + \beta \cos(x + \omega t).$$

This equation is explicitely integrable. Properties of the solution depend in a crucial way on whether

$$\alpha^2 + \beta^2 \geq \omega^2 \text{ (a particle is trapped) or } \alpha^2 + \beta^2 < \omega^2 \text{ (lack of trapping)}.$$

In the first case the particle moves with the wave, and for large time the only effect of the non-linear dynamics is a phase shift (the phase shift tends to a constant with an exponential speed). In the second case the particle oscillates with its own frequency

$$\mu = \sqrt{\omega^2 - \alpha^2 - \beta^2}. \tag{79}$$

The stochastic equation (73) takes the form

$$dq = \alpha \sin(q + \omega t)\, dt + \beta \cos(q + \omega t)\, dt + \sigma\, db. \tag{80}$$

Using the exponential representation of the trigonometric functions we can transform equation (80) into the Riccati equation

$$\frac{dz}{dt} = f_0(t) + f_1(t)z + f_2(t)z^2. \tag{81}$$

The Riccati equation can subsequently be linearized by the substitution

$$W(t) = \exp\left(-\int_0^t f_2(s)z(s)\, ds\right). \tag{82}$$

Then, W fulfills a linear equation

$$\frac{d^2W}{dt^2} - \left(f_1(t) + \frac{d}{dt}\ln f_2(t)\right)\frac{dW}{dt} + f_0(t)f_2(t)W = 0. \tag{83}$$

In the model (80) we may choose

$$W = \exp\left(-\frac{1}{2}(\alpha - i\beta)\int_0^t \exp(-i\omega s - iq(s))\, ds\right) \tag{84}$$

or

$$W_* = \exp\left(-\frac{1}{2}(\alpha + i\beta)\int_0^t \exp(i\omega s + iq(s))\, ds\right). \tag{85}$$

Both fulfil linear equations. Let us denote $\frac{dW}{dt} = W'$, then equation (83) can be expressed as a system of two Itô equations

$$dW = W' \, dt,$$

$$dW' = -i\left(\omega - \frac{\hbar}{2m}\right)W' \, dt + \frac{1}{4}(\alpha^2 + \beta^2)W \, dt - i\sigma W \, db. \tag{86}$$

Similarly, for W_* we get

$$dW_* = W'_* \, dt,$$

$$dW'_* = i\left(\omega - \frac{\hbar}{2m}\right)W'_* \, dt + \frac{1}{4}(\alpha^2 + \beta^2)W_* \, dt + i\sigma W_* \, db. \tag{87}$$

Let $w = E[W]$, then w is the solution of a deterministic equation

$$\frac{d^2w}{dt^2} + i\left(\omega - \frac{\hbar}{2m}\right)\frac{dw}{dt} - \frac{1}{4}(\alpha^2 + \beta^2)w = 0. \tag{88}$$

We can see from equation (88) that the noise produces the shift of the frequency

$$\omega \to \omega - \frac{\hbar}{2m}.$$

Hence, from equation (79) we can see that if

$$|\omega^2 - \alpha^2 - \beta^2| > \left|\frac{\omega\hbar}{m} - \frac{\hbar^2}{4m^2}\right| \tag{89}$$

then the quantum noise has no effect on stability changing only the frequency (or the Lyapunov exponent). However, if the condition (89) is not fulfilled, then the noise can change the exponential increase of $w(t)$ into an oscillatory behaviour or vice versa.

We want to express the Feynman formula (74) also in terms of the W variables. Note that

$$W(t)W_*(t) = \exp\left(\frac{1}{m}\int_0^t \Delta S_\tau(q(\tau)) \, d\tau\right). \tag{90}$$

Hence, the Feynman formula can be expressed in the form

$$\psi_t(x) = \exp\left(\frac{i}{\hbar}S_t(x)\right) E\left[\left(\sqrt{W(t)W_*(t)}\right)^{-1}\phi(q_t(W))\right]. \tag{91}$$

This way the Feynman integral can be expressed by variables which undergo a phase transition from exponential to oscillatory behaviour. This transition has a physical meaning as a transition from a trapped to an untrapped state of a particle in an electromagnetic wave.

Let us mention some other probabilistic approaches to quantum mechanics. The Feynman integral has been expressed by the Wiener measure in references [8] and [9]. The theory of Feynman integrals in terms of a complex measure is developed in reference [10]. A different stochastic version of quantum mechanics in real time has been developed by Nelson [11]. A probabilistic formulation of quantum mechanics in imaginary time is discussed in references [12] and [13].

REFERENCES

1. R. P. Feynman, *Proceedings*, Second Berkley Symp. (J. Neyman, ed.), Univ. Calif. Press, 1951.
2. M. Sato, J. Fac. Sci. Tokyo, sect. I **8** (1959), 139.
3. H. Bremermann, *Distributions, Complex Variables and Fourier Transforms*, Addison-Wesley, 1965.
4. Z. Haba, Phys. Lett. **175A** (1993), 371.
5. Z. Haba, Wroclaw preprint (1993).
6. Z. Haba, Journ. Math. Phys. **32** (1991), 3463.
7. J. von Neumann, Zeitschr. Phys. **57** (1929), 30.
8. H. Doss, Commun. Math. Phys. **73** (1980), 247.
9. T. Hida, H. H. Kuo, J. Potthoff, and L. Streit, *White Noise Analysis*, Reidel, 1993.
10. S. Albeverio and R. Hoegh-Krohn, *Mathematical Theory of Feynman Path Integrals*, Springer, 1976.
11. E. Nelson, Phys. Rev. **150** (1966), 1079.
12. S. Albeverio, K. Yasue, and J. C. Zambrini, Ann. Inst. H. Poincaré **A49** (1989), 259.
13. A. B. Cruzeiro and J. C. Zambrini, J. Func. Anal. **96** (1991), 62.

Z. HABA, INSTITUTE OF THEORETICAL PHYSICS, UNIVERSITY OF WROCLAW, WROCLAW, POLAND

Progress in Probability, Vol. 36

CRITICAL DIMENSIONS FOR THE EXISTENCE
OF SELF-INTERSECTION LOCAL TIMES
OF THE BROWNIAN SHEET IN $\mathbb{R}^d$

PETER IMKELLER AND FERENC WEISZ

ABSTRACT. Fix two rectangles A, B in $[0,1]^2$. Then the size of the random set of double points of the Brownian sheet $(W_t)_{t\in[0,1]^2}$ in $\mathbb{R}^d$, i.e. the set of pairs (s,t), where $s \in A$, $t \in B$, and $W_s = W_t$, can be measured as usual by a self-intersection local time. If $A = B$, we show that the critical dimension below which self-intersection local time exists, is given by $d = 4$. If $A \cap B$ consists of an axial parallel line, it is 6, if it consists of a point or is empty, 8. In all cases, we derive the rate of explosion of canonical approximations of self-intersection local time for dimensions above the critical one, and determine its smoothness in terms of the canonical Dirichlet structure on Wiener space.

§1 INTRODUCTION

In Varadhan [16], Szymanzik [15] and Wolpert [19], the construction of certain Euclidean quantum fields was related to self-intersection local times of Brownian motion W in $\mathbb{R}^d$. This discovery created a whole avalanche of mathematical papers dealing with properties of these objects which can be seen as measuring the size of the random set of double points of W, i.e. the set of pairs (s,t) where $W_s = W_t$. To quote just a few authors: Rosen [9], Yor [20], Le Gall [7], Dynkin [3]. See Dynkin [3] for an extensive bibliography.

In Rosen and Yor [13], Rosen [12] and Yor [20] it is shown that self-intersection local times can be understood by means of certain stochastic integrals, which figure in formulae of the Tanaka type. Representations of this kind make self-intersection local times analytically accessible. One can go one step further and represent the integrands appearing in the stochastic integrals in a similar way. Repeating this over and over again leads to a series representation of self-intersection local time in terms of multiple Wiener–Itô integrals. They have one particular shortcoming: working with series one has to forget about stochastic intuition of the sample path picture. But besides offering a relatively easy access to questions of renormalization of self-intersection local times or of the asymptotic behaviour of more general local time functionals (see Nualart, Vives [8], Imkeller, Perez–Abreu, Vives [4], Imkeller,

1991 *Mathematics Subject Classification.* Primary 60G60, 60J55; Secondary 60G15, 60H05.

Key words and phrases. Brownian sheet; self-intersection local time; multiple stochastic integrals; canonical Dirichlet structure.

Weisz [5]), they live on the space of Meyer distributions over Wiener space, and as such make perfect sense even as pure distributions, as for example the self-intersection local time of four-dimensional Brownian motion, and easily keep track of the smoothness properties of the functionals represented in terms of quadratic Sobolev spaces on Wiener space. It should be mentioned that they correspond to functionals derived from the representation of Donsker's delta distribution in the framework of Hida calculus (see Watanabe [17], Kuo [6], Shieh [14] and others).

In this paper we use the series representation to investigate the self-intersection local time of the Brownian sheet in $\mathbb{R}^d$. Let A and B be two rectangles in $\mathbb{R}^2_+$ off the boundary (we exlude boundary effects). Define

$$\alpha_\varepsilon(x, \cdot) = \int_B \int_A p_\varepsilon^d(W_t - W_s - x)\, ds\, dt, \quad x \in \mathbb{R}^d,\ \varepsilon > 0,$$

to measure the integral "closeness of $W_s + x$ and W_t of order ε", if s is allowed to run in A and t in B. We are mainly interested in answering two questions:

(a) For which dimensions d and which constellations of A and B does $\alpha_\varepsilon(x, \cdot)$ converge as $\varepsilon \to 0$ to an object we call self-intersection local time?
(b) If it does not converge, what is its rate of divergence? More precisely, we seek for deterministic functions $f(\varepsilon)$ such that $\{\alpha_\varepsilon(x, \cdot)/f(\varepsilon) : \varepsilon > 0\}$ is bounded and has non-trivial cluster points as $\varepsilon \to 0$.

It turned out that the approach of these questions by means of series representations allowed rather complete answers. We both provide "critical dimensions" and "explosion rates" of $\alpha_\varepsilon(x, \cdot)$ in all of four different typical configurations of A and B. Here are the essentials of what we found.

If $A = B$, the tendency of W to propagate its behaviour along axial parallel lines has enough room to create sets of double points, too regular and "thin" to allow the existence of self-intersection local times above the critical dimension $d = 4$. For $d = 4$, explosion is due to the 0th term of the series and is of order $\ln 1/\varepsilon$, so that just by "renormalization", i.e. subtraction of the 0th term, we are left with a well-behaved object. This parallels Varadhan's renormalization of two-dimensional Brownian motion (see Le Gall [7]). But even after renormalization, for dimension 5 the rate of explosion of $\alpha_\varepsilon(x, \cdot) - E(\alpha_\varepsilon(x, \cdot))$ as $\varepsilon \to 0$ is $\sqrt{\ln 1/\varepsilon}$, for dimension $d > 6$ it is $\varepsilon^{(5-d)/2}$.

If A and B have just one line (the upper boundary of A, the lower one of B) in common, the behaviour of admissible W_s and W_t has gained enough "independence" to create sets of double points much more chaotic and "equally distributed" in $\mathbb{R}^d$, that the critical dimension below which self-intersection local times exist, increases to 6. For $d = 6$, we see that the rate of explosion is dominated by the 0th term and is given by $\ln 1/\varepsilon$, whereas the remaining terms diverge at a rate of $\sqrt{\ln 1/\varepsilon}$. So in this case we obtain both a "law of large numbers" type result as well as a weak form of a "central limit" type statement. For $d > 6$ this dichotomy disappears. Explosion of all terms in the representation is governed by $\varepsilon^{(6-d)/2}$.

If the intersection of A and B shrinks to just one point, the critical dimension for the reasons indicated above, again increases by 2. The explosion properties of $\alpha_\varepsilon(x, \cdot)$ exactly parallel the ones of the preceding case, just with $d = 8$ replacing $d = 6$.

If A and B are finally disjoint, $\alpha_\varepsilon(x, \cdot)$ has a limit as $\varepsilon \to 0$ for any $d \in \mathbb{N}$. No explosion is possible any more. The limit is a function if $d < 8$, and a distribution if $d \geq 8$. This reflects a general pattern of our findings: convergence results for $\alpha_\varepsilon(x, \cdot)$ and boundedness results for $\alpha_\varepsilon(x, \cdot)/f(\varepsilon)$, as $\varepsilon \to 0$, hold with respect to Sobolev spaces of any order below $4 - d/2$ with respect to the canonical Dirichlet structure on Wiener space.

A special case of our last one $(A = [\alpha, \beta], \, B = [\gamma, \delta]$ with $\beta < \gamma)$ has been treated in Rosen [10], [11].

§2 PRELIMINARIES AND NOTATIONS

We shall deal with the canonical Brownian sheet indexed by $[0, 1]^2$ with values in $\mathbb{R}^d$ on the canonical Wiener space $(\Omega, \mathcal{F}, P)$. P is the probability measure under which W_t possesses the probability density

$$p_{t_1 t_2}^d(x) = \frac{1}{\sqrt{2\pi t_1 t_2}^d} \exp\left(-\frac{|x|^2}{2t_1 t_2}\right), \quad x \in \mathbb{R}^d, \; t = (t_1, t_2) \in [0, 1]^2.$$

The ordering of the parameter space is supposed to be coordinatewise linear ordering on $\mathbb{R}_+$. Intervals with respect to this partial ordering are defined in the usual way, and $s < t$ means $s_i < t_i$ for $i = 1, 2$.

Suppose now $d = 1$. It is well-known that $L^2(\Omega, \mathcal{F}, P)$ possesses an orthogonal decomposition by the eigenspaces of the Ornstein–Uhlenbeck operator on Wiener space, which are generated by the multiple Wiener–Itô integrals I_n, defined on $L^2(([0, 1]^2)^n)$ for $n \in \mathbb{N}_0$ (see for example Bouleau, Hirsch [2, pp. 78–80]). The multiple integrals fulfill the important orthogonality relation

$$\mathrm{E}(I_n(f)I_m(g)) = \begin{cases} 0, & \text{if } n \neq m, \\ n! \int_{([0,1]^2)^n} fg \, d\lambda, & \text{if } n = m, \end{cases}$$

where λ denotes Lebesgue measure without reference to the dimension of the space on which it is defined. If H_n is the nth Hermite polynomial defined by

$$H_n(x) = \frac{(-1)^n}{\sqrt{n!}} \exp\left(\frac{x^2}{2}\right) \left(\frac{d}{dx}\right)^n \left[\exp\left(-\frac{x^2}{2}\right)\right],$$

$x \in \mathbb{R}$, $n \geq 0$, and if

$$W(h) = \int_{[0,1]^2} h \, dW$$

denotes the Gaussian stochastic integral of a function $h \in L^2([0,1]^2)$ (we write $W(D) = W(1_D)$ for $D \in \mathcal{B}(\mathbb{R}_+^2)$, so $W_t = W(R_t)$ for $R_t = [0,t]$), the relation

$$H_n(W(h)) = \frac{1}{\sqrt{n!}} I_n(h^{\otimes n})$$

holds true whenever $\|h\| = 1$. Here $h^{\otimes n}$ denotes the n-fold tensor product of h with itself, while $\|\cdot\|$ is the norm in $L^2([0,1]^2)$. For $\varrho \in \mathbb{R}$ we may define the "Sobolev space of order ϱ" on Wiener space by introducing the norm

$$\|F\|_{2,\varrho} = \left(\sum_{n=0}^{\infty} (1+n)^{\varrho} \|I_n(f_n)\|_2^2 \right)^{1/2}$$

on the space

$$\left\{ F = \sum_{i=0}^{n} I_i(f_i) : f_i \in L^2([0,1]^{2i}), \, 0 \leq i \leq n, \, n \in \mathbb{N} \right\}$$

which is dense in $L^2(\Omega, \mathcal{F}, P)$ and completing with respect to $\|\cdot\|_{2,\varrho}$. We denote this space by $\mathbb{D}_{2,\varrho}$. In the case $\varrho = 1$ we just recover the domain of the gradient operator of the canonical Dirichlet form on Wiener space, for $\varrho < 0$ we obtain a space of distributions over Wiener space (see Watanabe [18], Bouleau, Hirsch [2]).

To denote multiple Wiener–Itô integrals with respect to the independent components W^i of W in $\mathbb{R}^d$, we use the symbol I_n^i, $1 \leq i \leq d$, $n \in \mathbb{N}_0$. Corresponding Sobolev spaces are defined for functionals of the d-dimensional Brownian sheet (see Watanabe [18]). We finally remark that $A \triangle B = (A \setminus B) \cup (B \setminus A)$ denotes the symmetric difference of the sets A and B.

§3 Two series representations and the characteristic integrals

In this section we shall consider a canonical approximation $\alpha_\varepsilon(x, \cdot)$, $\varepsilon > 0$, of self-intersection local time of the Brownian sheet in $\mathbb{R}^d$. We hereby fix two rectangles A and B off the boundary of $\mathbb{R}_+^2$ in which the time points s and t, at which intersections take place, are allowed to vary. We derive two series representations of $\alpha_\varepsilon(x, \cdot)$ in terms of multiple Wiener–Itô integrals. In computing the norms of these expansions in quadratic Sobolev spaces with respect to the canonical Dirichlet structure on Wiener space, we shall exhibit integrals $I(k, \varepsilon)$, $k \in \mathbb{N}_0$, $\varepsilon > 0$, the behaviour of which is characteristic for convergence of $\alpha_\varepsilon(x, \cdot)$ as $\varepsilon \to 0$, or, if convergence does not take place, for the rate of explosion of $\alpha_\varepsilon(x, \cdot)$ as $\varepsilon \to 0$. In addition, their dependence on k determines the order of smoothness of self-intersection local time in terms of the Dirichlet structure. It will be the task of the following section to investigate the behaviour of $I(k, \varepsilon)$ for different configurations of A and B and thus to characterize the properties of the respective self-intersection local times.

Let $A = [\alpha, \beta]$ and $B = [\gamma, \delta] \subset [0,1]^2$ be such that $\alpha, \gamma > 0$. For $\varepsilon > 0$ and $x \in \mathbb{R}^d$ let

$$\alpha_\varepsilon(x, \cdot) = \int_B \int_A p_\varepsilon^d(W_t - W_s - x)\, ds\, dt.$$

The first representation of this approximation of self-intersection local time we give is the straightforward one following from the analysis in Imkeller, Perez–Abreu, Vives [4, Proposition 1 and its corollary] (see also Kuo [6], Nualart, Vives [8]).

Proposition 1. *For $x \in \mathbb{R}^d$ and $\varepsilon > 0$ we have*

$$\alpha_\varepsilon(x, \cdot) = \sum_{n_i=0}^{\infty} \int_B \int_A \prod_{i=1}^{d} \frac{1}{\sqrt{n_i!}} I_{n_i}^i \left(\left[\frac{1_{R_t} - 1_{R_s}}{\sqrt{\varepsilon + \lambda(R_t \triangle R_s)}} \right]^{\otimes n_i} \right)$$

$$\times H_{n_i} \left(\frac{x_i}{\sqrt{\varepsilon + \lambda(R_t \triangle R_s)}} \right) p_{\varepsilon + \lambda(R_t \triangle R_s)}^d(x)\, ds\, dt.$$

Working with the representation of Proposition 1 alone would mean that we have to distinguish too many cases, due to the many different possible configurations of R_t and R_s. The following decomposition of R_t will simplify this a lot.

Choose $\eta > 0$ such that $\underline{\eta} = (\eta, \eta) < \alpha, \gamma$. For $t \in [\underline{\eta}, 1]$ let then

$$R_t = C_t \cup D_t,$$

where

$$C_t = R_t \setminus [\underline{\eta}, t], \quad D_t = [\underline{\eta}, t].$$

Then we have

$$W_t = W(C_t) + W(D_t).$$

For $s \in A$, $t \in B$ and $x \in \mathbb{R}^d$ let now

$$\xi(s, t, x) = x - (W(D_t) - W(D_s)).$$

Then obviously the families

$$(\xi(s, t, x) \colon s \in A, t \in B, x \in \mathbb{R}^d) \quad \text{and} \quad (W(C_t) - W(C_s) \colon s \in A, t \in B)$$

are independent. Consequently, we obtain the following representation of $\alpha_\varepsilon(x, \cdot)$.

Proposition 2. *For $x \in \mathbb{R}^d$ and $\varepsilon > 0$ we have*

$$\alpha_\varepsilon(x, \cdot) = \sum_{n_i=0}^{\infty} \int_B \int_A \prod_{i=1}^{d} \frac{1}{\sqrt{n_i!}} I_{n_i}^i \left(\left[\frac{1_{C_t} - 1_{C_s}}{\sqrt{\varepsilon + \lambda(C_t \triangle C_s)}} \right]^{\otimes n_i} \right)$$

$$\times H_{n_i} \left(\frac{\xi(s, t, x)_i}{\sqrt{\varepsilon + \lambda(C_t \triangle C_s)}} \right) p_{\varepsilon + \lambda(C_t \triangle C_s)}^d(x)\, ds\, dt.$$

Proof. By the independence stated above, Proposition 1 and its corollary in Imkeller, Perez–Abreu, Vives [4] may be applied to give, for $s \in A$, $t \in B$, $x \in \mathbb{R}^d$, and $\varepsilon > 0$,

$$p_\varepsilon^d(W_t - W_s - x) = p_\varepsilon^d[W(C_t) - W(C_s) - (x - W(D_t) + W(D_s))]$$

$$= p_\varepsilon^d[W(C_t) - W(C_s) - \xi(s,t,x)]$$

$$= \sum_{n_i=0}^{\infty} \prod_{i=1}^{d} \frac{1}{\sqrt{n_i!}} I_{n_i}^i \left(\left[\frac{1_{C_t} - 1_{C_s}}{\sqrt{\varepsilon + \lambda(C_t \triangle C_s)}} \right]^{\otimes n_i} \right)$$

$$\times H_{n_i}\left(\frac{\xi(s,t,x)_i}{\sqrt{\varepsilon + \lambda(C_t \triangle C_s)}} \right) p_{\varepsilon+\lambda(C_t \triangle C_s)}^d(\xi(s,t,x)).$$

It remains to integrate in s and t. $\square$

Let us now, for some $\varrho \in \mathbb{R}$, take the norm of $\alpha_\varepsilon(x,\cdot)$ in the two representations with respect to the quadratic Sobolev space of order ϱ. We have

$$\|\alpha_\varepsilon(x,\cdot)\|_{2,\varrho}^2$$

$$= \sum_{k=0}^{\infty} (1+k)^\varrho \int_B \int_A \int_B \int_A \frac{\left[\int_{[0,1]^2} (1_{R_t} - 1_{R_s})(1_{R_v} - 1_{R_u}) \, d\lambda \right]^k}{\left[(\varepsilon + \lambda(R_t \triangle R_s))(\varepsilon + \lambda(R_v \triangle R_u)) \right]^{k/2}}$$

$$\times \sum_{n_1+\cdots+n_d=k} \prod_{i=1}^{d} H_{n_i}\left(\frac{x_i}{\sqrt{\varepsilon + \lambda(R_t \triangle R_s)}} \right) H_{n_i}\left(\frac{x_i}{\sqrt{\varepsilon + \lambda(R_v \triangle R_u)}} \right) \quad (1)$$

$$\times p_{\varepsilon+\lambda(R_t \triangle R_s)}^d(x) \, p_{\varepsilon+\lambda(R_v \triangle R_u)}^d(x) \, ds \, dt \, du \, dv.$$

If in a similar computation for the representation of Proposition 2 we employ in addition the estimates of Proposition 2 and 3 of Imkeller, Perez–Abreu, Vives [4], for $\alpha = \beta = 1/2$, we find a constant c such that, for $x \in \mathbb{R}^d$ and $\varepsilon > 0$,

$$\|\alpha_\varepsilon(x,\cdot)\|_{2,\varrho}^2$$

$$\leq c \sum_{k=0}^{\infty} (1+k)^\varrho k^{d/2-1} \quad (2)$$

$$\times \int_B \int_A \int_B \int_A \frac{\lambda((C_t \triangle C_s) \cap (C_v \triangle C_u))^k}{\left[(\varepsilon + \lambda(C_t \triangle C_s))(\varepsilon + \lambda(C_v \triangle C_u)) \right]^{(k+d)/2}} \, ds \, dt \, du \, dv.$$

To investigate the behaviour of $\alpha_\varepsilon(x,\cdot)$ as $\varepsilon \to 0$ in the following section, we shall argue as follows.

If we want to establish convergence of $\alpha_\varepsilon(x,\cdot)$ to

$$\alpha(x,\cdot) = \sum_{n_i=0}^{\infty} \int_B \int_A \prod_{i=1}^{d} \frac{1}{\sqrt{n_i!}} I_{n_i}^i \left(\left[\frac{1_{R_t} - 1_{R_s}}{\sqrt{\lambda(R_t \triangle R_s)}} \right]^{\otimes n_i} \right)$$

$$\times H_{n_i}\left(\frac{x_i}{\sqrt{\lambda(R_t \triangle R_s)}} \right) p_{\lambda(R_t \triangle R_s)}^d(x) \, ds \, dt \quad (3)$$

in $\mathbb{D}_{2,\varrho}$, then, due to dominated convergence, all we have to do is prove that, denoting by

$$\tilde{I}(k,\varepsilon) = \int_B \int_A \int_B \int_A \frac{\lambda((C_t \,\triangle\, C_s) \cap (C_v \,\triangle\, C_u))^k}{\left[(\varepsilon + \lambda(C_t \,\triangle\, C_s))(\varepsilon + \lambda(C_v \,\triangle\, C_u))\right]^{(k+d)/2}} \, ds \, dt \, du \, dv,$$

$k \in \mathbb{N}_0$, $\varepsilon \geq 0$, the characteristic integrals, we have

$$\tilde{I}(k,0) \leq c_k \quad \text{and} \quad \sum_{k=0}^{\infty} c_k (1+k)^\varrho k^{d/2-1} < \infty. \tag{4}$$

If, on the other side, in case $\alpha_\varepsilon(x,\cdot)$ does not converge as $\varepsilon \to 0$, we want to find its rate of explosion, we shall argue as follows in two steps. We first determine a deterministic function $f(\varepsilon)$ such that

$$\sup_{\varepsilon>0} \frac{\tilde{I}(k,\varepsilon)}{f(\varepsilon)^2} \leq c_k \quad \text{and} \quad \sum_{k=0}^{\infty} c_k (1+k)^\varrho k^{d/2-1} < \infty. \tag{5}$$

With (5) we will have proved that $\{\alpha_\varepsilon(x,\cdot)/f(\varepsilon) : \varepsilon > 0\}$ is bounded in $\mathbb{D}_{2,\varrho}$. To show in the second step that $f(\varepsilon)$ gives the right order of explosion, we then consider the characteristic integrals

$$\tilde{J}(k,\varepsilon) = \int_B \int_A \int_B \int_A \frac{\left[\int_{[0,1]^2} (1_{R_t} - 1_{R_s})(1_{R_v} - 1_{R_u}) \, d\lambda\right]^k}{\left[(\varepsilon + \lambda(R_t \,\triangle\, R_s))(\varepsilon + \lambda(R_v \,\triangle\, R_u))\right]^{(k+d)/2}} \, ds \, dt \, du \, dv,$$

$k \in \mathbb{N}_0$, $\varepsilon \geq 0$, appearing in (1). We show that at least for some k

$$\lim_{\varepsilon \to 0} \frac{\tilde{J}(k,\varepsilon)}{f(\varepsilon)^2} > 0. \tag{6}$$

Since the functions

$$\varepsilon \mapsto H_n\left(\frac{x}{\sqrt{\varepsilon+s}}\right) \exp\left(-\frac{x^2}{2(\varepsilon+s)}\right), \quad x \in \mathbb{R}^d, \ s \geq 0,$$

are continuous and uniformly bounded, (6) will be sufficient for proving

$$\lim_{\varepsilon \to 0} \frac{\|\alpha_\varepsilon(x,\cdot)\|_{2,\varrho}}{f(\varepsilon)} > 0, \tag{7}$$

i.e. that $f(\varepsilon)$ is the correct rate of divergence.

Let us start by performing some general estimates on $\tilde{I}(k,\varepsilon)$. First of all, note that by choice of $\underline{\eta}$ we have, for $s \in A$ and $t \in B$,

$$\lambda(C_t \,\triangle\, C_s) = \eta(|t_1 - s_1| + |t_2 - s_2|),$$

and, for $u, s \in A$ and $t, v \in B$,

$$
\lambda((C_t \bigtriangleup C_s) \cap (C_v \bigtriangleup C_u)) = \eta \big[\lambda([s_1 \wedge t_1, s_1 \vee t_1] \cap [u_1 \wedge v_1, u_1 \vee v_1]) \\
+ \lambda([s_2 \wedge t_2, s_2 \vee t_2] \cap [u_2 \wedge v_2, u_2 \vee v_2])\big].
$$

Hence by symmetry for $k \in \mathbb{N}_0$ and $\varepsilon \geq 0$,

$$
\tilde{I}(k, \varepsilon) = 16 \int_B \cdots \int_A 1_{\{s_1 < t_1, u_1 < v_1, s_2 < t_2, u_2 < v_2\}}
$$
$$
\times \frac{\big[\lambda([s_1, t_1] \cap [u_1, v_1]) + \lambda([s_2, t_2] \cap [u_2, v_2])\big]^k}{\big[(\frac{\varepsilon}{\eta} + t_1 - s_1 + t_2 - s_2)(\frac{\varepsilon}{\eta} + v_1 - u_1 + v_2 - u_2)\big]^{(k+d)/2}} \, ds \ldots dv. \quad (8)
$$

To avoid the distinction of too many cases in (8), we shall use the following general estimate by which we still do not lose too much, as we shall see later.

Lemma 1. *Let* $S_1, S_2, T_1, T_2 \in \mathcal{B}([0,1]^2)$ *and* $\varepsilon \geq 0$. *Then*

$$
[\lambda(S_1 \cap T_1) + \lambda(S_2 \cap T_2) + \varepsilon][\lambda(S_1 \cup T_1) + \lambda(S_2 \cup T_2) + \varepsilon] \\
\leq 2[\lambda(S_1) + \lambda(S_2) + \varepsilon][\lambda(T_1) + \lambda(T_2) + \varepsilon].
$$

Proof. We have by additivity

$$
[\lambda(S_1 \cap T_1) + \lambda(S_2 \cap T_2) + \varepsilon][\lambda(S_1 \cup T_1) + \lambda(S_2 \cup T_2) + \varepsilon]
$$
$$
= [\lambda(S_1 \cap T_1) + \lambda(S_2 \cap T_2) + \varepsilon]
$$
$$
\times [\lambda(S_1) + \lambda(T_1) - \lambda(S_1 \cap T_1) + \lambda(S_2) + \lambda(T_2) - \lambda(S_2 \cap T_2) + \varepsilon]
$$
$$
\leq [\lambda(S_1 \cap T_1) + \lambda(S_2 \cap T_2) + \varepsilon][\lambda(S_1) + \lambda(S_2) + \varepsilon]
$$
$$
+ [\lambda(S_1 \cap T_1) + \lambda(S_2 \cap T_2) + \varepsilon][\lambda(T_1) + \lambda(T_2)]
$$
$$
\leq [\lambda(T_1) + \lambda(T_2) + \varepsilon][\lambda(S_1) + \lambda(S_2) + \varepsilon]
$$
$$
+ [\lambda(S_1) + \lambda(S_2) + \varepsilon][\lambda(T_1) + \lambda(T_2)].
$$

This obviously implies the desired inequality. $\square$

Now observe that $\eta > 0$ is a fixed number, and take

$$
S_1 = [s_1, t_1], \quad T_1 = [u_1, v_1], \quad \text{etc.}
$$

to see that instead of (8) it is enough to estimate the integrals

$$
I(k, \varepsilon) = \int_B \int_A \int_B \int_A 1_{\{s_1 < t_1, u_1 < v_1, s_2 < t_2, u_2 < v_2\}}
$$
$$
\times \frac{[\lambda(S_1 \cap T_1) + \lambda(S_2 \cap T_2)]^k \, ds \, dt \, du \, dv}{\big[(\varepsilon + \lambda(S_1 \cap T_1) + \lambda(S_2 \cap T_2))(\varepsilon + \lambda(S_1 \cup T_1) + \lambda(S_2 \cup T_2))\big]^{(k+d)/2}}. \quad (9)
$$

Decomposing $I(k, \varepsilon)$ further according to the three possible constellations

$$S_1 \cap T_1 \neq \emptyset \neq S_2 \cap T_2; \quad S_1 \cap T_1 \neq \emptyset, S_2 \cap T_2 = \emptyset \quad \text{and} \quad S_1 \cap T_1 = \emptyset, S_2 \cap T_2 \neq \emptyset,$$

and taking into account the symmetry of the latter two contributions, we see that we have to finally investigate the integrals

$$I(k, \varepsilon, 1) = \int_B \cdots \int_A 1_{\{a_i < b_i < c_i < d_i, i=1,2\}}$$
$$\times \frac{(c_1 - b_1 + c_2 - b_2)^k \, da_1 \ldots dd_2}{\left[(\varepsilon + c_1 - b_1 + c_2 - b_2)(\varepsilon + d_1 - a_1 + d_2 - a_2) \right]^{(k+d)/2}} \qquad (10)$$

and

$$I(k, \varepsilon, 2) = \int_B \cdots \int_A 1_{\{a_i < b_i < c_i < d_i, i=1,2\}}$$
$$\times \frac{(c_2 - b_2)^k \, da_1 \ldots dd_2}{\left[(\varepsilon + c_2 - b_2)(\varepsilon + b_1 - a_1 + d_1 - c_1 + d_2 - a_2) \right]^{(k+d)/2}} \qquad (11)$$

for $\varepsilon \geq 0$ and $k \in \mathbb{N}_0$. In the case $k = 0$ things are much simpler. So here we shall consider occasionally

$$I(0, \varepsilon) = \left(\int_B \int_A 1_{\{s_1 < t_1, s_2 < t_2\}} \frac{1}{(\varepsilon + t_1 - s_1 + t_2 - s_2)^{d/2}} \, ds \, dt \right)^2, \quad \varepsilon \geq 0. \qquad (12)$$

We are ready to start studying particular configurations of the intervals A and B now. We shall concentrate on four configurations, remarking that hereby we have covered all interesting cases. By cutting of rectangles, switching the axes of $[0,1]^2$ or time reversal in one time direction we shall always end up with the configurations considered.

§4 SELF-INTERSECTIONS
OF THE BROWNIAN SHEET IN DIFFERENT RECTANGLES

The four configurations of A and B we shall study can be described in the following way. For the first one, we take $A = B$. In the second, A and B are adjacent rectangles with one vertex in common (parallel to the x-axis), in the third, A and B are "almost disjoint", and have just one edge $\gamma = \beta$ in common. Finally, in the fourth configuration, A and B are disjoint. We shall see that as a consequence of the increase of "disjointness" of A and B, the "independence" of W_s and W_t will increase since s runs in A and t in B. This leads to a stepwise increase of the critical dimension below which self-intersection local times exist. More formally, we shall speak of

    case 1, if $A = B$,
    case 2, if $\alpha_1 = \gamma_1$, $\beta_1 = \delta_1$, $\beta_2 = \gamma_2$,
    case 3, if $\beta = \gamma$,
    case 4, if $\alpha_1 = \gamma_1$, $\beta_1 = \delta_1$, $\beta_2 < \gamma_2$ or $\beta < \gamma$.

Before we discuss the behaviour of $\alpha_\varepsilon(x, \cdot)$ as $\varepsilon \to 0$ in the four cases, we have to look at the characteristic integrals. We start with order zero.

Proposition 3.

(1) *We have $I(0,0) < \infty$ for $d < 4$ in case 1, $d < 6$ in case 2, $d < 8$ in case 3 and all $d \in \mathbb{N}$ in case 4.*

(2) *Moreover, there is a constant c such that, for $\varepsilon > 0$,*

$$\frac{I(0,\varepsilon)}{f(\varepsilon)} \le c,$$

where $f(\varepsilon) = (\ln 1/\varepsilon)^2$ for $d = 4$ in case 1, $d = 6$ in case 2, $d = 8$ in case 3, $f(\varepsilon) = \varepsilon^{4-d}$ for $d > 4$ in case 1, $f(\varepsilon) = \varepsilon^{6-d}$ for $d > 6$ in case 2 and $f(\varepsilon) = \varepsilon^{8-d}$ for $d > 8$ in case 3.

Proof. We concentrate on the more difficult argument needed to prove part (2). Indeed, in case 1 for $d \ge 4$,

$$\int_B \int_A 1_{\{s_1 < t_1, s_2 < t_2\}} \frac{1}{(\varepsilon + t_1 - s_1 + t_2 - s_2)^{d/2}} \, ds \, dt$$

is asymptotically equivalent with

$$\int_{\alpha_1}^{\beta_1} \int_{\alpha_1}^{\beta_1} 1_{\{s_1 < t_1\}} \frac{1}{(\varepsilon + t_1 - s_1)^{d/2-1}} \, ds_1 \, dt_1.$$

If $d = 4$, this is equivalent with $\ln \frac{1}{\varepsilon}$, if $d > 4$, with $\varepsilon^{2-d/2}$, as $\varepsilon \to 0$. In case 2 and for $d \ge 6$ we observe by integrating in $s_2 < \beta_2 < t_2$ first that

$$\int_B \int_A 1_{\{s_1 < t_1, s_2 < t_2\}} \frac{1}{(\varepsilon + t_1 - s_1 + t_2 - s_2)^{d/2}} \, ds \, dt$$

is asymptotically equivalent with

$$\int_{\alpha_1}^{\beta_1} \int_{\alpha_1}^{\beta_1} 1_{\{s_1 < t_1\}} \frac{1}{(\varepsilon + t_1 - s_1)^{d/2-2}} \, ds_1 \, dt_1,$$

and this in turn with $\ln \frac{1}{\varepsilon}$ if $d = 6$, and with $\varepsilon^{3-d/2}$ if $d > 6$. It is clear by now how case 3 has to be treated. Here we integrate in $s_2 < \beta_2 < t_2$ first, and then in $s_1 < \beta_1 < t_1$ to find out that the critical dimension has again increased by 2. Now consult (12) to conclude. □

We now consider the higher degree integrals $I(k, \varepsilon)$ with $k \in \mathbb{N}$.

Proposition 4.

(1) *There exists a constant c such that, for all $k \in \mathbb{N}$, $I(k,0)k^4 \leq c$ for $d < 5$ in case 1, for $d < 6$ in case 2, for $d < 8$ in case 3, and for all $d \in \mathbb{N}$ in case 4.*

(2) *Moreover, there exists a constant c' such that, for all $k \in \mathbb{N}$ and $\varepsilon > 0$,*

$$\frac{I(k,\varepsilon)k^4}{f(\varepsilon)} \leq c',$$

where $f(\varepsilon) = \ln 1/\varepsilon$ for $d = 5$ in case 1, $d = 6$ in case 2, $d = 8$ in case 3, $f(\varepsilon) = \varepsilon^{5-d}$ for $d > 5$ in case 1, $f(\varepsilon) = \varepsilon^{6-d}$ for $d > 6$ in case 2, and $f(\varepsilon) = \varepsilon^{8-d}$ for $d > 8$ in case 3.

Proof. Again we concentrate on proving the more difficult second statement. We also concentrate on combinations $\frac{k+d}{2} \geq 5$ resp. 6 resp. 8 in the three cases to deal with, remarking that the remaining integrals are similar to deal with, except for the appearance of a logarithm in the process of integration. Note first that in cases 1 and 2 $S_1 \cap T_1 = \varnothing$ is possible. Hence we have to deal with $I(k,\varepsilon,1)$ and $I(k,\varepsilon,2)$ in these cases. In case 3, however, it is enough to treat $I(k,\varepsilon,1)$, since there $S_1 \cap T_1 \neq \varnothing$ and $S_2 \cap T_2 \neq \varnothing$ is the only possibility, due to the relative position of A and B. In all cases, we integrate in d_1, a_1, d_2, a_2 first to obtain

$$\begin{aligned}
I(k,\varepsilon,1) &\leq \frac{c''}{k^4} \int_B \int_A 1_{\{b_1 < c_1, b_2 < c_2\}} \frac{(c_1 - b_1 + c_2 - b_2)^k}{(\varepsilon + c_1 - b_1 + c_2 - b_2)^{k+d-4}} \, db \, dc \\
&\leq \frac{c''}{k^4} \int_B \int_A 1_{\{b_1 < c_1, b_2 < c_2\}} \frac{1}{(\varepsilon + c_1 - b_1 + c_2 - b_2)^{d-4}} \, db \, dc \qquad (13) \\
&= \frac{c''}{k^4} H(\varepsilon).
\end{aligned}$$

For the remaining version, we integrate in $a_2, d_2, b_1, a_1, c_1, d_1$ first to obtain the inequality

$$\begin{aligned}
I(k,\varepsilon,2) &\leq \frac{c^{(3)}}{k^4} \int_{\alpha_2}^{\beta_2} \int_{\gamma_2}^{\delta_2} 1_{\{b_2 < c_2\}} \frac{(c_2 - b_2)^k}{(\varepsilon + c_2 - b_2)^{k+d-4}} \, dc_2 \, db_2 \\
&\leq \frac{c^{(3)}}{k^4} \int_{\alpha_2}^{\beta_2} \int_{\gamma_2}^{\delta_2} 1_{\{b_2 < c_2\}} \frac{1}{(\varepsilon + c_2 - b_2)^{d-4}} \, dc_2 \, db_2 \qquad (14) \\
&= \frac{c^{(3)}}{k^4} K(\varepsilon).
\end{aligned}$$

Here c'', $c^{(3)}$ are constants not depending on $\varepsilon > 0$ or $k \in \mathbb{N}$. It remains just to discuss the functions H and K in the different cases. But obviously, $H(\varepsilon) \leq K(\varepsilon)$ for $\varepsilon > 0$, and therefore it remains to discuss K in cases 1 and 2, H in case 3.

Let us consider case 1 now. Suppose that $d \geq 5$. Then obviously $K(\varepsilon)$ is asymptotically equivalent with $\ln \frac{1}{\varepsilon}$ if $d = 5$ and with ε^{5-d} if $d > 5$.

In case 2, supposing $d \geq 6$, we integrate in b_2, c_2 and observe that $b_2 < \beta_2 < c_2$. This yields easily that $K(\varepsilon)$ is asymptotically equivalent with $\ln \frac{1}{\varepsilon}$ if $d = 6$, and with ε^{6-d} if $d > 6$.

Finally, in case 3, we just have to study H. Here we integrate in b_1, c_1, b_2, c_2 and observe that $b_1 < \beta_1 < c_1$ and $b_2 < \beta_2 < c_2$. Hence we get the asymptotic equivalents of $H(\varepsilon)$ as $\ln \frac{1}{\varepsilon}$ if $d = 8$, and ε^{8-d} if $d > 8$. This completes the proof. □

As mentioned before (6), we also have to verify that the orders of divergence obtained in the preceding propositions are correct. We therefore investigate the integrals $\tilde{J}(k, \varepsilon)$ for particular k, and on particular domains D of integration contained in $B \times A \times B \times A$. If we call these integrals $J(k, \varepsilon)$, then

$$\lim_{\varepsilon \to 0} \frac{J(k, \varepsilon)}{f(\varepsilon)} > 0$$

will imply that

$$\lim_{\varepsilon \to 0} \frac{\tilde{J}(k, \varepsilon)}{f(\varepsilon)} > 0,$$

and we will have proved that $f(\varepsilon)$ is the correct order of divergence.

Proposition 5. *For $\varepsilon > 0$ let*

$$J(0, \varepsilon) = \left(\int_B \int_A 1_{\{s<t\}} \frac{1}{(\varepsilon + t_1 t_2 - s_1 s_2)^{d/2}} \, ds \, dt \right)^2 .$$

Then we have

$$\lim_{\varepsilon \to 0} \frac{J(0, \varepsilon)}{f(\varepsilon)} > 0,$$

where $f(\varepsilon) = (\ln \frac{1}{\varepsilon})^2$ for $d = 4$ in case 1, $d = 6$ in case 2, $d = 8$ in case 3, $f(\varepsilon) = \varepsilon^{4-d}$ for $d > 4$ in case 1, $f(\varepsilon) = \varepsilon^{6-d}$ for $d > 6$ in case 2, $f(\varepsilon) = \varepsilon^{8-d}$ for $d > 8$ in case 3.

Proof. We have

$$t_1 t_2 - s_1 s_2 = (t_1 - s_1)t_2 + s_1(t_2 - s_2) \begin{cases} \leq t_1 - s_1 + t_2 - s_2, \\ \geq \eta[t_1 - s_1 + t_2 - s_2]. \end{cases}$$

Hence $I(0, \varepsilon)$ and $J(0, \varepsilon)$ are asymptotically equivalent and the desired result follows from the proof of Proposition 3. □

Proposition 6. *For $\varepsilon > 0$ and $k \in \mathbb{N}$ let*

$$J(k, \varepsilon) = \int_B \int_A \int_B \int_A 1_{\{s_1 < t_1 < u_1 < v_1, s_2 < u_2 < v_2 < t_2\}}$$

$$\times \frac{\left[\int_{[0,1]^2} (1_{R_t} - 1_{R_s})(1_{R_v} - 1_{R_u}) \, d\lambda\right]^k}{\left[(\varepsilon + \lambda(R_t \bigtriangleup R_s))(\varepsilon + \lambda(R_v \bigtriangleup R_u))\right]^{(k+d)/2}} \, ds \, dt \, du \, dv.$$

Then

$$\lim_{\varepsilon \to 0} \frac{J(k, \varepsilon)}{f(\varepsilon)} > 0$$

for $\frac{k+d}{2} > 6$, where $f(\varepsilon) = \ln \frac{1}{\varepsilon}$ for $d = 5$ in case 1 and for $d = 6$ in case 2, $f(\varepsilon) = \varepsilon^{5-d}$ for $d > 5$ in case 1, and $f(\varepsilon) = \varepsilon^{6-d}$ for $d > 6$ in case 2.

Proof. The estimate of the preceding proof shows that $J(k, \varepsilon)$ is asymptotically equivalent with

$$\int_B \int_A \int_B \int_A 1_{\{s_1 < t_1 < u_1 < v_1, s_2 < u_2 < v_2 < t_2\}}$$

$$\times \frac{(v_2 - u_2)^k}{\left[(\varepsilon + t_1 - s_1 + t_2 - s_2)(\varepsilon + v_1 - u_1 + v_2 - u_2)\right]^{(k+d)/2}} \, ds \, dt \, du \, dv.$$

Now integrate in t_1, s_1, t_2, s_2 and v_1, u_1 to see that this integral is asymptotically equivalent with

$$\int_{\gamma_2}^{\delta_2} \int_{\alpha_2}^{\beta_2} 1_{\{u_2 < v_2\}} \frac{(v_2 - u_2)^k}{(\varepsilon + v_2 - u_2)^{k+d-4}} \, du_2 \, dv_2$$

$$= \sum_{l=0}^{k} (-\varepsilon)^l \binom{k}{l} \int_{\gamma_2}^{\delta_2} \int_{\alpha_2}^{\beta_2} 1_{\{u_2 < v_2\}} (\varepsilon + v_2 - u_2)^{4-l-d} \, du_2 \, dv_2.$$

In case 1 for $d \geq 5$, integration in v_2 and u_2 shows that this is asymptotically equivalent with $\ln \frac{1}{\varepsilon}$ if $d = 5$, and with ε^{5-d} if $d > 5$. In case 2 for $d \geq 6$, integration in v_2 and u_2, observing that $u_2 < \beta_2 < v_2$, leads to the same conclusion with $d = 6$ instead of 5. This completes the proof. $\square$

Proposition 7. *For $\varepsilon > 0$ and $k \in \mathbb{N}$ let*

$$J(k, \varepsilon) = \int_B \int_A \int_B \int_A 1_{\{u_1 < s_1 < v_1 < t_1, u_2 < s_2 < v_2 < t_2\}}$$

$$\times \frac{\left[\int_{[0,1]^2} (1_{R_t} - 1_{R_s})(1_{R_v} - 1_{R_u}) \, d\lambda\right]^k}{\left[(\varepsilon + \lambda(R_t \bigtriangleup R_s))(\varepsilon + \lambda(R_v \bigtriangleup R_u))\right]^{(k+d)/2}} \, ds \, dt \, du \, dv.$$

Then

$$\lim_{\varepsilon \to 0} \frac{J(k, \varepsilon)}{f(\varepsilon)} > 0$$

for $\frac{k+d}{2} > 8$, where $f(\varepsilon) = \ln \frac{1}{\varepsilon}$ if $d = 8$ and $f(\varepsilon) = \varepsilon^{8-d}$ if $d > 8$ in case 3.

Proof. The arguments are completely analogous to the preceding proof. The only difference is that $s_1 < \beta_1 < v_1$ and $s_2 < \beta_2 < v_2$. $\square$

We have now developed everything needed for the proof of the main results.

4.1 The case $A = B$. We first state the results on the existence of self-intersection local times, then on the rates of explosion in case of non-existence.

Theorem 1.

(1) *Let $d < 4$. Then, for any $x \in \mathbb{R}^d$,*

$$\alpha(x, \cdot) = \lim_{\varepsilon \to 0} \alpha_\varepsilon(x, \cdot)$$

exists in $\mathbb{D}_{2,\varrho}$ for any $\varrho < 4 - d/2$ and is given by (3).
(2) *Let $d < 5$. Then, for any $x \in \mathbb{R}^d$,*

$$\gamma(x, \cdot) = \lim_{\varepsilon \to 0} [\alpha_\varepsilon(x, \cdot) - \mathrm{E}(\alpha_\varepsilon(x, \cdot))]$$

exists in $\mathbb{D}_{2,\varrho}$ for $\varrho < 4 - d/2$.

Proof. Consult statements (1) in Propositions 3 and 4 for case 1, and remember that $\mathrm{E}(\alpha_\varepsilon(x, \cdot))$ is the first term in the development of $\alpha_\varepsilon(x, \cdot)$. The order of smoothness follows from these propositions as well and the fact that in (4) resp. (5) we can take $c_k = ck^{-4}$, and

$$\sum_{k=1}^{\infty} (1 + k)^\varrho k^{d/2-1} k^{-4} < \infty \iff \varrho + \frac{d}{2} - 5 < -1,$$

i. e. if $\varrho < 4 - d/2$. $\square$

Remarks.

(1) The second part of Theorem 1 claims the existence of a "renormalized" intersection local time in case $d = 4$. This is an analogue of Varadhan's renormalization for the intersection local time of two-dimensional Brownian motion (see Varadhan [16], Le Gall [7]).
(2) With arguments as in Imkeller, Perez–Abreu, Vives [4], we could have deduced a sharper existence result of $\lim_{\varepsilon \to 0} \alpha_\varepsilon(x, \cdot)$ than the one given in Theorem 1. Indeed, it could be shown that this limit exists for $d < 8$ as a function, and for $d \geq 8$ as a distribution on Wiener space. We shall not go into these refinements in the present paper.

Theorem 2.

(1) *Let $d \geq 4$. Then for any $x \in \mathbb{R}^d$ there is a constant $c_x > 0$ such that*

$$\lim_{\varepsilon \to 0} \frac{\alpha_\varepsilon(x, \cdot)}{f(\varepsilon)} = c_x$$

in $\mathbb{D}_{2,\varrho}$ for $\varrho < 4 - d/2$, where

$$f(\varepsilon) = \begin{cases} \ln 1/\varepsilon & \text{for } d = 4, \\ \varepsilon^{(4-d)/2} & \text{for } d > 4. \end{cases}$$

(2) *Let $d \geq 5$. Then, for any $x \in \mathbb{R}^d$,*

$$\left\{ \frac{\alpha_\varepsilon(x, \cdot) - \mathrm{E}(\alpha_\varepsilon(x, \cdot))}{f(\varepsilon)} : \varepsilon > 0 \right\}$$

is bounded in $\mathbb{D}_{2,\varrho}$ for any $\varrho < 4 - d/2$, where

$$f(\varepsilon) = \begin{cases} \sqrt{\ln 1/\varepsilon} & \text{for } d = 5, \\ \varepsilon^{(5-d)/2} & \text{for } d > 5. \end{cases}$$

Moreover, the limit of the $(2, \varrho)$-norms of these random variables as $\varepsilon \to 0$ is non-trivial.

Proof. This time, we have to quote the second parts of Propositions 3 and 4, as far as they refer to case 1. To see that the rates are sharp, consult Propositions 5 and 6. □

4.2 The case: A and B have a common vertex. Here the critical dimensions increase already by two, as our theorems show.

Theorem 3. *Let $d < 6$. Then, for any $x \in \mathbb{R}^d$,*

$$\alpha(x, \cdot) = \lim_{\varepsilon \to 0} \alpha_\varepsilon(x, \cdot)$$

exists in $\mathbb{D}_{2,\varrho}$ for any $\varrho < 4 - d/2$ and is given by (3).

Proof. Use the same arguments as in the proof of Theorem 1 and replace case 1 by case 2. □

The dichotomy of Theorem 2 exists in the case considered only in dimension 6.

Theorem 4. *Let $d = 6$.*

(1) *Then for any $x \in \mathbb{R}^d$ there exists a constant $c_x > 0$ such that*

$$\lim_{\varepsilon \to 0} \frac{\alpha_\varepsilon(x, \cdot)}{\ln 1/\varepsilon} = c_x$$

in $\mathbb{D}_{2,\varrho}$ for $\varrho < 4 - d/2$.

(2) *For any $x \in \mathbb{R}^d$ the set*

$$\left\{ \frac{\alpha_\varepsilon(x, \cdot) - \mathrm{E}(\alpha_\varepsilon(x, \cdot))}{\sqrt{\ln 1/\varepsilon}} : \varepsilon > 0 \right\}$$

is bounded in $\mathbb{D}_{2,\varrho}$ for $\varrho < 4 - d/2$. Moreover, the limit of the $(2, \varrho)$-norms of these random variables as $\varepsilon \to 0$ is non-trivial.

Proof. Use statement (2) of Propositions 3 and 4 for case 2 with $d = 6$, and also Propositions 5 and 6. □

Theorem 5. *Let $d > 6$. Then for any $x \in \mathbb{R}^d$ the set*

$$\left\{ \varepsilon^{d-6} \alpha_\varepsilon(x, \cdot) : \varepsilon > 0 \right\}$$

is bounded in $\mathbb{D}_{2,\varrho}$ for any $\varrho < 4 - d/2$. The limit of the $(2, \varrho)$-norms of these random variables as $\varepsilon \to 0$ is non-trivial.

Proof. See the appropriate case in Propositions 3 to 6. □

Remark. In case $d = 6$ we have a "law of large numbers" statement (part (1) of Theorem 4), as well as a weak form of a "central limit theorem" type statement (part (2) of Theorem 4). This is not the case for $d > 6$, as Theorem 5 shows. In our purely analytic and computational approach, we have no intuitive explanation for this behaviour.

4.3 The case: A and B have a common edge. The decrease of the dimension of $A \cap B$ by one again entails an increase of the critical dimension by two. The theorems we shall give are complete analogues of the theorems of the preceding subsection, and so are their proofs.

Theorem 6. *Let $d < 8$. Then, for any $x \in \mathbb{R}^d$,*

$$\alpha(x, \cdot) = \lim_{\varepsilon \to 0} \alpha_\varepsilon(x, \cdot)$$

exists in $\mathbb{D}_{2,\varrho}$ for any $\varrho < 4 - d/2$ and is given by (3).

Theorem 7. *Let $d = 8$.*

(1) *Then for any $x \in \mathbb{R}^d$ there exists a constant $c_x > 0$ such that*

$$\lim_{\varepsilon \to 0} \frac{\alpha_\varepsilon(x, \cdot)}{\ln 1/\varepsilon} = c_x$$

in $\mathbb{D}_{2,\varrho}$ for $\varrho < 4 - d/2$.
(2) *For any $x \in \mathbb{R}^d$ the set*

$$\left\{ \frac{\alpha_\varepsilon(x, \cdot) - \mathrm{E}(\alpha_\varepsilon(x, \cdot))}{\sqrt{\ln 1/\varepsilon}} : \varepsilon > 0 \right\}$$

is bounded in $\mathbb{D}_{2,\varrho}$ for $\varrho < 4 - d/2$. Moreover, the limit of the $(2, \varrho)$-norms of these random variables as $\varepsilon \to 0$ is non-trivial.

Theorem 8. *Let $d > 8$. Then for any $x \in \mathbb{R}^d$ the set*

$$\{\, \varepsilon^{d-8}\alpha_\varepsilon(x, \cdot) \colon \varepsilon > 0 \,\}$$

is bounded in $\mathbb{D}_{2,\varrho}$ for any $\varrho < 4 - d/2$. The limit of the $(2, \varrho)$-norms of these random variables as $\varepsilon \to 0$ is non-trivial.

4.4 The case: A and B are disjoint. We finally touch a situation of which a special case has been considered in Rosen [10], [11]. Here no explosions are possible any more.

Theorem 9. *For any $x \in \mathbb{R}^d$ we have that*

$$\alpha(x, \cdot) = \lim_{\varepsilon \to 0} \alpha_\varepsilon(x, \cdot)$$

exists in $\mathbb{D}_{2,\varrho}$ for any $\varrho < 4 - d/2$, hence is a function if and only if $d < 8$.

Proof. This is a consequence of Propositions 3 and 4, part (1), and the fact that

$$\sum_{k=1}^{\infty} (1 + k)^\varrho k^{d/2-1} k^{-4} < \infty \iff \varrho < 4 - d/2,$$

which has been discussed in the proof of Theorem 1. □

Remark. Theorem 9 corresponds to a result of Rosen [10], [11] in the case $\beta < \gamma$.

REFERENCES

1. R. Bass and D. Koshnevisan, *Intersection local times and Tanaka formulas*, Preprint, Univ. of Washington, 1992.
2. N. Bouleau and F. Hirsch, *Dirichlet forms and analysis on Wiener space*, W. de Gruyter, Berlin, 1991.
3. E. Dynkin, *Self-intersection gauge for random walks and for Brownian motion*, Ann. Probab. **16** (1988), 1–57.
4. P. Imkeller, V. Perez-Abreu, and J. Vives, *Chaos expansions of double intersection local time of Brownian motion in $\mathbb{R}^d$ and renormalization* (to appear in: Stoch. Proc. Appl.).
5. P. Imkeller and F. Weisz, *The asymptotic behaviour of local times and occupation integrals of the N-parameter Wiener process in $\mathbb{R}^d$*, Probab. Th. Rel. Fields **98** (1994), 47–75.
6. H. H. Kuo, *Donsker's delta function as a generalized Brownian functional and its application*, Lecture Notes in Control and Information Sciences, vol. 49, Springer-Verlag, Berlin, 1983, pp. 167–178.
7. J. F. Le Gall, *Sur le temps local d'intersection du mouvement brownien plan et la méthode de renormalisation de Varadhan*, Sém. de Prob. XIX 1983/84. LNM 1123, Springer-Verlag, Berlin, 1985, pp. 314–331.
8. D. Nualart and J. Vives, *Chaos expansion and local times*, Publicacions Matematiques **36**, 2 (1992), 827–836.
9. J. Rosen, *A local time approach to the self-intersections of Brownian paths in space*, Comm. Math. Phys. **88** (1983), 327–338.
10. J. Rosen, *Stochastic integrals and intersections of Brownian sheets*, Preprint, Univ. of Massachusetts.

11. J. Rosen, *Self-intersections of random fields*, Ann. Probab. **12** (1984), 108–119.
12. J. Rosen, *Tanaka's formula and renormalization for intersections of planar Brownian motion*, Ann. Probab. **14** (1986), 1245–1251.
13. J. Rosen and M. Yor, *Tanaka formulae and renormalization for triple intersections of Brownian motion in the plane*, Preprint.
14. N. R. Shieh, *White noise analysis and Tanaka formula for intersections of planar Brownian motion*, Nagoya Math. J. **122** (1991), 1–17.
15. K. Szymanzik, *Euclidean quantum field theory*, In: Local Quantum Theory (R. Jost, ed.), Academic Press, New York, 1969.
16. S. R. S. Varadhan, *Appendix to "Euclidean quantum field theory" by K. Szymanzik*, In: Local Quantum Theory (R. Jost, ed.), Academic Press, New York, 1969.
17. H. Watanabe, *The local time of self-intersections of Brownian motions as generalized Brownian functionals*, Letters in Math. Physics **23** (1991), 1–9.
18. S. Watanabe, *Lectures on stochastic differential equations and Malliavin calculus*, Springer-Verlag, Berlin, 1984.
19. R. Wolpert, *Wiener path intersection and local time*, J. Func. Anal. **30** (1978), 329–340.
10. M. Yor, *Sur la représentation comme intégrales stochastiques des temps d'occupation du mouvement Brownian dans $\mathbb{R}^d$*, Sém. de Prob. XX. 1984/85. LNM 1204, Springer-Verlag, Berlin, 1986, pp. 543–552.

PETER IMKELLER, LABORATOIRE DE MATHÉMATIQUE, UNIVERSITÉ DE FRANCHE-COMTÉ, 16, ROUTE DE GRAY, F-25030 BESANÇON CÉDEX, FRANCE

FERENC WEISZ, DEPARTMENT OF NUMERICAL ANALYSIS, EÖTVÖS L. UNIVERSITY, BOGDÁNFY U. 10/B, H-1117 BUDAPEST, HUNGARY

Progress in Probability, Vol. 36
© 1995 Birkhäuser Verlag Basel/Switzerland

DENSITY ESTIMATES FOR STOCHASTIC
PARTIAL DIFFERENTIAL EQUATIONS

RÉMI LÉANDRE AND FRANCESCO RUSSO

ABSTRACT. A survey is given on the density estimates for the law density of the solution to a stochastic PDE at a fixed point. A new result is constituted by the Davies type estimates for the stochastic wave equation.

§0 INTRODUCTION

Let L be a second order differential elliptic operator over $\mathbb{R}^n$ of the following type

$$\sum_{i,j=1}^{m} a_{ij}(t)\frac{\partial^2}{\partial x_i \partial x_j}, \quad a_{ij} \in C_b^\infty.$$

The ellipticity means that the matrix (a_{ij}) is strictly positive defined and so it has an inverse $(g_{ij}) = (a_{ij})^{-1}$. Let $(x,y) \mapsto d(x,y)$ be the distance associated with this matrix, that is to say:

$$d^2(x,y) = \sum_{i,j=1}^{m} g_{ij}(x_i - y_i)(x_j - y_j).$$

L is the generator of the diffusion solving the following stochastic differential equation:

$$dX_t = \sum_{j=1}^{m} A_j(X_t)\,dW_t^j,$$

$$X_0 = x, \tag{0.1}$$

where $(W^1, \ldots, W^m)$ is a m-dimensional Brownian motion, and $A_1, \ldots, A_m$ are vector fields $\mathbb{R}^n \to \mathbb{R}^n$, $m \geq n$, such that

$$(a_{ij}) = AA^T,$$

1991 *Mathematics Subject Classification.* 35C20, 35L05, 60F10, 60G60, 60H07.

Key words and phrases. Stochastic partial differential equations, large deviations, stochastic calculus of variations, density estimates, asymptotic expansion.

A being the $n \times m$ matrix whose columns are A_j vectors. In fact d can be also characterized as solution of the following problem

$$d^2(x,y) = \inf_{y_1(x,h)=y} \int_0^1 \dot{h}^2(s)\,ds \qquad (0.2)$$

where $y(x,h)$ solves the following skeleton problem

$$y_t(x,h) = x + \int_{[0,t]} \left\{ \sum_{j=1}^m A_j(y_r(x,h))\dot{h}_j(r) \right\} dr. \qquad (0.3)$$

The semigroup associated with L with x as a starting point, has a density $(p_t(x,\cdot))$. There are three types of estimates in relation with the distance d.

0.1 The Varadhan estimates. By mean of large deviations phenomena, the heat kernel $(p_t(x,\cdot))$ has an exponential decay when $t \to 0$: in this case one is interested in the behaviour of $\log p_t(x,\cdot)$ when $t \to 0$ and one is not bothered by the explosion of the heat kernel in small time. The result is that $t \log p_t$ behaves as $-d^2(x,y)$. This translates the fact that in small time, the diffusion concentrates on the geodesic going from x to y. These estimates are uniform even when L is hypoelliptic. We refer to [3], [4], [16], [17], [18] for such kind of estimates: the proof makes use of Malliavin calculus.

0.2 The Aronson estimates. They are upper and lower bounds of the form: $1/\sqrt{t}$ with a certain power times an exponential term of the type $-d^2(x,y)/(ct)$ where c is close to 2. One approach has been proposed by [10] using PDE's methods. The upper bound is also called Davies estimate.

The most complete approach by using Malliavin calculus (see [24]) is given by [14]. Concerning the hypoelliptic case there is no complete work in this direction when there is a bad drift inside the operator; however [18], [20], [33] study the blow up in terms of volume of the hypoelliptic balls related to d.

0.3 The estimates of Greiner–Minakshisundaram–Pleijel. These are in fact the product of a term of the type $\exp(-d^2(x,y)/(2t))$ times one term of the type $1/\sqrt{t}$ at a certain power $N(x,y)$ times a Taylor expansion. They concern a neighborhood of the diagonal: $y = x$ so that $d^2(x,y) = 0$.

These estimates have been first obtained by analytical tools by [13], [26]. The pionneer work using probabilistic methods has been performed by [25] in the elliptic case; successively, we have [37].

For the hypoelliptic case we refer to [3], [19], [32] (on the diagonal) and [4], [17], [34] (outside the diagonal). On the diagonal the behaviour is not exponential. Another non-exponential behaviour can be found in [12] where the authors illustrate a non-exponential decay in time for $p_t(x,y)$ where y belongs to a neighborhood of y_0 such that $p_t(x,y_0) = 0$.

Instead of operating in a small time framework, it is possible to formulate the discussion in terms of a parameter describing small perturbations of dynamical systems.

The goal of this communication is to discuss some extension of previous estimates to solutions of stochastic partial differential equations. We have considered in particular

(i) The case of two-parameter diffusions.
(ii) The one-dimensional wave equation weakly perturbed by a white noise.
(iii) The Zakai robust equation of non-linear filtering.
(iv) The heat equation perturbed by white noise.

Sections 1 to 4 are a review. However Section 5 is new: it is devoted to Davies estimates for the stochastic wave equation.

§1 THE TWO-PARAMETER DIFFUSIONS

Let $W_t = (W_t^1, \ldots, W_t^m)$, $t = (t_1, t_2) \in \mathbb{R}_+^2$, $[0, t] = [0, t_1] \times [0, t_2]$ be a n-dimensional Brownian sheet and $X = (X^1, \ldots, X^m) = (X(x))$ be the solution of the following Itô two-parameter integral equation. The stochastic integral will be understood in the sense of [35], [9].

$$X_t = x + \int_{[0,t]} \left\{ \sum_{j=1}^{n} A_j(X_s) \, dW_s^j + A_0(X_s) \, ds \right\}, \tag{1.1}$$

where A_j, $0 \le j \le m$, are smooth vector fields with bounded derivatives for each order. As for one-parameter diffusions, it is possible to show that X has a smooth version in x.

We set $\varepsilon = \sqrt{t_1 t_2}$. The law of $X_t(x)$ equals the law of $Y_{1,1} = Y_{1,1}(\varepsilon, x)$, where Y solves

$$Y_t = x + \int_{[0,t]} \left\{ \varepsilon \sum_{j=1}^{m} A_j(Y_s) \, dW_s^j + \varepsilon^2 A_0(Y_s) \, ds \right\}. \tag{1.2}$$

Since we are interested in law densities, W can be considered as the canonical Brownian sheet related to the abstract Wiener space Ω of the continuous function defined on $[0, 1]^2$ vanishing on the axes; this space is equipped with the Cameron–Martin space H constituted by functions $h(t) = \int_{[0,t]} \dot{h}(s) \, ds$, where $\dot{h} \in L^2([0, 1]^2)$. The inner product is given by $\langle h, k \rangle_H$ where

$$\langle k, k \rangle_H = \int_{[0,1]^2} \dot{h} \dot{k}(s) \, ds;$$

$\| \cdot \|_H$ symbolizes the associated norm.

The skeleton of (1.2) which is the two-parameter version of (0.3) is given by

$$y_t = x + \int_{[0,t]} \left\{ \sum_{j=1}^{n} A_j(y_s) \dot{h}^j(s) \right\} ds \tag{1.3}$$

where $y = (y_t(x, h))$, $x \in \mathbb{R}^n$, $h \in H$. The factor ε^2 obliges A_0 not to play any role for the Varadhan estimates.

Since two-parameter diffusions are not associated with any reasonable semi-group, we will define the analogous of the distance d through the two-parameter analogous of (0.3), that is to say (1.3). So, for $x, y \in \mathbb{R}^n$, we define

$$d^2(x, y) = \inf_{y_{1,1}(x,h)=y} \|h\|_H^2. \tag{1.4}$$

Let us formulate the following global coercive property. There exists $k > 0$ such that

$$\sum_{j=1}^{n} |A_j(x) \cdot y| \geq k|y|, \quad \forall x, y \in \mathbb{R}^n. \tag{1.5}$$

Under previous assumption we have the following result.

Proposition 1.1. *The application $(x, y) \mapsto d(x, y)$ is finite and continuous. More-over $d^{2/3}$ is a non-symmetric distance. This means that all the axioms for having a distance are fulfilled excepted the symmetry property.*

In fact $Y_{1,1}(\varepsilon, x)$ is smooth in Malliavin sense (it belongs to $\mathcal{D}^\infty$, see [36], [37]). The Gran–Malliavin matrix $\langle DY_{1,1}(\varepsilon, x), DY_{1,1}(\varepsilon, x) \rangle$ has an inverse which belongs to every L^p, $p \geq 1$. This implies that the law of $Y_{1,1}(\varepsilon, x)$ has a smooth density $y \mapsto q_\varepsilon(x, y)$, see [28].

The proof of the following theorem consists in a mixture of large deviations theory [11] and Malliavin calculus [28].

Theorem 1.2.
$$\lim_{\varepsilon \to 0+} 2\varepsilon^2 \log q_\varepsilon(x, y) = -d^2(x, y).$$

§2 THE CASE OF A ONE-DIMENSIONAL WAVE EQUATION

Let $a, b \in C_b^\infty(\mathbb{R})$, f, g Lipschitz functions. Let us consider the following stochastic wave equation.

$$\square X(t, x) = \varepsilon a(X(t, x))\dot{W}(t, x) + b(X(t, x)), \quad (t, x) \in \mathbb{R}_+ \times \mathbb{R}, \tag{2.1}$$

with initial condition

$$X(0, \cdot) \equiv f, \quad \frac{\partial X}{\partial t}(0, \cdot) \equiv g'. \tag{2.2}$$

$(\dot{W}(t, x), t \geq 0, x \in \mathbb{R})$ is a Gaussian white noise, ε a small parameter. This equation can be understood in distributional sense, see [35], [8], [22]. It is possible to prove that (2.1)+(2.2) has the following equivalent integral form:

$$X(\mathbf{z}) = X_0(\mathbf{z}) + \int_{D(\mathbf{z})} [a(X(\mathbf{r})) \, dW(\mathbf{r}) + b(X(\mathbf{r})) \, d\mathbf{r}], \quad \mathbf{z} \in \mathbb{R}_+ \times \mathbb{R} \tag{2.3}$$

where $\mathbf{z} = (t, x)$. $D(t, x)$ is the triangle whose vertices are $(0, t+x)$, $(0, x-t)$, and (t, x). X_0 is the deterministic process

$$X_0(t, x) = \frac{1}{2} \left\{ f(t+x) + f(x-t) + g(x+t) - g(x-t) \right\}.$$

The study of equation (2.3) is equivalent to the study of (2.3) on each compact triangle $T_0 = D(t_0, x_0)$ which will be from now on fixed. By fixed point methods, (2.3) can be shown to have a continuous process as unique solution.

Let Ω_0 be the Banach space of continuous function defining on T_0 but vanishing for $t = 0$; it is equipped with the uniform norm. H_0 is the Hilbert subspace composed by $h \in \Omega_0$ such there is $\dot{h} \in L^2(T_0)$ with $h(\mathbf{z}) = \int_{D(\mathbf{z})} \dot{h}(\mathbf{r}) \, d\mathbf{r}$. The Hilbert norm of H_0 is defined by

$$\|h\|_0^2 = \int_{D(\mathbf{z})} \dot{h}^2(\mathbf{r}) \, d\mathbf{r}.$$

By replacing $\varepsilon \dot{W}$ in (2.3) by $\dot{h} \in L^2(T_0)$, we obtain the skeleton equation

$$\psi(\mathbf{z}) = X_0(\mathbf{z}) + \int_{D(\mathbf{z})} \left\{ a(\psi(\mathbf{r}))\dot{h}(\mathbf{r}) + b(\psi(\mathbf{r})) \right\} d\mathbf{r}, \tag{2.4}$$

where $\psi = \psi(h)$. If A is a Borel subset of Ω_0, we set

$$\Lambda(A) = \inf_{\psi(h) \in A} \|h\|_D^2. \tag{2.5}$$

In [22] we get the following result

Theorem 2.1. *Let* $X = X^\varepsilon$ *be the solution to* (2.3). *Then*

$$-\Lambda(\mathring{A}) \leq \lim_{\varepsilon \to 0} \inf 2\varepsilon^2 \log P(X^\varepsilon \in A) \leq \lim_{\varepsilon \to 0} 2\varepsilon^2 \log P(X^\varepsilon \in A) \leq -\Lambda(\bar{A}). \quad \square$$

$X^\varepsilon(\mathbf{z}_0)$ belongs to the space of smooth test functionals in the sense of Wiener analysis [36], [37], Malliavin calculus [24].

We introduce two applications which are the analogous of the "distances" d and d_R introduced in [5]. From now on $\mathbf{z}_0$ will denote (t_0, x_0). We set

$$\ell^2(y) = \inf_{\psi(h)(\mathbf{z}_0) = y} \|h\|_0^2. \tag{2.6}$$

ℓ^2 is naturally associated with the large deviations functional Λ and it will naturally intervene in the logarithmic density estimates. The other application is related with the *Bismut condition* [2]. $h_0 \in H_0$ will be said to fulfill the Bismut condition if $h \mapsto \psi(h)(\mathbf{z}_0)$ is a submersion; this means that $D\psi(h_0)(\mathbf{z}_0)$ is injective as a linear form on H_0; this is in fact equivalent to $D\psi(h_0)(\mathbf{z}_0) \neq 0$. We define $\ell_R^2(y)$

as the infimum of $\|h\|_0^2$ over $h \in H_0$ such that $\psi(h)(\mathbf{z}_0) = y$ and h satisfies the Bismut condition. Clearly $\ell_R^2(y) \geq \ell^2(y)$.

[8] gives sufficient conditions on a, b, f, g such that the law of $X^\varepsilon(z_0)$ has a density for any $\varepsilon \neq 0$. They state the following assumptions.

Assumptions.

(A1) $a(y) \neq 0$ for some y in the closed interval with endpoints $f(x_0 - t_0)$ and $f(x_0 + t_0)$.

(A2) $f(x_0 - t_0) = f(x_0 + t_0) = y_0$, $a(y_0) = 0$, $a^{(n)}(y_0) \neq 0$ for some $n \geq 1$;

(A21) $a(f(\xi_0)) \neq 0$ for some $\xi_0 \in]x_0 - t_0, x_0 + t_0[$,

(A22) $a(f(\xi)) = 0$ for all $\xi \in [x_0 - t_0, x_0 + t_0]$

and

either $g'(x_0 - t_0) \neq 0$ or $g'(x_0 + t_0) \neq 0$

and

either $g''(x_0 - t_0) + f(y_0) \neq 0$ or $g''(x_0 + t_0) + f(y_0) \neq 0$. $\square$

Under (A1), under (A2)+(A21) or (A2)+(A22), $X^\varepsilon(z_0)$ has a law density p_ε. Under (A1) and (A22) every $h \in H_0$ has the Bismut condition so that $\ell^2 = \ell_R^2$.

In [22], again Malliavin calculus and Theorem 2.1 give the following logarithmic estimates.

Theorem 2.2.

$$-\ell_R^2(y) \leq \liminf_{\varepsilon \to 0} 2\varepsilon^2 \log p_\varepsilon(y) \leq \limsup_{\varepsilon \to 0} 2\varepsilon^2 \log p_\varepsilon(y) \leq -\ell^2(y). \quad \square$$

In [22], we also get the equivalent of the estimates on the diagonal; they are of the type Greiner–Minakshisundaram–Pleijel, see [13], [26].

Let us suppose that $h_0 = 0$ has the Bismut condition; it happens for instance under assumption (A1) and (A22). Let $y \in \mathbb{R}$ such that $\ell^2(y) = 0$; this happens for $X^0(z_0) = y$. In the diffusion case, when $b \equiv 0$, y equals the initial condition x. Because of this, the result below extends the diagonal case.

Theorem 2.3. *For every $N \in \mathbb{N}$, there are $a_0 > 0$, $a_1, \ldots a_N \in \mathbb{R}$ such that*

$$p_\varepsilon(y) = \frac{1}{\varepsilon}\left[\sum_{i=0}^{N} a_i \varepsilon^i + o(\varepsilon^N)\right].$$

§3 THE CASE OF THE ROBUST ZAKAI EQUATION

Let us consider a differential operator having the form $A = \frac{1}{2}\sum_{i=1}^{m} X_i^2 + X_0 + V$ where X_i are smooth vector fields on $\mathbb{R}^d$ whose derivatives of each order are bounded. If there is no potential V, we make the weak Hörmander assumption. Let $x \in \mathbb{R}^d$: the Lie algebra spanned by vectors $X_i(x)$, $X_0(x)$ above excluded, coincides with $\mathbb{R}^d$ at each point x.

The heat semigroup associated has of course a density $y \mapsto p_t(x, y)$. If $X_0(x) = 0$, a theorem of [5] says that for any positive integer n,

$$p_t(x, x) = \frac{1}{(\sqrt{t})^{N(x)}} \sum_{i=0}^{n} a_i(x)(\sqrt{t})^i + o(\sqrt{t})^n \tag{3.1}$$

where $N(x)$ is a theoretical integer depending on the geometry of $X_0, \ldots, X_n$. There are two possibilities:

(1) either $a_0(x) > 0$,
(2) or $a_0(x) = 0$ and in this case all the $a_i(x)$ vanish.

In the second case [12] has estimated the decay by using a Gevrey type argument and controlling the remaining term.

Let us suppose that for any $M > 0$, and multi-index r over $\mathbb{R}^d$

$$\sup_{|x-y|<M} |D^{(r)} X_i(y)| \leq (Cr!)^{cr} \tag{3.2}$$

where c and C are positive constants. In such a case, there exists $\alpha > 1$ such that

$$p_t(x, x) \leq \text{const} \exp(-|\log t|^\alpha). \tag{3.3}$$

Let us consider now the case where there is a potential V and let us study the Zakai robust equation

$$dp_t(y) = A^* p_t(y) \, dt + \sum_{i=1}^{n} h_i(y) p_t(y) \circ d\Theta^i, \tag{3.4}$$

$$p_0(y) = \delta_x,$$

where Θ^i is an auxiliary continuous semimartingale. Following [29], we can explicitly solve the previous equation. Let us namely define

$$A_t^\theta = \exp[h(x)\Theta(t)] A \exp(-h\Theta(t)) \tag{3.5}$$

and let us denote by $q_t^\theta(y)$ the solution of the Fokker–Planck equation

$$\frac{\partial}{\partial t} q_t^\Theta(y) = (A_t^\Theta)^* q_t^\Theta(y), \tag{3.6}$$

$$q_0(y) = \delta_x.$$

We have in fact

$$p_t(y) = \exp(h(x)\Theta(t)) q_t^\Theta(y). \tag{3.7}$$

(q_t^Θ) can be obtained through a Feynman–Kac formula and Girsanov theorem, see [23] and so deduced from $\frac{1}{2} \sum_{i=1}^{n} X_i^2 + X_0$. In the case of flat decay, the main result of [23] is the following.

Theorem 3.1. *Let us assume* (3.2). *Moreover, we suppose the existence of an integer N such that*

$$\sup_{|x-y|\leq M} |D^{(r)}V(y)| \leq (Cr!)^{cr^N}, \tag{3.8}$$

$$\sup_{|x-y|\leq M} |D^{(r)}h(y)| \leq (Cr!)^{cr^N}. \tag{3.9}$$

Then, there is $\alpha > 1$ such that

$$q_t^{\Theta}(x,x) \leq \text{const} \exp(-|\log t|^{\alpha}) \exp\left(\text{const } Ct \sup_{s\leq t} |\Theta(s)|^2\right). \quad \Box \tag{3.10}$$

Remark that [38] obtains an asymptotic expansion for $p_t(y)$.

§4 THE STOCHASTIC HEAT EQUATION

We end up the review part with some remarks about the stochastic heat equation. Let us consider the following stochastic PDE. Let $\tau > 0$

$$\frac{\partial X}{\partial t}(t,x) = \frac{\partial^2 X}{\partial x^2}(t,x) + a\left(X(t,x)\right)\dot{W}(t,x) + b\left(X(t,x)\right), \tag{4.1}$$
$$(t,x) \in [0,\tau] \times [0,1],$$

with initial condition

$$X(0,x) = X_0(x), \quad X_0 \in C[0,1], \tag{4.2}$$

and boundary condition

$$\frac{\partial X}{\partial x}(t,0) = \frac{\partial X}{\partial x}(t,1) = 0, \quad t \in [0,\tau]. \tag{4.3}$$

a, b are Lipschitz real functions, $\dot{W}$ is a space-time white noise.

In fact (4.1)+(4.2)+(4.3) have a rigorous meaning after an integration by parts again a test function $\varphi(t,x)$: this problem has in this way a distributional meaning.

As in the case of the stochastic wave equation, this problem has the following equivalent integral form:

$$X(t,x) = \int_0^1 G_t(x,y)X_0(y)\,dy + \int_0^t ds \int_0^1 G_{t-s}(x,y)\,b(X(s,y))\,dy$$
$$+ \int_0^t ds \int_0^1 dy\, G_{t-s}(x,y)\,a(X(s,y))\,dW(s,y),$$

where $(G_t(x,y))$ is the fundamental solution of the heat equation on $\mathbb{R}_+ \times [0,1]$ with Neumann boundary conditions. Let us remark the following points.

- [35] has proved by Picard methods existence and uniqueness of a continuous process.

- If a, b are smooth real functions whose first order derivatives are bounded and the high order derivatives have polynomial growth then [7] gives the following result: For every $x_1, \ldots, x_d \in \,]0, 1[$ which are pointwise distinct, $t \in [0, T]$, the law of the vector $(X(t, x_1), \ldots, X(t, x_d))$ admits a smooth density ρ on $\{a \neq 0\}^d$. [30] give existence results for the density under much weaker assumptions.
- If a is upper and lower bounded, [31] gives Freidlin–Ventcell type estimates if $\dot{W}$ is replaced by $\varepsilon \dot{W}$ when $\varepsilon \to 0$.
- Making use of a cocktail of large deviations theory and Malliavin calculus, it is possible to establish Varadhan estimates.
- Another problem which has been studied is the support of the solutions of this equation, see [6]; they had also treated the case of the two-parameter diffusions, see [27]. This is strongly related with the problem of establishing a criterion for positivity of the density at one fixed point. These authors use a general criterion of [1] which extends the ideas performed by [5] in the case of diffusions.

§5 THE DAVIES INEQUALITIES

These inequalities should work for a large class of stochastic PDE's. Here we will concentrate on the stochastic wave equation, see Section 2. We recall that Davies estimates are the upper bound version of Aronson estimates. In this section, we will suppose, $|a|$ lower and upper bounded by a positive constant. In particular, the assumption (A1) is fulfilled.

Lemma 5.1. ℓ *is finite and globally Lipschitz on each compact of* $\mathbb{R}$.

Proof. For each $x \in \mathbb{R}$, we set

$$A(\mathbf{z}, x) = X_0(\mathbf{z}) + \frac{\text{Leb}(D(\mathbf{z}))}{\text{Leb}(D(\mathbf{z}_0))}(x - X_0(\mathbf{z})) \tag{5.1}$$

where Leb stands for Lebesque measure. $A(\cdot, x)$ is solution to

$$\Box A(\cdot, x) = \frac{x - X_0(\mathbf{z})}{\text{Leb}(D(\mathbf{z}_0))} \tag{5.2}$$

with initial condition (2.2). We set

$$\dot{h}(\mathbf{r}) = \frac{\frac{x - X_0(\mathbf{z})}{\text{Leb}(D(\mathbf{z}_0))} - b(A(\mathbf{r}, x))}{a(A(\mathbf{r}, x))}.$$

Of course $\dot{h}_x \in L^2(T_0)$, so that $h_x(\mathbf{z}) = \int_{D(\mathbf{z})} \dot{h}_x(\mathbf{r}) \, d\mathbf{r}$ belongs to H_0. Clearly $A(\mathbf{z}, x) = \psi(h_x)(\mathbf{z})$ and $\psi(h_x)(\mathbf{z}_0) = x$. Consequently, $\ell(x) \leq \|h_x\|$. This implies that ℓ is finite.

Let us consider now $x, y \in \mathbb{R}$. For $\tilde{h}_x, \tilde{h}_y$ such that $\psi(\tilde{h}_x)(\mathbf{z}) = x$, $\psi(\tilde{h}_y)(\mathbf{z}) = y$, we have

$$\|\tilde{h}_y\| \leq \|\tilde{h}_y - \tilde{h}_x\| + \|\tilde{h}_x\|.$$

Now by taking the infimum over $\tilde{h}_y, \tilde{h}_x$, we get

$$\ell(y) \leq \ell(x) + \inf \|\tilde{h}_y - \tilde{h}_x\|$$

where the infimum is taken over the $\tilde{h}_x, \tilde{h}_y$ so that $\psi(\tilde{h}_x)(\mathbf{z}) = x$ and $\psi(\tilde{h}_y)(\mathbf{z}) = y$. By symmetry, we obtain that $|\ell(y) - \ell(x)|$ is upper bounded by the same infimum quantity.

It remains to prove the existence of a constant K depending on $\mathbf{z}_0$, a, b such that $\|h_y - h_x\| \leq K|y - x|$. Taking in account (2.4), we need to prove the existence of $K \geq 0$ such that

$$\int_{D(\mathbf{z})} \left\{ \frac{x - y - [b(A(\mathbf{r}, x)) - b(A(\mathbf{r}, y))]}{a(A(\mathbf{r}, x))} \right. \tag{5.3}$$
$$\left. + \left(\frac{1}{a(A(\mathbf{r}, x))} - \frac{1}{a(A(\mathbf{r}, y))} \right) (y - b(A(\mathbf{r}, y))) \right\}^2 d\mathbf{r}$$

is inferior to $K^2|x - y|^2$. Now (5.3) is upper bounded by

$$\frac{4}{C^2} \operatorname{Leb}^2(T_0)|x - y|^2 + \frac{4\|b'\|_\infty^2}{C^2}|x - y|^2 + \|a\|_\infty \frac{4}{C^2}|x - y|^2 \int_{T_0} d\mathbf{r}\, (y - b(A(\mathbf{r}, y)))^2,$$

where $\|a\|_\infty$ denotes the sup norm and C is a lower bound for a. $\square$

We set $y_0 = X^0(\mathbf{z}_0)$. For $\alpha > 0$, we define ℓ_α as the following regularisation of ℓ:

$$\ell_\alpha(x) = \int dy\, \phi^\alpha(x - y)\ell(y) - \ell(y_0)$$

where ϕ is a Friedrich mollifier, $\phi_\alpha(x) = \frac{1}{\alpha}\phi(\frac{x}{\alpha})$.

Lemma 5.2. *Let V be a bounded interval of $\mathbb{R}$. Then, there is a constant c such that ℓ_α is Lipschitz on V with constant c for any $\alpha \in]0, 1]$.*

Proof. Let $W = (V - 1) \cup (V + 1)$ be a neighborhood of V. Let $x_1, x_2 \in V$:

$$|\ell_\alpha(x_1) - \ell_\alpha(x_2)| = \int dy\, \phi(y)\, [\ell(x_1 - \alpha y) - \ell(x_2 - \alpha y)]$$

$$\leq \int_W dy\, \phi(y)|\ell(x_1 - \alpha y) - \ell(x_2 - \alpha y)| \leq c|x_1 - x_2|$$

where c is the Lipschitz constant of ℓ on W. $\square$

The next technical result will concern a uniform bound with respect to α of $E\{\exp(\ell_\alpha^2(X^\varepsilon(\mathbf{z}_0))/(\delta\varepsilon^2))\}$ if X^ε is the solution of (2.1) and $\delta > 0$ large enough.

Lemma 5.3. *Let V be a bounded interval of $\mathbb{R}$, $0 \leq \chi \leq 1$ be a smooth function vanishing outside V. Then, there is $\delta > 0$ large enough with*

$$\sup_{\substack{|\alpha| \leq 1 \\ |\varepsilon| \leq 1}} E\left(\chi(X^\varepsilon(\mathbf{z}_0)) \exp\left(\frac{\ell_\alpha^2(X^\varepsilon(\mathbf{z}_0))}{\delta\varepsilon^2} \right) \right) < \infty. \tag{5.4}$$

Proof. According to Lemma 5.2, ℓ_α is Lipschitz on V with a constant k not depending on α. We try to bound $|X^\varepsilon(\mathbf{z}_0) - y_0|$, where $y_0 = X^0(\mathbf{z}_0)$. The expectation in (5.4) is bounded by

$$E\left(\chi(X^\varepsilon(\mathbf{z}_0)) \exp\left(\frac{k^2|X^\varepsilon(\mathbf{z}_0) - y_0|^2}{\delta\varepsilon^2} \right) \right) \tag{5.5}$$

because $|\ell(X^\varepsilon(\mathbf{z}_0)) - \ell(y_0)| \leq k|X^\varepsilon(\mathbf{z}_0) - y_0|$. X^ε solves

$$X^\varepsilon(\mathbf{z}) = X_0(\mathbf{z}) + \varepsilon \int_{D(\mathbf{z})} d\mathbf{r}\{a(X^\varepsilon(\mathbf{r})) \, dW(\mathbf{r}) + b(X^\varepsilon(\mathbf{r}))\};$$

consequently

$$|X^\varepsilon(\mathbf{z}) - X^0(\mathbf{z})| = I_1(\mathbf{z}) + I_2(\mathbf{z}) \tag{5.6}$$

where

$$I_1 = \varepsilon \int_{D(\mathbf{z})} d\mathbf{r} \, a\left(X^\varepsilon(\mathbf{r}) \right) dW(\mathbf{r}),$$

$$I_2 = \int_{D(\mathbf{z})} d\mathbf{r} \left(b(X^\varepsilon(\mathbf{r})) - b(X^0(\mathbf{r})) \right).$$

We recall that a is supposed to be bounded. For the estimate of I_1, we will make use of the exponential inequality for two-parameter strong martingales, see Proposition 5 of [11]. We set $M(\mathbf{z}) = \int_{D(\mathbf{z})} a(X^\varepsilon(\mathbf{r})) \, dW(\mathbf{r})$; its quadratic variation process equals

$$[M, M](\mathbf{z}) = \int_{D(\mathbf{z})} d\mathbf{r} \, a^2(X^\varepsilon(\mathbf{r})).$$

$[M, M]$ is a.s. bounded by $c_0 = \mathrm{Leb}(D(\mathbf{z}_0))\|a\|_\infty$. According to exponential inequality, there are universal constants λ_1, λ_2 such that

$$P\left(\sup_{\mathbf{z} \in D(\mathbf{z})} |M(\mathbf{z})| \geq x \right) \leq \lambda_1 \exp\left(\frac{-x^2}{\lambda_2 c_0} \right). \tag{5.7}$$

Concerning I_2, if c_2 is the Lipschitz constant of b, then the following upper bound holds:

$$|I_2(\mathbf{z})| \leq c_2 \int_{D(\mathbf{z})} |X^\varepsilon(\mathbf{r}) - X^0(\mathbf{r})| \, d\mathbf{r}.$$

Consequently (5.6) gives

$$|X^\varepsilon(\mathbf{z}) - X^0(\mathbf{z})| \le c_2 \int_{D(\mathbf{z})} |X^\varepsilon(\mathbf{r}) - X^0(\mathbf{r})| \, d\mathbf{r} + \varepsilon \sup_{\mathbf{z}_0 \in D(\mathbf{z})} |M(\mathbf{z})|.$$

Gronwall type lemma of [22] says that

$$|X^\varepsilon(\mathbf{z}_0) - y_0|^2 \le \sup_{\mathbf{z} \in D(\mathbf{z}_0)} |X^\varepsilon(\mathbf{z}) - X^0(\mathbf{z})|^2 \le c_3 \varepsilon^2 \sup_{\mathbf{z} \in D(\mathbf{z}_0)} |M(\mathbf{z})|^2.$$

Let F be the distribution function of $\sup_{\mathbf{z} \in D(\mathbf{z}_0)} |M(\mathbf{z})|$. Then (5.5) is bounded by

$$\mathrm{E}\left((\chi(X^\varepsilon(\mathbf{z})) \exp\left(\frac{c_4}{\delta} \sup_{\mathbf{z} \in D(\mathbf{z}_0)} |M(\mathbf{z})|^2 \right) \right).$$

c_3 and c_4 are suitable constants. Hence (5.5) is controlled by

$$\int_0^\infty d(1 - F(y)) \exp\left(\frac{c_4}{\delta} y^2 \right)$$
$$= (1 - F(y)) \exp\left(\frac{c_4}{\delta} y^2 \right) \Big|_0^\infty - \frac{2c_4}{\delta} \int_0^\infty y \exp(y^2)(1 - F)(y) \, dy. \quad (5.8)$$

According to (5.7), if $\delta > c_4 \lambda_2 / c_0$ then

$$(1 - F)(y) \exp\left(y^2 \frac{c_4}{\delta} \right) \Big|_0^\infty = \lambda_1 \exp\left[\left(\frac{c_4}{\delta} - \frac{1}{\lambda_2 c_0} \right) y \right] \Big|_0^\infty = -\lambda_1.$$

Consequently (5.8) is bounded by

$$\lambda_1 + \frac{2c_4}{\delta} \int_0^\infty y \exp\left(\left(\frac{c_4}{\delta} - \frac{1}{\lambda_2 c_0} \right) y^2 \right) dy$$

which is finite because $\frac{c_4}{\delta} - \frac{1}{\lambda_2 c_0} < 0$. $\square$

Let $p \ge 1$, $N \in \mathbb{N}$, F a Wiener functional. In [22], [36], [37] the spaces $\mathcal{D}_{p,s}$ and $\mathcal{D}_{p,s}(H_0)$ and the corresponding norm $\|\cdot\|_{p,s}$ have been defined for $p \ge 1$, $s \in \mathbb{R}$. If $p \ge 1$, $s = 1$, we recall that

$$\|F\|_{p,1} = (\mathrm{E}\,|F|^p)^{1/p} + \mathrm{E}\left(\int_{T_0} (D_r F)^2 \, d\mathbf{r} \right)^{p/2}.$$

Moreover $DF \in L^p(\Omega; H_0)$. Let $\alpha > 0$. We set

$$Z_\delta^\alpha(\varepsilon) = \chi(X^\varepsilon(\mathbf{z}_0)) \exp\left(\frac{\ell_\alpha^2(X^\varepsilon(\mathbf{z}_0))}{\delta \varepsilon^2} \right),$$

as in (5.4).

Lemma 5.4. *For any $p \geq 1$, there is $\delta > 0$ large enough such that*

(i) $Z_\delta^\alpha(\varepsilon) \in \mathcal{D}_{p,1}$ *for any* $\alpha, \varepsilon \in [-1,1] \setminus \{0\}$,

(ii) $\displaystyle\sup_{\substack{|\alpha|,|\varepsilon|\leq 1 \\ \alpha, \varepsilon \neq 0}} \|Z_\delta^\alpha(\varepsilon)\|_{p,1} < \infty$.

Proof. $Z_\delta^\alpha(\varepsilon) \in L^p$ and $\sup_{|\alpha|,|\varepsilon|\leq 1} \|Z_\delta^\alpha(\varepsilon)\|_{L^p} < \infty$ for $\delta > 0$ large enough. This because of Lemma 5.3. According to Meyer inequalities, it remains to check that

(a) $DZ_\delta^\alpha(\varepsilon) \in L^p(\Omega; H_0)$,

(b) $\sup_{|\alpha|,|\varepsilon|\leq 1} \mathrm{E}(\int_{T_0} (D_r Z_\delta^\alpha(\varepsilon))^2 \, dr)^{p/2} < \infty$ for $\delta > 0$ large enough.

Now $DZ_\delta^\alpha = I_1 + I_2$, where

$$I_1 = \chi'(X^\varepsilon(\mathbf{z}_0)) \exp\left(\frac{\ell_\alpha^2(X^\varepsilon(\mathbf{z}_0))}{\delta\varepsilon^2}\right) DX^\varepsilon(\mathbf{z}_0),$$

$$I_2 = \chi(X^\varepsilon(\mathbf{z}_0)) \exp\left(\frac{\ell_\alpha^2(X^\varepsilon(\mathbf{z}_0))}{\delta\varepsilon^2}\right) \frac{2}{\delta} \frac{\ell_\alpha(X^\varepsilon(\mathbf{z}_0))}{\varepsilon} \ell_\alpha'(X^\varepsilon(\mathbf{z}_0)) \frac{DX^\varepsilon(\mathbf{z}_0)}{\varepsilon}.$$

For any $p \geq 2$, there is $\delta > 0$ large enough such that $I_1 \in L^p(\Omega; H_0)$. For this we use Cauchy–Schwarz inequality and the fact that $X^\varepsilon(\mathbf{z})$ belongs to all the Wiener Sobolev spaces $\mathcal{D}_{p,s}$ uniformly with respect to ε.

Concerning I_2, we observe that

$$\frac{DX^\varepsilon(\mathbf{z}_0)}{\varepsilon} = D\widehat{X}^\varepsilon(\mathbf{z}_0), \tag{5.9}$$

where

$$\widehat{X}^\varepsilon(\mathbf{z}_0) = \begin{cases} \frac{X^\varepsilon(\mathbf{z}_0) - y_0}{\varepsilon}, & \text{if } \varepsilon \neq 0, \\ \frac{\partial X^\varepsilon}{\partial \varepsilon}(\mathbf{z}_0)\big|_{\varepsilon=0}, & \text{if } \varepsilon = 0. \end{cases}$$

We recall that $\widehat{X}^\varepsilon(\mathbf{z}_0)$ belongs uniformly with respect to ε, to all the Sobolev spaces, see Proposition 4.19 of [22]. Therefore, for any $p \geq 2$,

$$\sup_{\varepsilon \in (-1,1)} \mathrm{E}\left(\int_{T_0} D_r \widehat{X}^\varepsilon(\mathbf{z}_0)^2 \, dr\right)^{p/2} < \infty.$$

According to Lemma 5.2, ℓ_α' is bounded on V with a constant c_1 not depending on α. So $|\ell_\alpha'(X^\varepsilon(\mathbf{z}))\chi(X^\varepsilon(\mathbf{z}))| \leq c_1$. On the other hand, if c_2 is one Lipschitz constant of ℓ_α on V

$$\frac{|\ell_\alpha(X^\varepsilon(\mathbf{z}_0))\chi(X^\varepsilon(\mathbf{z}_0))|}{\varepsilon} \leq c_2 \frac{|X^\varepsilon(\mathbf{z}_0) - y_0|}{\varepsilon} = c_2\left|\int_0^1 d\beta \frac{\partial X^{\varepsilon\beta}}{\partial \varepsilon}(\mathbf{z}_0)\right|.$$

Again, Proposition 4.19 in [22] says that $\frac{\partial X^{\varepsilon\beta}}{\partial \varepsilon}(\mathbf{z}_0)$ belongs to all the Sobolev spaces $\mathcal{D}_{p,s}$ uniformly with respect to ε. Therefore for any $p \geq 2$,

$$\sup_{\substack{|\varepsilon|\leq 1 \\ |\alpha|\leq 1}} \left\|\ell_\alpha'(X^\varepsilon(\mathbf{z}_0)) \frac{\ell_\alpha(X^\varepsilon(\mathbf{z}_0))}{\varepsilon} D\widehat{X}^\varepsilon(\mathbf{z}_0)\right\|_{L^p(\Omega; H_0)} < \infty.$$

By Cauchy–Schwarz inequality and Lemma 5.3, for any $p \geq 2$, there is $\delta > 0$ large enough so that $Z_\delta^\alpha(\varepsilon)$ fulfills (a) and (b). □

Let $\alpha, \varepsilon \neq 0$. By classical arguments we notice that for any $p \geq 2$, $N \in \mathbb{N}$, there is δ large enough so that $Z_\delta^\alpha(\varepsilon)$ belongs to $\mathcal{D}_{p,N}$. We consider the following Radon measures:

$$
\begin{aligned}
\hat{\mu}_{\varepsilon,\delta}^\alpha &: f \longmapsto \mathrm{E}\big(f(\widehat{X}^\varepsilon(\mathbf{z}_0))Z_\delta^\alpha(\varepsilon)\big), \\
\mu_{\varepsilon,\delta}^\alpha &: f \longmapsto \mathrm{E}\big(f(X^\varepsilon(\mathbf{z}_0))Z_\delta^\alpha(\varepsilon)\big).
\end{aligned}
\tag{5.10}
$$

By an easy adaptation of Proposition 2.1 in [22], there is δ large enough such that $\hat{\mu}_{\varepsilon\delta}^\alpha$ (resp. $\mu_{\varepsilon\delta}^\alpha$) has a smooth density $\hat{q}_{\varepsilon,\delta}^\alpha$ (resp. $q_{\varepsilon,\delta}^\alpha$). By construction of $Z_\delta^\alpha(\varepsilon)$, these densities vanish outside V. Therefore, they are bounded.

Proposition 5.5. *For any $\delta > 0$ large enough there is a constant $C > 0$ not depending on α and ε such that*

(i) $\sup_y |\hat{q}_{\varepsilon,\delta}^\alpha(y)| \leq C$,
(ii) $\sup_y |q_{\varepsilon,\delta}^\alpha(y)| \leq C/\varepsilon$.

Proof. Let $\phi \in C_b^\infty(\mathbb{R})$. Proposition 6.8 of [22] says that

$$
\sup_{|\varepsilon| \leq 1} \mathrm{E}\left(\|D\widehat{X}^\varepsilon(\mathbf{z}_0)\|^{-p}\right) < \infty, \quad \forall p \geq 2.
\tag{5.11}
$$

This implies the existence of $\delta > 0$ large enough so that

$$
\sup_{\substack{|\alpha| \leq 1 \\ |\varepsilon| \leq 1}} \|\widetilde{Z}_\delta^\alpha(\cdot, \varepsilon)\|_{2,1} < \infty
\tag{5.12}
$$

where $\|\cdot\|$ is the H_0-norm and

$$
\widetilde{Z}_\delta^\alpha(\mathbf{r}, \varepsilon) = \frac{D_\mathbf{r}\widehat{X}^\varepsilon(\mathbf{z}_0)Z_\delta^\alpha(\varepsilon)}{\|D\widehat{X}^\varepsilon(\mathbf{z}_0)\|^2}.
$$

Let us consider now $G \in C_b^\infty(\mathbb{R})$. We have

$$
\begin{aligned}
\mathrm{E}\left(G'(\widehat{X}^\varepsilon(\mathbf{z}))Z_\delta^\alpha(\varepsilon)\right) &= \mathrm{E}\left\{\int_{T_0} d\mathbf{r}\, D_\mathbf{r}(G(\widehat{X}^\varepsilon(\mathbf{z})))\widetilde{Z}_\delta^\alpha(\mathbf{r}, \varepsilon)\right\} \\
&= \mathrm{E}\left\{G(\widehat{X}^\varepsilon(\mathbf{z}))\partial(\widetilde{Z}_\delta^\alpha(\cdot, \varepsilon))\right\}
\end{aligned}
\tag{5.13}
$$

where ∂ denotes Skorohod integration. Using the continuity properties of Skorohod integrals, we obtain that $\partial(\widetilde{Z}_\delta^\alpha(\cdot, \varepsilon)) \in L^2$ and

$$
\sup_{\substack{|\alpha| \leq 1 \\ |\varepsilon| \leq 1}} \|\partial\widetilde{Z}_\delta^\alpha(\cdot, \varepsilon)\|_{L^2}
\tag{5.14}
$$

is finite for δ large enough.

Let us consider now a Friedrich mollifier ϕ and $\phi_\lambda(y) = \frac{1}{\lambda}\phi(\frac{y}{\lambda})$, so that ϕ_λ converges to Dirac delta function when $\lambda \to 0$. Let $G_\lambda(x) = \int_{-\infty}^{x} \phi_\lambda(y)\,dx$. Now $|G_\lambda| \leq \int \phi_\lambda(y)\,dy = 1$, so that G_λ are uniformly bounded. Consequently for $x_0 \in \mathbb{R}$, (5.13) implies

$$
\begin{aligned}
\mathrm{E}\left(\phi_\lambda(\widehat{X}^\varepsilon(\mathbf{z}_0) - x_0)Z_\delta^\alpha(\varepsilon)\right) &= \mathrm{E}\left(G_\lambda'(\widehat{X}^\varepsilon(\mathbf{z}_0) - x_0)Z_\delta^\alpha(\varepsilon)\right) \\
&= \mathrm{E}\left(G_\lambda(\widehat{X}^\varepsilon(\mathbf{z}_0) - x_0)\partial\widetilde{Z}_\delta^\alpha(\varepsilon)\right) \\
&\leq \sqrt{\mathrm{E}\left(G_\lambda(\widehat{X}^\varepsilon(\mathbf{z}_0)) - x_0)\right)^2 \mathrm{E}(\partial\widetilde{Z}_\delta^\alpha(\varepsilon))^2} \\
&\leq \sqrt{\sup_{\substack{|\alpha|\leq 1 \\ |\varepsilon|\leq 1}} \mathrm{E}(\partial\widetilde{Z}_\delta^\alpha(\varepsilon))^2}.
\end{aligned}
\tag{5.15}
$$

Let us denote by C previous square root. We let now go λ to 0 in (5.13) and we obtain $\widehat{q}_{\varepsilon,\delta}^\alpha(x_0)$ and finally (i) is established.

Let us prove (ii). Since $D\widehat{X}^\varepsilon(\mathbf{z}_0) = \frac{1}{\varepsilon}DX^\varepsilon(\mathbf{z}_0)$, similarly to (5.13) and (5.15), we obtain

$$
\mathrm{E}\left(\phi_\lambda(X^\varepsilon(\mathbf{z}_0) - x_0)Z_\delta^\alpha(\varepsilon)\right) = \frac{1}{\varepsilon}\mathrm{E}(G_\lambda(X^\varepsilon(\mathbf{z}_0) - x_0)\partial\widetilde{Z}_\delta^\alpha(\varepsilon)) \leq \frac{C}{\varepsilon}
$$

and (ii) follows in an analogous manner as (i). $\square$

Finally, we can state the main theorem of the section. Let $\mathbf{z}_0 = (t_0, x_0)$, $t_0 > 0$, $\varepsilon \neq 0$, X^ε be the solution of the stochastic wave equation (2.1)+(2.2), p_ε be the law density of $X^\varepsilon(\mathbf{z}_0)$. We recall that

$$
\widehat{p}(y) = \frac{1}{\varepsilon}p(y - y_0),
$$

where $y_0 = X^0(\mathbf{z}_0)$.

Theorem 5.6. *Let V be a bounded neighborhood of y_0 where $\ell(y_0) = 0$. We suppose:*

(i) *$a, b \in C_b^\infty(\mathbb{R})$,*

(ii) *a is upper and lower bounded.*

Then there are constants δ, $c > 0$ such that

$$
p_\varepsilon(y) \leq \frac{c}{\varepsilon}\exp\left(\frac{-\ell^2(y)}{\delta\varepsilon^2}\right),
$$

for any $y \in V$.

Proof. For $\delta > 0$, we define the following Random measures.

$$
\nu_\varepsilon : f \longmapsto \mathrm{E}\left(f(X^\varepsilon(\mathbf{z}_0))\right),
$$

$$
\nu_{\varepsilon,\delta}^\alpha : f \longmapsto \mathrm{E}\left(f(X^\varepsilon(\mathbf{z}_0))\exp\frac{(\ell_\alpha^2(X^\varepsilon(\mathbf{z}_0)))}{\delta\varepsilon^2}\right).
$$

Through usual arguments of Malliavin calculus, for $\varepsilon \neq 0$, $\alpha \neq 0$, these measures have a law density denoted respectively by p_ε, $p_{\varepsilon,\delta}^\alpha$.

For $\alpha \in \,]0,1]$, let $0 \leq \chi \leq 1$ be a smooth function with value 1 on V and 0 outside $V_0 = (V-1) \cup (V+1)$. We recall that $\mu_{\varepsilon,\delta}^\alpha$ has been defined in (5.10) with the density $q_{\varepsilon,\delta}^\alpha$, for $\delta > 0$ large enough. Under these condition clearly

$$p_{\varepsilon,\delta}^\alpha(y) = q_{\varepsilon,\delta}^\alpha(y), \quad \forall y \in V, \tag{5.16}$$

for $\alpha, \varepsilon \neq 0$, $|\alpha|$, $|\varepsilon| \leq 1$.

According to (5.16) and Proposition 5.5, we get

$$\sup_{y \in V} p_{\varepsilon,\delta}^\alpha(y) \leq \frac{c}{\varepsilon},$$

for a suitable constant c. By a change of variables we get

$$p_{\varepsilon,\delta}^\alpha(y) = \exp\left(\frac{\ell_\alpha^2(y)}{\delta \varepsilon^2}\right) p_\varepsilon(y),$$

so that

$$p_\varepsilon(y) = \exp\left(\frac{\ell_\alpha^2(y)}{\delta \varepsilon^2}\right) p_{\varepsilon,\delta}(y) \leq \frac{c}{\varepsilon} \exp\left(\frac{\ell_\alpha^2(y)}{\delta \varepsilon^2}\right), \quad \forall y \in V.$$

Letting $\alpha \to 0$ and using the fact $\ell_\alpha \to \ell$ uniformly on each compact, the conclusion of the theorem can follow. $\qquad \square$

Acknowledgements. The second author has been partially supported by a grant of the "Fonds National Suisse de la Recherche Scientifique" for the first part and by a Human Capital and Mobility grant of the European Union for the second part.

References

1. Aida, S., Kusuoka, S., Stroock, D., *On the support of Wiener functionals*, in Asymptotic problems in probability theory (K. D. Elworthy and N. Ikeda, eds.), Pitman Research Notes, Longman, 1993.
2. Bismut, J. M., *Large deviations and the Malliavin calculus*, Progress in Math., vol. 45, Birkhäuser, Basel, 1984.
3. Ben Arous, G, *Développement asymptotique du noyau de la chaleur hypoelliptique sur la diagonale*, Ann. Institut Fourier (Grenoble) **39** (1989), 73–99.
4. Ben Arous, G, *Flots et séries de Taylor stochastiques*, Prob. Th. Rel. Fields **81** (1989), 29–77.
5. Ben Arous, G., Léandre, R., *Décroissance exponentielle du noyau de la chaleur sur la diagonale (II)*, Prob. Th. Rel. Fields **90** (1991), 377–402.
6. Bally, V., Millet, M., Sanz, M., *On the support of the solution to the stochastic heat equation*, preprint.
7. Bally, V., Pardoux, E., *Malliavin calculus for white noise driven SPDE's*, preprint.

8. Carmona, R., Nualart, D., *Random non-linear wave equations: smoothness of the solutions*, Prob. Th. Rel. Fields **79** (1988), 469–508.

9. Cairoli, R., Walsh, J. B., *Stochastic integrals in the plane*, Acta Math. **134** (1975), 111–183.

10. Davies, E. B., *Heat kernels and spectral theory*, Cambridge Univ. Press, 1989.

11. Doss, H., Dozzi, M., *Estimations de grandes déviations pour les processus à paramètre multidimensionnel*, Séminaire de Probabilités XX (Lect. Notes Math., vol. 1204, pp. 68–80), Springer-Verlag, 1986.

12. Florchinger, P., Léandre, R., *Décroissance non-exponentielle du noyau de la chaleur*, Prob. Th. Rel. Fields **95** (1993), 237–262.

13. Greiner, P., *An asymptotic expansion for the heat equation*, Arch. Rat. Mech. Anal. **41** (1971), 163–218.

14. Kusuoka, S., Stroock, D., *Applications of the Malliavin calculus*, Part III, J. Fac. Sci. Univ. Tokyo Sect. 1A Math. **34.2** (1987), 291–442.

15. Kusuoka, S. Stroock, D., *Applications of the Malliavin calculus*, Part I, Stochastic Analysis. Proc. Taniguchi Symp. Katata 1982, Kinokuniya, 1984, pp. 271–306.

16. Léandre, R., *Majoration de la densité d'une diffusion hypoelliptique*, Prob. Th. Rel. Fields **74** (1987), 429–451.

17. Léandre, R., *Intégration dans la fibre associée à une diffusion dégénérée*, Prob. Th. Rel. Fields **76** (1987), 341–358.

18. Léandre, R., *Volume de boules sous-riemaniennes et explosion du noyau de la chaleur au sens de Stein*, Séminaire de Probabilités XXIII. Lecture Notes in Math., vol. 1372, Springer-Verlag, 1990, pp. 426–448.

19. Léandre, R., *Développement asymptotique de la densité d'une diffusion dégénérée*, Forum Mathematicum **4** (1992), 341–358.

20. Léandre, R., *Uniform upper bound for hypoelliptic heat kernels with drift* (to appear: Journal of Mathematics of Kyoto University).

21. Léandre, R., Russo, F., *Estimation de Varadhan pour des diffusions à deux paramètres*, Prob. Th. Rel. Fields **84** (1990), 429–451.

22. Léandre, R., Russo, F., *Small stochastic perturbations of a one-dimensional wave equation*, Prob. Th. Rel. Fields **84** (1990), 429–451.

23. Léandre, R., Russo, F., *Estimation de la densité de la solution de l'équation de Zakai robuste* (to appear: Journal of Potential Analysis).

24. Malliavin, P., *Stochastic calculus of variations and hypoelliptic operators*, Proceedings International Symposium on SDE, Kyoto (K. Itô, ed.), Kinokuniya, Tokyo, 1978.

25. Molchanov, S. A., *Diffusion processes and Riemannian geometry*, Russian Math. Survey **30** (1975), 1–30.

26. Minakshisundaram, S., Pleijel, A., *Some properties of the eigenfunctions of the Laplace operator on Riemannian manifolds*, Canada J. Math. **1** (1949), 242–256.

27. Millet, A., Sanz, M., *The support of a hyperbolic partial differential equation*, Submitted: Prob. Th. Rel. Fields.

28. Nualart, D., Sanz, M., *Malliavin calculus for two-parameter Wiener functionals*, Wahrscheinlichkeitsth. Verw. Geb. **70** (1985), 573–590.

29. Ocone, D., *Stochastic calculus of variations for partial differential equations*, Journal of Functional Analysis **79** (1976), 247–320.

30. Pardoux, E., Tusheng, Z., *Absolute continuity of the law of the solution of a parabolic SPDE*, Journal of Funct. Analysis **112, 2** (1993), 447–458.

31. Sowers, R., *Large deviations for a reaction-diffusion equation with non-Gaussian perturbations*, Ann. of Prob. **20, 1** (1992), 504–537.

32. Takanobu, S., *Diagonal short time asymptotics of heat kernels for degenerate second order operator of Hoermander type*, Publ. Rims, vol. 24, Kyoto University, 1988, pp. 169–203.

33. Takanobu, S., *Diagonal estimates of transition probability density of certain degenerate diffusion processes*, Journal of Functional Analysis **91** (1990), 221–236.

34. Takanobu, S., Watanabe, S., *Asymptotic expansion formulas of Schilder type for a class of conditional Wiener functional integration*, Asymptotic problems in probability theory: Wiener functionals and asymptotics (K. Elworthy, N. Ikeda, eds.), Pitman 284, Longman, 1992, pp. 194–241.

35. Walsh, J. B., *An introduction to stochastic partial differential equations*, École d'Été de Probabilités de Saint-Flour XIV. (Lect. Notes Math., vol. 1180, pp. 266–437), Springer-Verlag, Berlin Heidelberg New York, 1986.

36. Watanabe, S., *Lectures on stochastic differential equations and Malliavin Calculus*, Tata Institute of Fundamental Research, Bombay, 1984.

37. Watanabe, S., *Analysis of Wiener Functionals (Malliavin Calculus) and its applications to heat kernels)*, Ann. Probab. **15** (1987), 1–39.

38. Zhang, H, *Développement en temps petit de la solution de l'équation de Zakai et résolution numérique par maillage adaptatif*, Thèse Université de Provence, 1992.

RÉMI LÉANDRE, DÉPARTEMENT DE MATHÉMATIQUES, UNIVERSITÉ DE NANCY I, B.P. 239, 54506 VANDOEUVRE-LES-NANCY CEDEX, FRANCE

FRANCESCO RUSSO, UNIVERSITÉ PARIS-NORD, DÉPARTEMENT DE MATHÉMATIQUES, INSTITUT GALILÉE, AV. JEAN BAPTISTE CLÉMENT, F-93430 VILLETANEUSE AND UNIVERSITÄT BIELEFELD, BIBOS, D-33615 BIELEFELD 1

Progress in Probability, Vol. 36

ALMOST SURE CONVERGENCE OF
STOCHASTIC DIFFERENTIAL EQUATIONS
OF JUMP-DIFFUSION TYPE

C. W. Li

ABSTRACT. We consider a class of stochastic differential equations of jump-diffusion type

$$dx_t = f(t, x_t)\, dt + G(t, x_t)\, dW_t + \int_U h(t-, u, x_{t-})\, N(dt, du),$$

where f, G, and h are càdlàg adapted processes; W_t is an m-dimensional standard Brownian motion and $N(dt, du)$ is a Poisson random counting measure with deterministic characteristic measure $\lambda(du)$ on the measurable space U. The pathwise unique strong solution and the stability problem for a sequence of stochastic systems with small perturbations in almost sure convergence are discussed by means of a stochastic version of Gronwall's inequality.

§1 INTRODUCTION

Stochastic differential equations of jump-diffusion type have been rapidly used to model a lot of physical systems with impulse phenomenons. Uniqueness of the strong solution in mean square convergence or in probability has been discussed in many text books e. g. [1–3] while that in almost sure convergence is investigated in Gal'chuk [4] by a stochastic version of Gronwall's inequality. We will follow this approach to show uniqueness, local boundedness, stability and related results of stochastic differential equations of jump-diffusion type. This provides a new technique in stochastic analysis.

Another approach, using Tanaka's formula for one-dimensional stochastic systems with Poisson jumps, is shown by Situ Rong [5]. A typical problem in optimal stochastic control using a stochastic Gronwall inequality to prove almost sure convergence is shown in Li and Blankenship [6].

§2 PRELIMINARY RESULTS

Let $(\Omega, \mathcal{F}, \mathcal{P})$ be a complete probability space equipped with a filtration $\{\mathcal{F}_t\}_{t\geq 0}$ of an increasing family of right-continuous sub-σ-algebras of $\mathcal{F}$ such that the stochastic basis $(\Omega, \mathcal{F}, \{\mathcal{F}_t\}, \mathcal{P})$ is a complete filtered probability space.

1991 *Mathematics Subject Classification.* 60G55, 60H20, 60J75, 93E15.

Key words and phrases. Jump-diffusion equation, Poisson random point measure, stochastic Gronwall inequality, almost sure convergence, stability.

The following lemmas concern the explicit unique solution of linear stochastic integral equations.

Lemma 2.1. *Let Z be a semimartingale with $Z_0 = 0$. Then the stochastic exponential equation*

$$X_t = 1 + \int_0^t X_{s-} \, dZ_s \qquad (2.1)$$

admits the unique solution

$$X_t = e^{Z_t - \frac{1}{2}[Z,Z]_t^c} \prod_{0 < s \le t} (1 + \Delta Z_s) e^{-\Delta Z_s}. \qquad (2.2)$$

Proof. The result is well-known. See Protter [3]. □

Denote the stochastic exponential (Doléans-Dade exponential) by $\mathcal{E}(Z)_t = X_t$, where X_t is given by expression (2.2).

Lemma 2.2. *Let H and Z be semimartingales with $Z_0 = 0$. Then the linear stochastic integral equation*

$$X_t = H_t + \int_0^t X_{s-} \, dZ_s \qquad (2.3)$$

admits the unique solution

$$
\begin{aligned}
X_t &= \mathcal{E}_H(Z)_t \\
&\triangleq \mathcal{E}(Z)_t \left\{ H_0 + \int_{0+}^t \mathcal{E}(Z)_{s-}^{-1} \, d(H_s - [H,Z]_s) \right. \\
&\quad \left. + \sum_{0 < s \le t} \mathcal{E}(Z)_{s-}^{-1} \frac{\Delta Z_s^2}{1 + \Delta Z_s} \Delta H_s \right\}
\end{aligned}
\qquad (2.4)
$$

where $[\,,\,]$ denotes the quadratic covariation. If Z is of finite variation, then

$$\mathcal{E}_H(Z)_t = \mathcal{E}(Z)_t \left\{ H_0 + \int_{0+}^t \mathcal{E}(Z)_{s-}^{-1} \, d(H_s - J_s) \right\} \qquad (2.5)$$

where

$$J_t = \sum_{0 < s \le t} \frac{\Delta Z_s}{1 + \Delta Z_s} \Delta H_s. \qquad (2.6)$$

Proof. Apply the method of variation of constants, we assume that $X_t = C_t Y_t$ and $Y_t = \mathcal{E}(Z)_t$. Integration by parts gives

$$
\begin{aligned}
dX_t &= C_{t-} \, dY_t + Y_{t-} \, dC_t + d[C,Y]_t = C_{t-} Y_{t-} \, dZ_t + Y_{t-} \, dC_t + Y_{t-} \, d[C,Z]_t \\
&= X_{t-} \, dZ_t + Y_{t-} \, d\{C_t + [C,Z]_t\}.
\end{aligned}
$$

Since
$$dX_t = dH_t + X_{t-} \, dZ_t,$$

we have
$$dH_t = Y_{t-} \, d\{C_t + [C, Z]_t\}$$

and $Y_t = \mathcal{E}(Z)_t \neq 0$ for all t, so that

$$\int_0^t Y_{s-}^{-1} \, dH_s = C_t + [C, Z]_t. \tag{2.7}$$

As $[C, Z]$ is of finite variation,

$$[[C, Z], Z] = [C, Z]_0 Z_0 + \sum_{0 < s \leq t} \Delta[C, Z]_s \Delta Z_s = \sum_{0 < s \leq t} \Delta C_s \Delta Z_s^2. \tag{2.8}$$

From (2.7)
$$Y_{t-}^{-1} \Delta H_t = \Delta C_t + \Delta C_t \Delta Z_t = \Delta C_t (1 + \Delta Z_t)$$

or
$$\Delta C_t = \frac{\Delta H_t}{Y_{t-}(1 + \Delta Z_t)}. \tag{2.9}$$

Using (2.8) and (2.9), the quadratic covariation of both sides of (2.7) with Z becomes

$$\int_{0+}^t Y_{s-}^{-1} \, d[H, Z]_s = [C, Z]_t + \sum_{0 < s \leq t} \frac{\Delta H_s \Delta Z_s^2}{Y_{s-}(1 + \Delta Z_s)}$$

or
$$[C, Z]_t = \int_{0+}^t Y_{s-}^{-1} \, d[H, Z]_s - \sum_{0 < s \leq t} \frac{\Delta H_s \Delta Z_s^2}{Y_{s-}(1 + \Delta Z_s)},$$

so that from (2.7),

$$C_t = H_0 + \int_{0+}^t Y_{s-}^{-1} \, d(H_s - [H, Z]_s) + \sum_{0 < s \leq t} \frac{\Delta H_s \Delta Z_s^2}{Y_{s-}(1 + \Delta Z_s)}. \tag{2.10}$$

The result (2.4) follows. Suppose that Z is of finite variation. Then

$$[H, Z]_t = \sum_{0 < s \leq t} \Delta H_s \Delta Z_s,$$

so that (2.10) becomes

$$C_t = H_0 + \int_{0+}^t Y_{s-}^{-1} \, d\left(H_s - \sum_{0 < \tau \leq s} \frac{\Delta H_\tau \Delta Z_\tau}{1 + \Delta Z_\tau}\right)$$

and the result (2.5) follows. $\square$

Lemma 2.3. *If H is not a semimartingale but only a càdlàg (it has sample paths which are right-continuous with left-hand limits almost surely), $\{\mathcal{F}_t\}$-adapted process, then the solution of (2.3) becomes*

$$X_t = H_t + \mathcal{E}(Z)_t \left\{ \int_{0+}^t \mathcal{E}(Z)_{s-}^{-1} H_{s-} \, d(Z_s - [Z, Z]_s) \right. \tag{2.11}$$

$$\left. + \sum_{0 < s \le t} \mathcal{E}(Z)_{s-}^{-1} H_{s-} \frac{\Delta Z_s^3}{1 + \Delta Z_s} \right\}.$$

If Z is of finite variation, then the solution of (2.3) becomes

$$X_t = H_t + \mathcal{E}(Z)_t \left\{ \int_{0+}^t \mathcal{E}(Z)_{s-}^{-1} H_{s-} \, d(Z_s - \tilde{J}_s) \right\}, \tag{2.12}$$

where

$$\tilde{J}_t = \sum_{0 < s \le t} \frac{\Delta Z_s^2}{1 + \Delta Z_s}. \tag{2.13}$$

Proof. Consider $\tilde{X}_t = X_t - H_t$. Then

$$\tilde{X}_t = \int_0^t X_{s-} \, dZ_s = K_t + \int_0^t \tilde{X}_{s-} \, dZ_s, \tag{2.14}$$

where

$$K_t = \int_0^t H_{s-} \, dZ_s$$

is a semimartingale. Applying Lemma 2.2 to (2.14), we get

$$\tilde{X}_t = \mathcal{E}(Z)_t \left\{ K_0 + \int_{0+}^t \mathcal{E}(Z)_{s-}^{-1} \, d(K_s - [K, Z]_s) + \sum_{0 < s \le t} \mathcal{E}(Z)_{s-}^{-1} \frac{\Delta Z_s^2}{1 + \Delta Z_s} \Delta K_s \right\}.$$

Since $K_0 = 0$, $\Delta K_t = H_{t-} \Delta Z_t$, and

$$[K, Z]_t = \int_0^t H_{s-} \, d[Z, Z]_s$$

the result (2.11) follows. If Z is of finite variation, then

$$\tilde{X}_t = \mathcal{E}(Z)_t \left\{ K_0 + \int_{0+}^t \mathcal{E}(Z)_{s-}^{-1} \, d\left(K_s - \sum_{0 < \tau \le s} \frac{\Delta Z_\tau}{1 + \Delta Z_\tau} \Delta K_\tau \right) \right\}$$

$$= \mathcal{E}(Z)_t \int_{0+}^t \mathcal{E}(Z)_{s-}^{-1} H_{s-} \, d\left(Z_s - \sum_{0 < \tau \le s} \frac{\Delta Z_\tau^2}{1 + \Delta Z_\tau} \right)$$

and the result (2.12) follows. □

Remark. The proofs of Lemma 2.2 and Lemma 2.3 are just adapted from the approach in Protter [3] in which the case when Z is a continuous semimartingale is shown. See also Jacod [2].

Remark. Expressions (2.4) and (2.5) are equivalent to those of (2.11) and (2.12), respectively, if H is a semimartingale.

Theorem 2.4 (Stochastic Gronwall Inequality). *Let Z be an increasing process that is continuous from the right with $Z_0 = 0$ and let H be an $\{\mathcal{F}_t\}$-adapted càdlàg process. Suppose that V is an $\{\mathcal{F}_t\}$-adapted càdlàg process which satisfies the stochastic integral inequality*

$$V_t \leq H_t + \int_0^t V_{s-}\, dZ_s \qquad (2.15)$$

almost surely for all t. Then

$$V_t \leq H_t + \mathcal{E}(Z)_t \left\{ \int_{0+}^t \mathcal{E}(Z)_{s-}^{-1} H_{s-}\, d(Z_s - \tilde{J}_s) \right\} \quad a.\,s. \qquad (2.16)$$

with $\tilde{J}_t$ defined by (2.13). If H is a semimartingale, then

$$V_t \leq \mathcal{E}_H(Z)_t = \mathcal{E}(Z)_t \left\{ H_0 + \int_{0+}^t \mathcal{E}(Z)_{s-}^{-1}\, d(H_s - J_s) \right\} \quad a.\,s. \qquad (2.17)$$

with J_t defined by (2.6).

Proof. Let

$$G_t = V_t - \int_0^t V_{s-}\, dZ_s.$$

Then

$$V_t = G_t + \int_0^t V_{s-}\, dZ_s$$

admits the unique solution

$$V_t = G_t + \mathcal{E}(Z)_t \int_{0+}^t \mathcal{E}(Z)_{s-}^{-1} G_{s-}\, d(Z_s - \tilde{J}_s) \quad \text{a.\,s.} \qquad (2.18)$$

by Lemma 2.3. As

$$\tilde{J}_t = \sum_{0 < s \leq t} \frac{\Delta Z_s^2}{1 + \Delta Z_s} \leq \sum_{0 < s \leq t} \Delta Z_s \leq Z_t,$$

$Z_t - \tilde{J}_t$ is an increasing process, $\mathcal{E}(Z)_t^{-1} > 0$, and $G_t \leq H_t$, so that (2.16) follows from (2.18). Note that the right-hand side of (2.16) and that of (2.17) are equivalent if H is a semimartingale. $\square$

Remark. (2.17) is also shown in Gal'chuk [4] by a different approach.

Corollary 2.5. *If H is a local continuous martingale with $H_0 = 0$ and V is a non-negative $\{\mathcal{F}_t\}$-adapted càdlàg process that satisfies (2.15), then $V \equiv 0$ a.\,s.*

Proof. Here $J_t = 0$. The stochastic Gronwall inequality (2.17) implies that

$$0 \leq E\{\mathcal{E}(Z)_t^{-1} V_t\} \leq E\left\{ \int_{0+}^t \mathcal{E}(Z)_{s-}^{-1}\, dH_s \right\} = 0.$$

Therefore, $\mathcal{E}(Z)_t^{-1} V_t \equiv 0$ and the result follows as $\mathcal{E}(Z)_t^{-1} > 0$. $\square$

§3 MAIN RESULTS

We consider a class of stochastic differential equations of jump-diffusion type

$$dx_t = f(t, x_t)\, dt + G(t, x_t)\, dW_t + \int_U h(t-, u, x_{t-})\, N(dt, du) \qquad (3.1)$$

almost surely for all $t \geq 0$. Here W_t is an $\{\mathcal{F}_t\}$-adapted m-dimensional independent standard Brownian motion; $N(dt, du)$ is a Poisson random counting measure which is adapted to $\mathcal{F}_t \otimes \mathcal{B}(U)$ with deterministic finite characteristic measure $\lambda(du)$ on the measurable space $(U, \mathcal{B}(U))$ so that $\tilde{N}(dt, du) = N(dt, du) - \lambda(du)\, dt$ is a Poisson random martingale measure and so multiple, simultaneous jumps are of probability zero and there are only a finite number of jumps within each bounded interval almost surely. We also assume that W_t and $N(dt, du)$ are independent. Suppose further that for each $t \geq 0$, $f(t, x)$ is an n-dimensional càdlàg $\mathcal{F}_t \otimes \mathcal{B}(\mathbb{R}^n)$ jointly measurable vector process, $G(t, x)$ is an $n \times m$ càdlàg $\mathcal{F}_t \otimes \mathcal{B}(\mathbb{R}^n)$ jointly measurable matrix process while $h(t, u, x)$ is an n-dimensional càdlàg $\mathcal{F}_t \otimes \mathcal{B}(U) \otimes \mathcal{B}(\mathbb{R}^n)$ jointly measurable vector process. Denote

$$\|x\| = (x_1^2 + \cdots + x_n^2)^{1/2}, \quad x \in \mathbb{R}^n,$$

$$\||A\|| = \left(\sum_{k=1}^{m} \sum_{i=1}^{n} A_{ik}^2 \right)^{1/2}, \quad A \in \mathbb{R}^{n \times m}.$$

The existence of a local strong solution of (3.1) is easy to obtain if f, G, and h are of locally at most linear growth, i.e., for each $t \geq 0$, $u \in U$, and all $x \in \mathbb{R}^n$,

$$\|f(t, x)\|^2 \leq \alpha_t (1 + \|x\|^2),$$
$$\||G(t, x)\||^2 \leq \beta_t (1 + \|x\|^2), \qquad (3.2)$$
$$\|h(t, u, x)\|^2 \leq \gamma_t(u)(1 + \|x\|^2)$$

for some non-negative locally bounded $\{\mathcal{F}_t\}$-adapted processes α_t, β_t, and $\gamma_t(u)$ which are independent of the Brownian motion W_t and the Poisson random point measure $N(dt, du)$. See [1, 2] for details. Here we only concentrate on the pathwise uniqueness of a strong solution of the jump-diffusion stochastic differential equation (3.1).

Assume further that f, G, and h satisfy the local Lipschitz conditions, i.e., for each $t \geq 0$, $u \in U$, and all $x, y \in \mathbb{R}^n$,

$$\|f(t, x) - f(t, y)\| \leq \hat{\alpha}_t \|x - y\|,$$
$$\||G(t, x) - G(t, y)\|| \leq \hat{\beta}_t \|x - y\|, \qquad (3.3)$$
$$\|h(t, u, x) - h(t, u, y)\| \leq \hat{\gamma}_t(u) \|x - y\|$$

for some non-negative locally bounded $\{\mathcal{F}_t\}$-adapted processes $\hat{\alpha}_t$, $\hat{\beta}_t$, and $\hat{\gamma}_t(u)$ which are independent of the Brownian motion W_t and the Poisson random measure $N(dt, du)$. Suppose that x_t and y_t are strong solutions of (3.1) in the sense that x_t and y_t are $\{\mathcal{F}_t\}$-adapted càdlàg processes that satisfy almost surely the integral form of stochastic differential equation (3.1) with given initial condition x_0 and y_0, respectively. Apply Itô's differential rule to obtain

$$
\begin{aligned}
d\|x_t - y_t\|^2 &= 2(x_t - y_t)^{\mathrm{T}} \{ (f(t, x_t) - f(t, y_t))\, dt + (G(t, x_t) - G(t, y_t))\, dW_t \} \\
&\quad + \|(G(t, x_t) - G(t, y_t))^{\mathrm{T}} (1 \ldots 1)^{\mathrm{T}}\|^2\, dt \\
&\quad + \int_U \{ \|x_{t-} - y_{t-} + h(t-, u, x_{t-}) - h(t-, u, y_{t-})\|^2 \\
&\qquad\quad - \|x_{t-} - y_{t-}\|^2 \}\, N(dt, du) \\
&\leq 2(x_t - y_t)^{\mathrm{T}} \{ f(t, x_t) - f(t, y_t) + n\||G(t, x_t) - G(t, y_t)\||^2 \}\, dt \\
&\quad + 2(x_t - y_t)^{\mathrm{T}} (G(t, x_t) - G(t, y_t))\, dW_t \\
&\quad + \int_U \{ 2(x_{t-} - y_{t-})^{\mathrm{T}} (h(t-, u, x_{t-}) - h(t-, u, y_{t-})) \\
&\qquad\quad + \|h(t-, u, x_{t-}) - h(t-, u, y_{t-})\|^2 \}\, N(dt, du).
\end{aligned}
$$

After simplification and using the Lipschitz conditions (3.3),

$$
\|x_t - y_t\|^2 \leq \|x_0 - y_0\|^2 + B_t + \int_0^t \|x_{s-} - y_{s-}\|^2\, dZ_s, \tag{3.4}
$$

where

$$
B_t = 2 \int_0^t (x_s - y_s)^{\mathrm{T}} (G(s, x_s) - G(s, y_s))\, dW_s, \tag{3.5}
$$

$$
Z_t = \int_0^t (2\hat{\alpha}_s + n\hat{\beta}_s^2)\, ds + \int_0^t \int_U (2\hat{\gamma}_{s-}(u) + \hat{\gamma}_{s-}^2(u))\, N(ds, du) \tag{3.6}
$$

are a continuous local martingale and a local increasing process, respectively. By Corollary 2.5, we know that $\|x_t - y_t\|^2 \equiv 0$ a. s. if $x_0 = y_0$ a. s. Hence we have proved the following theorem.

Theorem 3.1 (Uniqueness). *If the at most linear growth conditions (3.2) and the local Lipschitz conditions (3.3) hold, then the jump-diffusion stochastic differential equation (3.1) has a local pathwise unique strong solution almost surely.*

Remark. The local result is justified if we define $\{\mathcal{F}_t\}$-adapted stopping times for $r \geq 0$ by

$$
T_r = \inf\{ t \geq 0 \colon B_t > r,\ Z_t > r \} \tag{3.7}
$$

such that (3.4) is defined locally and the result follows if we let $r \to +\infty$.

Now, suppose that $\{x_{k,t}\}_{k\geq 1}$ are the almost sure local pathwise unique strong solutions corresponding to the jump-diffusion stochastic differential equations

$$dx_t = f_k(t, x_t)\, dt + G_k(t, x_t)\, dW_t + \int_U h_k(t-, u, x_{t-})\, N(dt, du) \quad \text{a. s.} \qquad (3.8)$$

Assume that the following stability conditions hold, i. e., for each $t \geq 0$,

$$\lim_{k\to\infty} \int_0^t \sup_{x\in\mathbb{R}^n} \|f_k(s, x) - f(s, x)\|^2\, ds = 0 \quad \text{a. s.,}$$

$$\lim_{k\to\infty} \int_0^t \sup_{x\in\mathbb{R}^n} |||G_k(s, x) - G(s, x)|||^2\, ds = 0 \quad \text{a. s.,} \qquad (3.9)$$

$$\lim_{k\to\infty} \int_0^t \int_U \sup_{x\in\mathbb{R}^n} \|h_k(s-, u, x) - h(s-, u, x)\|^2\, N(ds, du) = 0 \quad \text{a. s.}$$

Without loss of generality, we can define stopping times similar to (3.7) such that the following results hold locally:

$$d\|x_{k,t}\|^2 = 2x_{k,t}^T\{f_k(t, x_{k,t})\, dt + G_k(t, x_{k,t})\, dW_t\} + \left\|G_k(t, x_{k,t})^T(1\ldots 1)^T\right\|^2 dt$$

$$+ \int_U \{\|x_{k,t-} + h_k(t-, u, x_{k,t-})\|^2 - \|x_{k,t-}\|^2\}\, N(dt, du)$$

$$\leq (\|x_{k,t}\|^2 + 2\|f(t, x_{k,t})\|^2 + 2\|f_k(t, x_{k,t}) - f(t, x_{k,t})\|^2)\, dt$$

$$+ 2n|||G(t, x_{k,t})|||^2\, dt + 2n|||G_k(t, x_{k,t}) - G(t, x_{k,t})|||^2\, dt$$

$$+ 2x_{k,t}^T G_k(t, x_{k,t})\, dW_t + \int_U \{\|x_{k,t-}\|^2 + 4\|h(t-, u, x_{k,t-})\|^2$$

$$+ 4\|h_k(t-, u, x_{k,t-}) - h(t-, u, x_{k,t-})\|^2\}\, N(dt, du).$$

After simplification and using the at-most-linear-growth conditions (3.2) and stability conditions (3.9), we have

$$\|x_{k,t}\|^2 \leq \|x_{k,0}\|^2 + B_{k,t} + P_t + Q_{k,t} + \int_0^t \|x_{k,s-}\|^2\, d\hat{Z}_s, \qquad (3.10)$$

where

$$B_{k,t} = 2\int_0^t x_{k,s}^T G_k(s, x_{k,s})\, dW_s$$

is a continuous local martingale and

$$P_t = \int_0^t 2(\alpha_s + n\beta_s)\, ds + \int_0^t \int_U 4\gamma_{s-}(u)\, N(ds, du),$$

$$Q_{k,t} = \int_0^t \{2\|f_k(s, x_{k,s}) - f(s, x_{k,s})\|^2 + 2n|||G_k(s, x_{k,s}) - G(s, x_{k,s})|||^2\}\, ds$$

$$+ \int_0^t \int_U 4\|h_k(s-, u, x_{k,s-}) - h(s-, u, x_{k,s-})\|^2\, N(ds, du), \qquad (3.11)$$

$$\hat{Z}_t = \int_0^t (1 + 2\alpha_s + 2n\beta_s)\, ds + \int_0^t \int_U (1 + 4\gamma_{s-}(u))\, N(ds, du)$$

are local increasing processes. By the stochastic Gronwall inequality (Theorem 2.4),

$$\|x_{k,t}\|^2 \le \mathcal{E}(\hat{Z})_t \left\{ \|x_{k,0}\|^2 + \int_{0+}^{t} \mathcal{E}(\hat{Z})_{s-}^{-1} \, d(B_{k,s} + P_s + Q_{k,s} - \hat{J}_s) \right\} \quad \text{a. s.} \quad (3.12)$$

with

$$\hat{J}_t = \sum_{a < s \le t} \frac{\Delta \hat{Z}_s}{1 + \Delta \hat{Z}_s} (\Delta B_{k,s} + \Delta P_s + \Delta Q_{k,s})$$

$$= 2 \int_{0+}^{t} \int_{U} \left\{ \gamma_{s-}(u) + \|h_k(s-, u, x_{k,s-}) - h(s-, u, x_{k,s-})\|^2 \right\}$$

$$\times \frac{1 + 4\gamma_{s-}(u)}{1 + 2\gamma_{s-}(u)} \, N(ds, du). \quad (3.13)$$

As B_k, P, Q_k, $\hat{J}$, and $\mathcal{E}(\hat{Z})$ are a. s. locally bounded by the stability conditions (3.9), then so is $\|x_{k,t}\|^2$ by (3.12).

Let $x_{\infty,t}$ be the almost surely local pathwise unique strong solution of (3.1). Then

$$d\|x_{k,t} - x_{\infty,t}\|^2$$
$$= 2(x_{k,t} - x_{\infty,t})^{\mathrm{T}} \{ (f_k(t, x_{k,t}) - f(t, x_{\infty,t})) \, dt + (G_k(t, x_{k,t}) - G(t, x_{\infty,t})) \, dW_t \}$$
$$+ \left\| (G_k(t, x_{k,t}) - G(t, x_{\infty,t}))^{\mathrm{T}} (1 \ldots 1)^{\mathrm{T}} \right\|^2 dt$$
$$+ \int_{U} \{ \|x_{k,t-} - x_{\infty,t-} + h_k(t-, u, x_{k,t-}) - h(t-, u, x_{\infty,t-})\|^2$$
$$- \|x_{k,t-} - x_{\infty,t-}\|^2 \} \, N(dt, du)$$

$$\le \{ \|x_{k,t} - x_{\infty,t}\|^2 + 2\|f_k(t, x_{k,t}) - f(t, x_{k,t})\|^2 + 2\|f(t, x_{k,t}) - f(t, x_{\infty,t})\|^2 \} \, dt$$
$$+ \{ 2n\|\|G_k(t, x_{k,t}) - G(t, x_{k,t})\|\|^2 + 2n\|\|G(t, x_{k,t}) - G(t, x_{\infty,t})\|\|^2 \} \, dt$$
$$+ 2(x_{k,t} - x_{\infty,t})^{\mathrm{T}} (G_k(t, x_{k,t}) - G(t, x_{\infty,t})) \, dW_t$$
$$+ \int_{U} \{ \|x_{x,t-} - x_{\infty,t-}\|^2 + 4\|h_k(t-, u, x_{k,t-}) - h(t-, u, x_{k,t-})\|^2$$
$$+ 4\|h(t-, u, x_{k,t-}) - h(t-, u, x_{\infty,t-})\|^2 \} \, N(dt, du).$$

Then, after simplification using the local Lipschitz conditions (3.3),

$$\|x_{k,t} - x_{\infty,t}\|^2 \le \|x_{k,0} - x_{\infty,0}\|^2 + B_{k,t}^* + Q_{k,t}^* + \int_0^t \|x_{k,s-} - x_{\infty,s-}\|^2 \, dZ_s^*, \quad (3.14)$$

where

$$B_{k,t}^* = 2 \int_0^t (x_{k,s} - x_{\infty,s})^{\mathrm{T}} (G_k(s, x_{k,s}) - G(s, x_{\infty,s})) \, dW_s$$

is a local continuous martingale and

$$Q_{k,t}^* = \int_0^t \left\{ 2\|f_k(t, x_{k,t}) - f(t, x_{k,t})\|^2 + 2n\|\|G_k(t, x_{k,t}) - G(t, x_{k,t})\|\|^2 \right\} dt$$

$$+ \int_0^t \int_U 4\|h_k(t-, u, x_{k,t-}) - h(t-, u, x_{k,t-})\|^2 \, N(ds, du),$$

$$Z_t^* = \int_0^t (1 + 2\hat{\alpha}_s^2 + 2n\hat{\beta}_s^2) \, ds + \int_0^t \int_U (1 + 4\hat{\gamma}_{s-}^2(u)) \, N(ds, du)$$

$$(3.15)$$

are local increasing processes. Then, by the stochastic Gronwall inequality (Theorem 2.4),

$$\mathcal{E}(Z^*)_t^{-1} \|x_{k,t} - x_{\infty,t}\|^2$$

$$\leq \|x_{k,0} - x_{\infty,0}\|^2 + \int_{0+}^t \mathcal{E}(Z^*)_{s-}^{-1} \, d(B_{k,s}^* + Q_{k,s}^* - J_{k,s}^*) \quad \text{a. s.} \quad (3.16)$$

where

$$J_{k,t}^* = \sum_{0 < s \leq t} \frac{\Delta Z_s^*}{1 + \Delta Z_s^*} (\Delta B_{k,s}^* + \Delta Q_{k,s}^*)$$

$$= 2 \int_{0+}^t \int_U \frac{1 + 4\hat{\gamma}_s^2(u)}{1 + 2\hat{\gamma}_s^2(u)} \|h_k(s-, u, x_{k,s-}) - h(s-, u, x_{k,s-})\|^2 \, N(ds, du). \quad (3.17)$$

If we assume for the initial conditions

$$\lim_{k \to \infty} \|x_{k,0} - x_{\infty,0}\| = 0 \quad \text{a. s.,} \quad (3.18)$$

then, by the stability conditions (3.9), for each $t \geq 0$,

$$\lim_{k \to \infty} Q_{k,t}^* = 0, \quad \text{and} \quad \lim_{k \to \infty} J_{k,t}^* = 0 \quad \text{a. s.}$$

Applying the dominated convergence theorem to the right-hand side of (3.16), we know that the limit of the left-hand side of (3.16) exists locally. Thus,

$$0 \leq \mathrm{E}\left\{ \lim_{k \to \infty} \mathcal{E}(Z^*)_t^{-1} \|x_{k,t} - x_{\infty,t}\|^2 \right\} \leq \mathrm{E}\left\{ \lim_{k \to \infty} \int_{0+}^t \mathcal{E}(Z^*)_{s-}^{-1} \, dB_{k,s}^* \right\}$$

$$\leq \lim_{k \to \infty} \mathrm{E}\left\{ \int_{0+}^t \mathcal{E}(Z^*)_{s-}^{-1} \, dB_{k,s}^* \right\} = 0,$$

where the last inequality is justified by Fatou's lemma as the limit exists locally, so that

$$\lim_{k \to \infty} \|x_{k,t} - x_{\infty,t}\|^2 = 0 \quad \text{a. s.}$$

We summarize the above results in the following theorem:

Theorem 3.2 (Stability). *Suppose that $x_{k,t}$ for $k \geq 1$ are the respective almost sure local pathwise unique strong solutions of* (3.8) *and that conditions* (3.2), (3.3), *and* (3.9) *hold. Then* $\|x_{k,t}\|^2$ *are locally bounded. Let $x_{\infty,t}$ be the almost sure local pathwise unique strong solution of* (3.1) *such that* (3.18) *holds. Then almost surely* $\|x_{k,t} - x_{\infty,t}\|^2 \to 0$ *locally as $k \to \infty$.*

Remark. Similar results are obtained for the one-dimensional case by Situ Rong [5] using Tanaka's formula.

REFERENCES

1. N. Ikeda and S. Watanabe, *Stochastic Differential Equations and Diffusion Processes*, (2nd ed.), North-Holland, Amsterdam, 1989.
2. J. Jacod, *Stochastic Calculus and Martingale Problems*, Lecture Notes in Math., vol. 714, Springer-Verlag, 1979.
3. P. Protter, *Stochastic Integration and Differential Equations*, Springer-Verlag, 1990.
4. L. I. Gal'chuk, *Strong uniqueness of solution of stochastic integral equations for semimartingale components*, English translation, 1984, Math Notes **35, 1–2**, 157–161.
5. Situ Rong, *On strong solution, uniqueness, stability and comparison theorems for a stochastic system with Poisson jumps*, Distributed Parameter Systems, Springer-Verlag, Berlin, 1985, pp. 352–381.
6. C. W. Li and G. L. Blankenship, *Optimal stochastic control of linear stochastic systems with Poisson process disturbances*, Control-Theory and Advanced Technology **9, 4** (1993), 799–818.

C. W. Li, Department of Mathematics, City Polytechnic of Hong Kong, 83 Tat Chee Avenue, Kowloon, Hong Kong

Progress in Probability, Vol. 36
© 1995 Birkhäuser Verlag Basel/Switzerland

APPLICATIONS AND FOUNDATIONS
OF QUASI SURE ANALYSIS

PAUL MALLIAVIN

ABSTRACT. A short overview is given on foundations and applications of quasi sure analysis

§1 APPLICATIONS

We denote by X the classical Wiener space: X is the probability space of the $\mathbb{R}^d$-valued Brownian motion, starting from zero and running on the interval time $[0, 1]$. It is possible to define on X a whole scale of Sobolev spaces: $D_r^p(X)$. Let us denote by $\mathcal{I}$ the Itô functionals which are defined on X as solutions of stochastic differential equations with smooth coefficients. Now $\mathcal{I}$ are not continuous for the uniform norm topology, nevertheless they belong to all the Sobolev spaces. In finite dimensions the classical Sobolev imbedding says that $D_r^p(\mathbb{R}^n)$ is contained in the space of continuous functions if and only if $pr > n$. Therefore, (p, r) being fixed, the immersion into the continuous functions fails for n large enough. This fact explains the paradoxical regularity of $\mathcal{I}$. It could be shortly said that: *Quasi-sure analysis is an infinite-dimensional substitute to the Sobolev imbedding.*

It is also possible to give purely probabilistic motivations. The stochastic calculus of variations [17] has transferred to probability theory the tools of differential geometry; regularity criteria of the laws of Itô functionals have been obtained by this approach. An elementary fact is the existence of regular desintegration along a coordinate function of an $\mathbb{R}^d$-valued smooth random variable; the quasi-sure analysis will provide ([2], [27]) *regular desintegration of the Wiener measure along a smooth Itô functional.*

In classical probability theory, negligible events are events of probability zero; then the conditioning for a prescribed value of a random variable has no meaning; this fact excludes in this context the possibility to construct regular desintegrations. A smaller class of negligible sets, the *slim* sets will be introduced; this class will be stable under conditioning by an $\mathbb{R}^n$-valued smooth Itô functionnal. We shall get ([2], [27], [21]) the *descent principle: properties being true quasi-surely remain true a. s. under conditioning.* Typical examples of conditioning are the

1991 *Mathematics Subject Classification.* 31A15, 31B35, 31C15, 46F25, 60H07, 60J45.

Key words and phrases. Quasi sure analysis, regular desintegration of the Wiener measure, descent principle, quasisure theory of stochastic flews, capacities, slimsets.

Brownian bridge and its generalization to an arbitrary diffusion (the final value of the path is prescribed). There exist two classical approaches to this situation: Doob's h-theory and the Itô–Yor theory of enlargement of filtrations. The quasi-sure analysis needs differentiability hypothesisses which are not required in the classical approach. Under those more stringent hypotheses, it leads to exponential estimates which cannot be otherwise easily reached. When the conditioning consists in the equality of the inital and the final value, we get some loop spaces. In particular, using a bi-invariant elliptic operator on a compact Lie group G, we get [16] in the loop group $L(G)$ through those exponential estimates the following non-commutative Cameron–Martin type result: *The left (or the right) action of $L^1(G)$ on $L(G)$ leaves the measure quasi-invariant.*

All the results in probability theory which are formulated with the quantifier "almost sure" can be eventually strenghten in "quasi sure" results. Geometric theory of the Brownian curves can be considered in this way ([7], [28]); for instance the Wiener theorem telling that the quadratic variation of a Brownian curve is equal to 1, holds true quasi surely. In the same way Fatou–Doob almost-sure convergence of martingales can be strenghten into quasi-sure analogues; the same happens for Kolmogorov's version theorem. In this way ([23], [21], [13], [25]), appears to be a *theory of quasi-sure convergence (or version, or continuity).*

Outside of its own realm, quasi-sure analysis is appearing as an efficient tool in other branchs of probability: anticipative Girsanov transformations [4], anticipative SDE [22]. This methodology works in the case where the anticipative σ-field can be contained in a σ-field generated by a finite number of "independent" smooth random variables; in order to realize this situation it could be useful to extend the Wiener space into a larger probability space. For instance to solve a Stratanovitch SDE in which the non-adapteness comes only from the initial condition, we extend the Wiener space X into Y obtained by a direct product of X with the initial condition; the original problem on X becomes equivalent to a transferred problem on Y raised under an appropriate conditioning; making the quasi-sure analysis working on Y, the descent principle will solve the original problem. In order to make quasi-sure analysis working on Y we need ([23], [10], [22], [5]) *a quasi-sure theory of stochastic flows.*

In the finite-dimensional case, conditional laws of smooth random variables can be computed using the theory of differential forms, their restriction to a sub-manifold, and their integration. In order to carry out the same program in the infinite-dimensional case, the following prerequisites appear: to define a submanifold of finite codimension on the Wiener space, to define on such a manifold an area, to define the differential forms with their usual operations of restriction and integration. This program can be fulfilled; it leads to the expression of conditional laws as the product of the area by the inverse of the determinant of the covariance matrix. Conditonal divergence could be reached by the computation of the Ricci curvature of those submanifolds; the classical formalism of differential geometry leads here to divergent series which have to be properly renormalized. Conditional large deviations will depend upon appropriate implicit function theorems.

All these facts ([18], [9], [15], [1], [29], [6], [19], [30]) constitute the *quasi-sure differential geometry.*

§2 FOUNDATIONS

2.1 Universal Wiener space [11], [20]. We need to have the facility of extension by product which exists in the framework of abstract probability spaces. The Gross theory of Gaussian measures carried by Banach spaces satisfies this prerequisite. Nevertheless we shall work at the more intrinsic level of *universal Wiener spaces.* A numerical Wiener space will be the data of an abstract probability space on which we have chosen a separating sequence of independent Gaussian normal random variables g_i; a unitary transformation on the g_i defines a new sequence and therefore a new numerical Wiener space; the class of *all* numerical Wiener spaces constructed in this way will constitute the data of a *single* universal Wiener space. The stability by product of the class of numerical (or universal) Wiener spaces is clear.

2.2 Topology on a numerical Wiener space. A numerical Wiener space can be identified through $g = (g_i)$, $i \in \mathbb{N}$, to $\mathbb{R}^{\mathbb{N}}$; therefore it inherits the product topology of $\mathbb{R}^{\mathbb{N}}$; this topology is separable and metrisable.

2.3 Capacities on a numerical Wiener space [18], [28], [12], [8], [14]. Given $p > 1$, $r \in \mathbb{N}$, we consider the Sobolev space $D_r^p(\mathbb{R}^{\mathbb{N}})$ and its associated capacity $c_{p,r}$ defined for an open set O by

$$c_{p,r}(O) = \inf \|f\|_{p,r}, \quad f > 1_O \text{ a. s.,}$$

and prolonged to any set A as the infimum of the capacities of all the open sets containing A.

The (p,r)-capacity is *tight,* which means that there exists an increasing sequence of compacts K_n such that $c_{p,r}(K_n^c)$ converges to zero.

A function f defined on $\mathbb{R}^{\mathbb{N}}$ is *quasi-continuous* if there exists such sequence of compacts such that, furthermore, the restriction of f to K_n is continuous. A random variable $h \in D_r^p$ has a quasi-continuous version h^*; two such versions differ on a set of (p,r)-capacity zero.

Given a positive linear form l on D_r^p there exists a positive Borelian measure λ on $\mathbb{R}^{\mathbb{N}}$, which does not charge any Borelian of (p,r)-capacity zero, such that

$$l(h) = \int h^* \, d\lambda.$$

We call the family of such measures λ the family of *measures of (p,r)-finite energy.* A Borelian is of (p,r)-capacity zero if and only if it is not charged by all measures of (p,r)-finite energy ([27], [14]).

2.4 Capacities on a universal Wiener space. An unitary transformation U of the sequence (g_i) realizes the passage from one numerical model to another one. A capital fact [26] is that the quasi-continuous version U^* of U realizes a quasi-homeomorphism between the corresponding two numerical models preserving the class of measures of (p,r)-finite energy. Therefore the theory of (p,r)-capacities on two numerical models associated with the same universal Wiener space are canonically isomorphic: we have constructed in this way the theory of the (p,r)-capacity on a universal Wiener space.

The same invariance property holds true in the framework of the Gross theory of Gaussian measures on Banach spaces [3].

2.5 Quasi-sure analysis on a universal Wiener space. A set will be called *slim* if all its (p,r)-capacities are equal to zero.

Negligible sets in quasi-sure analysis are the slim sets.

REFERENCES

1. H. Airault, *Differential calculus on finite codimensional submanifold of the Wiener space*, J. Funct. Anal. **100** (1991), 291–316.
2. H. Airault and P. Malliavin, *Intégration géométrique sur l'espace de Wiener*, Bull. Sci. Math. **112** (1988), 3–52.
3. S. Albeverio, M. Fukushima, W. Hansen, Z. Ma, M. Röckner, *Invariance result for capacities on the Wiener space*, J. Func. Anal. **106** (1992), 35–49.
4. R. Buckdahn and H. Föllmer, *A conditional approach to the anticipating Girsanov transformation*, Probab. Theor. Rel. Fields **95** (1993), 311–330.
5. L. Denis, *Analyse quasi-sure de l'approximation d'Euler du flot des EDS*, C. R. Acad. Sci., Paris, 1993.
6. S. Fang, *Le calcul différentiel quasi-sure et son application à l'estimation du noyau de la chaleur*, Trans. Amer. Math. Soc. **339** (1993), 221–241.
7. M. Fukushima, *Basic properties of the Brownian motion and capacity on the Wiener space*, J. Math. Soc. Japan **36** (1984), 171–176.
8. M. Fukushima and H. Kaneko, *On (p,r)-capacities for general Markovian semi-group*, in: Inf. Dim. Ana. and Stoch. Proc. (S. Albeverio, ed.), Pitman, 1985.
9. E. Getzler, *Degree theory for Wiener map*, J. Func. Ana. **68** (1986), 388–403.
10. Z. Huang and J. Ren, *Quasi-sure stochastic flow*, Stoch. **33** (1990), 149–158.
11. K. Itô, *On Malliavin Tensor Fields*, Comm. Pure and Appl. Math **47** (1994), 1–27.
12. H. Kaneko, *On (p,r)-capacities for Markov processes*, Osaka J. Math. **23** (1984), 307–319.
13. T. Kazumi, *Refinements in terms of capacities of certain limit theorems on an abstract Wiener space*, J. Math. Kyoto Univ. **32** (1992), 1–33.
14. T. Kazumi and I. Shigekawa, *Measures of finite (p,r)-energy and potential on a separable metric space*, Sem. Prob. 26, Lecture Notes in Math., vol. 1526, 1992, pp. 415–444.
15. S. Kusuoka, *Analysis on the Wiener space, II, Differential forms*, J. Funct. Anal. **103** (1992), 229–274.
16. M. P. Malliavin and P. Malliavin, *Integration on loop groups, I, Quasi invariant measures*, J. Funct. Ana. **93** (1990), 207–237.
17. P. Malliavin, *Stochastic calculus of variations and hypoelliptic operators*, Int. Symp. on SDE (K. Itô, ed.), Kyoto, 1976, pp. 195–274.
18. P. Malliavin, *Implicit function in finite corank on the Wiener space*, Stochastic Analysis (K. Itô, ed.), Katata 1982, North-Holland, 1984, pp. 369–386.

19. P. Malliavin, *Infinite dimensional analysis*, Saint Chéron Round Tables, Bull. Sci. Math. (M. Yor, ed.), vol. 117, 1993, pp. 63–90.

20. P. Malliavin, *Universal Wiener Spaces*, San Felipe Conference (1993), Birkhäuser, 77–102.

21. P. Malliavin and D. Nualart, *Quasi-sure analysis and Stratanovitch anticipative SDE*, Prob. Th. and Rel. Fields **96** (1993), 45–55.

22. P. Malliavin and D. Nualart, *Quasi-sure analysis of stochastic flows*, J. Funct. Anal. **112** (1993), 287–317.

23. J. Ren, *Analyse quasi-sure des équations différentielles stochastiques*, Bull. Sci. Math. **112** (1990), 187–214.

24. J. Ren, *Topologie p-fine sur l'espace de Wiener et théorie des fonctions implicites*, Bull. Sci. Math. **114** (1990), 99–114.

24. J. Ren, *Analyse quasi-sure des martingales régulières*, C. R. Acad. Sci. Paris **317** (1993), 299–303.

26. I. Shigekawa, *A quasi homeomorphism of the Wiener space*, AMS Summer Institute, Cornell, July, 1993.

27. H. Sugita, *Positive generalized functions and potential theory over an abstract Wiener space*, J. Math. Kyoto Univ. **25** (1985), 717–725.

28. M. Takeda, *(p,r)-capacity on the Wieneer space and properties of the Brownian motion*, Z. Wahrsch. Verw. Gebiete **68** (1984), 149–162.

29. J. Van Biesen, *The divergence on submanifold of the Wiener space*, J. Funct. Ana. **113** (1993), 426–461.

30. N. Yoshida, *A large deviation principle for capacities on the Wiener space*, Prob. Th. Rel. Fields **94** (1993), 473–488.

PAUL MALLIAVIN, 10 RUE SAINT LOUIS EN L'ISLE, 75004 PARIS

Progress in Probability, Vol. 36

A DUALITY FORMULA ON THE POISSON SPACE
AND SOME APPLICATIONS

DAVID NUALART AND JOSEP VIVES

ABSTRACT. We establish a duality formula for the chaotic derivative operator on the canonical Poisson space. The adjoint of this operator is proved to coincide with the stochastic integration on predictable processes.

§1 INTRODUCTION

The stochastic calculus of variations for functionals of the Wiener process is based on the properties of the derivative operator D on the Wiener space and its adjoint δ which is called the divergence operator or the Skorohod integral ([7]).

In the case of a Poisson process different approaches have been developed in order to define a version of the operator D. A first method (see, for instance, Bismut [3], Bichteler, Gravereaux and Jacod [2], and Bass and Cranston [1]) is based on a perturbation of the jump sizes. On the other hand, in the approach introduced by Carlen and Pardoux [4], the derivative operator is introduced by means of a variation of the jump times. In both methods one deduces a type of integration by parts formula which allows to establish general criteria for regularity of probability laws.

In the case of the Brownian motion one way to define the derivative and the divergence operators is to use the Wiener chaos expansion of a square-integrable functional. Dermoune, Krée and Wu [5] have adapted this approach to the Poisson case using the expansion obtained through the Lévy–Khintchine representation. On the other hand, Privault [10] has studied the operators which arise from the Fock space structure associated to the infinite sequence of independent exponential interjump times. The derivative operator obtained in this way is related to the gradient which appears in Carlen and Pardoux's work.

In a previous paper [8] we have studied the annihilation and creation operators corresponding to the Fock space associated with the chaotic development of a square-integrable functional of the Poisson space. Unlike the Wiener space the annihilation operator D cannot be interpreted as a derivative operator but as a difference operator, namely, $D_t F(\omega) = F(\omega + \delta_t) - F(\omega)$, where ω is a point measure on the parameter space. The purpose of this work is to establish a duality

---

1991 *Mathematics Subject Classification.* 60H07, 60G55.

Key words and phrases. Poisson process, Fock space, orthogonal expansions, duality formula.

formula for this difference operator and its adjoint, which is proved in the canonical probability space of the Poisson process. As a consequence we can identify these operators with those defined by means of the chaotic expansion, and obtain their closure in L^2.

§2 THE POISSON PROCESS

Let $(T, \mathcal{B}, \lambda)$ be a measure space such that T is a locally compact space with countable basis, $\mathcal{B}$ is the Borel σ-field of T and λ is a continuous Radon measure.

Let $M_{\mathrm{p}}(T)$ be the space of Radon point measures over T. In other words, $m \in M_{\mathrm{p}}(T)$ if and only if $m = \sum_i \delta_{t_i}(\cdot)$, and any compact subset of T has only a finite number of points t_i. We can introduce on this space the σ-field $\mathcal{M}_{\mathrm{p}}$ generated by the maps $\phi_F(m) = m(F)$, $F \in \mathcal{B}$.

Definition 1. The *Poisson process* is a measurable mapping N from a probability space $(\Omega, \mathcal{F}, P)$ to $(M_{\mathrm{p}}(T), \mathcal{M}_{\mathrm{p}})$ which induces a probability law over $M_{\mathrm{p}}(T)$ such that:

(i) For all $F \in \mathcal{B}$ with $\lambda(F) < \infty$, $N(F)$ is a Poisson random variable with parameter $\lambda(F)$. If $F \in \mathcal{B}$ and $\lambda(F) = \infty$ then $N(F) = \infty$ a. s.

(ii) If $F_1, \ldots, F_k$ are sets of $\mathcal{B}$, pairwise disjoint, then $N(F_1), \ldots, N(F_k)$ are independent random variables.

We can introduce multiple stochastic integrals with respect to the Poisson process ([9]). Let us briefly recall their construction.

We will say that $f \in L^2(T^p)$ is a simple function if it has the form

$$f(t_1, \ldots, t_p) = 1_{B_1}(t_1) \cdots 1_{B_p}(t_p),$$

where $B_1, \ldots, B_p$ are pairwise disjoint sets in $\mathcal{B}$ with finite measure. The linear space S_p generated by simple functions is dense in $L^2(T^p)$. Given a simple function $f(t_1, \ldots, t_p)$, we can define its multiple stochastic integral with respect to the compensated Poisson process $\tilde{N} = N - \lambda$ in the following way

$$I_p(f) = \int_T \cdots \int_T f(t_1, \ldots, t_p) \, d\tilde{N}(t_1) \cdots d\tilde{N}(t_p) = \tilde{N}(B_1) \cdots \tilde{N}(B_p).$$

By convenience we take $I_0(c) = c$, for all $c \in \mathbb{R}$.

This definition can be extended by linearity to S_p. The linear map $I_p : S_p \longrightarrow L^2(\Omega)$ has the following properties:

(i)
$$I_p(f) = I_p(\tilde{f}),$$

where $\tilde{f}$ is the symmetrization of f, and

(ii)
$$E[I_p(f)I_q(g)] = \begin{cases} p!\langle \tilde{f}, \tilde{g} \rangle_{L^2(T^p)} & \text{if } p = q, \\ 0 & \text{if } p \neq q. \end{cases}$$

Notice that $\|I_p(f)\|_2 = \sqrt{p!}\|\tilde{f}\|_2 \leq \sqrt{p!}\|f\|_2$. Hence I_p is a bounded linear operator on S_p whose norm is bounded by $\sqrt{p!}$, and we can extend this operator to $L^2(T^p)$. In particular I_p is an isometry between the subspace $L_s^2(T^p)$ of symmetric functions of $L^2(T^p)$ with the modified norm $\sqrt{p!}\|f\|_2$, and the subspace C_p of $L^2(\Omega)$ generated by the $I_p(f)$, which is called the chaos of order p.

Theorem 1. *Assume that* $\mathcal{F} = \sigma\{N(A), A \in \mathcal{B}\}$. *Then we have the following orthogonal decomposition:*

$$L^2(\Omega) = \bigoplus_{n=0}^{\infty} C_n.$$

This theorem allows us to represent the space $L^2(\Omega)$ as a Fock space, and we can consider in this framework the operators D and δ that are in duality (see [8]).

In the sequel we will introduce the canonical Poisson space over $(T, \mathcal{B}, \lambda)$. In order to simplify the exposition we will assume that λ is a finite measure. Consider the disjoint union $\Omega = \bigcup_{n=0}^{\infty} T^n$, where $T^0 = \{a\}$, and a is a distinguished point. We can introduce on Ω a σ-field $\mathcal{F}$ defined as follows: $G \subset \Omega$ belongs to $\mathcal{F}$ if and only if $G \cap T^n \in \mathcal{B}^{\otimes n}$ for all $n \geq 1$. If λ^n is the product measure in T^n, for all $n \geq 1$, and λ^0 is δ_a, we define over $(\Omega, \mathcal{F})$ the probability P given by

$$P(G) = \sum_{n=0}^{\infty} e^{-\lambda(T)} \frac{1}{n!} \lambda^n(G \cap T^n).$$

With these ingredients we will introduce a Poisson process on this probability space $(\Omega, \mathcal{F}, P)$. Consider the mapping N from Ω into $M_p(T)$ such that for any $n \geq 1$ we have $N(s_1, \ldots, s_n) = \sum_{i=1}^{n} \delta_{s_i}$, and $N(a) = 0$. This mapping has the following properties:

   (i) $P\{N(T) = k\} = e^{-\lambda(T)} \frac{1}{k!} \lambda(T)^k$.
   (ii) $P\{G|N(T) = m\} = \lambda^m(G \cap T^m)/\lambda(T)^m$. In particular, P restricted to T^n is a measure equivalent to λ^n, for all $n \geq 1$.

From these two properties it follows that N is a Poisson process (see [6]). An essential property proved in [6] is the following:

Lemma 1. *Let* $\mathcal{B}_s^{\otimes n}$ *be the sub-σ-field of symmetric Borel sets of* T^n. *Let* $\mathcal{F}_s$ *be the σ-field generated by the sets* $G \in \mathcal{F}$ *such that* $G \cap T^n \in \mathcal{B}_s^{\otimes n}$. *Then, we have* $N^{-1}(\mathcal{M}_p(T)) = \mathcal{F}_s$.

This lemma means that measurable functionals with respect to the Poisson process will be functionals with symmetric projections over every T^n.

We will recall two additional properties of this probability measure P.

   (1) $P\{N(\{t\}) > 1 \text{ for some } t \in T\} = 0$.
   (2) For any fixed $t \in T$, $P\{N(\{t\}) \geq 1\} = 0$.

These properties are immediate using the fact that on every T^n the measures P and λ^n are equivalent.

§3 THE TRANSLATION OPERATOR AND ITS ADJOINT

We recall that the σ-field $\mathcal{F}_s$ introduced in Lemma 1 coincides with the σ-field generated by the Poisson process N. Consider a random variable F which is $\mathcal{F}_s$-measurable and fix $t \in T$. We define the random variable $\Psi_t F$ as follows

$$\Psi_t F(s_1, \ldots s_n) = F(s_1, \ldots s_n, t) - F(s_1, \ldots, s_n), \quad n \geq 1,$$
$$\Psi_t F(a) = F(t) - F(a).$$

On the other hand, if $u = \{u_s, s \in T\}$ is a measurable stochastic process such that $\int_T |u_s| \lambda(ds) < \infty$, we define a random variable Φu as follows

$$(\Phi u)(s_1, \ldots s_n) = \sum_{j=1}^n u_{s_j}(s_1, \ldots, \hat{s}_j, \ldots s_n) - \int_T u_s(s_1, \ldots s_n) \lambda(ds), \quad n \geq 2,$$

$$(\Phi u)(s_1) = u_{s_1}(a) - \int_T u_s(s_1) \lambda(ds),$$

and

$$(\Phi u)(a) = - \int_T u_s(a) \lambda(ds).$$

The operator Ψ is well defined from $L^0(\Omega)$ into $L^0(\Omega \times T)$, and, in the same way, the operator Φ maps $L^0(\Omega, L^1(T))$ in $L^0(\Omega)$. This is true because P restricted to T^n is a measure equivalent to λ^n, for all $n \geq 1$.

The duality relation between these two operators is given by the following integration by parts formula:

Theorem 2. *Consider $F \in L^0(\Omega)$ and $u \in L^0(\Omega \times T)$, such that $\int_T |u_s| \lambda(ds) < \infty$. Suppose that $E(\int_T |Fu_t| \lambda(dt)) < \infty$. Then $F\Phi u \in L^1(\Omega)$ if and only if $E(\int_T |\Psi_t Fu_t| \lambda(dt)) < \infty$, and in this case*

$$E[F\Phi u] = E\left(\int_T \Psi_t Fu_t \, \lambda(dt) \right). \tag{1}$$

Proof. We have

$$E[F\Phi u] = \sum_{n=2}^\infty \int_{T^n} \left[F(s_1, \ldots, s_n) \left(\sum_{j=1}^n u_{s_j}(s_1, \ldots, \hat{s}_j, \ldots, s_n) \right. \right.$$

$$\left. - \int_T u_s(s_1, \ldots s_n) \lambda(ds) \right) \bigg] dP$$

$$+ \int_T F(s_1) \left(u_{s_1}(a) - \int_T u_s(s_1) \lambda(ds) \right) dP + F(a) \int_T u_s(a) \lambda(ds).$$

On the other hand,

$$E\left(\int_T \Psi_t F u_t \, \lambda(dt)\right) = \sum_{n=1}^{\infty} \int_{T^n} \int_T [F(s_1, \ldots, s_n, t)$$

$$- F(s_1, \ldots, s_n)] u_t(s_1, \ldots, s_n) \, \lambda(dt) \, dP$$

$$+ \int_T [F(t) - F(a)] u_t(a) \, \lambda(dt).$$

Then it is sufficient to see for every $n \geq 2$

$$\int_{T^n} F(s_1, \ldots, s_n) \sum_{j=1}^n u_{s_j}(s_1, \ldots, \hat{s}_j, \ldots, s_n) \, dP$$

$$= \int_{T^{n-1}} \int_T F(s_1, \ldots, s_{n-1}, t) u_t(s_1, \ldots, s_{n-1}) \, \lambda(dt) \, dP.$$

The restriction to T^n of dP is given by

$$dP = \frac{e^{-\lambda(T)}}{n!} \lambda(ds_1) \cdots \lambda(ds_n).$$

Hence, we have to check that

$$\int_{T^n} F(s_1, \ldots, s_n) \sum_{j=1}^n u_{s_j}(s_1, \ldots, \hat{s}_j, \ldots, s_n) \frac{e^{-\lambda(T)}}{n!} \lambda(ds_1) \cdots \lambda(ds_n)$$

$$= \int_{T^{n-1}} \int_T F(s_1, \ldots, s_{n-1}, t) u_t(s_1, \ldots, s_{n-1}) \lambda(dt) \frac{e^{-\lambda(T)}}{(n-1)!} \lambda(ds_1) \cdots \lambda(ds_{n-1}),$$

and this is clear because F is a symmetric functional. $\square$

Now we will establish the relation between the operators Ψ and Φ introduced above and the operators D and δ associated with the structure of Fock space underlying to $L^2(\Omega)$. Let

$$T_*^m = \{ (s_1, \ldots, s_m) \in T^m : s_i \neq s_j, \text{ if } i \neq j \}.$$

Lemma 2. *For any function $g_m \in L_s^2(T^m)$ we have for all $\omega = (s_1, \ldots, s_n) \in T^n$ a. e., $n \geq 1$,*

$$I_m(g_m)(t_1, \ldots, t_m)(s_1, \ldots, s_n)$$

$$= \int_{T_*^m} g_m(t_1, \ldots, t_m) \left(\sum_{i=1}^n \delta_{s_i} - \lambda\right)(dt_1) \cdots \left(\sum_{i=1}^n \delta_{s_i} - \lambda\right)(dt_m).$$

Furthermore, $I_m(g_m)(t_1, \ldots, t_m)(a) = (-1)^m \int_{T_*^m} g_m(t_1, \ldots, t_m) \lambda(dt_1) \cdots \lambda(dt_m)$.

Proof. Both expressions coincide when g_m is a simple function, and define bounded linear operators over $L_s^2(T^m)$. $\square$

Lemma 3. *If* $F = I_m(g_m)$ *where* $g_m \in L^2_s(T^m)$, *we have* $\Psi_t F(\omega) = D_t F(\omega)$ *for almost all* $(\omega, t) \in \Omega \times T$.

Proof. Fix $n \geq 1$. For all $\omega = (s_1, \ldots, s_n)$, we define the measure $\tilde{N}(\omega) = \sum_{j=1}^{n} \delta_{t_j} - \lambda$. Then, using Lemma 2 we have for almost all $(\omega, t) \in T^n \times T$,

$$
\begin{aligned}
(\Psi_t F)(\omega) \\
&= I_m(g_m)(\omega, t) - I_m(g_m)(\omega) \\
&= \sum_{k=1}^{m} \binom{m}{k} \int_{T^m_*} g_m(t_1, \ldots, t_m) \, \delta_t(dt_1) \cdots \delta_t(dt_k) \tilde{N}(\omega)(dt_{k+1}) \cdots \tilde{N}(\omega)(dt_m) \\
&= m \int_{T^{m-1}_*} g_m(t_1, \ldots, t_{m-1}, t) \, \tilde{N}(\omega)(dt_1) \cdots \tilde{N}(\omega)(dt_{m-1}) \\
&= m I_m(g_m(\cdot, t)) \\
&= (D_t F)(\omega).
\end{aligned}
$$

The last equality follows from the definition of the operator D on multiple stochastic integrals. In the case $\omega = a$, by the definition of the operator Ψ we have $(\Psi_t F)(a) = I_m(g_m)(t) - I_m(g_m)(a)$. Moreover, $\tilde{N}(a) = -\lambda$, and by the same argument we obtain $\Psi_t F(a) = D_t F(a)$. $\square$

Lemma 4. *Let* g_m *be a function on* $L^2(T^{m+1})$ *such that for any* $t \in T$, $g_m(\cdot, t) \in L^2_s(T^m)$. *Set* $u_t = I_m(g_m(\cdot, t))$, *then* $\Phi u = \delta u$ *a. s.*

Proof. We recall that $\delta(u) = I_{m+1}(\tilde{g}_m(\cdot, t))(\omega)$, where $\tilde{g}_m$ denotes the symmetrization of the function g_m in all its variables. Using Lemma 2 we obtain for almost all $\omega = (s_1, \ldots, s_n)$, $n \geq 1$,

$$
\begin{aligned}
I_{m+1}(\tilde{g}_m(\cdot, t))(\omega) \\
&= \int_{T^{m+1}_*} \tilde{g}_m(t_1, \ldots, t_m, t) \, \tilde{N}(\omega)(dt_1) \cdots \tilde{N}(\omega)(dt_m) N(\omega)(dt) - \int_T u_t(\omega) \, \lambda(dt) \\
&= \sum_{j=1}^{n} \int_{T^m_*} \tilde{g}_m(t_1, \ldots, t_m, s_j) 1_{\{t_i \neq s_j, i=1, \ldots, m\}} \, \tilde{N}(\omega)(dt_1) \cdots \tilde{N}(\omega)(dt_m) \\
&\quad - \int_T u_t(\omega) \, \lambda(dt).
\end{aligned}
$$

Define $\tilde{N}^j(\omega) = \sum_{l=1, l \neq j}^{n} \delta_{s_l} - \lambda$ for any $j = 1, \ldots, n$. We can decompose $\tilde{N}$ as $\tilde{N}(\omega) = \delta_{s_j} + \tilde{N}^j(\omega)$. Using these notations, the above expression can be written as follows

$$
\begin{aligned}
\sum_{j=1}^{n} \sum_{k=0}^{m} \binom{m}{k} \int_{T^m_*} \tilde{g}_m(t_1, \ldots, t_m, s_j) 1_{\{t_i \neq s_j, i=1, \ldots, m\}} \\
\times \, \delta_{s_j}(dt_1) \cdots \delta_{s_j}(dt_k) \tilde{N}^j(\omega)(dt_{k+1}) \cdots \tilde{N}^j(\omega)(dt_m) - \int_T u_t(\omega) \, \lambda(dt).
\end{aligned}
$$

The terms with $k = 1, \ldots, m$ in the above sum are zero because we avoid the diagonals. Hence, we obtain, taking into account the definition of the operator Φ,

$$\sum_{j=1}^{n} \int_{T_*^m} \tilde{g}_m(t_1, \ldots, t_m, s_j) 1_{\{t_i \neq s_j, i=1,\ldots,m\}} \tilde{N}^j(\omega)(dt_1) \cdots \tilde{N}^j(\omega)(dt_m)$$

$$- \int_T u_t(\omega)\, \lambda(dt) = (\Phi u)(\omega).$$

$\square$

The following propositions characterize the domain of the operator D in terms of the square integrability of the random variable F and the process $\Psi_t F$. A similar result holds for the operator δ.

Proposition 1. *Let F be a random variable in $L^2(\Omega)$. Then $\Psi F \in L^2(\Omega \times T)$ if and only if $F \in \mathrm{Dom}(D)$ and in this case $DF = \Psi F$.*

Proof. Suppose first that $\Psi F \in L^2(\Omega \times T)$. It is sufficient to show that there exists a process $\eta \in L^2(\Omega \times T)$ such that for any multiple stochastic integral of the form $u_t = I_m(g_m(\cdot, t))$, where $g_m \in L^2(T^{m+1})$ and is symmetric in the first m variables, one has

$$E[F\Phi(u)] = E[\langle \eta, u \rangle_{L^2(T)}].$$

By the duality formula (1) we have

$$E[F\Phi(u)] = E[\langle \Psi F, u \rangle_{L^2(T)}].$$

Hence, $\eta = \Psi F \in L^2(\Omega \times T)$ satisfies the desired condition.

Conversely, if $F \in \mathrm{Dom}(D)$, for any $u_t = I_m(g_m(\cdot, t))$ as above, we have, using the duality between the operators D and δ, Lemma 4, and Theorem 2,

$$E[\langle DF, u \rangle] = E[F\delta(u)] = E[F\Phi u] = E[\langle \Psi F, u \rangle].$$

Hence $\Psi F = DF$, and this completes the proof of the proposition. $\square$

A similar result for the operator δ is the following:

Proposition 2. *Let $u = \{u_t, t \in T\}$ be a stochastic process in $L^2(\Omega \times T)$. Then $\Phi u \in L^2(\Omega)$ if and only if $u \in \mathrm{Dom}(\delta)$ and in this case $\delta u = \Phi u$.*

Proof. Suppose first that $\Phi u \in L^2(\Omega)$. For any $F = I_m(g_m)$, using Lemma 3 and Theorem 2, we obtain

$$E[\langle u, DF \rangle_{L^2(T)}] = E[\langle u, \Psi F \rangle_{L^2(T)}] = E[F\Phi u].$$

As a consequence, $\delta(u) = \Phi(u)$.

Conversely, if $u \in \mathrm{Dom}(\delta)$ for any $F = I_m(g_m)$, using the duality between the operators D and δ, Lemma 3, and Theorem 2, we obtain

$$E[F\delta(u)] = E[\langle DF, u \rangle] = E[\langle \Psi F, u \rangle] = E[F\Phi u].$$

Hence $\delta(u) = \Phi(u)$, and this completes the proof of the proposition. $\square$

As an application of these results we will compute the derivative of the jump times of the Poisson process. Let $T = [0, 1]$, and let $S_1, S_2, \ldots, S_n, \ldots$ be the jump times of the Poisson process over T. We will first compute the transformation $\Psi_t S_i = S_i(\omega + \delta_t) - S_i(\omega)$. We have

$$\begin{aligned}
\Psi_t S_i &= 0, & \text{if } t > S_i, \\
\Psi_t S_i &= t - S_i, & \text{if } S_{i-1} < t < S_i, \\
\Psi(S_i) &= S_{i-1} - S_i, & \text{if } t < S_{i\ 1}.
\end{aligned}$$

Observe that the random variables S_i satisfy the conditions of Proposition 1 because they have moments of all orders. Hence $S_i \in \mathrm{Dom}(D)$ for any $i \geq 1$ and we have $\Psi_t S_i = D_t S_i$ almost everywhere on $\Omega \times T$.

Observe that the operator Ψ is not local. Indeed, consider the following example. Assume that F and G are functionals such that they vanish on the subset $\{N(1/2) = N(1)\}$ and take the values 1 and 2 respectively over the complementary subset. For any $t > 1/2$ we have $\Psi_t F = 1$ and $\Psi_t G = 2$ although F and G coincide on a set of positive probability.

We close this note with the following theorem which allows to interpretate the operator Φ as a generalization of the stochastic integral with respect to the Poisson process, in the sense that it coincides with this integral over predictable processes.

Theorem 3. *Suppose that $T = [0, 1]$, and let $u = \{u_t, t \in [0, 1]\}$ be a predictable process such that $\int_0^1 |u_t| \, dt < \infty$. Then we have*

$$\Phi u = \int_0^1 u(t) \, d\tilde{N}(t).$$

Proof. For any $t \in [0, 1]$ we will denote by $\mathcal{F}_t$ the σ-field generated by the random variables $\{N_s, 0 \leq s \leq t\}$. Suppose first that u is a simple process such that $u(t) = F(\omega)1_{(a,b]}(t)$ and F is $\mathcal{F}_a$-measurable, then for any $\omega = (s_1, \ldots, s_n)$, $n \geq 1$, we have

$$\Phi(u)(\omega) = \sum_{j=1}^{n} F(s_1, \ldots, \hat{s}_j, \ldots, s_n)1_{(a,b]}(s_j) - \int_0^1 u_t(\omega) \, dt.$$

The fact that F is $\mathcal{F}_a$-measurable implies that for each $n \geq 1$ and for any $k = 0, \ldots, n$, $F(s_1, \ldots, s_n)$ depends only on the coordinates $\{s_1, \ldots, s_k\}$ on the set $\{s_k < a < s_{k+1}\}$. Consequently, we obtain

$$\Phi(u)(\omega) = F(\omega)(N_b - N_a) - F(\omega)(b - a) = \int_0^1 u_t(\omega) \, d\tilde{N}_t.$$

Finally this equality can be extended to all predictable processes verifying $\int_0^1 |u_t| \, dt < \infty$ by an approximation argument. In fact, we can find a sequence of elementary predictable processes (linear combinations of processes of the form $= F(\omega)1_{(a,b]}(t)$, where F is $\mathcal{F}_a$-measurable) u^n such that $\int_0^1 |u_t - u_t^n|\, dt$ converges in probability to zero as n tends to infinity. Then we have

$$\Phi u^n = \int_0^1 u_t^n \, d\tilde{N}(t),$$

and taking the limit as n tends to infinity, and using the definiton of the operator Φ we obtain the desired result. □

Remark. If λ is a Radon measure which is not necessarily finite, then we can also consider a canonical Poisson space (see [6]), and all the preceding results can be properly extended with minor modifications.

REFERENCES

1. R. F. Bass, M. Cranston, *The Malliavin calculus for pure jump processes and applications to local time*, Annals of Probability **14** (1986), 490–532.
2. K. Bichteler, J. B. Gravereaux and J. Jacod, *Malliavin Calculus for Processes with Jumps*, Gordon and Breach, 1987.
3. J. M. Bismut, *Calcul des variations stochastiques et processus de sauts*, Z. für Wahrschein. verw. Gebiete **63** (1983), 147–235.
4. E. A. Carlen, E. Pardoux, *Differential calculus and integration by parts on Poisson space*, In: Stochastics, algebra and analysis in classical and quantum dynamics, Kluwer Academic, Dordrecht, 1990, pp. 63–73.
5. A. Dermoune, P. Krée, L. Wu, *Calcul stochastique non adapté par rapport à la mesure aléatoire de Poisson*, Lecture Notes in Math., vol. 1321, 1980, pp. 477–484.
6. J. Neveu, *Processus Ponctuels*, Lecture Notes in Math, vol. 598, Springer, 1976.
7. D. Nualart, E. Pardoux, *Stochastic calculus with anticipating integrands*, Probability Theory and Related Fields **78** (1988), 535–581.
8. D. Nualart, J. Vives, *Anticipative calculus for the Poisson process based on the Fock space*, *Séminaire de Probabilités XXIV*, Lecture Notes on Math, vol. 1426, 1990, pp. 154–165.
9. H. Ogura, *Orthogonal functionals of the Poisson processes*, Trans. IEEE Inf. Theory **IT-18**, 4 (1972), 473–481.
10. N. Privault, *Chaotic and variational calculus for the Poisson process*. Thesis.

DAVID NUALART, DEPARTAMENT D'ESTADÍSTICA, UNIVERSITAT DE BARCELONA, GRAN VIA 585, 08007 BARCELONA, SPAIN

JOSEP VIVES, DEPARTAMENT DE MATEMÀTIQUES, UNIVERSITAT AUTÒNOMA DE BARCELONA, 08193 BELLATERRA, SPAIN

Progress in Probability, Vol. 36
© 1995 Birkhäuser Verlag Basel/Switzerland

GENERALIZED FUNCTIONS AND STOCHASTIC PROCESSES

MICHAEL OBERGUGGENBERGER

ABSTRACT. Differential algebras of generalized stochastic processes are constructed, which contain irregular processes as e. g. white noise. Solutions to nonlinear stochastic differential equations are obtained in these algebras. The methods extend the nonlinear theories of generalized functions as developed by J. F. Colombeau, Yu. V. Egorov, E. E. Rosinger to the stochastic setting. Both an approach based on regularized sample paths as well as on sequences of smooth L^0-valued functions are presented.

§1 INTRODUCTION

The purpose of this note is to explore and indicate how the "nonlinear theories of generalized functions" that have been developed over the past decades by J. F. Colombeau [3–5], Yu. V. Egorov [7], D. Laugwitz [11], E. E. Rosinger [17–20], and others, can be adapted to produce a new approach to stochastic differential equations involving nonlinearities and irregular stochastic processes.

Let us briefly describe the scope of the nonlinear theories of generalized functions refered to (in the deterministic setting): These theories provide differential algebras of generalized functions and are capable of handling nonlinear operations, differentiation, and singular objects (distributions in the sense of L. Schwartz) simultaneously. In particular, nonlinear partial differential equations with singular data and/or discontinuous terms can be treated. The basic ingredients are regularization, infinite families of smooth objects, and factorization with respect to certain ideals (so as to process the inherent information). Typical examples are the Colombeau ideal (giving optimal consistency of algebraic operations with the corresponding classical, pointwise operations), the ultrapower constructions in nonstandard analysis (producing the powerful logical transfer – and permanence principles), and the nowhere dense ideal of Rosinger (adapted to deal with solutions defined off nowhere dense, closed sets). For a presentation of details and recent applications we have to refer to the literature, e. g. [15] and Colombeau [3] as well as the review articles [16] and Colombeau [4].

1991 *Mathematics Subject Classification.* 46F10, 60G20, 60H10.

Key words and phrases. Algebras of generalized functions and stochastic processes, generalized solutions to nonlinear stochastic differential equations.

In this paper we shall translate the Colombeau approach into the setting of generalized stochastic processes and apply it to ordinary stochastic differential equations

$$X(t) = F\big(X(t), Y(t)\big)$$

where Y is any generalized stochastic process, e. g. white noise, and F can be a nonlinear, discontinuous or even generalized function. Thus we present our results in the framework of one-parameter processes. However, it will be obvious that the basic constructions can be carried out for multi-dimensional processes, random fields and stochastic partial differential equations in a similar fashion.

We should note that various authors have already studied stochastic partial differential equations involving white noise by means of regularization (see the notion of "functional processes" of Holden, Lindstrøm, Øksendal, Ubøe, and Zhang [9] and the study of semilinear wave equations with white noise excitation of Albeverio, Haba, and Russo [1]). The new ingredient in the present paper is the stressing of *factorization* and the construction of *algebras* of generalized stochastic processes. We should also like to point out that our approach is *not* related to the hyperfinite constructions customary in nonstandard analysis (see e. g. Nelson [14]), but rather views generalized processes as "internal smooth processes" in the nonstandard terminology.

The plan of exposition is as follows: In Section 2 we recall the basic features of the existing nonlinear theories of generalized functions, focussing on the Colombeau approach. In Section 3 we develop a corresponding theory of stochastic processes, based on a sample path perspective: We construct differential algebras of generalized stochastic processes, containing all processes with distribution-valued paths, and admitting the algebra of smooth random processes as a subalgebra. Section 4 serves to elaborate applications to stochastic differential equations. Finally, in Section 5, we present an alternative construction of algebras of generalized stochastic processes, starting from topological properties of the classical algebra of measurable functions.

§2 MULTIPLICATION OF DISTRIBUTIONS

In this section we discuss some of the basic ideas from the existing nonlinear theories of generalized functions. Letting Δ be an open subset of $\mathbb{R}^n$, $C^\infty(\Delta)$ denotes the algebra of infinitely differentiable functions on Δ, $\mathcal{D}(\Delta)$ the subalgebra of compactly supported smooth functions; $\mathcal{D}'(\Delta)$ the space of Schwartz distributions on Δ, $\mathcal{E}'(\Delta)$ the subspace of distributions of compact support; finally, $\mathcal{S}(\mathbb{R}^n)$ denotes the algebra of smooth, rapidly decreasing functions on $\mathbb{R}^n$.

Our goal is to construct an associative, commutative differential algebra $\mathcal{G}(\Delta)$, possessing derivations $\partial_1, \ldots, \partial_n : \mathcal{G}(\Delta) \to \mathcal{G}(\Delta)$ with the properties:

(a) there is a linear imbedding $\mathcal{D}'(\Delta) \to \mathcal{G}(\Delta)$;

(b) $\partial_j|_{\mathcal{D}'(\Delta)}$ coincides with the usual distributional derivative on $\mathcal{D}'(\Delta)$ for $j = 1, \ldots, n$;

(c) the multiplication of $\mathcal{G}(\Delta)$ induces the usual pointwise product on $C^\infty(\Delta)$.

Owing to Colombeau [3, 4], such an algebra can indeed be constructed. We note that the indicated list of properties is optimal in as much as given (a) and (b), point (c) cannot be improved to consistency with the product on function algebras larger than $C^\infty(\Delta)$. In addition, the algebra $\mathcal{G}(\Delta)$ will have good localization properties, i.e. it constitutes a sheaf.

To specify the construction, we start with the algebra of sequences of smooth functions

$$\mathcal{E}[\Delta] = \left(C^\infty(\Delta)\right)^{\mathbb{N}}.$$

A sequence $f = (f_\nu)_{\nu \in \mathbb{N}} \in \mathcal{E}[\Delta]$ will be called *moderate* if for all compact $K \subset \Delta$ and all $\alpha \in \mathbb{N}_0^n$ there exists an $N > 0$ such that

$$\sup_{x \in K} |\partial^\alpha f_\nu(x)| = O(\nu^N) \quad \text{as} \quad \nu \to \infty.$$

Further, f is said to be *null* if, for all compact $K \subset \Delta$, all $\alpha \in \mathbb{N}_0^n$, and all $N > 0$,

$$\sup_{x \in K} |\partial^\alpha f_\nu(x)| = O(\nu^{-N}) \quad \text{as} \quad \nu \to \infty.$$

Denoting the algebra of moderate sequences by $\mathcal{E}_M[\Delta]$ and the ideal of null sequences by $\mathcal{N}(\Delta)$, the algebra $\mathcal{G}(\Delta)$ is defined as the factor algebra

$$\mathcal{G}(\Delta) = \mathcal{E}_M[\Delta]/\mathcal{N}(\Delta).$$

For the sake of a coherent presentation, we have chosen to work with sequences here; this will be convenient in the context of measurability questions in the next sections. However, families of larger cardinality are also possible and often useful.

Any given $\varphi \in \mathcal{S}(\mathbb{R}^n)$ with $\int_{\mathbb{R}^n} \varphi(x)\,dx = 1$ produces a mollifying sequence $\varphi_\nu(x) = \nu^n \varphi(\nu x)$ and an imbedding of the compactly supported distributions. Indeed, to any $g \in \mathcal{E}'(\Delta)$ we can assign the class of $(g * \varphi_\nu\!\restriction_\Delta)_{\nu \in \mathbb{N}}$ in $\mathcal{G}(\Delta)$. This imbedding apparently respects partial derivatives. In addition, if we take φ with all moments vanishing, i.e. $\int_{\mathbb{R}^n} x^\alpha \varphi(x)\,dx = 0$ for all $\alpha \in \mathbb{N}_0^n$ satisfying $|\alpha| \geq 1$, then we obtain by Taylor expansion and the defining property of $\mathcal{N}(\Delta)$ that $(g - g * \varphi_\nu\!\restriction_\Delta)_{\nu \in \mathbb{N}}$ is a null sequence for every $g \in \mathcal{D}(\Delta)$. This produces property (c) for compactly supported smooth functions. The extension of the indicated imbedding to all of $\mathcal{D}'(\Delta)$, together with property (c) for all of $C^\infty(\Delta)$ is achieved by localization using the respective sheaf structures.

Various other algebras containing the distributions are obtained by varying the choice of the ideal. Prominent cases are ideals defined by vanishing properties, as for example the algebra $^*C^\infty(\Delta)$ of nonstandard analysis and the algebra $\mathcal{E}[\Delta]/\mathcal{I}_{\mathrm{nd}}(\Delta)$ of Rosinger, where $\mathcal{I}_{\mathrm{nd}}(\Delta)$ denotes the ideal of sequences vanishing eventually off some nowhere dense, closed subset of Δ. However, the algebra $\mathcal{G}(\Delta)$ or its variants are currently the only known ones enjoying all of the properties (a)–(c).

A further important notion of the theory is the association relation. Two elements $f, g \in \mathcal{G}(\Delta)$ are called *associated*, denoted by $f \approx g$, if for any representatives the difference $f_\nu - g_\nu$ converges to zero in $\mathcal{D}'(\Delta)$ as $\nu \to \infty$. Obviously, the association relation reduces to equality in $\mathcal{D}'(\Delta)$, thus it can be viewed as bringing the information on the objects down to the customary level of distribution theory. And indeed, one obtains a range of coherence results with classical operations in the association sense: For example, the product of two continuous functions in $\mathcal{G}(\Delta)$ is associated with their classical product.

Finally, we shall make use of the ring of constants in $\mathcal{G}(\Delta)$, defined as the zero-dimensional generalized functions $\overline{\mathbb{R}} = \mathcal{G}(\mathbb{R}^0)$. The field of real numbers $\mathbb{R}$ is a subring of $\overline{\mathbb{R}}$. Generalized constants arise as point values of arbitrary elements, or as integrals of compactly supported elements of $\mathcal{G}(\Delta)$.

§3 ALGEBRAS OF GENERALIZED STOCHASTIC PROCESSES

The purpose of this section is to introduce algebras of Colombeau-type generalized stochastic processes. Here and in the sequel, we fix a probability space (Ω, Σ, μ) where Σ denotes the measurable subsets of Ω and μ a probability measure (the terms "measurable" and "almost surely" will refer to Σ and μ). The space $\mathcal{L}^p$, $0 \leq p \leq \infty$, comprises the measurable functions X such that $\mathrm{E}\,|X|^p$ is finite. As noted in the introduction, we take $\mathbb{R}$ as our time interval; a (classical) stochastic process is thus a map from $\mathbb{R}$ to $\mathcal{L}^0$. The algebras of (classical) stochastic processes with almost surely continuous (respectively smooth) paths will be denoted by $\mathcal{CP}$ (respectively $\mathcal{C}^\infty\mathcal{P}$).

We are now in the position to introduce differential algebras of generalized stochastic processes, sample path style (a variant based on $\mathcal{C}^\infty(\mathbb{R}, \mathcal{L}^0)$ will be presented in Section 5). The basic objects are sequences of processes with almost surely smooth paths,

$$\mathcal{EP} = (\mathcal{C}^\infty\mathcal{P})^{\mathbb{N}}.$$

This is the space of all $X : \mathbb{N} \times \mathbb{R} \times \Omega \to \mathbb{R}$ such that

$\omega \mapsto X_\nu(t, \omega)$ is measurable, for all $\nu \in \mathbb{N}$ and $t \in \mathbb{R}$;

$t \to X_\nu(t, \omega)$ belongs to $\mathcal{C}^\infty(\mathbb{R})$, for all $\nu \in \mathbb{N}$ and almost all $\omega \in \Omega$.

The elements of $\mathcal{EP}$ will be denoted by $X = (X_\nu)_{\nu \in \mathbb{N}}$. The product of two elements X, Y in $\mathcal{EP}$ is defined componentwise

$$XY = (X_\nu Y_\nu)_{\nu \in \mathbb{N}}$$

and similarly the derivative

$$\frac{d}{dt}X = \left(\frac{d}{dt}X_\nu\right)_{\nu \in \mathbb{N}}$$

and in this way $\mathcal{EP}$ becomes a differential algebra. In order to achieve consistency properties with operations on classical processes, analogous to Section 2, we introduce the differential subalgebra

$$\begin{aligned}
\mathcal{EP}_M = \{(X_\nu)_{\nu \in \mathbb{N}} \in \mathcal{EP} : \text{ for almost all } \omega \in \Omega, \\
\forall T > 0 \; \forall j \in \mathbb{N}_0 \; \exists N(\omega), \, C(\omega), \, \eta(\omega) > 0 \text{ with} \\
\sup_{t \in [-T,T]} |\partial^j X_\nu(t,\omega)| \le C(\omega)\nu^{N(\omega)}, \, \nu \ge \eta(\omega)\}
\end{aligned} \tag{1}$$

and the differential ideal

$$\begin{aligned}
\{(X_\nu)_{\nu \in \mathbb{N}} \in \mathcal{EP} : \text{ for almost all } \omega \in \Omega, \\
\forall T > 0 \; \forall j \in \mathbb{N}_0 \; \forall N > 0 \; \exists C(\omega), \, \eta(\omega) > 0 \text{ with} \\
\sup_{t \in [-T,T]} |\partial^j X_\nu(t,\omega)| \le C(\omega)\nu^{-N}, \, \nu \ge \eta(\omega)\}.
\end{aligned} \tag{2}$$

The *differential algebra* $\mathcal{GP}$ of *generalized stochastic processes* is introduced as the factor algebra

$$\mathcal{GP} = \mathcal{EP}_M / \mathcal{NP}.$$

This concludes our construction. Thus a *generalized stochastic process* is an equivalence class of a three-parameter family

$$(\nu, t, \omega) \mapsto X_\nu(t,\omega)$$

where t signifies time, ω the sample realization, and, in view of the imbedding of irregular processes below, ν may be considered as measuring the degree of singularity. Due to the fact that $\mathcal{GP}$ is a differential algebra, generalized stochastic processes have (generalized) derivatives of any order and can be multiplied. In addition, superposition with polynomially bounded functions or even generalized functions is possible.

The algebra of smooth functions $F : \mathbb{R}^2 \to \mathbb{R}$ which, together with all derivatives, grow at most polynomially at infinity, is denoted by $\mathcal{O}_M(\mathbb{R}^2)$. A sequence $G = (G_\nu)_{\nu \in \mathbb{N}}$ will be called *tempered* if $G_\nu \in \mathcal{O}_M(\mathbb{R}^2)$ for all $\nu \in \mathbb{N}$ and

$$\begin{aligned}
\forall \alpha \in \mathbb{N}_0^2 \; \exists N, \, C > 0 \text{ such that} \\
|\partial^\alpha G_\nu(x,y)| \le C\nu^N(1 + |x| + |y|)^N, \quad \text{for all } (x,y) \in \mathbb{R}^2 \text{ and } \nu \in \mathbb{N}.
\end{aligned} \tag{3}$$

Thus the sequence G defines a tempered generalized function in the sense of Colombeau [4; Chapter 4].

Proposition 3.1. *If $X, Y \in \mathcal{GP}$ are generalized stochastic processes and G is a tempered sequence, then $G(X,Y)$ is a well-defined element of $\mathcal{GP}$.*

Proof. Letting $(X_\nu)_{\nu \in \mathbb{N}}$ and $(Y_\nu)_{\nu \in \mathbb{N}}$ be representatives of X and Y, it follows easily from property (3) that $(G_\nu(X_\nu, Y_\nu))_{\nu \in \mathbb{N}}$ belongs to $\mathcal{EP}_M$. Its class in $\mathcal{GP}$

defines the superposition $G(X, Y)$, and it is readily checked that the class does not depend on the choice of representative for X and Y. □

As a particular case we see that $F(X, Y)$ is well-defined if $F \in \mathcal{O}_M(\mathbb{R}^2)$. Of course, more than two variables can be dealt with in the same manner.

We now specify an imbedding of the continuous path processes $\mathcal{CP}$ into $\mathcal{GP}$. To this end we choose an element $\varphi \in \mathcal{S}(\mathbb{R})$ with $\int_{\mathbb{R}} \varphi(s) \, ds = 1$ and all moments vanishing, $\int_{\mathbb{R}} s^j \varphi(s) \, ds = 0$ for all $j \geq 1$. Further, we take $\psi \in \mathcal{D}(\mathbb{R})$ such that $\psi(s) \equiv 1$ in a neighborhood of $s = 0$. Given a process Z with almost surely continuous paths, we define the corresponding element of $\mathcal{GP}$ as

$$\iota(Z) = \text{class of } (Z * (\psi \varphi_\nu))_{\nu \in \mathbb{N}} \tag{4}$$

where $\varphi_\nu(s) = \nu \varphi(\nu s)$ and the convolution is effected pathwise

$$(Z * \psi \varphi_\nu)(t, \omega) = \int_{\mathbb{R}} Z(s, \omega) \psi(t - s) \varphi_\nu(t - s) \, ds. \tag{5}$$

Since $Z(\cdot, \omega)$ is uniformly bounded on compact time intervals, for fixed $\omega \in \Omega$, it follows that (5) yields an element of $\mathcal{EP}_M$ and so we can indeed take its class in $\mathcal{GP}$. We note that the cut-off function ψ could have been dropped, had we taken φ with compact support; but then φ could not have all moments vanishing, as is needed in Proposition 3.4.

Example 3.2. Let B be a *Brownian motion*. Then $B \in \mathcal{CP}$ and $W = \frac{d}{dt} \iota(B)$ represents *white noise* as an element of $\mathcal{GP}$.

Remark 3.3.
  (a) Formula (4) could also serve to imbed generalized random processes in the sense of Gel'fand and Vilenkin [8; Chapter III] into $\mathcal{GP}$. However, it suffices to work with derivatives of continuous path processes to accommodate the irregular processes commonly in use.
  (b) In case the process Z belongs to $\mathcal{C}(\mathbb{R}, \mathcal{L}^2)$, as for instance Brownian motion, the integral in (5) can be interpreted in the stronger sense of a Bochner integral with values in $\mathcal{L}^2$. This will become relevant in Section 5.
  (c) From any second order stationary process one can construct an approximating sequence of smooth $\mathcal{L}^2$-valued processes, by regularizing the autocorrelation function by means of mollifiers of positive type (see e. g. Loève [13; 37.1, 37.2]). However, the autocorrelation function alone does not sufficiently specify the process for our purposes and hence this approach will not be pursued here.

The processes $\mathcal{C}^\infty \mathcal{P}$ with almost surely smooth paths can be imbedded into $\mathcal{GP}$ by means of the standard imbedding

$$\sigma(Z) = \text{class of } (Z)_{\nu \in \mathbb{N}} \tag{6}$$

as well. The crucial consistency property, to be proven now, is that $\iota \restriction_{\mathcal{C}^\infty \mathcal{P}} = \sigma$. From there it follows immediately that $\iota(\mathcal{C}^\infty \mathcal{P})$ is a differential subalgebra of $\mathcal{GP}$.

Proposition 3.4. *If Z is a process with almost surely smooth paths, then $\iota(Z) = \sigma(Z)$ in $\mathcal{GP}$.*

Proof. Suppressing $\omega \in \Omega$ in our notation and assuming, for convenience, that $\psi(s) \equiv 1$ for $|s| \leq 1$ and $\psi(s) \equiv 0$ for $|s| \geq 2$, we have

$$(Z * \psi\varphi_\nu)(t) - Z(t)$$
$$= \int_{-2\nu}^{2\nu} \left(Z\left(t - \frac{s}{\nu}\right) - Z(t) \right) \psi\left(\frac{s}{\nu}\right) \varphi(s)\, ds + Z(t) \int_{\mathbb{R}} \left(\psi\left(\frac{s}{\nu}\right) - 1 \right) \varphi(s)\, ds.$$

The second integral actually involves only $|s| \geq \nu$ and hence is $\mathcal{O}(\nu^{-M})$ for any $M > 0$, since φ is rapidly decreasing. In the first integral, we develop $Z(t - \frac{s}{\nu})$ in a Taylor series around t up to order $N - 1$. The remainder term is of order ν^{-N}; the terms involving $\int_{-2\nu}^{2\nu} s^j \psi(\frac{s}{\nu}) \varphi(s)\, ds$ are small of order ν^{-M} for all M, since $\int_{\mathbb{R}} s^j \varphi(s)\, ds = 0$, $\psi(\frac{s}{\nu}) \equiv 1$ for $|s| \leq \nu$, and again φ is rapidly decreasing. The estimates are uniform with respect to $t \in [-T, T]$ for any $T > 0$, and so we get

$$\sup_{t \in [-T,T]} |(Z * \psi\varphi_\nu)(t) - Z(t)| = O(\nu^{-N})$$

as $\nu \to \infty$, for every $N > 0$ (at almost all $\omega \in \Omega$). This is the zero-order estimate required in (2). The derivatives are handled similarly. We conclude that

$$(Z * (\psi\varphi_\nu) - Z)_{\nu \in \mathbb{N}} \in \mathcal{NP},$$

thus $\iota(Z) = \sigma(Z)$ in the factor algebra $\mathcal{GP}$. $\square$

In order to be able to evaluate a generalized stochastic process at fixed points of time, we introduce the concept of a *generalized random variable*. To this end let $\mathcal{ER} = (\mathcal{L}^0)^{\mathbb{N}}$ be the space of sequences of measurable functions on Ω. We set

$$\mathcal{ER}_M = \Big\{ (X_\nu)_{\nu \in \mathbb{N}} \in \mathcal{ER} : \text{for almost all } \omega \in \Omega,$$
$$\exists\, N(\omega), C(\omega), \eta(\omega) > 0 \text{ such that}$$
$$|X_\nu(\omega)| \leq C(\omega)\nu^{N(\omega)}, \nu \geq \eta(\omega) \Big\},$$
$$\mathcal{NR} = \Big\{ (X_\nu)_{\nu \in \mathbb{N}} \in \mathcal{ER} : \text{for almost all } \omega \in \Omega,$$
$$\forall N > 0\ \exists\, C(\omega), \eta(\omega) > 0 \text{ such that}$$
$$|X_\nu(\omega)| \leq C(\omega)\nu^{-N}, \nu \geq \eta(\omega) \Big\},$$

and define

$$\mathcal{GR} = \mathcal{ER}_M / \mathcal{NR}.$$

Clearly, $\mathcal{GR}$ is an algebra. If $X \in \mathcal{GP}$ is a generalized process and $t_0 \in \mathbb{R}$, then $X(t_0)$ is a generalized random variable, i.e. belongs to $\mathcal{GR}$.

§4 STOCHASTIC DIFFERENTIAL EQUATIONS

In this section we shall obtain an existence theory for the stochastic differential equation

$$\dot{X}(t) = F(X(t), Y(t)), \quad t \in \mathbb{R},$$
$$X(0) = A,$$

$$(7)$$

in the setting of generalized stochastic processes developed in Section 3. The solution is sought as an element of $\mathcal{GP}$,

$$X = \text{class of } (X_\nu)_{\nu \in \mathbb{N}} \text{ in } \mathcal{GP},$$

the driving term is a generalized stochastic process

$$Y = \text{class of } (Y_\nu)_{\nu \in \mathbb{N}} \text{ in } \mathcal{GP};$$

the initial value is a generalized random variable

$$A = \text{class of } (A_\nu)_{\nu \in \mathbb{N}} \text{ in } \mathcal{GR};$$

finally, the function F may be given by a tempered sequence

$$F = (F_\nu)_{\nu \in \mathbb{N}} \in \mathcal{E}[\mathbb{R}^2]$$

satisfying condition (3) of Section 3. The equation itself is understood in sense of the differential algebra $\mathcal{GP}$, that is, for chosen representatives we require

$$\left(\dot{X}_\nu - F_\nu(X_\nu, Y_\nu) \right)_{\nu \in \mathbb{N}} \in \mathcal{NP} \quad \text{and} \quad \left(X_\nu(0) - A_\nu \right)_{\nu \in \mathbb{N}} \in \mathcal{NR}. \tag{8}$$

Due to Proposition 3.1, this is a valid solution concept not depending on the choice of representative; more precisely, (8) holds for all representatives if it holds for one.

In addition to being given by a tempered sequence, we shall require of F that each of its defining terms F_ν has a globally bounded gradient, together with the logarithmic estimate

$$\exists C > 0 : |\nabla F_\nu(x, y)| \leq C \log \nu, \quad \forall (x, y) \in \mathbb{R}^2, \nu \in \mathbb{N}. \tag{9}$$

We note that these hypotheses hold in particular when F is a classical globally Lipschitz continuous function belonging to $\mathcal{O}_M(\mathbb{R}^2)$, $F_\nu \equiv F$ for all $\nu \in \mathbb{N}$.

Theorem 4.1. (Existence and uniqueness) *Let F be as described above, satisfying (3) and (9). Given $Y \in \mathcal{GP}$, $A \in \mathcal{GR}$, problem (7) has a unique solution $X \in \mathcal{GP}$.*

Proof: (Existence). At fixed $\nu \in \mathbb{N}$, the hypotheses on F_ν and Y_ν are such that we can invoque the classical existence theorems for pathwise solutions (see e. g. Bunke [2; Satz 1.2]), to obtain a unique solution X_ν with smooth paths, i. e.

$(X_\nu)_{\nu\in\mathbb{N}} \in \mathcal{EP}$. To prove existence of the generalized solution, it suffices to show that $(X_\nu)_{\nu\in\mathbb{N}}$ belongs to $\mathcal{EP}_M$; its class in $\mathcal{GP}$ then yields a solution. We have, at fixed $\omega \in \Omega$,

$$X_\nu(t) = A_\nu + t\,F_\nu(0,0) + \int_0^t \partial_y F_\nu\big(\theta(X_\nu(s), Y_\nu(s))\big)\,Y_\nu(s)\,ds$$

$$+ \int_0^t \partial_x F_\nu\big(\theta(X_\nu(s), Y_\nu(s))\big)\,X_\nu(s)\,ds$$

for some function θ. Using that $F_\nu(0,0)$ and $Y_\nu|_{[-T,T]}$ are of order ν^N on any fixed time interval $[-T,T]$, while $\partial_x F_\nu$ and $\partial_y F_\nu$ are of order $\log\nu$, Gronwall's inequality gives

$$\sup_{t\in[-T,T]} |X_\nu(t)| \le C\,\nu^N \exp(T\log\nu) = C\,\nu^{N+T}.$$

This is the desired zero-order estimate for X_ν. Similar estimates on the derivatives of X_ν follow immediately from the equation using the properties of F_ν and induction.

(Uniqueness) Let X and $\widetilde{X}$ be two solutions. Then there are $I \in \mathcal{NR}$, $J \in \mathcal{NP}$ such that

$$(X - \widetilde{X})^{\boldsymbol{\cdot}} = F(X,Y) - F(\widetilde{X},Y) + J,$$

$$X(0) - \widetilde{X}(0) = I.$$

Thus

$$X_\nu(t) - \widetilde{X}_\nu(t) = I_\nu + \int_0^t J_\nu(s)\,ds + \int_0^t \partial_x F_\nu(\theta, Y_\nu)\,(X_\nu(s) - \widetilde{X}_\nu(s))\,ds$$

for some function $\theta = \theta(X_\nu, \widetilde{X}_\nu)$. By Gronwall's inequality, we have

$$\sup_{t\in[-T,T]} |X_\nu(t) - \widetilde{X}_\nu(t)| \le C(1+T)\nu^{-M} \exp(T\log\nu) = C(1+T)\nu^{T-M}$$

and this proves the zero-order property in (2) for $X - \widetilde{X}$. From there one shows inductively as above that

$$(X_\nu - \widetilde{X}_\nu)_{\nu\in\mathbb{N}} \in \mathcal{NP}$$

thus $X = \widetilde{X}$ in $\mathcal{GP}$. $\square$

In case $F_\nu \equiv F$ is a classical globally Lipschitz continuous function and the driving process is smooth, say given by $S \in \mathcal{C}^\infty\mathcal{P}$, and $A_\nu \equiv A \in \mathcal{L}^0$ is a classical random variable, then the problem

$$\dot{Z} = F(Z,S) \quad \text{with} \quad Z(0) = A \tag{10}$$

has a classical, smooth solution $Z \in \mathcal{C}^\infty\mathcal{P}$ as well. A major point of our approach is the validity of the assertion that the generalized solution $X \in \mathcal{GP}$ coincides with the classical solution Z in this case:

Proposition 4.2. (Consistency) *Let* $F_\nu \equiv F \in \mathcal{O}_M(\mathbb{R}^2)$, $|\nabla F|$ *bounded, and* $A_\nu \equiv A \in \mathcal{L}^0$. *Assume that* $S \in C^\infty\mathcal{P}$, *let* $Z \in C^\infty\mathcal{P}$ *be the classical solution to* (10), *and let* $X \in \mathcal{G}\mathcal{P}$ *be the generalized solution to* (7) *with* $Y = \iota(S)$. *Then* $X = \iota(Z)$ *in* $\mathcal{G}\mathcal{P}$.

Proof. As shown in Proposition 3.4, $Y_\nu \equiv S$ gives a representative of $Y = \iota(S)$. Similarly, the classical solution Z, more precisely the sequence $(Z)_{\nu \in \mathbb{N}}$, may serve as a representative of the unique generalized solution X. $\square$

Remark 4.3. As noted above, Theorem 4.1 is certainly applicable in case the function F is a classical, globally Lipschitz continuous member of $\mathcal{O}_M(\mathbb{R}^2)$. However, the generalized setting makes it possible to deal with smooth F of arbitrary growth as well, namely by means of a suitable cut-off procedure. For example, if $F(X,Y) = XY$ we can produce a sequence $F_\nu(X,Y) = M_\nu(X)M_\nu(Y)$ satisfying the required hypotheses (3) and (9) by taking $M_\nu(x) = x$ for $|x| \leq \log \nu - 1$; $M_\nu(x) = \log \nu$ for $|x| \geq \log \nu$, and smooth in between. A faster growth of F at infinity can be counterbalanced by a corresponding slower growth with $\nu \to \infty$. On the other hand, if further conditions are put on the driving term Y, the cut-off in F may become unnecessary (see Example 4.4 below). Further, discontinuous F can be accommodated as well. For example, $F(X) = \text{sign } X$ can be mollified by taking $F_\nu(X) = \tanh(X \log \nu)$ to satisfy (9).

Example 4.4. A few phenomena can be demonstrated by means of the *multiplicative noise equation*

$$\dot{X}(t) = \alpha X(t)W(t) \quad \text{with} \quad X(0) = A \tag{11}$$

where $W \in \mathcal{G}\mathcal{P}$ denotes white noise (Example 3.2), $\alpha \in \mathbb{R}$, and $A \in \mathcal{G}\mathcal{R}$ is a generalized random variable. Though the right-hand side $F(X,W) = \alpha XW$ does not directly (i.e. without cut-off) satisfy the hypotheses of Theorem 4.1, we nevertheless have a unique solution $X \in \mathcal{G}\mathcal{P}$, due to the local boundedness of Brownian motion. Indeed, letting $(B_\nu)_{\nu \in \mathbb{N}}$ be a representative of $\iota(B)$ according to (4), we have that $(\dot{B}_\nu)_{\nu \in \mathbb{N}}$ is a representative of W. If $(X_\nu)_{\nu \in \mathbb{N}}$ is a representative of a solution $X \in \mathcal{G}\mathcal{P}$ to (11) then

$$\dot{X}_\nu(t) = \alpha X_\nu(t)\dot{B}_\nu(t) + J_\nu(t) \quad \text{with} \quad X_\nu(0) = A_\nu + I_\nu$$

for some $J \in \mathcal{N}\mathcal{P}$ and $I \in \mathcal{N}\mathcal{R}$, where $(A_\nu)_{\nu \in \mathbb{N}}$ is a representative of A. Thus

$$X_\nu(t) = (A_\nu + I_\nu) \exp\big(\alpha B_\nu(t) - \alpha B_\nu(0)\big)$$
$$+ \int_0^t \exp\big(\alpha B_\nu(t) - \alpha B_\nu(s)\big) J_\nu(s)\, ds.$$

Due to the pathwise boundedness of $B_\nu(t)$ on any interval $[-T, T]$, independently of $\nu \in \mathbb{N}$, the sequence $(X_\nu)_{\nu \in \mathbb{N}}$ belongs to $\mathcal{E}\mathcal{P}_M$ and hence defines a solution.

Uniqueness is evident from the fact that the terms involving I_ν and J_ν belong to $\mathcal{NP}$. One representative of the solution is actually

$$\widetilde{X}_\nu(t) = A_\nu \exp\big(\alpha B_\nu(t) - \alpha B_\nu(0)\big).$$

In case $A_\nu \equiv A$ is a classical random variable, $X_\nu(t)$ or $\widetilde{X}_\nu(t)$ converge, pathwise uniformly on $t \in [-T, T]$, as $\nu \to \infty$, to

$$Z(t) = A \exp\big(\alpha B(t)\big).$$

This can be interpreted as saying that the generalized solution $X \in \mathcal{GP}$ is *associated* with the classical process $Z \in \mathcal{CP}$. As is to be expected from classical convergence results (Wong and Zakai [21]), $Z(t)$ is the *Stratonovich solution* to problem (11), see e.g. Kloeden and Platen [10; Section 6.1]. In general, when white noise is replaced by an arbitrary generalized process in (11), it will be a difficult (and important) question to decide whether the generalized solution $X \in \mathcal{GP}$ is associated with some classical process or not. A complete positive answer exists only for smooth driving processes, where X actually equals the classical solution (Proposition 4.2).

§5 $\mathcal{L}^0$-VALUED GENERALIZED FUNCTIONS

While in Sections 3 and 4 we have based our constructions of generalized stochastic processes on sample path properties, we shall now outline an approach employing the topological-algebraic properties of $\mathcal{L}^0$ and sequences of smooth $\mathcal{L}^0$-valued functions on $\mathbb{R}$.

As in the previous section we work on a fixed probability space (Ω, Σ, μ). We need a few preliminary observations concerning the space $\mathcal{L}^0$ of measurable functions on Ω. As is well-known (see e.g. Loève [13; §9.3]), $\mathcal{L}^0$ equipped with convergence in probability is a complete metrizable topological vector space. With respect to multiplication and superposition defined pointwise almost everywhere, we have the following assertions.

Lemma 5.1.

  (a) $\mathcal{L}^0$ is a topological algebra.
  (b) Let $F \in C^1(\mathbb{R}^m)$ such that $|\nabla F|$ is polynomially bounded. Then F defines, by superposition, a continuous map from $(\mathcal{L}^0)^m$ to $\mathcal{L}^0$.

Proof. (a) We work with the pseudonorms $\pi_a(X) = \mu(|X| > a)$ which define the topology of $\mathcal{L}^0$. Recall that $\pi_{2a}(X+Y) \le \pi_a(X) + \pi_a(Y)$. To prove the continuity of multiplication on $\mathcal{L}^0$ it therefore suffices to show that $X_j \to 0$ and $Y_j \to Y$ in probability imply $\pi_a(X_j Y_j) \to 0$ as $j \to \infty$ for all $a > 0$. For this we take $\epsilon > 0$ and first choose j so large that $\mu(|Y_j - Y| > a) < \epsilon/4$. By the triangle inequality it follows that $\mu(|Y_j| > a + |Y|) < \epsilon/4$ as well. Next we take b so large that $\mu(|Y| > b) < \epsilon/4$. Then

$$\mu(|Y_j| > a + b) = \mu(|Y_j| > a + b, |Y| > b) + \mu(|Y_j| > a + b, |Y| \le b)$$
$$\le \mu(|Y| > b) + \mu(|Y_j| > a + |Y|) < \frac{\epsilon}{2}.$$

Finally,

$$\mu(|X_j Y_j| > a) = \mu(|X_j Y_j| > a, |Y_j| \le a + b) + \mu(|X_j Y_j| > a, |Y_j| > a + b)$$
$$\le \mu\left(|X_j| > \frac{a}{a+b}\right) + \mu(|Y_j| > a + b) < \epsilon$$

for sufficiently large $j \in \mathbb{N}$.

(b) Let $X_j = (X_{1j}, \ldots, X_{mj}) \in (\mathcal{L}^0)^m$ with $X_j \to X$ in probability as $j \to \infty$. Then

$$\pi_a\big(F(X_j) - F(X)\big) = \pi_a\left(\left[\int_0^1 \nabla F(\sigma X_j + (1 - \sigma)X)\, d\sigma [X_j - X]\right)\right.$$
$$\le \pi_a\big(C(1 + |X_j| + |X|)^k |X_j - X|\big)$$

for some $k \in \mathbb{N}$ and $C > 0$, where the integral is understood pointwise at fixed $\omega \in \Omega$. Using that $\mathcal{L}^0$ is a topological algebra from (a), we get that this last expression converges to zero as $j \to \infty$. □

As $\mathcal{L}^0$ is a topological vector space, we can talk about $\mathcal{C}^\infty(\mathbb{R}, \mathcal{L}^0)$, the space of smooth functions of $t \in \mathbb{R}$ with values in $\mathcal{L}^0$. Due to Lemma 5.1 we have:

Corollary 5.2. *Let* $F \in \mathcal{O}_M(\mathbb{R}^m)$ *and* $X \in (\mathcal{C}^\infty(\mathbb{R}, \mathcal{L}^0))^m$. *Then* $F(X) \in \mathcal{C}^\infty(\mathbb{R}, \mathcal{L}^0)$ *as well. In particular,* $\mathcal{C}^\infty(\mathbb{R}, \mathcal{L}^0)$ *is an algebra.*

Proof. It follows from Lemma 5.1(b) that $F(X) \in \mathcal{C}(\mathbb{R}, \mathcal{L}^0)$: Applying the mean value theorem twice we get, at fixed $\omega \in \Omega$,

$$\frac{1}{h}\big(F(X(t + h)) - F(X(t))\big)$$
$$= \left(\frac{1}{h}\int_0^1 \int_0^1 \sigma \nabla^2 F\big(\sigma \tau X(t + h) + (1 - \sigma\tau)X(t)\big)\, d\sigma\, d\tau\right)$$
$$[X(t + h) - X(t), X(t + h) - X(t)]$$
$$+ \nabla F(X(t))\left[\frac{1}{h}(X(t + h) - X(t))\right].$$

Due to the polynomial bounds of $\nabla^2 F$ we can estimate the integral by a polynomial in $X(t + h)$ and $X(t)$. Hence, by the topological-algebra property of $\mathcal{L}^0$, the first term on the right-hand side converges to zero while the second converges to $\nabla F(X(t))[\frac{d}{dt}X(t)]$ in probability. Thus $F(X(t)) \in \mathcal{C}^1(\mathbb{R}, \mathcal{L}^0)$, and the chain rule holds. Similarly, one derives the Leibnitz product rule. Taking successively higher derivatives and employing the chain – and Leibnitz rules together with the polynomial bounds on all derivatives of F, one finally obtains that $F(X) \in C^\infty(\mathbb{R}, \mathcal{L}^0)$. $\square$

What we have done so far already enables us to consider a very basic algebra of generalized stochastic processes, or rather $\mathcal{L}^0$-valued generalized functions: the sequence algebra

$$\mathcal{E}[\mathbb{R}, \mathcal{L}^0] = \left(C^\infty(\mathbb{R}, \mathcal{L}^0)\right)^{\mathbb{N}}.$$

The space $\mathcal{C}(\mathbb{R}, \mathcal{L}^2)$ of classical second order processes is imbedded in $\mathcal{E}[\mathbb{R}, \mathcal{L}^0]$ via

$$\iota : Z \to (Z * \psi \varphi_\nu)_{\nu \in \mathbb{N}} \tag{12}$$

with ψ and φ_ν as in (4), the convolution integral understood as an $\mathcal{L}^2$-valued Bochner integral. Actually, (12) makes even sense for $Z \in \mathcal{D}'(\mathbb{R}, \mathcal{L}^2)$. We shall say that an element $(Y_\nu)_{\nu \in \mathbb{N}} \in \mathcal{E}[\mathbb{R}, \mathcal{L}^0]$ is of L^2-type, if each Y_ν belongs to $\mathcal{C}(\mathbb{R}, \mathcal{L}^2)$. Of course, $\iota(Z)$ in (12) is of L^2-type, as are all of its (generalized) derivatives. We come to an existence result concerning the stochastic differential equation (7) in $\mathcal{E}[\mathbb{R}, \mathcal{L}^0]$.

Proposition 5.3. *Let F be given by a sequence $(F_\nu)_{\nu \in \mathbb{N}}$ such that each F_ν is in $\mathcal{O}_M(\mathbb{R}^2)$ and $|\nabla F_\nu|$ is bounded (no growth restrictions with respect to ν). Let $Y \in \mathcal{E}[\mathbb{R}, \mathcal{L}^0]$ be of L^2-type and $A \in (\mathcal{L}^2)^{\mathbb{N}}$. Then problem (7) has a solution $X \in \mathcal{E}[\mathbb{R}, \mathcal{L}^0]$ such that X and $\dot{X}$ are of L^2-type.*

Proof. The result is evident from the classical $\mathcal{L}^2$-existence theory for $\dot{X}_\nu = F_\nu(X_\nu, Y_\nu)$ with $X_\nu(0) = A_\nu$, giving a solution $X_\nu \in \mathcal{C}^1(\mathbb{R}, \mathcal{L}^2)$ under the above hypotheses on F_ν. The fact that $X_\nu \in \mathcal{C}^\infty(\mathbb{R}, \mathcal{L}^0)$ follows from the differential equation by arguments similar to those in the proof of Corollary 5.2. $\square$

We remark that the solution X has generalized first and second moments, given by the sequence of real numbers $(\mathrm{E}(X_\nu^p))_{\nu \in \mathbb{N}}$ for $p = 1, 2$. By contrast, the definition of generalized moments in the algebra $\mathcal{GP}$ of Section 3 poses difficulties due to the presence of the ideal $\mathcal{NP}$, which prevents well-definedness on representatives. On the other hand, we have no uniqueness result so far in the situation of Proposition 5.3. Consistency with smooth random processes in $\mathcal{E}[\mathbb{R}, \mathcal{L}^0]$ similar to Section 3 can be obtained by a suitable factorization, though, as follows.

Owing to the lack of good multiplicative properties of the pseudonorms π_a on $\mathcal{L}^0$, we shall turn to estimates in $\mathcal{L}^p$, $0 < p \leq 1$, in the pseudonorms

$$\|X\|_p = \int_\Omega |X(\omega)|^p \, d\mu(\omega).$$

Lemma 5.4.

(a) *If $0 < q \leq p \leq 1$ and $X \in \mathcal{L}^p$, then $X \in \mathcal{L}^q$ and $\|X\|_q \leq \|X\|_p^{q/p}$.*

(b) *If $p \leq \frac{1}{2}$ and $X, Y \in \mathcal{L}^p$, then $XY \in \mathcal{L}^{p^2}$ and $\|XY\|_{p^2} \leq \|X\|_p^p \|Y\|_p^p$.*

Proof. Easy and left to the reader. $\square$

We thus define

$$\mathcal{E}_M[\mathbb{R}, \mathcal{L}^0] = \big\{ (X_\nu)_{\nu \in \mathbb{N}} \in \mathcal{E}[\mathbb{R}, \mathcal{L}^0] :$$
$$\forall T > 0 \; \forall j \in \mathbb{N}_0 \; \exists N > 0 \; \exists p \leq 1 \text{ with}$$
$$\sup\nolimits_{t \in [-T,T]} \|\partial^j X_\nu(t)\|_p = O(\nu^N) \text{ as } \nu \to \infty \big\}$$

$$\mathcal{N}[\mathbb{R}, \mathcal{L}^0] = \big\{ (X_\nu)_{\nu \in \mathbb{N}} \in \mathcal{E}[\mathbb{R}, \mathcal{L}^0] :$$
$$\forall T > 0 \; \forall j \in \mathbb{N}_0 \; \exists p \leq 1 \; \forall N > 0 :$$
$$\sup\nolimits_{t \in [-T,T]} \|\partial^j X_\nu(t)\|_p = O(\nu^{-N}) \text{ as } \nu \to \infty \big\}.$$

From Lemma 5.4 it is clear that $\mathcal{E}_M[\mathbb{R}, \mathcal{L}^0]$ is a differential algebra and $\mathcal{N}[\mathbb{R}, \mathcal{L}^0]$ is an ideal therein. We thus may define a differential algebra of generalized $\mathcal{L}^0$-valued functions as

$$\mathcal{G}(\mathbb{R}, \mathcal{L}^0) = \mathcal{E}_M[\mathbb{R}, \mathcal{L}^0] / \mathcal{N}(\mathbb{R}, \mathcal{L}^0).$$

By going over to the respective equivalence classes, (12) defines an imbedding of $\mathcal{C}(\mathbb{R}, \mathcal{L}^2)$ or more generally $\mathcal{D}'(\mathbb{R}, \mathcal{L}^2)$ into $\mathcal{G}(\mathbb{R}, \mathcal{L}^2)$. In addition, one proves as in Proposition 3.4 that

$$(Z * (\psi\varphi_\nu) - Z)_{\nu \in \mathbb{N}} \in \mathcal{N}(\mathbb{R}, \mathcal{L}^0)$$

for $Z \in \mathcal{C}^\infty(\mathbb{R}, \mathcal{L}^2)$, and this is the desired consistency result.

Conclusion. We have presented various approaches to a framework capable of handling generalized stochastic processes in nonlinear, possibly nonsmooth differential equations. The basic elements were sequences of smooth random processes and factorization with respect to certain ideals. The ideals constructed were geared to establishing maximal consistency with smooth random processes (a number of other possibilities have not been pursued, like factorization involving ultrafilters and leading to the nonstandard spaces $^*\mathcal{C}^\infty\mathcal{P}$, respectively $^*\mathcal{C}^\infty(\mathbb{R}, \mathcal{L}^0)$). Our approach offers a generalized solution concept and gives a meaning to differential-algebraic and more general nonlinear operations with stochastic processes. On the other hand, we have not investigated probabilistic properties and the transfer of probabilistic arguments into this setting; this important and (due to various regularization procedures that appear) difficult task remains to be done. Finally, we point out that nonlinear operations in white noise calculus can also be achieved with the help of the *Wick product* (see e. g. Lindstrøm, Øksendal, Ubøe [12]). A comparison with this theory remains open as well.

REFERENCES

1. S. Albeverio, Z. Haba, F. Russo, *Trivial solutions for a non-linear two-space dimensional wave equation perturbed by space-time white noise*, Prépublication 93–17, Université de Provence, Marseille, 1993.
2. H. Bunke, *Gewöhnliche Differentialgleichungen mit zufälligen Parametern*, Akademie-Verlag, Berlin, 1972.
3. J. F. Colombeau, *New Generalized Functions and Multiplication of Distributions*, North-Holland, Amsterdam, 1984.
4. J. F. Colombeau, *Elementary Introduction to New Generalized Functions*, North-Holland, Amsterdam, 1985.
5. J. F. Colombeau, *Multiplication of distributions*, Lecture Notes Math. 1532, Springer-Verlag, Berlin, 1992.
6. J. F. Colombeau, *Multiplication of distributions*, Bull. Am. Math. Soc. **23** (1990), 251–268.
7. Yu. V. Egorov, *A contribution to the theory of generalized functions*, Russian Math. Surveys **45:5** (1990), 1–49, Translated from: Uspekhi Mat. Nauk **45:5** (1990), 3–40.
8. I. M. Gel'fand, N. Ya. Vilenkin, *Generalized Functions, Vol. 4: Applications of Harmonic Analysis*, Academic Press, New York, 1964.
9. H. Holden, T. Lindstrøm, B. Øksendal, J. Ubøe, T.-S. Zhang, *The Burgers equation with a noisy force and the stochastic heat equation*, Comm. Part. Diff. Eqs. **19** (1994), 119–141.
10. P. E. Kloeden, E. Platen, *Numerical Solution of Stochastic Differential Equations*, Springer-Verlag, Berlin, 1992.
11. D. Laugwitz, *Anwendung unendlich kleiner Zahlen. I. Zur Theorie der Distributionen*, J. reine angew. Math. **207** (1961), 53–60.
12. T. Lindstrøm, B. Øksendal, J. Ubøe, *Wick multiplication and Itô–Skorohod stochastic differential equations*, In: Ideas and Methods in Mathematical Analysis, Stochastics, and Applications (S. Albeverio, J. E. Fenstad, H. Holden, T. Lindstrøm, eds.), vol. I, Cambridge University Press, 1992.
13. M. Loève, *Probability Theory*, vol. I and II. Fourth Ed., Springer-Verlag, New York, 1977/1978.
14. E. Nelson, *Radically Elementary Probability Theory*, Annals Math. Studies, vol. 117, Princeton University Press, 1987.
15. M. Oberguggenberger, *Multiplication of Distributions and Applications to Partial Differential Equations*, Pitman Research Notes Math., vol. 259, Longman, Harlow, 1993.
16. M. Oberguggenberger, *Nonlinear theories of generalized functions*, In: Advances in Analysis, Probability, and Mathematical Physics – Contribution from Nonstandard Analysis (S. Albeverio, W. A. J. Luxemburg, M. P. H. Wolff, eds.), Kluwer, Dordrecht, 1994.
17. E. E. Rosinger, *Distributions and Nonlinear Partial Differential Equations*, Lecture Notes Math. 684, Springer-Verlag, Berlin, 1978.
18. E. E. Rosinger, *Nonlinear Partial Differential Equations. Sequential and Weak Solutions*, North Holland, Amsterdam, 1980.
19. E. E. Rosinger, *Generalized Solutions of Nonlinear Partial Differential Equations*, North-Holland, Amsterdam, 1987.
20. E. E. Rosinger, *Non-linear Partial Differential Equations. An Algebraic View of Generalized Solutions*, North Holland, Amsterdam, 1990.
21. E. Wong, M. Zakai, *On the convergence of ordinary integrals to stochastic integrals*, Ann. Math. Statistics **36** (1965), 1560–1564.

MICHAEL OBERGUGGENBERGER, INSTITUT FÜR MATHEMATIK UND GEOMETRIE, UNIVERSITÄT INNSBRUCK, A-6020 INNSBRUCK, AUSTRIA

Progress in Probability, Vol. 36
© 1995 Birkhäuser Verlag Basel/Switzerland

ON THE GEOMETRY DEFINED BY DIRICHLET FORMS

KARL-THEODOR STURM

ABSTRACT. Every regular, local Dirichlet form on a locally compact, separable space X defines in an intrinsic way a pseudo metric ρ on the state space. Assuming that this is actually a complete metric (compatible with the original topology), we prove that (X, ρ) is a geodesic space. That is, any two points in X are joined by a minimal geodesic.

Also analogues of the Hopf–Rinow Theorem and of the Cartan–Hadamard Theorem are obtained. The latter requires the notion of curvature which is defined by means of the CAT-inequality.

§1 THE DIRICHLET SPACE $(\mathcal{E}, \mathcal{F})$ AND THE ENERGY MEASURE Γ

The basic object for the sequel is a fixed regular Dirichlet form $\mathcal{E}$ with domain $\mathcal{F} = \mathcal{F}(X)$ on a real Hilbert space $L^2(X, m)$ with norm $\|u\| = (\int_X u^2 \, dm)^{1/2}$. $\mathcal{F}$ is again a real Hilbert space with norm $\|u\|_{\mathcal{F}} := \sqrt{\mathcal{E}(u, u) + \|u\|}$.

The underlying topological space X is a locally compact, separable Hausdorff space and m is a positive Radon measure with $\operatorname{supp}[m] = X$. The initial Dirichlet form $\mathcal{E}$ is always assumed to be *symmetric* (i. e. $\mathcal{E}(u, v) = \mathcal{E}(v, u)$) and of *diffusion type* (i. e. $\mathcal{E}(u, v) = 0$ whenever $u \in \mathcal{F}$ is constant on a neighborhood of the support of $v \in \mathcal{F}$ or, in other words, $\mathcal{E}$ has no killing measure and no jumping measure). The selfadjoint operator associated with the initial form $\mathcal{E}$ is denoted by L.

Any such form can be written as

$$\mathcal{E}(u, v) = \int_X d\Gamma(u, v)$$

where Γ is a positive semidefinite, symmetric bilinear form on $\mathcal{F}$ with values in the signed Radon measures on X (the so-called *energy measure*). It can be defined by the formulae

$$\int_X \phi \, d\Gamma(u, u) = \mathcal{E}(u, \phi u) - \frac{1}{2}\mathcal{E}(u^2, \phi)$$

$$= \lim_{t \to 0} \frac{1}{2t} \int_X \int_X \phi(x) \cdot [u(x) - u(y)]^2 \, T_t(x, dy) \, m(dx)$$

1991 *Mathematics Subject Classification.* 60J60, 31C25, 58G32, 35J70.

Key words and phrases. Dirichlet forms, symmetric diffusion, Markov process, Carathéodory metric, geodesic space.

for every $u \in \mathcal{F}(X) \cap L^\infty(X, m)$ and every $\phi \in \mathcal{F}(X) \cap \mathcal{C}_0(X)$. Since $\mathcal{E}$ is assumed to be of diffusion type, the energy measure Γ is local and satisfies the Leibniz rule as well as the chain rule, cf. [8], [13], [2], [3] and [18]. As usual we extend the quadratic forms $u \mapsto \mathcal{E}(u, u)$ and $u \mapsto \Gamma(u, u)$ to the whole spaces $L^2(X, m)$ resp. $L^2_{\text{loc}}(X, m)$ in such a way that $\mathcal{F}(X) = \{ u \in L^2(X, m) : \mathcal{E}(u, u) < \infty \}$ and $\mathcal{F}_{\text{loc}}(X) = \{ u \in L^2_{\text{loc}}(X, m) : \Gamma(u, u) \text{ is a Radon measure} \}$.

§2 THE INTRINSIC METRIC ρ AND ASSUMPTION (A)

The energy measure Γ defines in an intrinsic way a pseudo metric ρ on X by

$$\rho(x, y) = \sup\{ u(x) - u(y) : u \in \mathcal{F}_{\text{loc}}(X) \cap \mathcal{C}(X), \, d\Gamma(u, u) \le dm \text{ on } X \}, \quad (1)$$

called intrinsic metric or Carathéodory metric (cf. [2], [3], [5], [22]). The condition $d\Gamma(u, u) \le dm$ in (1) means that the energy measure $\Gamma(u, u)$ is absolutely continuous w. r. t. the reference measure m with Radon–Nikodym derivative $\frac{d}{dm}\Gamma(u, u) \le 1$ on X (m-almost everywhere). The density $\frac{d}{dm}\Gamma(u, u)(z)$ should be interpreted as the square of the (length of the) gradient of u at $z \in X$. In general, ρ may be degenerate (i. e. $\rho(x, y) = \infty$ or $\rho(x, y) = 0$ for some $x \ne y$). Throughout this paper we make the

Assumption (A). *ρ is a complete metric on X which is compatible with the original topology on X.*

This assumption in particular implies that ρ is non-degenerate and that for any $y \in X$ the function $x \mapsto \rho(x, y)$ is continuous on X. It is discussed in more details in the paper [18]. There we also compared it with the weaker

Assumption (A'). *The topology induced by ρ is equivalent to the original topology.*

Note that under (A') the following assertions are equivalent:

- ρ is a metric (i.e. it is non-degenerate),
- $\rho(x, y) < \infty$ for all $x, y \in X$,
- X is connected.

In [18] we proved that under (A') the following basic property of the distance function holds true.

Lemma 1. *For every $y \in X$ the distance function $\rho_y : x \mapsto \rho(x, y)$ satisfies $\rho_y \in \mathcal{F}_{\text{loc}}(X) \cap \mathcal{C}(X)$ and*

$$d\Gamma(\rho_y, \rho_y) \le dm. \quad (2)$$

Hence, the distance function ρ_y can be used to construct cut-off functions on intrinsic balls $B_r(y)$ of the form $\rho_{y,r} : x \mapsto (r - \rho(x, y))_+$.

Let us list up some important facts on intrinsic balls (which need not be true in general metric spaces).

Proposition 1. *Under (A') the following properties are satisfied for any ball* $B_r(x) = \{ y \in X : \rho(x,y) < r \}$ *resp. its closure* $\overline{B}_r(x)$:

(i) $B_r(x)$ *is connected;*

(ii) $\overline{B}_r(x) = \{ y \in X : \rho(x,y) \leq r \}$.

Proof. (i) Put $B = B_r(x)$. Assume that B is not connected. Let A be a nonempty subset of B which is open and closed in B. Put $C = B \setminus A$. Without restriction $x \in C$. Define the function ψ to be $\equiv 0$ on $X \setminus C$ and to be $r - \rho(x, \cdot)$ on C. Then ψ satisfies $\psi \in \mathcal{F}_{\text{loc}}(X) \cap \mathcal{C}(X)$ and $d\Gamma(\psi, \psi) \leq dm$. Moreover, $\psi(x) - \psi(y) = r$ for all $y \in A$. Hence, by the definition of ρ we must have $\rho(x,y) \geq r$ for all $y \in A$, i.e. $A \cap B = \emptyset$.

(ii) Put $K_r(x) = \{ y \in X : \rho(x,y) \leq r \}$. Obviously, $K_r(x)$ is closed and contains $\overline{B}_r(x)$. Assume that $K_r(x) \neq \overline{B}_r(x)$. Then there exists $y \in K_r(x)$ and $\epsilon > 0$ such that $B_\epsilon(y) \cap B_r(x) = \emptyset$. Now consider the following function ψ: on $B_r(x)$ we put $\psi = r - \rho(x, \cdot)$, on $B_\epsilon(y)$ we put $\psi = \rho(y, \cdot) - \epsilon$ and on the rest we put $\psi = 0$. Then ψ satisfies $\psi \in \mathcal{F}_{\text{loc}}(X) \cap \mathcal{C}(X)$ and $d\Gamma(\psi, \psi) \leq dm$. Moreover, $\psi(x) - \psi(y) = r + \epsilon$. Hence, by the definition of ρ this is a contradiction to $\rho(x,y) = r$. □

In Theorem 2 we shall see that under Assumption (A) all balls $B_r(x)$ are relatively compact.

§3 EXAMPLES

The main examples which we have in mind are:

- L is the Laplace–Beltrami operator on a Riemannian manifold and m is the Riemannian volume; in this case, ρ is just the Riemannian distance. More generally, L can be chosen to be a uniformly elliptic operator on a Riemannian manifold (cf. [16]).

- L is a uniformly elliptic operator with a nonnegative weight ϕ on $\mathbb{R}^N$, i.e. $\mathcal{E}(u,v) = \sum_{i,j=1}^{N} \int a_{ij} \cdot \frac{\partial}{\partial x_i} u \cdot \frac{\partial}{\partial x_j} v \cdot \phi \, dx$ and $(u,v) = \int uv\phi \, dx$ with (a_{ij}) symmetric and uniformly elliptic and ϕ as well as $\phi^{-1} \in L^1_{\text{loc}}(\mathbb{R}^N, dx)$; in this case, ρ is equivalent to the Euclidean distance (cf. [3], [14], [21]).

- L is a subelliptic operator on $\mathbb{R}^N$, i.e.

$$\mathcal{E}(u,v) = \sum_{i,j=1}^{N} \int a_{ij} \cdot \frac{\partial}{\partial x_i} u \cdot \frac{\partial}{\partial x_j} v \, dx$$

and $(u,v) = \int uv \, dx$ with (a_{ij}) symmetric and such that $\mathcal{E}(u,u) \geq \delta \cdot \|u\|^2_{H^\epsilon} - \|u\|^2$ for some $\delta, \epsilon > 0$; in this case, ρ is equal to the metric used e.g. by Fefferman/Phong [6], Fefferman/Sanchez-Calle [7], Jerison [11], Jerison/Sanchez-Calle [12], Nagel/Stein/Wainger [15]; it can locally be estimated by the Euclidean distance $|\cdot|$ as follows

$$\frac{1}{C} \cdot |x - y| \leq \rho(x,y) \leq C \cdot |x - y|^\epsilon.$$

This includes also Hörmander type operators with bounded measurable coefficients in the sense of Saloff–Coste/Stroock [17].

§4 CONSTRUCTION OF GEODESICS

We always take Assumption (A) for granted. A *curve* (or *path*) on X is a continuous map $\gamma : I \to X$ where I is an interval (=connected set) in $\mathbb{R}$. The *length* $L(\gamma)$ of a curve $\gamma : I \to X$ is defined as

$$L(\gamma) = \sup\left\{ \sum_{i=1}^{n} \rho\big(\gamma(t_i), \gamma(t_{i-1})\big) : n \in \mathbb{N}, \, I \ni t_0 < t_1 < \cdots < t_n \in I \right\}. \quad (3)$$

Obviously, the length of a curve $\gamma : [a, b] \to X$ dominates the distance of its endpoints, i.e.

$$L(\gamma) \geq \rho\big(\gamma(a), \gamma(b)\big)$$

and the length of a composed curve $\gamma : [a, b] \cup [b, c] \to X$ is the sum of the lengths of the parts $\gamma_1 = \gamma|_{[a,b]}$ and $\gamma_2 = \gamma|_{[b,c]}$, i.e.

$$L(\gamma) = L(\gamma_1) + L(\gamma_2).$$

Let us state (without proof) the following elementary properties of curves on X.

Lemma 2. *For a curve* $\gamma : I \to X$ *the following properties are equivalent:*

(i) *for all* $r, s, t \in I$ *with* $r < s < t$

$$\rho\big(\gamma(r), \gamma(t)\big) = \rho\big(\gamma(r), \gamma(s)\big) + \rho\big(\gamma(s), \gamma(t)\big); \quad (4)$$

(ii) *for all compact intervals* $J = [s, t] \subset I$

$$L(\gamma|_{[s,t]}) = \rho\big(\gamma(s), \gamma(t)\big); \quad (5)$$

(iii) *for an increasing sequence of compact intervals* $J_n = [s_n, t_n]$ *with* $I = \bigcup_{n \in \mathbb{N}} J_n$ *(e. g. for* $J_n \equiv I$ *if* I *itself is compact) property (5) holds true;*

(iv) *there exists a reparametrization* $\tilde{\gamma} : \tilde{I} \to X$ *with* $\tilde{\gamma}(\tilde{I}) = \gamma(I)$ *and*

$$\rho\big(\tilde{\gamma}(s), \tilde{\gamma}(t)\big) = |t - s| \quad \text{for all } s, t \in I. \quad (6)$$

Definitions.

(i) *A minimal geodesic on* X *is a curve* $\gamma : I \to X$ *with one (hence all) of the properties mentioned in the Lemma 2.*

(ii) *A subunit curve on* X *is a curve* $\gamma : I \to X$ *with* $|\dot{\gamma}| \leq 1$ *on* I, *where the speed* $|\dot{\gamma}|$ *of* γ *is defined by* $|\dot{\gamma}(t)| = \limsup_{\epsilon \to 0} \frac{\rho(\gamma(t), \gamma(t \pm \epsilon))}{\epsilon}$.

Property (6) in Lemma 2 states that the reparametrized curve $\tilde{\gamma}$ is an isometry from $\tilde{I}$ to X. Such a curve is also called minimal geodesic *parametrized by arc length*. A *geodesic* on X is a curve $\gamma : I \to X$ with the property that for every $t \in I$ there exists an $\epsilon > 0$ such that the restriction of γ on $I \cap]t - \epsilon, t + \epsilon[$ is a minimal geodesic. Obviously, every geodesic which is parametrized by arc length is a subunit curve.

Lemma 3. *For any points $x, y \in X$ with $\rho(x, y) = R$ and for any $r \in]0, R[$ there exists an intermediate point $z \in X$ with*

$$\rho(x, z) = r \quad and \quad \rho(z, y) = R - r.$$

In general, this intermediate point z between x and y is of course not unique.

Proof. If there exists a point z with $\rho(x, z) \leq r$ and $\rho(z, y) \leq R - r$ then by the triangle inequality this point z already must satisfy $\rho(x, z) = r$ and $\rho(z, y) = R - r$. (Otherwise $\rho(x, y) < R$!)

If $\rho(x, y) = 0$ then $x = y$ and thus one can choose $z = x$. Assume that $0 < \rho(x, y) < \infty$ and that there exists no such point z. That is, $K_x = \overline{B_r}(x)$ and $K_y = \overline{B_{R-r}}(y)$ are disjoint closed sets and K_x is compact. In this case, these two sets must have a strictly positive distance, say $\rho(K_x, K_y) \geq 3\delta > 0$. But then also the larger sets $L_x = \overline{B_{r+\delta}}(x)$ and $L_y = \overline{B_{R-r+\delta}}(y)$ have a strictly positive distance $\rho(L_x, L_y) \geq \delta > 0$.

Now let us consider the following function

$$\psi_0 = \begin{cases} \rho(x, \cdot) - (r + \delta) & \text{in } L_x, \\ 0 & \text{in } X \setminus (L_x \cup L_y), \\ -\rho(y, \cdot) + R - r + \delta & \text{in } L_y. \end{cases}$$

From Lemma 1 and the truncation property it follows that $\psi_0 \in \mathcal{C}(X) \cap \mathcal{F}_{\text{loc}}(X)$ with

$$d\Gamma(\psi_0, \psi_0) = 1_{L_x} \, d\Gamma(\rho_x, \rho_x) + 0 + 1_{L_y} \, d\Gamma(\rho_y, \rho_y) \leq dm.$$

Moreover, one obviously has $\psi_0(y) - \psi_0(x) = R + 2\delta > R$. But this is a contradiction to $R = \sup\{ \psi(y) - \psi(x) : \psi \in \mathcal{C}(X) \cap \mathcal{F}_{\text{loc}}(X) \text{ with } d\Gamma(\psi, \psi) \leq dm \}$. □

Theorem 1. (X, ρ) *is a geodesic space. That is, any two points $x, y \in X$ are joined by a minimal geodesic. In other words, there exists a continuous map $\gamma : [0, 1] \to X$ with $\gamma(0) = x$, $\gamma(1) = y$ and*

$$\rho(\gamma(r), \gamma(t)) = \rho(\gamma(r), \gamma(s)) + \rho(\gamma(s), \gamma(t)) \quad for \ all \ 0 \leq r < s < t \leq 1.$$

Proof. (i) We fix points $x, y \in X$ with $\rho(x, y) = R > 0$. By Assumption (A), the metric space (X, ρ) is complete and locally compact. Hence, for the point $x_0 = x$ there exists a maximal radius $R_0 \in]0, R]$ with the property that $B_r(X_0)$ is relatively compact for all $r < R_0$. First of all, we construct the desired geodesic γ on the interval $[0, R_0]$ such that its graph lies inside the closed ball $\overline{B_{R_0}}(x_0)$.

(ii) According to Lemma 3, there exists an intermediate point $\gamma(R_0/2)$ between $\gamma(0) = x_0$ and $\gamma(R) = y$ with the properties $\rho(\gamma(0), \gamma(R_0/2)) = R_0/2$ and $\rho(\gamma(R_0/2), \gamma(R)) = R - R_0/2$. Applying the same argument to the pairs of points $\gamma(0), \gamma(R_0/2)$ and $\gamma(R_0/2), \gamma(R)$ one obtains intermediate points $\gamma(R_0/4)$ (between $\gamma(0)$ and $\gamma(R_0/2)$) and $\gamma(3/4R_0)$ (between $\gamma(R_0/2)$ and $\gamma(R)$). Doing this iteratively, one finally gets a countable set of points $\{\gamma(\alpha R_0) : \alpha \in [0, 1[\text{ dyadic} \}$ with the property

$$\rho(x, \gamma(\alpha R_0)) = \alpha R_0 = R - \rho(\gamma(\alpha R_0), y) \quad \text{for all dyadic numbers } \alpha \in [0, 1[.$$

The closure of this countable set is the minimal geodesic γ restricted to $[0, R_0]$.

(iii) Denote the point $\gamma(R_0)$ on $\partial B_{R_0}(x_0)$ by x_1. Either $R_0 = R$ then $x_1 = y$ and we are finished, or $R_0 < R$.

In the latter case, similarly as in (i), we fix a maximal radius $R_1 \in]0, R - R_0]$ such that $B_r(x_1)$ is relatively compact for all $r < R_1$, and (as in (ii)) we construct the geodesic γ on the interval $[R_0, R_0 + R_1]$. The graph of this part of γ lies in $\overline{B_{R_1}}(x_1) \subset \overline{B_{R_0+R_1}}(x_0)$. Its endpoint $\gamma(R_0 + R_1)$ on $\partial B_{R_1}(x_1)$ (which lies also on $\partial B_{R_0+R_1}(x_0)$) will be denoted by x_2.

Doing this successively, one obtains sequences $\{x_n\}_n$ of points in X and $\{R_n\}_n$ of radii in $]0, R]$. Put $S_n = \sum_{k<n} R_k$ and $S_\infty = \sup_n S_n$. Then $x_n = \gamma(S_n)$. The above construction yields the desired geodesic γ restricted to the interval $[0, S_\infty] \subset [0, R]$.

(iv) If $S_\infty = R$ we are already finished. Hence, suppose $S_\infty < R$. Then $\{R_n\}_n$ converges to 0 and $\{x_n\}_n$ is a Cauchy sequence in $B_R(x)$ which converges to the point $x_\infty := \gamma(S_\infty)$. By Assumption (A), there exists a radius $R_\infty \in]0, R - S_\infty]$ such that $B_{R_\infty}(x_\infty)$ is relatively compact. For finally all $n \in \mathbb{N}$, the balls $B_{2R_n}(x_n)$ are contained in that ball $B_{R_\infty}(x_\infty)$ which implies that they are also relatively compact. This, however, contradicts the maximality of R_n. Therefore, S_∞ must equal R and we are finished. $\square$

Remark. Assume that instead of (A) only (A') is satisfied and that $\rho(x, y) < \infty$ for two points under consideration. Instead of requiring that (X, ρ) is complete it suffices to assume in Lemma 3 and Theorem 1 that there exists a complete (w. r. t. the metric ρ) subset $Y \subset X$ containing the convex hull of x and y. The latter means that all $z \in X$ satisfying the equality $\rho(x, z) + \rho(z, y) = \rho(x, y)$ lie in Y.

For instance, the closed ball $\overline{B_R}(x)$ contains the convex hull of x and $y \in \partial B_R(x)$. Thus, in Lemma 3 and Theorem 1 it suffices to assume that either $\overline{B_R}(x)$ or $\overline{B_R}(y)$ is complete.

Let us again assume that (A) is satisfied.

Corollary 1. *The following distances on X coincide:*

- $\rho(x, y) = \sup\{ \psi(x) - \psi(y) : \psi \in \mathcal{C}(X) \cap \mathcal{F}_{\text{loc}}(X), d\Gamma(\psi, \psi) \leq dm \}$,
- $\rho_0(x, y) = \sup\{ \psi(x) - \psi(y) : \psi \in \mathcal{C}_0(X) \cap \mathcal{F}(X), d\Gamma(\psi, \psi) \leq dm \}$,
- $\rho_1(x, y) = \inf\{ L(\gamma) : \gamma \text{ is a geodesic in } X \text{ joining } x \text{ and } y \}$,
- $\rho_2(x, y) = \inf\{ L(\gamma) : \gamma \text{ is a curve in } X \text{ joining } x \text{ and } y \}$,
- $\rho_3(x, y) = \inf\{ R > 0 : \text{ there exists a subunit curve } \gamma : [0, R] \to X$
  $\text{with } \gamma(0) = x \text{ and } \gamma(R) = y \}$.

Proof. The equality "$\rho = \rho_0$" was already proven in [18]. For the inequality "$\rho \geq \rho_1$" note that the previous Theorem states that $\rho(x, y)$ is the length of a suitable (minimal) geodesic joining x and y. The inequalities "$\rho_1 \geq \rho_2$" and "$\rho_2 \geq \rho$" are obvious.

In order to see "$\rho_1 \geq \rho_3$", note that in the definition of ρ_1 we may replace "geodesic" by "geodesic parametized by arc length" and that for the latter the length equals the length of its defining interval. Finally, note that every geodesic parametrized by arc length is a subunit curve.

For the remaining inequality "$\rho_3 \leq \rho$" note that every subunit curve $\gamma : [0, R] \rightarrow X$ satisfies $\rho(\gamma(s), \gamma(t)) \leq |t - s|$ for all $s, t \in [0, R]$. In particular, $\rho(\gamma(0), \gamma(R)) \leq R$. $\square$

We close this section with some analogue to the Hopf–Rinow Theorem.

Theorem 2. *Under Assumption (A') and assuming that X is connected the following are equivalent:*

   (i) *the metric space (M, ρ) is complete;*
   (ii) *every ball $B_r(x)$ is relatively compact;*
   (iii) *every (minimal) geodesic $\gamma : I \rightarrow X$ defined on an interval $I \subset \mathbb{R}$ can be extended to a (minimal) geodesic defined on $\bar{I}$.*

Proof. The implication (i) $\Rightarrow$ (ii) was proved by Gromov ([10], Thme. 1.10, cf. also [9], Lemme III. 18). It is based on the result of Theorem 1 saying that under (i) the metric space (X, ρ) is a geodesic space. The implications (ii) $\Rightarrow$ (i) and (i) $\Rightarrow$ (iii) are trivial. In order to prove the converse implication, assume that (X, ρ) is not complete, say, the ball $B_{R_0}(x)$ is not relatively compact for some fixed $x \in X$ and some $R_0 < \infty$. Let $R = R(x) = \inf\{r > 0 : B_r(x)$ is not relatively compact $\}$. By assumption (A'), $B_R(x)$ is not relatively compact whereas $B_r(x)$ is relatively compact for all $r < R$.

Now consider a Cauchy sequence $\{y_n\}_n$ in $(B_R(x), \rho)$ which does not converge in $\overline{B_R}(x)$ but to an abstract point y. Take $\hat{B} = \overline{B_R}(x) \cup \{y\}$. Lemma 3 allows to construct a midpoint z between x and y. Namely, for any $n \in \mathbb{N}$ there exists a midpoint z_n between x and y_n satisfying $\rho(x, z_n) = \rho(z_n, y_n) = 1/2\rho(x, y_n)$. Obviously, $|\rho(x, z_n) - 1/2\rho(x, y)| = 1/2|\rho(x, y_n) - \rho(x, y)| \leq 1/2\rho(y_n, y) \rightarrow 0$ and $|\rho(z_n, y) - 1/2\rho(x, y)| \leq |\rho(z_n, y_n) - 1/2\rho(x, y_n)| + 3/2\rho(y_n, y) = 3/2\rho(y_n, y) \rightarrow 0$ (for $n \rightarrow \infty$). All these points z_n, $n \in \mathbb{N}$, lie in the compact set $\overline{B_{r/2}}(x)$. Hence, there exists a cluster point z of $\{z_n\}_n$ in $\overline{B_{r/2}}(x)$ which (by continuiuty of ρ) must satisfy $\rho(x, z) = 1/2\rho(x, y)$ and $\rho(z, y) = 1/2\rho(x, y)$.

Therefore, a slight modification of Theorem 1 allows to construct a (minimal) geodesic $\gamma : [0, R] \rightarrow \hat{B}$ from x to y. The curve $\gamma : [0, R[\rightarrow B_R(x)$ is a (minimal) geodesic in X defined on the parameter interval $[0, R[$ which can not be extended to a curve in X defined on $[0, R]$. $\square$

Even if the state space (X, ρ) is complete, it is in general not possible to extend a geodesic $\gamma : I \rightarrow X$ to a geodesic defined on the whole interval $\mathbb{R}$. For instance, it is not possible if X is the closed unit ball in $\mathbb{R}^N$ and $\mathcal{E}$ is the classical Dirichlet form with Neumann boundary conditions (i. e. $\mathcal{F} = H^1(B_1(0))$). In this case, no (!) geodesic can be extended to a geodesic defined on the whole $\mathbb{R}$.

Remark. Let (A) be satisfied. The following two conditions are equivalent:
- every geodesic $\gamma : I \to X$ can be extended to a geodesic defined on $\mathbb{R}$;
- every point $x \in X$ has a neighborhood Y such that for every $y \in Y$ there exists a point $z \in X \setminus \{x\}$ satisfying

$$\rho(y, z) = \rho(y, x) + \rho(x, z).$$

§5 AN ALTERNATIVE APPROACH TO THE INTRINSIC METRIC

Up to now we can measure the distance ρ between two points x and y on X only by means of functions ψ which are defined on the whole space. Our aim is to localize this procedure. In particular, we want to measure distances by moving along paths.

Of course, this can already be done according to Corollary 1. Note, however, that our previous definition of the length of a curve requires the knowledge of the distance of points. In the sequel, we will reverse the procedure. We define an alternative notion L^* of length of curves without referring to the metric ρ and by means of L^* we define an alternative notion ρ^* of distance of points.

Definitions.

(i) For a curve $\gamma : [a, b] \to X$ we define

$$L^*(\gamma) = \sup\{ u(\gamma(a)) - u(\gamma(b)) : Y \text{ is an open neighborhood of}$$
$$\gamma([a, b]) \subset X,\ u \in \mathcal{F}_{\mathrm{loc}}(Y) \cap \mathcal{C}(Y),\ d\Gamma(u, u) \le dm \text{ on } Y \}$$

and, generally, for a curve $\gamma : I \to X$ we define

$$L^*(\gamma) = \limsup_{n \in \mathbb{N}} L^*(\gamma|_{I \cap [-n, n]}).$$

(ii) For $x, y \in X$ we define

$$\rho^*(x, y) = \inf\{ L^*(\gamma) :\ \gamma \text{ is a curve joining } x \text{ and } y \}.$$

Note that this notion L^* of length ignores loops. In particular, the L^*-length of a closed curve is 0.

The main advantage of this alternative definition $L^*(\gamma)$ is that it depends only on quantities in an arbitrary close neighborhood of $\gamma(I)$. Obviously, $L^*(\gamma) \ge \rho(\gamma(a), \gamma(b))$ for any curve $\gamma : [a, b] \to X$ and thus

$$\rho^* \ge \rho. \tag{7}$$

This holds true without Assumption (A) or (A').

Lemma 4. *Let* $\gamma : [a, c] \to X$ *be a curve without selfintersections and let* $b \in \,]a, c[$. *Define* $\gamma_1 = \gamma|_{[a,b]}$ *and* $\gamma_2 = \gamma|[b, c]$. *Then*

$$L^*(\gamma) = L^*(\gamma_1) + L^*(\gamma_2).$$

Proof. Let $x = \gamma(a)$, $y = \gamma(b)$ and $z = \gamma(c)$. The inequality "$\leq$" is obvious since every function u used in the definition of $L^*(\gamma)$ can simultaneously be used in the definition of $L^*(\gamma_1)$ and $L^*(\gamma_2)$ and since $u(x) - u(z) = [u(x) - u(y)] + [u(y) - u(z)]$.

For the converse inequality, choose $\epsilon > 0$. We have to construct a neighborhood Y of $\gamma([a, c])$ and a suitable function u on it with $u(x) - u(z) \geq L^*(\gamma_1) + L^*(\gamma_2) - 4\epsilon$ and $d\Gamma(u, u) \leq dm$ on Y. For $i = 1, 2$ there exist neighborhoods Y_i of the graphs of γ_i and functions $u_i \in \mathcal{F}_{\text{loc}}(Y_i) \cap C(Y_i)$ with $u_1(x) - u_1(y) \geq L^*(\gamma_1) - \epsilon$ and $u_2(y) - u_2(z) \geq L^*(\gamma_2) - \epsilon$ and $d\Gamma(u_i, u_i) \leq dm$ on Y_i.

Let $U \subset Y_1 \cap Y_2$ be a neighborhood of y such that $u_1 < u_1(y) + \epsilon$ on U and $u_2 > u_2(y) - \epsilon$. Since γ is without selfintersections, we may assume without restriction that $U = Y_1 \cap Y_2$. Now choose an open neighborhood Y of the graph of γ with $Y \subset Y_1 \cup Y_2$. Define

$$u = \begin{cases} (u_1 - u_1(y) - \epsilon) \vee 0 & \text{in } Y \cap Y_1, \\ (u_2(y) - u_2 + \epsilon) \wedge 0 & \text{in } Y \cap Y_2. \end{cases}$$

Then $u(x) - u(z) \geq L^*(\gamma_1) + L^*(\gamma_2) - 4\epsilon$. Moreover, $u \equiv 0$ in U. Therefore (by means of the truncation property), we deduce that $u \in \mathcal{F}_{\text{loc}}(Y) \cap C(Y)$ with $d\Gamma(u, u) \leq dm$ on Y. $\square$

Theorem 3. *Assume (A') and let* $\gamma : I \to X$ *be a curve without selfintersections and such that* $\gamma(I)$ *is relatively compact in* (X, ρ). *Then*

$$L^*(\gamma) = L(\gamma).$$

Proof. Let $L^*(\gamma) = L^*$ and, without restriction, $I = [a, b]$. Choose $\epsilon > 0$, an open neighborhood Y of $\gamma(I)$ and an admissible function u on Y with $u(\gamma(a)) - u(\gamma(b)) \geq L^* - \epsilon$. (Here and below we call a function u on an open set $Y \subset X$ admissible on Y if $u \in \mathcal{F}_{\text{loc}}(Y) \cap C(Y)$ and $d\Gamma(u, u) \leq dm$ on Y.)

Let $\delta = 1/4 \cdot \rho(\gamma(I), X \setminus Y)$. The relative compactness of $\gamma(I)$ implies $\delta > 0$. Choose $a = t_0 < t_1 < \ldots < t_n = b$ with $\delta_i := \rho(\gamma(t_i), \gamma(t_{i-1})) \leq \delta$. Then for every $i = 1, \ldots, n$ the function u is defined and admissible on the whole ball $B_{4\delta_i}(\gamma(t_i))$. Hence, $\tilde{v}_i = [3\delta_i - \rho(\gamma(t_i), \cdot)] \wedge [u - u(\gamma(t_{i-1}))]$ is defined and admissible on $B_{4\delta_i}(\gamma(t_i))$. It immediately follows that $\tilde{v}_i \leq 0$ on $B_{4\delta_i}(\gamma(t_i)) \setminus B_{3\delta_i}(\gamma(t_i))$. Hence,

$$v_i = \begin{cases} \tilde{v}_i \vee 0, & \text{on } B_{3\delta_i}(\gamma(t_i)), \\ 0, & \text{else,} \end{cases}$$

is defined and admissible on the whole space X. From the definition of ρ it follows now that

$$v_i\big(\gamma(t_i)\big) - v_i\big(\gamma(t_{i-1})\big) \leq \rho\big(\gamma(t_i), \gamma(t_{i-1})\big) = \delta_i$$

and thus

$$u\big(\gamma(t_i)\big) - u\big(\gamma(t_{i-1})\big) \le \delta_i.$$

This implies

$$L^*(\gamma) - \epsilon \le u\big(\gamma(a)\big) - u\big(\gamma(b)\big) = \sum_{i=1}^{n} u\big(\gamma(t_i)\big) - u\big(\gamma(t_{i-1})\big)$$

$$\le \sum_{i=1}^{n} \rho\big(\gamma(t_i), \gamma(t_{i-1})\big) \le L(\gamma)$$

and, hence, $L^*(\gamma) = L(\gamma)$. $\square$

Corollary 2. *Assume that (X, ρ) satisfies Assumption (A). Then*

$$\rho^* = \rho. \tag{8}$$

§6 CURVATURE AND THE CARTAN-HADAMARD THEOREM

In the following section we briefly sketch some ideas and results from the general theory of geodesic spaces and the method of "comparison of geometries". This approach goes back to A. D. Aleksandrov (cf. [1]) and was further elaborated among many others by M. Gromov (cf. [10]).

We again make Assumption (A) which implies that (X, ρ) is a locally compact, complete, geodesic metric space. The basic idea is to define upper bounds for the "curvature" on X by comparing geodesic triangles in X with isometric triangles in spaces of constant sectional curvature.

For $\kappa \in \mathbb{R}$, we denote by H_κ the two-dimensional complete, simply connected Riemannian manifold of constant sectional curvature κ. For $\kappa = 0$ this is the Euclidean plane, for $\kappa > 0$ it is a two-dimensional sphere of radius $1/\sqrt{\kappa}$ and for $\kappa < 0$ it is the two-dimensional hyperbolic plane (homothetic to the Poincaré disc).

A geodesic triangle T in X consists of three points in X and three minimal geodesics connecting them. A comparison triangle T_κ for T in H_κ is a geodesic triangle in H_κ with the same edge lenghts as T. It is clear that T_κ is unique up to an isometry of H_κ and such a triangle T_κ exists if $\kappa \le 0$ or if $\kappa > 0$ and the perimeter of T is less than $2\pi/\sqrt{\kappa}$. There is a unique map $T \to T_\kappa$ which takes each edge of T isometrically onto the corresponding edge of T_κ. For each $x \in T$ let x_κ denote its image in T_κ under this map.

Definitions.

(i) A triangle T in X (of perimeter $\le 2\pi/\sqrt{\kappa}$ if $\kappa > 0$) *satisfies the CAT(κ)-inequality iff for any vertex $y \in T$ and any point $x \in T$ on the side opposite to y we have*

$$\rho(x, y) \le \rho_\kappa(x_\kappa, y_\kappa), \tag{9}$$

where x_κ and y_κ are the corresponding points of a comparison triangle in H_κ and ρ_κ denotes the distance in H_κ. (The abbreviation CAT comes from C=comparison, A=Aleksandrov, T=Toponogov.)

(ii) The geodesic space X *has curvature* $\leq \kappa$ iff any $x \in X$ has a neighborhood Y such that any geodesic triangle in Y (of perimeter $\leq 2\pi/\sqrt{\kappa}$ in the case $\kappa > 0$) satisfies the CAT(κ)-inequality.

Remark (cf. [1], [9]). Let (M, g) be a complete Riemannian manifold. Then M has curvature $\leq \kappa$ (in the sense of the above definition) if and only if it has sectional curvature $K \leq \kappa$.

Having at hand the notion of curvature, one can derive for our geodesic spaces many properties which are known to hold for complete Riemannian manifolds with sectional curvature $K \leq \kappa$. We pick out one of these results, namely the famous Cartan–Hadamard Theorem. A complete proof of this result in the full generality of geodesic spaces was given by W. Ballmann ([9], Chap. 10, Thm. 14).

Theorem 4. *If X is simply connected and has curvature ≤ 0, then it is contractible. Any two points $x, y \in X$ are joined by exactly one geodesic and this geodesic is minimal and depends continuously on x and y.*

Acknowledgement. The author is grateful to Professor T. Lyons for stimulating discussions on the subject of this paper.

REFERENCES

1. A. D. Aleksandrov, V. N. Berestovvskii and I. G. Nikolaev, *Generalized Riemannian spaces*, Russian Math. Surveys **41**, 3 (1986), 1–54.

2. M. Biroli and U. Mosco, *Formes de Dirichlet et estimations structurelles dans les milieux discontinus*, C. R. Acad. Sci. Paris **313** (1991), 593–598.

3. M. Biroli and U. Mosco, *A Saint-Venant Principle for Dirichlet forms on discontinuous media*, Elenco Preprint, Rome, 1992.

4. E. A. Carlen, S. Kusuoka and D. W. Stroock, *Upper bounds for symmetric Markov transition functions*, Ann. Inst. H. Poincaré **2** (1987), 245–287.

5. E. B. Davies, *Heat kernels and spectral theory*, Cambridge University Press, 1989.

6. C. L. Fefferman and D. Phong, *Subelliptic eigenvalue problems*, in: Conf. on Harmonic Analysis, Chicago (W. Beckner et al., ed.), Wadsworth, 1981, pp. 590–606.

7. C. L. Fefferman and A. Sanchez-Calle, *Fundamental solutions for second order subelliptic operators*, Ann. of Math. **124** (1986), 247–272.

8. M. Fukushima, *Dirichlet forms and Markov processes*, North Holland and Kodansha, Amsterdam, 1980.

9. E. Ghys and P. de la Harpe, *Sur les groupes hypberboliques d'aprés Mikhael Gromov*, Prog. Math., vol. 83, Birkhäuser, Boston, 1990.

10. M. Gromov, *Structures métriques pour les variétés riemanniennes*, Rédigé par J. Lafontaine et P. Pansu., Cedic/F.Nathan, 1981.

11. D. Jerison, *The Poincaré inequality for vector fields satisfying an Hörmander's condition*, Duke J. Math. **53** (1986), 503–523.

12. D. Jerison and A. Sanchez-Calle, *Estimates for the heat kernel for a sum of squares of vector fields*, Indiana Univ. J. of Math. **35** (1986), 835–854.

13. Y. Lejan, *Mesures associées a une forme de Dirichlet. Applications*, Bull. Soc. Math. France **106** (1978), 61–112.

14. Z. Ma and M. Röckner, *Introduction to the theory of (non-symmetric) Dirichlet forms*, Springer Universitext, 1992.

15. A. Nagel, E. Stein and S. Wainger, *Balls and metrics defined by vector fields. I: Basic properties*, Acta Math. **155** (1985), 103–147.

16. L. Saloff-Coste, *Uniformly elliptic operators on Riemannian manifolds*, J. Diff. Geom. **36** (1992), 417–450.

17. L. Saloff-Coste and D. W. Stroock, *Operateurs uniformement sous-elliptiques sur des groupes de Lie*, J. Funct. Anal. **98** (1991), 97–121.

18. K.-Th. Sturm, *Analysis on local Dirichlet spaces – I. Recurrence, conservativeness and L^p-Liouville properties* (to appear in J. Reine Angew. Math.).

19. K.-Th. Sturm, *Analysis on local Dirichlet spaces – II. Upper Gaussian estimates for the fundamental solutions of parabolic equations* (to appear in Osaka J. Math.).

20. K.-Th. Sturm, *Analysis on local Dirichlet spaces – III. The parabolic Harnack inequality*, Preprint, Erlangen, 1994.

21. K.-Th. Sturm, *Sharp estimates for capacities and applications to symmetric diffusions*, Preprint, Erlangen, 1994.

22. N. Varopoulos, L. Saloff-Coste and T. Coulhon, *Analysis and geometry on groups*, Cambridge University Press, 1992.

K.-T. STURM, MATHEMATISCHES INSTITUT, UNIVERSITÄT ERLANGEN-NÜRNBERG, BISMARCK-STRASSE 1 1/2, D-91054 ERLANGEN

Progress in Probability, Vol. 36

RANDOM BROWNIAN SCALING
AND SOME ABSOLUTE CONTINUITY RELATIONSHIPS

Marc Yor

ABSTRACT. Consider $(X_t, \ t \geq 0)$ a one-dimensional, or d-dimensional Brownian motion, or d-dimensional Bessel process, starting from 0, and let a and b be two random times with $a < b$, almost surely. One finds, in the literature, a number of examples of such random couples (a, b) such that the law of the process

$$X_t^{[a,b]} \equiv \frac{1}{\sqrt{b-a}} X_{a+t(b-a)}, \quad t \leq 1,$$

which is the transform of X by *random Brownian scaling* on the interval $[a, b]$, is absolutely continuous with respect to the law of $(X_t, \ t \leq 1)$, or the law of another interesting process $(Y_t, \ t \leq 1)$, with a particularly simple density. In the following notes, I shall

(a) present a list of such results; this is done in Section 1;
(b) attempt to unify their proofs; this is partly done in Section 2;
(c) show how these absolute continuity results may be applied to give a deep insight into P. Lévy's arc sine law for Brownian motion; this is done in Section 3;
(d) develop some examples of applications to Bessel processes, in Section 4.

§1 A LIST OF EXAMPLES

The following examples are somewhat scattered in the literature, and admit various extensions. To my knowledge, the fourth example is new.

Example 1. *A representation of the Brownian bridge.* If $(B_u, \ u \geq 0)$ is a one-dimensional Brownian motion starting from the origin, and if, for $s > 0$, we define $g_s = \sup\{ t < s \colon B_t = 0 \}$, then

$$\left(\frac{1}{\sqrt{g_s}} B_{ug_s}; \ u \leq 1 \right) \quad \text{is a standard Brownian bridge.}$$

This representation of the Brownian bridge goes back, at least, to Lévy [8].

1991 *Mathematics Subject Classification.* primary 60E07; secondary 60G55, 60J30, 60J55.
Key words and phrases. Scaling property, Brownian meander, Brownian bridge.

Example 2. *The pseudo-Brownian bridge* ([4], [9]). The pseudo-Brownian bridge is defined, for a fixed $s > 0$, as the process

$$\left(B_u^{\#} = \frac{1}{\sqrt{\tau_s}} B_{u\tau_s}; u \leq 1 \right),$$

where $(\tau_s = \inf\{ u : \ell_u > s \}; s \geq 0)$ is the inverse of $(\ell_u; u \geq 0)$, the local time at 0 of $(B_u; u \geq 0)$. The distribution of $(b(u), u \leq 1)$, the standard Brownian bridge, is equivalent to that of $(B_u^{\#}; u \leq 1)$; more precisely, one has

$$\mathrm{E}[F(b(u); u \leq 1)] = \mathrm{E}\left[\sqrt{\frac{\pi}{2}} \frac{s}{\sqrt{\tau_s}} F(B_u^{\#}; u \leq 1) \right] \tag{1.a}$$

for every measurable functional $F \colon C([0,1]; \mathbb{R}) \to \mathbb{R}_+$. This relation may be used to compute the distributions of certain functionals of the Brownian bridge, using excursion theory.

Example 3. *The Brownian meander* (Imhof [7], Biane–Yor [2]). Using the notation in Example 1, we introduce, for fixed $s > 0$,

$$\left(m(u) = \frac{1}{\sqrt{s - g_s}} |B_{g_s + u(s - g_s)}|; u \leq 1 \right),$$

the so-called Brownian meander. It was discovered by Imhof [7] that the law of $(m(u); u \leq 1)$ is equivalent to that of $(R_u; u \leq 1)$, the three-dimensional Bessel process; more precisely, one has:

$$\mathrm{E}[F(m(u); u \leq 1)] = \mathrm{E}\left[\left(\sqrt{\frac{\pi}{2}} \frac{1}{R_1} \right) F(R_u; u \leq 1) \right] \tag{1.b}$$

for every measurable functional $F \colon C([0,1]; \mathbb{R}_+) \to \mathbb{R}_+$.

A number of results about the Brownian meander have been gathered in [2]; a proof of the absolute continuity relation combining enlargement of filtrations formulae and Girsanov's theorem is given in Azéma–Yor ([1], p. 294).

Example 4. *Brownian motion conditioned to be in $\mathbb{R}_+$ at time 1.* Define $\alpha_s^+ = \inf\{ u : \int_0^u dt \, 1_{(B_t > 0)} > s \}$, the inverse of the process of (occupation) time spent in $\mathbb{R}_+$ by $(B_t; t \geq 0)$. Then, as will be shown in the next section, one has

$$\mathrm{E}\left[1_{(B_1 > 0)} F(B_u; u \leq 1) \right] = \mathrm{E}\left[\frac{1}{\alpha_1^+} F\left(\frac{1}{\alpha_1^+} B_{u\alpha_1^+}; u \leq 1 \right) \right] \tag{1.c}$$

for every measurable functional $F \colon C([0,1]; \mathbb{R}) \to \mathbb{R}_+$.

§2 A GENERAL ABSOLUTE CONTINUITY RESULT

For simplicity, I shall consider here a one-dimensional Brownian motion $(B_t,\ t \geq 0)$ starting from 0, and a continuous increasing process $(A_t,\ t \geq 0)$ which enjoys a scaling property, jointly with $(B_t,\ t \geq 0)$, i.e., there exists $r \in \mathbb{R}$ such that

$$\left(B_{ct},\ A_{ct};\ t \geq 0\right) \overset{\text{(law)}}{=} \left(\sqrt{c}B_t,\ c^{r+1}A_t;\ t \geq 0\right) \quad \text{for } c > 0. \tag{2.a}$$

We shall assume that

$$\mathrm{E}[A_1] < \infty; \tag{2.b}$$

hence, we have: $\mathrm{E}[A_t] \equiv t^{r+1}\,\mathrm{E}[A_1] < \infty$ for every t.

The main object we shall consider in this section is the *deterministic* measure $\mu^{A,F}(dt)$ on $\mathbb{R}_+$, which we associate with the pair (A, F), where $F \colon C([0, 1]; \mathbb{R}_+) \to \mathbb{R}_+$ is a bounded measurable functional; $\mu^{A,F}(dt)$ is defined via the formula

$$\int_0^\infty \mu^{A,F}(dt)\,\varphi(t) = \mathrm{E}\left[\int_0^\infty dA_t\, F\left(\frac{1}{\sqrt{t}}B_{ut};\ u \leq 1\right)\varphi(t)\right],$$

where $\varphi \colon \mathbb{R}_+ \to \mathbb{R}_+$ is a generic Borel function. We shall also write, slightly abusively

$$\mu^{A,F}(dt) = \mathrm{E}\left[dA_t\, F\left(\frac{1}{\sqrt{t}}B_{ut};\ u \leq 1\right)\right].$$

Here is now our key result:

Theorem 2.1. *Under the above hypothesis, the measure* $\mu^{A,F}(dt)$ *is a multiple of* $t^r\, dt$, *i.e. there exists a constant* $c_{A,F}$ *such that*

$$\mu^{A,F}(dt) = c_{A,F}\, t^r\, dt.$$

More precisely, one has:

$$\mathrm{E}\left[dA_t\, F\left(\frac{1}{\sqrt{t}}B_{ut};\ u \leq 1\right)\right] = t^r\, dt\, \mathrm{E}\left[\frac{r+1}{\alpha_1^{r+1}}F\left(\frac{1}{\sqrt{\alpha_1}}B_{s\alpha_1};\ s \leq 1\right)\right] \tag{2.c}$$

where $(\alpha_u,\ u \geq 0)$ *is the inverse of* $(A_t,\ t \geq 0)$, *i.e.* $\alpha_u \equiv \inf\{\, t \colon A_t > u\,\}$.

Proof. Let $\varphi \colon \mathbb{R}_+ \to \mathbb{R}_+$ be a Borel function; then, we have

$$I_\varphi \overset{\text{def}}{=} \mathrm{E}\left[\int_0^\infty dA_t\, F\left(\frac{1}{\sqrt{t}}B_{st};\ s \leq 1\right)\varphi(t)\right]$$

$$= \mathrm{E}\left[\int_0^\infty du\, \varphi(\alpha_u)F\left(\frac{1}{\sqrt{\alpha_u}}B_{s\alpha_u};\ s \leq 1\right)\right],$$

by time-changing. From the scaling property (2.a), we deduce that for fixed $u > 0$, we have

$$\left\{\left(\frac{1}{\sqrt{\alpha_u}}B_{s\alpha_u};\ s \leq 1\right);\ \alpha_u\right\} \overset{\text{(law)}}{=} \left\{\left(\frac{1}{\sqrt{\alpha_1}}B_{s\alpha_1};\ s \leq 1\right);\ u^m\alpha_1\right\}$$

with $\frac{1}{m} = r + 1$. Hence, we have

$$I_\varphi = \mathrm{E}\left[\int_0^\infty du\, \varphi(u^m \alpha_1) F\left(\frac{1}{\sqrt{\alpha_1}} B_{s\alpha_1}; s \leq 1\right)\right]$$

$$= \mathrm{E}\left[(r+1) \int_0^\infty dt\, t^r \frac{\varphi(t)}{\alpha_1^{r+1}} F\left(\frac{1}{\sqrt{\alpha_1}} B_{s\alpha_1}; s \leq 1\right)\right],$$

after making the change of variables $t = u^m \alpha_1$. The identity (2.c) follows. □

We now give two important consequences of Theorem 2.1.

Corollary 2.1.1. *Let $r > -1$ and $(\theta_t, t \geq 0)$ be an $\mathbb{R}_+$-valued measurable process, which satisfies the joint scaling property*

$$(B_{ct}, \theta_{ct}; t \geq 0) \overset{(\text{law})}{=} (\sqrt{c}B_t, c^r \theta_t; t \geq 0), \quad \text{for } c > 0, \tag{2.a'}$$

and the integrability condition

$$\mathrm{E}[\theta_1] < \infty. \tag{2.b'}$$

Define

$$A_t = \int_0^t ds\, \theta_s \quad \text{and} \quad \alpha_u \equiv \inf\{t: A_t > u\}.$$

Then, we have

$$\mathrm{E}[\theta_1 F(B_u; u \leq 1)] = \mathrm{E}\left[\frac{r+1}{\alpha_1^{r+1}} F\left(\frac{1}{\sqrt{\alpha_1}} B_{s\alpha_1}; s \leq 1\right)\right] \tag{2.d}$$

for every measurable functional $F: C([0,1]; \mathbb{R}) \to \mathbb{R}_+$.

Proof. We first remark that, from the hypotheses (2.a') and (2.b'), we have

$$\mathrm{E}[A_1] = \int_0^1 ds\, \mathrm{E}[\theta_s] = \int_0^1 ds\, s^r\, \mathrm{E}[\theta_1] < \infty,$$

so that (2.b) is satisfied, and, as a consequence, A_t is almost surely finite, for every t, and satisfies the scaling property (2.a). Moreover, in this particular case, the left-hand side of (2.c) is equal to

$$dt\, \mathrm{E}\left[\theta_t F\left(\frac{1}{\sqrt{t}} B_{ut}; u \leq 1\right)\right] = dt\, t^r\, \mathrm{E}[\theta_1 F(B_u; u \leq 1)],$$

from (2.a'). The identity (2.d) now follows from (2.c). □

As a particular case of (2.d), we obtain the identity (1.c), by taking

$$A_t = \int_0^t ds\, 1_{(B_s > 0)}.$$

In Section 4, we shall develop some applications of (2.d) to Bessel processes, when $\theta_s = R_s^{2r}$, and R is a Bessel process.

We now consider another interesting case, when $(A_t \equiv \ell_t; t \geq 0)$ is the local time at 0 of $(B_t; t \geq 0)$, from which we shall deduce the identity (1.a).

Corollary 2.1.2 ([4]). *Let $(b(u); u \leq 1)$ be a standard Brownian bridge, and let $\tau_t = \inf\{u: \ell_u > t\}$, where $(\ell_u; u \geq 0)$ is the local time at 0 of the one-dimensional Brownian motion $(B_u; u \geq 0)$. Then, we have*

$$E[F(b(u); u \leq 1)] = \sqrt{\frac{\pi}{2}} E\left[\frac{1}{\sqrt{\tau_1}} F\left(\frac{1}{\sqrt{\tau_1}} B_{u\tau_1}; u \leq 1\right)\right] \qquad (1.a)$$

for every measurable functional $F \colon C([0,1]; \mathbb{R}_+) \to \mathbb{R}_+$.

Proof. We take in (2.c) $A_t \equiv \ell_t$, for $t \geq 0$; then, we have $\alpha_t \equiv \tau_t$ and from Revuz–Yor [12, Exercise 1.16, p. 378], or, more generally, from Fitzsimmons–Pitman–Yor [6], the left-hand side of (2.c) is equal to

$$E\left[(d\ell_t) E\left[F\left(\frac{1}{\sqrt{t}} B_{ut}; u \leq 1\right) \Big| B_t = 0\right]\right]$$

$$= E[(d\ell_t)] E[F(b(u); u \leq 1)]$$

$$= \sqrt{\frac{2}{\pi}} \left(\frac{dt}{2\sqrt{t}}\right) E[F(b(u); u \leq 1)],$$

since $\ell_t \overset{(\text{law})}{=} |B_t|$ for fixed t. Comparing this last formula with the right-hand side of (2.c), in which we have, for this particular example, $r = -1/2$, we find (1.a). □

§3 AN APPLICATION TO P. LÉVY'S ARC SINE LAW

In this section, I will use the notation introduced in Example 4 above; moreover, $(A_t^{\pm}; t \geq 0)$ denotes the processes

$$A_t^{\pm} = \int_0^t ds \, 1_{(B_s \in \mathbb{R}_{\pm})}, \quad t \geq 0.$$

Some remarkable features of Lévy's arc sine law are the following "invariance" results, which have been proved and discussed in Pitman–Yor [10], using excursion theory, and later extended in [5], using the arguments below, to reflecting Brownian motion perturbed by its local time at 0.

Theorem 3.1. *Consider the triple $\frac{1}{T}(A_T^+, A_T^-, \ell_T^2)$, where T is a random time. Then, the law of this triple is the same for*

(i) $T = t$, *a constant time;*
(ii) $T = \alpha_s^+ = \inf\{u: A_u^+ = s\}$;
(iii) $T = \tau(u) = \inf\{h: \ell_h > u\}$.

The proof of the theorem will be done in two steps.

Step 1. We first prove the identity in law for the triples corresponding to (ii) and (iii). For this, we note the relations

$$A_{\alpha^+(1)}^- = A_\tau^-(\ell_{\alpha_1^+}) \quad \text{and} \quad \alpha_1^+ = A_{\alpha_1^+}^+ + A_{\alpha_1^+}^- = 1 + A_\tau^-(\ell_{\alpha_1^+}). \qquad (3.a)$$

Then, thanks to the independence of the two processes $(A_\tau^+(t); t \geq 0)$ and $(A_\tau^-(t); t \geq 0)$, we deduce from (3.a) that

$$\left(A_{\alpha_1^+}^-, \ell_{\alpha_1^+}^2, \alpha_1^+\right) \stackrel{\text{(law)}}{=} \left(\ell_{\alpha_1^+}^2 A_\tau^-(1); \ell_{\alpha_1^+}^2; 1 + \ell_{\alpha_1^+}^2 A_\tau^-(1)\right)$$

$$\stackrel{\text{(law)}}{=} \left(\frac{A_\tau^-(1)}{A_\tau^+(1)}; \frac{1}{A_\tau^+(1)}; \frac{\tau(1)}{A_\tau^+(1)}\right)$$

from which we easily obtain the desired identity in law.

Step 2. We now prove the identity in law between the triples corresponding to (i) and (iii), as a consequence of formula (1.c), from which we deduce

$$E\left[f(A_1^+, A_1^-, \ell_1^2)1_{(B_1 > 0)}\right] = E\left[\frac{1}{\alpha_1^+} f\left(\frac{A_{\alpha_1^+}^+}{\alpha_1^+}, \frac{A_{\alpha_1^+}^-}{\alpha_1^+}, \frac{\ell_{\alpha_1^+}^2}{\alpha_1^+}\right)\right]$$

where $f : [0,1]^2 \times \mathbb{R}_+ \to \mathbb{R}_+$ is a Borel function; now, the last written expectation is equal, thanks to the identity in law between the triples in (ii) and (iii), to

$$E\left[\frac{A_{\tau(1)}^+}{\tau(1)} f\left(\frac{A_{\tau(1)}^+}{\tau(1)}, \frac{A_{\tau(1)}^-}{\tau(1)}, \frac{1}{\tau(1)}\right)\right]. \tag{3.b_+}$$

Changing B into $-B$, we also obtain

$$E\left[f(A_1^+, A_1^-, \ell_1^2)1_{(B_1 < 0)}\right] = E\left[f\left(\frac{A_{\tau(1)}^+}{\tau(1)}, \frac{A_{\tau(1)}^-}{\tau(1)}, \frac{1}{\tau(1)}\right)\frac{A_{\tau(1)}^-}{\tau(1)}\right], \tag{3.b_-}$$

and, adding up $(3.b_+)$ and $(3.b_-)$, we obtain

$$E\left[f(A_1^+, A_1^-, \ell_1^2)\right] = E\left[f\left(\frac{A_{\tau(1)}^+}{\tau(1)}, \frac{A_{\tau(1)}^-}{\tau(1)}, \frac{1}{\tau(1)}\right)\right],$$

which proves the desired identity in law. □

The main consequence of Theorem 3.1 is that A_1^+ follows the arc sine law since, from the identity in law between the triples corresponding to (i) and (iii), one has

$$A_1^+ \stackrel{\text{(law)}}{=} \frac{A_{\tau(u)}^+}{A_{\tau(u)}^+ + A_{\tau(u)}^-}, \quad \text{for every fixed } u > 0, \tag{3.c}$$

with $A_{\tau(u)}^+$ and $A_{\tau(u)}^-$ two independent, identically distributed, stable $\left(\frac{1}{2}\right)$, $\mathbb{R}_+$-valued random variables.

Taking even more advantage of Theorem 3.1, one may complete the identity in law (3.c) as follows:

$$\frac{1}{\ell_t^2}(A_t^+, A_t^-) \stackrel{\text{(law)}}{=} \frac{1}{u^2}(A_{\tau(u)}^+, A_{\tau(u)}^-). \tag{3.d}$$

A number of variants of these results are found in [10] and [5].

§4 SOME APPLICATIONS TO BESSEL PROCESSES

As announced in the introduction, I now give some examples of application of Corollary 2.1.1 to Bessel processes. Here, $(R_t, \ t \geq 0)$ denotes a δ-dimensional Bessel process starting from 0, and $\theta_t = R_t^{2r}$, for $r > -1$; $P^{(\delta)}$ denotes the law of R. Then, under the further hypothesis

$$\delta + 2r > 0, \tag{4.a}$$

which ensures that $\int_0 \varrho^{\delta-1}\varrho^{2r}\,d\varrho < \infty$, the integrability condition

$$\mathrm{E}^{(\delta)}[\theta_1] < \infty \tag{2.b'}$$

is satisfied. Hence, the identity (2.d) may be applied to R, instead of B, to obtain

$$\mathrm{E}^{(\delta)}\left[f\left(\int_0^1 ds\, R_s^{2r}\right)R_1^{2r}\right] = \mathrm{E}^{(\delta)}\left[\frac{r+1}{\alpha_1^{r+1}}f\left(\frac{1}{\alpha_1^{r+1}}\right)\right] \tag{4.b}$$

for every Borel function $f : \mathbb{R}_+ \to \mathbb{R}_+$, and

$$\alpha_t = \inf\left\{u : \int_0^u ds\, R_s^{2r} > t\right\}.$$

As we shall now discuss, the identity (4.b) is particularly interesting in the case $r = -1/2$ (hence, in order that (4.a) be satisfied, we need $\delta > 1$). In this particular case, we obtain, from (4.b):

$$\mathrm{E}^{(\delta)}\left[\frac{1}{R_1}f\left(\int_0^1 \frac{ds}{R_s}\right)\right] = \mathrm{E}^{(\delta)}\left[\frac{1}{2\sqrt{\alpha_1}}f\left(\frac{1}{\sqrt{\alpha_1}}\right)\right]. \tag{4.c}$$

We shall be able to exploit this identity thanks to the following elementary

Lemma 4.1. *Let X be an $\mathbb{R}_+$-valued random variable such that $P(X = 0) = 0$. Assume that $\mathrm{E}[\frac{1}{\sqrt{X}}] < \infty$. Then, there exists a unique normalizing constant $c > 0$ such that*

$$c\int_{-\infty}^{\infty} dx\,\varphi(x) = 1, \quad \text{where} \quad \varphi(x) = \mathrm{E}\left[\exp\left(-\frac{x^2}{2}X\right)\right].$$

If we denote $\widehat{\varphi}(\lambda) \overset{\text{def}}{=} c\int_{-\infty}^{\infty} dx\,\varphi(x)\exp(i\lambda x)$, we have

$$\widehat{\varphi}(\lambda) = c\,\mathrm{E}\left[\sqrt{\frac{2\pi}{X}}\exp\left(-\frac{\lambda^2}{2X}\right)\right]. \tag{4.d}$$

Proof. We deduce from Fubini's theorem that

$$
\begin{aligned}
\widehat{\varphi}(\lambda) &= c\,\mathrm{E}\left[\int_{-\infty}^{\infty} dx\,\exp\left(-\frac{x^2 X}{2}\right)\exp(i\lambda x)\right] \\
&= c\,\mathrm{E}\left[\int_{-\infty}^{\infty} \frac{dy}{\sqrt{X}}\,\exp\left(-\frac{y^2}{2}\right)\exp\left(i\frac{\lambda y}{\sqrt{X}}\right)\right] \\
&= c\,\mathrm{E}\left[\sqrt{\frac{2\pi}{X}}\,\exp\left(-\frac{\lambda^2}{2X}\right)\right].
\end{aligned}
$$

$\square$

As a consequence of Lemma 4.1, $\{\widehat{\varphi}(\lambda),\ \lambda \in \mathbb{R}\}$ now appears as the Laplace transform in $\lambda^2/2$ of the random variable $1/X$ under the probability measure $Q = \left(c\sqrt{2\pi/X}\right)P$. The following examples are particularly relevant. Let $\beta > 0$; then, if we define

$$
\varphi_\beta(x) = \frac{1}{(\cosh x)^\beta},
$$

we get

$$
c \equiv c_\beta = \frac{\Gamma\left(\frac{\beta+1}{2}\right)}{\sqrt{\pi}\,\Gamma\left(\frac{\beta}{2}\right)} \quad \text{and} \quad \widehat{\varphi}_\beta(\lambda) = \frac{\left|\Gamma\left(\frac{\beta}{2} + \frac{i\lambda}{2}\right)\right|^2}{\left(\Gamma\left(\frac{\beta}{2}\right)\right)^2}. \tag{4.e}
$$

Moreover, it is well-known that

$$
\varphi_\beta(x) = \mathrm{E}\left[\exp\left(-\frac{x^2}{2}X_\beta\right)\right], \quad \text{where} \quad X_\beta \stackrel{\text{(law)}}{=} \int_0^1 ds\,R_{2\beta}^2(s), \tag{4.f}
$$

and $(R_{2\beta}(s),\ s \geq 0)$ is a 2β-dimensional Bessel process starting from 0.

Before we proceeed with the application to Bessel processes, we recall that in the particular case where β is an integer, the formulae (4.e) simplify thanks to the fundamental identities (such as the complement formula) satisfied by the gamma function.

Lemma 4.2 (see, e. g., Mehta [9, p. 117]).

(i) *For $\beta = 1$, we have*

$$
\widehat{\varphi}_1(\lambda) = \frac{1}{\cosh(\pi\lambda/2)},
$$

and more generally, for every $k > 0$,

$$
\widehat{\varphi}_{2k+1}(\lambda) = \frac{1}{\cosh(\pi\lambda/2)}\prod_{j=1}^{k}\left(\frac{\lambda^2 + (2j-1)^2}{(2j-1)^2}\right).
$$

(ii) *For $\beta = 2$, we have*

$$
\widehat{\varphi}_2(\lambda) = \frac{(\pi\lambda/2)}{\sinh(\pi\lambda/2)},
$$

and more generally, for every integer $k > 0$,

$$\widehat{\varphi}_{2k+2}(\lambda) = \frac{(\pi\lambda/2)}{\sinh(\pi\lambda/2)} \prod_{j=1}^{k}\left(1 + \frac{\lambda}{j}\right)^2.$$

Now, in order to apply Lemma 4.1 to Bessel processes, we recall that a power of a Bessel process is another Bessel process, up to a time-change; more precisely, if, in order to avoid confusion with the previous notation R_δ, which was used – following the identity (4.f) – for a Bessel process with dimension δ, we now write $(R_{[\nu]}(t); t \geq 0)$ for a Bessel process with dimension $\delta = 2(\nu + 1)$, starting from 0, then, we have

$$q R_{[\nu]}^{1/q}(t) = R_{[\nu q]}\left(\int_0^t ds\, R_{[\nu]}^{-2/p}(s)\right), \quad t \geq 0, \tag{4.g}$$

where p and q denote two conjugate numbers, and $\nu > -1/q$. As a consequence of (4.g), we obtain, with $p = q = 2$,

$$\frac{1}{4}\left(\int_0^1 \frac{ds}{R_{[\nu]}(s)}\right)^2 \overset{\text{(law)}}{=} \left(\int_0^1 ds\, R_{[2\nu]}^2(s)\right)^{-1} \tag{4.h}$$

(for details concerning both identities (4.g) and (4.h), see [12, Chapter XI, Proposition (1.11) and Corollary (1.12)]). On the other hand, we deduce from (4.b), where we take $r = -1/2$ and $\delta = 2(\nu + 1)$,

$$\mathrm{E}^{(\delta)}\left[\exp\left(-\frac{\lambda^2}{2}\left(\int_0^1 \frac{ds}{R_{[\nu]}(s)}\right)^2\right)\frac{1}{R_{[\nu]}(1)}\right] = \frac{1}{2}\,\mathrm{E}^{(\delta)}\left[\frac{1}{\sqrt{\alpha_1}}\exp\left(-\frac{\lambda^2}{8\alpha_1}\right)\right]. \tag{4.i}$$

Now, from (4.g), it follows that

$$\alpha_t = \frac{1}{4}\int_0^t ds\, R_{[2\nu]}^2(s).$$

We then apply Lemma 4.1 to

$$X = 4\alpha_1 \equiv \int_0^1 du\, R_{[2\nu]}^2(u),$$

which gives, with the notation introduced in (4.d),

$$c\sqrt{2\pi}\left(\frac{1}{2}\,\mathrm{E}^{(\delta)}\left[\frac{1}{\sqrt{\alpha_1}}\exp\left(-\frac{\lambda^2}{8\alpha_1}\right)\right]\right) = \widehat{\varphi}(\lambda), \tag{4.j}$$

where

$$\varphi(x) = \mathrm{E}\left[\exp\left(-\frac{x^2}{2}X\right)\right] = \frac{1}{(\cosh x)^\beta} \quad \text{and} \quad \beta = 2\nu + 1.$$

Finally, we deduce from (4.i) and (4.j) the following formula:

$$\left(c_{2\nu+1}\sqrt{2\pi}\right) E^{(\delta)}\left[\exp\left(-\frac{\lambda^2}{8}\left(\int_0^1 \frac{ds}{R_{[\nu]}(s)}\right)^2\right)\frac{1}{R_{[\nu]}(1)}\right] = \frac{\left|\Gamma\left(\frac{2\nu+1+i\lambda}{2}\right)\right|^2}{\left(\Gamma\left(\frac{2\nu+1}{2}\right)\right)^2} \quad (4.k)$$

(the right-hand side of (4.k) is $\hat{\varphi}_{2\nu+1}(\lambda)$, with the notation used in (4.e)).

In particular, for $\nu = \frac{1}{2}$, $(R(t) \equiv R_{[1/2]}(t); \; t \geq 0)$ is the three-dimensional Bessel process starting from 0, and formula (4.k) may be translated into a result involving the meander, thanks to Imhof's relation (1.b) we obtain:

$$E\left[\exp-\frac{\lambda^2}{2}\left(\frac{1}{2}\int_0^1 \frac{du}{m(u)}\right)^2\right] = \frac{\pi\lambda/2}{\sinh(\pi\lambda/2)}. \quad (4.l)$$

REFERENCES

1. J. Azéma and M. Yor, *Sur les zéros des martingales continues*, Sém. Probas. XXVI, Lect. Notes in Maths. 1526, Springer, 1992, pp. 248–306.
2. Ph. Biane and M. Yor, *Quelques précisions sur le méandre brownien*, Bull. Sciences Maths., 2ème série, vol. 112, 1988, pp. 101–109.
3. Ph. Biane and M. Yor, *Valeurs principales associées aux temps locaux browniens*, Bull. Sciences Maths., 2ème série, vol. 111, 1987, pp. 23–101.
4. Ph. Biane, J. F. Le Gall, and M. Yor, *Un processus qui ressemble au pont brownien*, Sém. Probas. XXI, Lect. Notes in Maths. 1247, Springer, 1987, pp. 270–275.
5. Ph. Carmona, F. Petit, and M. Yor, *Some extensions of the arc sine law as partial consequences of the scaling property of Brownian motion*, 1994 (to appear in Prob. Th. and Rel. Fields).
6. P. Fitzsimmons, J. W. Pitman, and M. Yor, *Markovian Bridges: Construction, Palm interpretation, and Splicing*, Seminar on Stochastic Processes (R. Bass, K. Burdzy, eds.), Birkhäuser, 1993.
7. J. P. Imhof, *Density factorisation for Brownian motion and the three-dimensional Bessel processes and applications*, J. App. Proba. **21** (1984), 500–510.
8. P. Lévy, *Sur certains processus stochastiques homogènes*, Compositio Math. **7** (1939), 283–339.
9. M. L. Mehta, *Matrix Theory: Selected topics and useful results*, Les Editions de Physique, Les Ulis (France), 1989.
10. J. W. Pitman and M. Yor, *Arc sine laws and interval partitions derived from a stable subordinator*, Proc. London Math. Soc. **3, 65** (1992), 326–356.
11a. J. W. Pitman and M. Yor, *Random scaling of Brownian and Bessel bridges*, Preprint (University of California, Berkeley, August 1992).
11b. J. W. Pitman and M. Yor, *Dilatations d'espace-temps, réarrangements des trajectoires browniennes, et quelques extensions d'une identité de Knight*, Comptes Rendus Acad. Sci. Paris, t. 316, Série I, 1993, pp. 723–726.
12. D. Revuz and M. Yor, *Continuous martingales and Brownian motion*, Springer, 1991.
13. M. Yor, *Some aspects of Brownian motion, Part I*, Lectures in Mathematics, ETH Zürich, Birkhäuser, 1992.
14. Zhan Shi and M. Yor, *Sur la loi du supremum des temps locaux d'un pont de Bessel*, Preprint. Prépublication n° 179 du Laboratoire de Probabilités de Paris VI (Mai 1993).

MARC YOR, LABORATOIRE DE PROBABILITÉS ASSOCIÉ AU C.N.R.S., UNIVERSITÉ PARIS VI, 4, PLACE JUSSIEU, TOUR 56, 75252 PARIS CÉDEX 05

Progress in Probability, Vol. 36
© 1995 Birkhäuser Verlag Basel/Switzerland

RECENT PROGRESS
IN THE HYPERCONTRACTIVE SEMIGROUPS

BOGUSLAW ZEGARLINSKI*

ABSTRACT. Recent progress in the studying of the ergodicity problem of Markov semigroups on infinite dimensional spaces based on use of the hypercontractivity property is presented.

§1 THE ERGODICITY PROBLEM

Let μ be a probability measure on a Polish space (Ω, Σ) and let $P_t \equiv e^{t\mathcal{L}}$, $t \geq 0$ be a Markov semigroup on the space $(\mathcal{C}(\Omega), \|\cdot\|_u)$ of continuous real functions with the supremum norm, such that

$$\mu f \cdot P_t g = \mu P_t f \cdot g$$

for every $f, g \in \mathcal{C}(\Omega)$. In the present lecture we would like to consider the following question.

The Ergodicity Problem:

$$\exists \ ? \begin{cases} \text{a } P_t\text{-invariant probability measure} & \mu \text{ on } (\Omega, \Sigma) \\ \text{a dense subset} & \mathcal{A}_o \subset \mathcal{C}(\Omega), \ \overline{\mathcal{A}_o}^{\|\ \|_u} = \mathcal{C}(\Omega) \\ \text{a rate function} & \varepsilon : \mathbb{R}^+ \to \mathbb{R}^+, \ \varepsilon(t) \xrightarrow[t \to \infty]{} 0 \\ \text{and a norm} & \|\cdot\| \text{ on } \mathcal{A}_o \end{cases}$$

such that for any $f \in \mathcal{A}_o$ and $\eta \in \Omega$ we have

$$|P_t f(\eta) - \mu f| \leq C(\eta)\varepsilon(t)\|f - \mu f\|$$

with a constant $C(\eta) \in (0, \infty)$ (essentially) independent of the function f and the time t.

1991 *Mathematics Subject Classification.* primary 60K35, 60K, 60; secondary 82C22, 82C, 82.
Key words and phrases. Stochastic dynamics, infinite dimensional spaces, ergodicity problem, logarithmic Sobolev inequalities.
*Supported by SFB 237

§2 THE FINITE DIMENSIONAL CASE

It is known that if a probability measure μ satisfies the *Classical Sobolev inequality,* i. e. the inequality of the following form

$$(\mu|f|^q)^{\frac{2}{q}} \leq a\mu\Gamma(f,f) + b\mu f^2 \qquad\qquad \text{(CS)}$$

with some $q > 2$ and some constant $a \in (0,\infty)$, $b \in [0,\infty)$ independent of a function $f \in L_2(\mu)$ for which the square of the gradient quadratic form

$$\Gamma(f,f) \equiv \frac{1}{2}\left(\mathcal{L}f^2 - 2f\mathcal{L}f\right) \geq 0$$

is well defined and has finite μ-expectation, then there is a standard way of turning L_2-information into the pointwise information. This is because in this case we have *Ultracontractivity estimate*, i. e. we have for any $f \in L_2(\mu)$ and $T \in (0,\infty)$ the following bound

$$\|P_T f\|_u \leq D(T)^N \|f\|_2$$

with a constant $D(T) \in (0,\infty)$ independent of the function f and with N being the dimension of the space Ω, related to the biggest possible q for which (CS) is true by $N = 2q/(q-2)$. In this situation, if additionally the operator $(-\mathcal{L})$, as a selfadjoint operator in $L_2(\mu)$, has a gap $m \equiv \mathrm{gap}_2(\mathcal{L}) > 0$ at a bottom of its spectrum, or equivalently the following *Spectral Gap inequality* is satisfied

$$m\mu(f,f) \leq \mu\Gamma(f,f) \qquad\qquad \text{(SG)}$$

with $\mu(f,f)$ denoting the corresponding variance, then one easily gets the following

Theorem 2.1.

$$\text{(CS)} + \text{(SG)} \Longrightarrow \text{Strong Ergodicity,}$$

i. e. we have

$$\|P_t f - \mu f\|_u \leq D(1)^N e^{-m(t-1)}\|f - \mu f\|_2.$$

Thus the Classical Sobolev inequality and Spectral Gap inequality for one invariant measure (with respect to a semigroup $P_t \equiv e^{t\mathcal{L}}$, $t \geq 0$) allow us to solve completely the corresponding ergodicity problem. Unfortunately if the dimension N of the space Ω is infinite the corresponding Classical Sobolev inequality can only be true with $q = 2$ and therefore it does not provide us with nontrivial information contained in the ultracontractivity estimate. In this situation one has to seek for a more subtle strategy to solve our ergodicity problem.

§3 THE INFINITE DIMENSIONAL CASE

In the infinite dimensional case the best what we can have is the following property of the probability measure.

Definition 3.1. A probability measure μ satisfies the *Logarithmic Sobolev inequality with a coefficient* $c \in (0, \infty)$ iff

$$\mu f^2 \log f \leq c\mu\Gamma(f, f) + \mu f^2 \log(\mu f^2)^{1/2} \tag{LS}$$

for every positive function f for which the right hand side is finite.

Then one has the following classical result.

Theorem 3.2. ([3])

$$(\text{LS}) \iff P_t, \ t \geq 0 \ \textit{is hypercontractive},$$

i. e.

$$\|P_t f\|_q \leq \|f\|_2$$

for any $q \in [1, q(t)]$, *with*

$$q(t) \equiv 1 + e^{2t/c}.$$

Additionally then we have also the following spectral gap property.

Theorem 3.3. ([7], [8])

$$(\text{LS}) \implies \text{gap}_2(\mathcal{L}) \geq \frac{1}{c}.$$

We will show that the hypercontractivity property provides us with a sufficiently strong tool to solve our ergodicity problem in infinite dimensional case. The corresponding strategy will be presented later. In order to be able to use this strategy we need to have first an effective method for proving Logarithmic Sobolev inequality. Therefore we shall begin from presenting the corresponding technology.

In fact we plan to consider a slightly more general problem. Namely we will assume that by a suitable choice of coordinates in the infinite dimensional space Ω the corresponding generator of the Markov semigroup has the following representation

$$\mathcal{L} = \sum_{i \in \mathbb{L}} \mathcal{L}_i$$

with $\mathcal{L}_i$, for i in an infinite countable set $\mathbb{L}$, being a local Markov generator acting on a finite number of coordinates. Denoting the corresponding square of the gradient quadratic forms by Γ_i, we will consider a problem when a probability measure μ satisfies the *Logarithmic Sobolev inequality with the local coefficients* c_i, i. e. the inequality

$$\mu f^2 \log f \leq \sum_{i \in \mathbb{L}} c_i \mu \Gamma_i(f, f) + \mu f^2 \log(\mu f^2)^{1/2} \tag{LLS}$$

for every positive function f for which the right hand side is finite.

§4 A Method for Proving the Logarithmic Sobolev Inequalities

Let $\mathbb{L}$ be a countable infinite set and let $\mathcal{F}$ be the family of its all finite subsets. Suppose that for every $\Lambda \in \mathcal{F}$ there exists a σ-algebra $\Sigma_\Lambda \subset \Lambda$, such that

    (i) (Ω, Σ_Λ) is isomorphic to *a finite dimensional* Polish space $(M_\Lambda, \mathcal{B}(M_\Lambda))$,

    (ii) If $\Lambda_1 \subset \Lambda_2$, then $\Sigma_{\Lambda_1} \subset \Sigma_{\Lambda_2}$

and for every $\Lambda \in \mathcal{F}$

$$\Sigma_{\Lambda^c} = \overline{\bigcup_{\substack{\Lambda' \in \mathcal{F} \\ \Lambda' \subset \Lambda^c}} \Sigma_{\Lambda'}}$$

and

$$\Sigma = \overline{\bigcup_{\Lambda \in \mathcal{F}} \Sigma_\Lambda}.$$

Since (Ω, Σ) is a Polish space, there exists *a regular conditional expectation*

$$E_\mu(F|\Sigma_{\Lambda^c})(\omega) = E_\Lambda^\omega(F)$$

defined for every $\omega \in \Omega$ by a probability measure E_Λ^ω such that $E_{\Lambda|\Sigma_{\Lambda^c}}^\omega = \delta_{\omega|\Sigma_{\Lambda^c}}$. Clearly, by the property of conditional expectations, the following *compatibility condition* is satisfied

$$\Lambda_1 \subset \Lambda_2 \Longrightarrow E_{\Lambda_2}^\omega(E_{\Lambda_1}^\cdot(F)) = E_{\Lambda_2}^\omega(F).$$

Suppose the generator $\mathcal{L}$ of a Markov semigroup can be represented as follows

$$\mathcal{L} = \sum_{i \in \mathbb{L}} \mathcal{L}_i$$

with some Markov generators $\mathcal{L}_i$ and thus we have

$$\Gamma(f, f) = \sum_{i \in \mathbb{L}} \Gamma_i(f, f).$$

Suppose also that, for any $\Lambda \in \mathcal{F}$ and $\omega \in \Omega$, a semigroup

$$P_t^{\Lambda, \omega} \equiv e^{t\mathcal{L}_{\Lambda, \omega}}$$

with a generator

$$\mathcal{L}_{\Lambda, \omega} \equiv \delta_{\omega|\Sigma_{\Lambda^c}} \sum_{i \in \Lambda} \mathcal{L}_i$$

preserves the measure E_Λ^ω. Now let us consider the restriction $\mu_{|\Sigma_\Lambda}$ of the measure μ to the σ-algebra Σ_Λ. Since by our assumption (Ω, Σ_Λ) is isomorphic to a finite dimensional space, therefore in a good situation one can expect that the measure

$\mu_{|\Sigma_\Lambda}$ satisfies *a logarithmic Sobolev inequality with the local coefficients* $c_i(\mu, \Lambda) \in (0, \infty)$, $i \in \Lambda$, i.e. we have the following inequality

$$\mu_{|\Sigma_\Lambda} f^2 \log f \leq \sum_{i \in \Lambda} c_i(\mu, \Lambda) \mu_{|\Sigma_\Lambda} \Gamma_i(f, f) + \mu_{|\Sigma_\Lambda} f^2 \log(\mu_{|\Sigma_\Lambda} f^2)^{1/2}$$

for every positive Σ_Λ-measurable function f for which the right hand side is finite. Let us suppose also that the conditional measures $E_{\Lambda'}^\omega$, $\Lambda' \subset \Lambda$ also satisfies the similar logarithmic Sobolev inequalities with the corresponding local coefficients $c_i^{(\Lambda')}$ independent of the configuration $\omega \in \Omega$. We want to argue that *the knowledge of* $\{c_i^{(\Lambda_0)}\}$, *for some smaller set* Λ_0, *can be used to improve the estimates on the local coefficients* $\{c_i(\mu, \Lambda)\}$.

For this we note that, using the definition of conditional expectation and (LLS) for conditional measures, we have

$$\mu_{|\Sigma_\Lambda} f^2 \log f = \mu_{|\Sigma_\Lambda} E_{\Lambda_0}^\omega f^2 \log f$$
$$\leq \mu_{|\Sigma_\Lambda} \left(\sum_{i \in \Lambda_0} c_i^{(\Lambda_0)} E_{\Lambda_0}^\omega \Gamma_i(f, f) + E_{\Lambda_0}^\omega f^2 \log(E_{\Lambda_0}^\omega f^2)^{1/2} \right)$$
$$\leq \sum_{i \in \Lambda_0} c_i^{(\Lambda_0)} \mu_{|\Sigma_\Lambda} \Gamma_i(f, f) + \mu_{|\Sigma_\Lambda} \left(E_{\Lambda_0}^\omega f^2 \log(E_{\Lambda_0}^\omega f^2)^{1/2} \right).$$

Thus, assuming some smoothness of the operation of taking the conditional expectations and applying the (LLS) for the measure $\mu_{|\Sigma_\Lambda}$ to the last term on the right hand side, we obtain

$$\mu_{|\Sigma_\Lambda} f^2 \log f \leq \sum_{i \in \Lambda_0} c_i^{(\Lambda_0)} \mu_{|\Sigma_\Lambda} \Gamma_i(f, f) + \sum_{i \in \Lambda \setminus \Lambda_0} c_i(\mu, \Lambda) \mu_{|\Sigma_\Lambda} \Gamma_i(E_{\Lambda_0}^\omega f^2)^{1/2}$$
$$+ \mu_{|\Sigma_\Lambda} f^2 \log(\mu_{|\Sigma_\Lambda} f^2)^{1/2}.$$

To continue we need one more ingredient, namely let us suppose that the following *sweeping out relation* is true

$$\Gamma_j \left((E_{\Lambda_0}^\omega f^2)^{1/2}, (E_{\Lambda_0}^\omega f^2)^{1/2} \right) \leq \sum_{k \in \Lambda_0 \cup j} \gamma_{jk}^{\Lambda_0} E_{\Lambda_0}^\omega \Gamma_k(f, f)$$

with some constant $\gamma_{jk}^{\Lambda_0} \in (0, \infty)$ independent of $\omega \in \Omega$. Using this together with our previous inequality we arrive to the following new (LLS)

$$\mu_{|\Sigma_\Lambda} f^2 \log f \leq \sum_{i \in \Lambda} \tilde{c}_i(\mu, \Lambda) \mu_{|\Sigma_\Lambda} \Gamma_i(f, f) + \mu_{|\Sigma_\Lambda} f^2 \log(\mu_{|\Sigma_\Lambda} f^2)^{1/2}$$

with the new local coefficients given by

$$\tilde{c}_i(\mu, \Lambda) \equiv \begin{cases} c_i^{\Lambda_0} + \sum_{j \in \Lambda \setminus \Lambda_0} c_j(\mu, \Lambda) \gamma_{ji}^{\Lambda_0} & \text{for } i \in \Lambda_0 \\ c_i(\mu, \Lambda) \gamma_{ii}^{\Lambda_0} & \text{for } i \in \Lambda \setminus \Lambda_0. \end{cases}$$

We have the following theorem (proven essentially in [9], [10], [11], [14] and [15]).

Theorem 4.1. *If there are positive numbers L and v, $L \leq v$, such that*

$$\sup_{|\Lambda_0|=v} \sup_{i \in \Lambda_0^c} \sum_{j \in \Lambda_0} \gamma_{ij}^{\Lambda_0} < \infty$$

and

$$\sup_{|\Lambda_0|=v} \sup_{j \in \Lambda_0} \sum_{\substack{i \in \Lambda_0^c \\ d(i,j) \geq L}} \gamma_{ij}^{\Lambda_0} \leq \varepsilon$$

with some sufficiently small $\varepsilon \in (0,1)$. Then there is a constant $c \in (0,\infty)$ such that, for any $\Lambda \in \mathcal{F}$ and $\omega \in \Omega$, we have

$$E_\Lambda^\omega f^2 \log f \leq c E_\Lambda^\omega \Gamma_\Lambda(f,f) + E_\Lambda^\omega f^2 \log(E_\Lambda^\omega f^2)^{1/2}$$

for every positive function f for which the right hand side is finite.

Let us give few examples of applications of our method.

Example 4.2. Let $\Omega = M^{\mathbb{L}}$ with

$$M = \begin{cases} \text{continuous case:} & \text{\textit{a smooth, compact, connected finite dimensional}} \\ & \text{\textit{Riemannian manifold,}} \\ \text{discrete case:} & \text{\textit{a finite set}} \end{cases}$$

and $\mathbb{L} = \mathbb{Z}^d$. For $X \subset \mathbb{L}$, let $\mathcal{A}_X$ denotes a space of smooth real functions dependent only on the coordinates ω_i, $i \in X$. Let

$$E_\Lambda^\omega f \equiv \delta_\omega \frac{\mu_{0|\Lambda}(e^{-U_\Lambda} f)}{\mu_{0|\Lambda}(e^{-U_\Lambda})}$$

where

$$\mu_{0|\Lambda} \equiv \otimes_{i \in \Lambda} \rho(d\omega_i)$$

with ρ a uniform (Riemannian) probability measure on M. (Let us note that $\rho \in LS(c_0)$, and thus also the measures $\mu_{0|\Lambda}$ satisfy logarithmic Sobolev inequality with the same constant.) U_Λ is an additive functional given by

$$U_\Lambda \equiv \sum_{\substack{X \cap \Lambda \neq \emptyset \\ X \in \mathcal{F}}} \Phi_X$$

with a Gibbsian potential $\Phi \equiv \{\Phi_X \in \mathcal{A}_X\}_{X \in \mathcal{F}}$

$$\|\Phi\| \equiv \sup_{i \in \mathbb{Z}^d} \sum_{\substack{X \in \mathcal{F} \\ X \ni i}} \|\Phi_X\|_u \quad < \infty$$

of finite range, i. e.

$$\exists R > 0 \qquad \Phi_X \equiv 0 \qquad \text{if } \operatorname{diam} X > R.$$

We set

$$\Gamma_\Lambda(f,f) \equiv \sum_{i\in\Lambda} \Gamma_i(f,f) \equiv \begin{cases} \sum_{i\in\Lambda} |\nabla_i f|^2 & \text{in the continuous case,} \\ \sum_{i\in\Lambda} E_{x_0+i}(f,f) & \text{in the discrete case.} \end{cases}$$

Let us introduce also the following seminorm

$$\|\!|f|\!\| \equiv \sum_{i\in\mathbb{Z}^d} \|\Gamma_i(f,f)^{1/2}\|_u.$$

In the above described situation one has the following result:

Theorem 4.3. ([9], [10], [11])
The following conditions are equivalent:

(i) *Dobrushin-Shlosman mixing:*

$$|E_\Lambda^\omega(f,g)| \le Ce^{-Md(X,Y)}\|\!|f|\!\| \cdot \|\!|g|\!\|$$

for any $f \in \mathcal{A}_X$, $g \in \mathcal{A}_Y$ with some constants $C, M \in (0,\infty)$ independent of Λ, ω and f, g.

(ii) *Uniform Logarithmic Sobolev inequality:*
There is a constant $c \in (0,\infty)$ such that for any $\Lambda \in \mathcal{F}$ and $\omega \in \Omega$ we have

$$E_\Lambda^\omega f^2 \log f \le cE_\Lambda^\omega \Gamma_\Lambda(f,f) + E_\Lambda^\omega f^2 \log(E_\Lambda^\omega f^2)^{1/2}$$

for all positive functions f for which the right hand side is finite.

Remark. It is known that (i) is equivalent to the analyticity of the map

$$\Phi \longmapsto E_{\Lambda,\Phi}^\omega$$

uniformly with respect to $\Lambda \in \mathcal{F}$ and $\omega \in \Omega$.

Example 4.4. Let

$$\Omega = \mathcal{S}'(\mathbb{Z}^d) \equiv \text{ the set of tempered sequences in } \mathbb{R}^{\mathbb{Z}^d}.$$

Let

$$E_\Lambda^\omega f \equiv \delta_\omega \frac{E_{G,\Lambda}^\omega\left(e^{-U_\Lambda} f\right)}{E_{G,\Lambda}^\omega\left(e^{-U_\Lambda}\right)}$$

with $E_{G,\Lambda}^\omega$ denoting a regular conditional expectation with respect to Σ_{Λ^c} associated to a Gaussian measure μ_G with mean zero and a covariance G such that $G^{-1} \ge m_0^2 > 0$ and

$$G_{ij}^{-1} \equiv 0, \quad \text{if } d(i,j) > R$$

and

$$U_\Lambda(\omega) \equiv \sum_{i\in\Lambda} (V(\omega_i) + W(\omega_i))$$

with $V, W \in \mathcal{C}^2(\Omega)(\mathbb{R})$, V convex, $V''(x) \xrightarrow[x\to\infty]{} \infty$ and $\|W'\|_u, \|W''\|_u < \infty$.

Theorem 4.5. ([16]) (d=1)
There is a constant $c \in (0, \infty)$ such that for any $\Lambda \in \mathcal{F}$, $\omega \in \Omega$ we have

$$E_\Lambda^\omega f^2 \log f \le c E_\Lambda^\omega \Gamma_\Lambda(f, f) + E_\Lambda^\omega f^2 \log(E_\Lambda^\omega f^2)^{1/2}$$

for all positive f for which the rhs is finite.

Remark. The result extends to $\mathcal{S}'(\mathbb{R})$, and (with some restrictions on the local interactions) also to $\mathcal{S}'(\mathbb{Z}^d)$, $d > 1$.

§5 ERGODICITY OF HYPERCONTRACTIVE SEMIGROUPS

In the setting of the Examples 1 and 2 one has the following result.

Theorem 5.1. ([12], [16])
Suppose an invariant for P_t, $t \ge 0$ probability measure μ satisfies Logarithmic Sobolev inequality with some coefficient $c \in (0, \infty)$, i. e. we have

$$\mu f^2 \log f \le c \mu \Gamma(f, f) + \mu f^2 \log(\mu f^2)^{1/2}$$

for every positive function f for which the right hand side is finite.

    Then there is a set $\mathcal{A}_o$, $\bar{\mathcal{A}}_o = \mathcal{C}(\Omega)$ (containing all smooth functions dependent on a finite number of coordinates) and a norm $\| \cdot \|$ on it, such that for every $f \in \mathcal{A}_o$ and $\eta \in \Omega$ we have

$$|P_t f(\eta) - \mu f| \le C(\eta) e^{-mt} \|f - \mu f\|$$

for any $m \in (0, \mathrm{gap}_2 \, \mathcal{L})$ with some constant $C(\eta) \in (0, \infty)$ independent of the function f. (If $\Omega = M^{\mathbb{Z}^d}$ with M a compact space, the constant $C(\eta)$ can be chosen to be independent of $\eta \in \Omega$.)

Remark. Let us note that in the earlier proofs of the strong ergodicity result one used (LS) for a sequence of cubes invading all the lattice, see [10], [11] and also [5], [6].

Idea of the Proof. We use *the finite speed of propagation property*

$$|P_t f(\eta) - P_t^{\Lambda, \eta_0} f(\eta)| \le D(\eta) e^{-At} \|\|f\|\| \tag{1a}$$

valid for arbitrary $A \in (0, \infty)$ and some constant $D(\eta) \in (0, \infty)$ independent of $f \in \mathcal{A}_X$, $X \in \mathcal{F}$, provided

$$d(X, \partial\Lambda) \ge Ct \tag{1b}$$

for some $C \in (0, \infty)$ dependent only on A. Then for any $q \in (1, \infty)$ we have

$$\begin{aligned}
|P_t^{\Lambda, \eta_0}(f - \mu f)(\eta)| &= \left(|P_t^{\Lambda, \eta_0}(f - \mu f)(\eta)|^q\right)^{1/q} \\
&\le \left(P_T^{\Lambda, \eta_0} |P_{t-T}^{\Lambda, \eta_0}(f - \mu f)|^q(\eta)\right)^{1/q} \\
&\le I_{\Lambda, T}(\eta)^{\frac{1}{q}} \left(E_\Lambda^{\eta_0} |P_{t-T}^{\Lambda, \eta_0}(f - \mu f)|^q\right)^{1/q}
\end{aligned}$$

Since for some $p \in (1, \infty]$ we have

$$\left(\mu \left| \frac{dE_\Lambda^{\eta_0}}{d\mu_{|\Sigma_\Lambda}} \right|^p \right)^{1/p} \leq e^{B|\partial\Lambda|}$$

we get

$$(E_\Lambda^\eta |P_{t-T}(f - \mu f)|^q)^{1/q} \leq e^{B|\partial\Lambda|/q} \left(\mu |P_{t-T}(f - \mu f)|^{qs} \right)^{1/qs}$$

with $1/p + 1/s = 1$. Now we use hypercontractivity of the semigroup P_t with

$$qs = 1 + e^{2/c(\delta t - T)} \tag{2}$$

for some $\delta \in (0, 1)$. This gives

$$(\mu |P_{t-T}(f - \mu f)|^{qs})^{1/qs} \leq \left(\mu |P_{(1-\delta)t}(f - \mu f)|^2 \right)^{1/2}$$
$$\leq e^{-t(1-\delta) \, \mathrm{gap}_2 \, \mathcal{L}} \left(\mu(f - \mu f)^2 \right)^{1/2}$$

where in the last step we have used the fact that the generator of the hypercontractive semigroup has a spectral gap. Since for $q = q(t)$ satisfying (2) and $\Lambda = \Lambda(t)$ satisfying (1) we have

$$\lim_{t \to \infty} I_{\Lambda(t), T}(\eta)^{\frac{1}{q(t)}} = 1 \quad \text{for all } \eta \in \Omega$$

and

$$\lim_{t \to \infty} |\partial\Lambda(t)|/q(t) = 0.$$

Choosing $A \geq (1 - \delta) \, \mathrm{gap}_2 \, \mathcal{L}$ we obtain

$$|P_t f(\eta) - \mu f| \leq C(\eta) e^{-t(1-\delta) \, \mathrm{gap}_2 \, \mathcal{L}} \max \left((\mu(f - \mu f)^2)^{1/2}, \|f\| \right).$$

This ends the arguments necessary for the proof of our strong ergodicity result for the hypercontractive semigroup.

Remarks. One can get also the strong ergodicity result with a slower decay to equilibrium in the case when (LLS) is true with local coefficients growing not faster than linearly; see [4] and [12].

Acknowledgements:
The author would like to thank the organizers, and especially Erwin Bolthausen and Francesco Russo, for a very nice and interesting conference.

REFERENCES

1. Davies, E. B., Gross, L. and Simon, B., *Hypercontractivity: A bibliographical review*, in proceedings of the Hoegh–Krohn Memorial Conference.
2. Davies, E. B. and Simon, B., *Ultracontractivity and the Heat Kernel for Schrödinger Operators and Dirichlet Laplacians*, J. Func. Anal. **59** (1984), 335–395.
3. Gross, L., *Logarithmic Sobolev inequalities*, Amer. J. Math. **97** (1976), 1061–1083.
4. Holley, R. and Stroock, D. W., *Logarithmic Sobolev inequalities and stochastic Ising models*, J. Stat. Phys. **46** (1987), 1159–1194.
5. ShengLin, Lu, Horng-Tzer, Yau, *Spectral Gap and Logarithmic Sobolev Inequality for Kawasaki and Glauber Dynamics*, Preprint 1993.
6. Martinelli, F., Olivieri, E., *Approach to Equilibrium of Glauber Dynamics in the One Phase Region: I. The Attractive case, II. The General Case*, Preprints 1992/1993.
7. Rothaus, O. S., *Logarithmic Sobolev Inequalities and the Spectrum of Schrödinger Operators*, J. Func. Anal. **42** (1981), 110–378.
8. Simon, B., *A remark on Nelson's best hypercontractive estimates*, Proc. AMS **55** (1976), 376–378.
9. Stroock, D. W. and Zegarlinski, B., *The Logarithmic Sobolev Inequality for Continuous Spin Systems on a Lattice*, J. Func. Anal. **104** (1992), 299–326.
10. Stroock, D. W. and Zegarlinski, B., *The Equivalence of the Logarithmic Sobolev Inequality and the Dobrushin–Shlosman Mixing Condition*, Commun. Math. Phys. **144** (1992), 303–323.
11. Stroock, D. W. and Zegarlinski, B., *The Logarithmic Sobolev Inequality for Discrete Spin Systems on a Lattice*, Commun. Math. Phys. **149** (1992), 175–193.
12. Stroock, D. W. and Zegarlinski, B., *Some remarks about Glauber Dynamics and logarithmic Sobolev inequalities*, MIT preprint 1993.
13. Zegarlinski, B., *On log-Sobolev inequalities for infinite lattice systems*, Lett. Math. Phys. **20** (1990), 173–182.
14. _____, *Log-Sobolev inequalities for infinite one-dimensional lattice systems*, Comm. Math. Phys. **133** (1990), 147–162.
15. _____, *Dobrushin uniqueness theorem and logarithmic Sobolev inequalities*, J. Func. Anal. **105** (1992), 77–111.
16. _____, *Strong exponential decay to equilibrium for the Markov hypercontractive semigroups associated to unbounded spin systems on a lattice*, MIT Preprint 1992.

BOGUSLAW ZEGARLINSKI, MATHEMATISCHES INSTITUT, RUB, 4630 BOCHUM, GERMANY

FINANCIAL MODELS

Progress in Probability, Vol. 36
© 1995 Birkhäuser Verlag Basel/Switzerland

ALTERNATIVE ESTIMATORS OF A DIFFUSION MODEL OF THE TERM STRUCTURE OF INTEREST RATES. A MONTE CARLO COMPARISON

CARLO BIANCHI[1], RICCARDO CESARI[2] AND LORENZO PANATTONI[3]

ABSTRACT. We evaluate, through Monte Carlo experiments, the econometric performance of six alternative estimators of the basic parameters of the Cox–Ingersoll–Ross single-factor diffusion model of the term structure of interest rates. Different generating schemes are compared and the unobservability of the state-variable is taken into account. The effects of approximating interest rates, increasing frequency data and starting values are analyzed. A Monte Carlo evaluation of the effects on bond prices of biased parameter estimates is provided.

1. INTRODUCTION

Any asset has many characteristics: the issuer (with a certain probability of default), the coupon flow, the maturity date T, the taxation (of coupon and capital gains), other contractual provisions (options, etc.). The asset price $P(t)$ is clearly a function of these characteristics: $P(t, \text{risk, coupon, maturity, taxes}, \ldots)$. The relation, *ceteris paribus,* of the asset price with respect to maturity is called the term structure. The "ceteris paribus" clause is specifically expressed as a zero default risk, zero coupon, unit payment at maturity, fixed, uniform taxation. In this case the asset is a default-free zero-coupon unit discount bond, i. e., a security with time t (money) price $P(t, T)$, promising, with probability one, a payment of one (money) unit at time $T > t$. In other words, at time t, $P(t, T)$ is the present value (discount factor) of one (sure) money unit in T. Assuming continuous time, the price function $P(t, T)$ can be transformed into

$$R(t, T) = -\log(P)/(T - t)$$

1991 *Mathematics Subject Classification.* 90A20, 62M10.

Key words and phrases. Monte Carlo simulation, diffusion processes, term structure of interest rates, econometric estimation.

[1] Università di Pisa, Dipartimento di Scienze Economiche
[2] Banca d'Italia, Sede di Trieste
[3] IBM-Semea

where $R(t,T)$ is, by definition, the continuously compounded spot rate of return to maturity of the bond and $T - t$ is the bond time to maturity.[4]

As a function of maturity $\tau = T - t$, $R(t,\tau)$ is called the term structure of interest rates. A theory of the term structure is essentially an explanation of the difference (risk- or term-premium) between short-term and long-term interest rates or, equivalently, between spot and forward interest rates[5]. Classical examples of term structure theories are Fisher's [25] pure expectations theory, Hicks' [33] liquidity preference, Modigliani and Sutch's [50] preferred habitat. More recently, into the no-arbitrage asset pricing literature started up by Black and Scholes [9], the theory of the term structure has known a new period of successful innovation and implementation.[6] Not only the shape of the term structure but also the factors affecting its level and dynamics have been analyzed in a general equilibrium context, showing explicitly the link between agent preferences, uncertainty, production technology and term premia (see Cox, Ingersoll and Ross [16]). The new term structure models were developed in the typical Black and Scholes framework: continuous-time, frictionless markets, diffusion processes i. e. intertemporal dynamics given by stochastic differential equations, no-arbitrage condition i. e. prices constrained by the maintained hypothesis that equivalent assets (or portfolios) in terms of cash flows and other characteristics must earn the same return. These models, and in particular the general equilibrium model[7] of Cox, Ingersoll and Ross [17], have been applied to financial data in many countries.[8]

Considering this empirical literature, a first point to note is that notwithstanding a unified theoretical framework behind the various models, different estimation methods have been used by different authors. Secondly, in many cases, the elegant, formal derivation used in the theoretical development of the model is far from being used in the empirical counterpart, to develop and justify the adopted estimation method. Thirdly, almost no analytical or numerical comparison has been made to evaluate the various estimators, in particular in the case of discrete and small sample data.[9]

Our paper is addressed toward these critical points. Using the single-factor

---

[4]In discrete time, using the discrete-time interval as a natural choice for the time unit, the rate of return is $R(t,n) = (1/P)^{1/n} - 1$, n being the (possibly fractional) number of intervals (periods) between current time t and the bond maturity date T.

[5]The forward rate $R^F(t,S,T)$ is the rate implied in a contract written and completely specified at time t to buy a bond at time S with maturity date $T > S$. The no-arbitrage condition implies that forward rates (prices) can be specified in terms of spot rates (prices).

[6]On previous models and results see e. g. Masera [46] and Dobson, Sutch and Vanderford [20]. A recent survey is in Shiller [55].

[7]The equilibrium condition (supply equals demand) of the Cox–Ingersoll–Ross model implies the no-arbitrage condition. Different models, with alternative dynamic specifications, should be mentioned: e. g. Dothan [21], Vasicek [59], Brennan and Schwartz [10, 11].

[8]See Barone, Cuoco and Zautzik [6] for Italy; Cox, Ingersoll and Ross [15], Brennan and Schwartz [10], Brown and Dybvig [13], Gibbons and Ramaswamy [29] for the U. S.; Fischer and Zechner [24] for the Federal Republic of Germany; Brown and Schaefer [12] for the U. K.

[9]An important exception is Fournie and Talay [26] who consider only the asymptotic behaviour of estimators.

model developed by Cox, Ingersoll and Ross [17], we shall consider the foundations of different estimation methods for a diffusion model of the term structure as well as the evaluation, through Monte Carlo experiments, of their relative performance in the typical case of discrete, finite samples. The paper is as follows: in Section 2 the theoretical model is sketched; in Section 3 different estimation and testing methods are considered, in continuous or discrete time; Section 4 draws a set up for Monte Carlo simulation of continuous (diffusion) processes; Section 5 presents the main results and Section 6 concludes the paper.

§2 A UNIVARIATE MODEL OF THE TERM STRUCTURE

The model of Cox, Ingersoll and Ross [17] is an equilibrium model of a production/exchange economy with a single production/consumption good,[10] in which the state variable describing the system is a univariate diffusion, agents have rational expectations and logarithmic preferences, markets are perfectly competitive and frictionless.

Given that the instantaneous rate of return $r(t)$ on instantaneously maturing bonds is, in equilibrium, a linear function of the single state variable, it is possible to use this very short-term rate as an instrument for the system state variable. It turns out that the short-rate dynamics are given by the Feller [23] stochastic differential equation

$$dr(t) = \kappa(\theta - r)\,dt + \sigma\sqrt{r}\,dw(t) \tag{2.1}$$

where dw is the Itô differential of the one-dimensional standard Wiener process (Brownian motion), defined by

$$w(t) \sim \mathcal{N}(0, t), \tag{2.2a}$$

$$\mathrm{E}(w(t)w(s)) = \min(t, s), \tag{2.2b}$$

$$w(0) = 0 \quad \text{a.s.} \tag{2.2c}$$

Notice, in particular, that

$$\mathrm{E}\big(w(s)[w(u) - w(t)]\big) = 0 \quad \text{for every } s \leq t \leq u, \tag{2.2d}$$

i.e., non-overlapping increments are independent.

For $\sigma, \kappa, \theta > 0$ the process $r(t)$, solution of (2.1), is pathwise unique (strong solution), non-negative, and it displays mean-reversion toward the long-run value θ, the parameter κ denotes the speed of the expected short-term adjustment. According to Feller [23, Lemma 4], if $2\kappa\theta \geq \sigma^2$, then the origin is inaccessible (entrance boundary) and the process is ergodic.[11]

---

[10]The economic system is essentially a real (non-monetary) economy, with prices expressed in terms of the single good (or bundle).

[11]See Ikeda and Watanabe [34, p. 223]. On entrance boundaries and ergodicity see Karlin and Taylor [36, p. 241].

Solving Kolmogorov's backward differential equation, the transition probability density from $r(t)$ to $r(s)$ is obtained as (Karlin and McGregor [35])

$$p(r(s)|r(t)) = ce^{-(u+v)/2}\left(\frac{v}{u}\right)^{(q-1)/2}I_{q-1}\left(\sqrt{uv}\right), \qquad (2.3)$$

where

$$c = \frac{2\kappa}{\sigma^2(1 - e^{-\kappa(s-t)})}, \qquad u = 2cr(t)e^{-\kappa(s-t)}, \qquad v = 2cr(s), \qquad q = \frac{2\kappa\theta}{\sigma^2}$$

and I_α is the modified Bessel function of the first kind of order α:

$$I_\alpha(z) = \sum_{k=0}^{\infty}\frac{(z/2)^{2k+\alpha}}{k!\,\Gamma(k+\alpha+1)}. \qquad (2.4)$$

The transition distribution is therefore a stationary[12] non-central chi-square distribution $\chi^2(v; 2q, u)$ with $2q$ degrees of freedom and non-centrality parameter u. Conditional expected value[13] and variance are given by

$$E(r(s)|r(t)) = r(t)e^{-\kappa(s-t)} + \theta(1 - e^{-\kappa(s-t)}), \qquad (2.5a)$$

and

$$\mathrm{Var}(r(s)|r(t)) = r(t)\left(\frac{\sigma^2}{\kappa}\right)\left(e^{-\kappa(s-t)} - e^{-2\kappa(s-t)}\right) + \theta\left(\frac{\sigma^2}{2\kappa}\right)\left(1 - e^{-\kappa(s-t)}\right)^2. \qquad (2.5b)$$

Note that

$$\lim_{\kappa\uparrow\infty} E(r(s)|r(t)) = \theta, \qquad \lim_{\kappa\uparrow\infty}\mathrm{Var}(r(s)|r(t)) = 0,$$

and

$$\lim_{\kappa\downarrow 0} E(r(s)|r(t)) = r(t), \qquad \lim_{\kappa\downarrow 0}\mathrm{Var}(r(s)|r(t)) = \sigma^2 r(t)(s-t). \qquad (2.6)$$

The invariant (steady-state) distribution is obtained as $s \uparrow \infty$ and it is a gamma distribution with positive parameters q and q/θ and density[14,15]

$$p(r) = \frac{(q/\theta)^q}{\Gamma(q)}r^{q-1}e^{-rq/\theta}. \qquad (2.7a)$$

---

[12] A Markov process with stationary transition distributions is time-homogeneous. In the case of diffusions this property follows from the independence of drift and diffusion coefficient with respect to time (autonomous stochastic differential equation).

[13] The expected value solves the ordinary differential equation (deterministic version of (2.1)): $dr = k(\theta - r)\,du$, $r(t) = r_t$.

[14] It solves the stationary form of Kolmogorov's forward differential equation.

[15] Time-homogeneity and the existence of an invariant distribution are necessary and sufficient conditions for the Markov process $r(t)$ to the stationary (Arnold [1, p. 33]).

The first moments are

$$\bar{E}(r) = \theta, \qquad \overline{\text{Var}}(r) = \frac{\sigma^2\theta}{2\kappa}, \qquad \bar{E}(1/r) = \frac{2\kappa}{2\kappa\theta - \sigma^2}, \qquad (2.7b)$$

where $\bar{E}$ denotes expectations with respect to the invariant distribution. Note that, in steady state, κ and σ^2 cannot be separately identified.

In this economy any asset price with no coupon or dividend flow must satisfy the following partial differential (valuation) equation:

$$\frac{1}{2}\sigma^2 r P_{rr} + [\kappa(\theta - r) - \lambda r]P_r + P_t - rP = 0, \qquad (2.8)$$

where λr is the "market price of risk" (covariance between percentage change in wealth and change in the short interest rate).[16]

If $P(r,t,T)$ is the price at time t of a unit discount bond, the appropriate terminal condition is

$$P(r,T,T) = 1, \qquad (2.9)$$

so that the solution, dependent only on r and $\tau = T - t$, is[17]

$$P(r,\tau) = F(\tau)\exp(-rG(\tau)), \qquad (2.10)$$

where

$$F(\tau) = \left[\frac{\phi_1 \exp(\phi_2\tau)}{\phi_2(\exp(\phi_1\tau) - 1) + \phi_1}\right]^{\phi_3} \in \,]0,1[,$$

$$G(\tau) = \left[\frac{\exp(\phi_1\tau) - 1}{\phi_2(\exp(\phi_1\tau) - 1) + \phi_1}\right] > 0, \qquad (2.11)$$

$$\phi_1 = \sqrt{(\kappa + \lambda)^2 + 2\sigma^2}, \qquad \phi_2 = (\kappa + \lambda + \phi_1)/2, \qquad \phi_3 = 2\kappa\theta/\sigma^2 > 0.$$

Note that

$$\sigma^2 = 2\phi_2(\phi_1 - \phi_2), \qquad \phi_1 > \phi_2 > 0. \qquad (2.12)$$

The term structure is given by

$$R(r,\tau) \equiv -\log(P)/\tau = r(t)G(\tau)/\tau - \log(F(\tau))/\tau \qquad (2.13)$$

with a constant infinite-maturity interest rate (consol rate)

$$\lim_{\tau\to\infty} R(r,\tau) \equiv R_\infty = \frac{2\kappa\theta}{\kappa + \lambda + \phi_1} = (\phi_1 - \phi_2)\phi_3. \qquad (2.14)$$

---

[16]If λ is negative (positive), then the rate of return contains a risk premium (discount) the market requires to compensate for holding a bond whose price has a positive (negative) covariance with wealth.

[17]See Friedman [27, p. 147], for a mean-value representation of the unique solution of a general parabolic boundary value problem.

If

$$\frac{2\kappa\theta}{\kappa+\lambda+\phi_1} < r(t) < \frac{\kappa\theta}{\kappa+\lambda}, \tag{2.15}$$

then the term structure is humped. It is rising for values of $r(t)$ below this interval, falling for values above it.[18] Clearly,

$$\lim_{\tau\to 0} R(r,\tau) = r(t). \tag{2.16}$$

Note that $P(r,\tau) = \exp(-R(r,\tau)\tau)$ so that a change of the time unit (e.g. time from years to months) implies a corresponding change of the interest rates (e.g. from annual to monthly rates: $\tau \cdot 12$, $R/12$). The price formula is invariant to a scale change in time and rates if τ is multiplied (divided) and r, κ, θ, σ and λ are divided (multiplied) by the scale constant (e.g. 12).[19]

§3 ESTIMATION METHODS

3.1 Continuous-time maximum-likelihood estimation of SDE. Let us assume, for the moment, that the process $r(t)$ is observable and that a continuous record of observations is available: $\{r(t),\ t \in [0,T]\}$. Our purpose is to estimate the stochastic differential equation (2.1).

From the quadratic variation property of semimartingales (Shiryayev [56]) we obtain

$$\int_0^T dr(t)\, dr(t) = \int_0^T \sigma^2 r(t)\, dt,$$

that is

$$\sigma^2 = \int_0^T [dr(t)]^2 \bigg/ \int_0^T r(t)\, dt. \tag{3.1.1}$$

The integral in the numerator[20] is defined through the constructive characterization of the quadratic variation process: if $T_n = (t_0^{(n)}, t_1^{(n)}, \ldots, t_{N_n}^{(n)})$ is an increasing (i.e. nested) partition of $[0,T]$, i.e., $T_n \subset T_{n+1}$, $0 = t_0^{(n)} < t_1^{(n)} < \cdots < t_{N_n}^{(n)} = T$ and

$$\lim_{n\to\infty} \max_{p\in\{1,\ldots,N_n\}} \left(t_p^{(n)} - t_{p-1}^{(n)}\right) = 0,$$

---

[18]Using the approximation $R_\infty \approx \theta$ for $\lambda < 0$, Barone and Cuoco [5] obtain $\phi_1 = k - \lambda$, $\phi_2 = k$, and $\phi_3 = -\theta/\lambda$.

[19]To show this take the differential of $r^\circ = r/m$ and make a deterministic time change $t' = mt$ using $dt' = m\, dt$ and $dz(t') = \sqrt{m}\, dz(t)$ obtaining $dr^\circ(t') = \kappa^\circ(\theta^\circ - r^\circ)\, dt' + \sigma^\circ\sqrt{r^\circ}\, dz(t')$. Alternatively, make a change of variables in the partial differential equation (2.8).

[20]Remember that a diffusion is a Markov process with continuous but nowhere differentiable (and therefore of unbounded variation) sample paths. The Stieltjes definition of integrals does not apply.

then

$$\lim_{n\to\infty} \sum_{p=1}^{N_n} \left[r\big(t_p^{(n)}\big) - r\big(t_{p-1}^{(n)}\big)\right]^2 = \int_0^T [dr(t)]^2$$

$$= \sigma^2 \int_0^T r(t)\, dt \tag{3.1.2}$$

$$= \sigma^2 \lim_{n\to\infty} \sum_{p=1}^{N_n} r\big(t_{p-1}^{(n)}\big)\big(t_p^{(n)} - t_{p-1}^{(n)}\big)$$

and the convergence is almost sure as n goes to infinity for any given T.[21] This means that, when considering more and more frequent observations over a given interval $[0, T]$, the diffusion coefficient σ becomes known with probability one.

In order to estimate the drift coefficients we use the probability measure induced by $r(t, \alpha)$, where $\alpha = (\kappa, \theta)$ is the parameter vector. If $L_T(r, \alpha)$ is the likelihood ratio (Liptser and Shiryayev [41, Chapter 7]), the maximum-likelihood estimators are obtained by setting the score vector $q_T(r, \alpha)$ equal to 0:

$$q_T(r, \alpha) \equiv \frac{\partial}{\partial\alpha} \log L_T(r, \alpha)$$

$$= \int_0^T \left(\frac{\partial}{\partial\alpha} A(r, \alpha)\right) G^{-2}(r)\, dr_s \tag{3.1.3}$$

$$- \int_0^T \left(\frac{\partial}{\partial\alpha} A(r, \alpha)\right) G^{-2}(r) A(r, \alpha)\, ds$$

$$= 0$$

where $A(r, \alpha) = \kappa(\theta - r)$ and $G(r) = \sigma\sqrt{r}$. It can be easily seen that, in the case of Wiener noises (exponential families), the maximum-likelihood estimator is equivalent to the minimum distance (weighted least squares) estimator

$$\min_{\alpha} \int_0^T [dr - A(r, \alpha)\, ds]' G^{-2}(r)[dr - A(r, \alpha)\, ds]. \tag{3.1.4}$$

Using martingale limit theorems it can be shown that the maximum-likelihood estimator $\hat{\alpha}_T$ is strongly consistent for α and asymptotically normal in the sense that[22]

$$I_T^{1/2}(r, \alpha)(\hat{\alpha}_T - \alpha) \overset{aD}{\sim} \mathcal{N}(0, 1) \quad \text{as } T \to \infty \tag{3.1.5}$$

---

[21] An example of nested partitions is $T_n = (pT/2^n,\ p = 0, 1, \ldots, 2^n)$. If partitions are not nested as for $T_n = (pT/n,\ p = 0, 1, \ldots, n)$ the first convergence is only in quadratic mean. See Wong and Hajek [60, p. 53].

[22] McKeague [47] shows that for stationary ergodic processes strong consistency and asymptotic normality (with a loss in efficiency) still hold even in the case of a misspecified diffusion function ($\sigma\sqrt{r}$ in our case).

where the process I_T (quadratic covariation of q_T) is given by

$$I_T(r,\alpha) = \int_0^T \left(\frac{\partial}{\partial\alpha}A(r,\alpha)\right)G^{-2}(r)\left(\frac{\partial}{\partial\alpha}A(r,\alpha)\right)ds \qquad (3.1.6)$$

and, under regularity conditions, it can be considered a random Fisher conditional information matrix for dependent observations.[23]

In the case of an ergodic process with "smooth" drift function we have (see Kutoyants [39])

$$\sqrt{T}(\hat\alpha_T - \alpha) \overset{aD}{\sim} \mathcal{N}(0, D^{-1}(\alpha)) \quad \text{as } T \to \infty$$

where

$$D(\alpha) = \bar{\mathbb{E}}\left(\frac{\partial}{\partial\alpha}A(r,\alpha)G^{-2}(r)\frac{\partial}{\partial\alpha}A(r,\alpha)\right)$$
$$= \lim_{T\to\infty}\frac{1}{T}I_T(\alpha) \quad \text{a.s.} \quad \text{(by ergodicity).} \qquad (3.1.7)$$

In our case the maximum-likelihood estimators are[24]

$$\hat\kappa = \frac{\int_0^T \frac{ds}{r(s)}\int_0^T dr(s) - T\int_0^T \frac{dr(s)}{r(s)}}{T^2 - \int_0^T r(s)\,ds\int_0^T \frac{ds}{r(s)}} \qquad (3.1.8)$$

and

$$\hat\theta = \frac{T\int_0^T dr(s) - \int_0^T r(s)\,ds\int_0^T \frac{dr(s)}{r(s)}}{\int_0^T \frac{ds}{r(s)}\int_0^T dr(s) - T\int_0^T \frac{dr(s)}{r(s)}} \qquad (3.1.9)$$

where

$$\int_0^T dr(s) = r(T) - r(0), \qquad \int_0^T \frac{dr(s)}{r(s)} = \log\left(\frac{r(T)}{r(0)}\right) + \frac{\sigma^2}{2}\int_0^T \frac{ds}{r(s)}$$

and the (symmetric) Fisher matrix is given by

$$I_T(\kappa,\theta) = \begin{bmatrix} \int_0^T \frac{(\theta-r(s))^2}{\sigma^2 r(s)}\,ds & \int_0^T \frac{\kappa(\theta-r(s))}{\sigma^2 r(s)}\,ds \\ \int_0^T \frac{\kappa(\theta-r(s))}{\sigma^2 r(s)}\,ds & \int_0^T \frac{\kappa^2}{\sigma^2 r(s)}\,ds \end{bmatrix}. \qquad (3.1.10)$$

Strong consistency of $\hat\alpha_T$ can be easily checked using (2.7b). Moreover, the Rao–Cramér bound, for large T, is obtained using (3.1.7) as

$$I_T^{-1}(\alpha) \simeq \frac{1}{T}D^{-1}(\alpha) = \frac{1}{T}\begin{bmatrix} 2\kappa & -\sigma^2/\kappa \\ -\sigma^2/\kappa & \theta\sigma^2/\kappa^2 \end{bmatrix} \qquad (3.1.11)$$

---

[23]See Feigin [22]. The optimal asymptotic results depend on the factorization of the score (as conditional families): $q_T(\alpha) = I_T(\alpha)(\hat\alpha_T - \alpha)$ in which case $I_T^{-1}(\alpha)$ identifies a Rao–Cramér minimum variance bound.

[24]We used the fact that $\max L(\kappa,\theta) = \max L(\kappa,\kappa\theta/\kappa) \equiv \max M(\kappa,\kappa\theta)$.

going to zero, by consistency, as T goes to infinity. Notice that, from Itô's formula, the transformation $y = \sqrt{r}$ has dynamics[25]

$$dy(t) = \left[\frac{(4\kappa\theta - \sigma^2)}{8y} - \frac{ky}{2} \right] dt + \frac{\sigma}{2} \, dw \qquad (3.1.12)$$

so that

$$\sigma^2 = \frac{4}{T} \int_0^T [dy(t)]^2.$$

Comparing the last equation with (3.1.1) we obtain:

$$T = \frac{4 \int_0^T \left[d\sqrt{r(t)} \right]^2 \int_0^T r(t) \, dt}{\int_0^T [dr(t)]^2}. \qquad (3.1.13)$$

In practice, the assumption of a continuous record of observations is not satisfied and, usually, only a set of discrete observations is available at the partition $0 = t_0 < t_1 < \cdots < t_n = T$ of $[0, T]$, where $\delta_n = t_{p+1} - t_p$ is independent of p (constant step) and refining to zero, i. e., $\lim_{n\to\infty} \delta_n = 0$.

The suggestion is then to replace continuous integrals with discrete sums (e. g. Cauchy approximations). In particular, take $t_p = pT/n \equiv p\delta_n$ for $p = 0, 1, \ldots, n$ and define $\hat{\sigma}^2_{n,T,r}$ as the discretized version of (3.1.1):

$$\hat{\sigma}^2_{n,T,r} = \frac{\sum_{p=1}^n [r(t_p) - r(t_{p-1})]^2}{\delta_n \sum_{p=1}^n r(t_{p-1})}. \qquad (3.1.14)$$

We have

$$L^2\text{-}\lim_{n\to\infty} (\hat{\sigma}^2_{n,T,r} - \sigma^2) = 0.$$

At the same time, from (3.1.12),

$$\hat{\sigma}^2_{n,T,\sqrt{r}} = \frac{4}{T} \sum_{p=1}^n \left[\sqrt{r(t_p)} - \sqrt{r(t_{p-1})} \right]^2 \qquad (3.1.15)$$

which is consistent (in mean square) and recursive (Banon [3, p. 392]):

$$\hat{\sigma}^2_{n+1,T+\delta,\sqrt{r}} = \frac{n}{n+1} \hat{\sigma}^2_{n,T,\sqrt{r}} + \frac{4}{T+\delta} \left(\sqrt{r(t_{n+1})} - \sqrt{r(t_n)} \right)^2. \qquad (3.1.16)$$

Analogously, define $\hat{\alpha}_{n,T}$ as the discretized version of the maximum-likelihood estimator $\hat{\alpha}_T$ (Cauchy approximating sums instead of integrals). Then, as in Le Breton [40]:

$$P\text{-}\lim_{n\to\infty} (\hat{\alpha}_{n,T} - \hat{\alpha}_T) = 0, \qquad (\hat{\alpha}_{n,T} - \hat{\alpha}_T) = O_P(\delta_n^{1/2}), \qquad (3.1.17)$$

---

[25]It can be shown that, for regular transformations $y = h(r)$ the maximum-likelihood estimators are unaffected, being $q_T(r, \alpha) = q_T(h(r), \alpha)$ and $I_T(r, \alpha) = I_T(h(r), \alpha)$.

where $O_P(\cdot)$ means "same order in probability as". This means that as $n \to \infty$, for given T, i.e. as the frequency of observations per unit time increases ($\delta_n \downarrow 0$) toward the limit of a continuous record, the discretized estimator $\hat{\alpha}_{n,T}$ is consistent for $\hat{\alpha}_T$ (but not for α).[26]

Note that in the case of sample-path integrals, more refined approximations (trapezoidal, Simpson's, etc.) can be used. In particular, for any smooth function f:

$$\int_0^T f(t)\,dt \simeq \begin{cases} \delta_n \sum_{p=0}^{n-1} f(t_p) & \text{Cauchy,} \\[2em] \delta_n \sum_{p=0}^{n-1} \dfrac{f(t_p) + f(t_{p+1})}{2} & \text{trapezoidal,} \\[2em] \delta_n \sum_{\substack{p=0 \\ p \text{ even}}}^{n-2} \dfrac{f(t_p) + 4f(t_{p+1}) + f(t_{p+2})}{3} & \text{Simpson.} \end{cases}$$

In the first case the approximation error is of order $O(\delta)$ (we omit the subscript n of δ_n), in the second case it is of order $O(\delta^2)$ and for Simpson's formula it is of order $O(\delta^4)$.[27]

3.2 Estimation of the discrete equivalent of linearized SDE.
It is well-known that the linear stochastic differential equation

$$dx(t) = (Ax + a)\,dt + b\,dw(t) \tag{3.2.1}$$

has the solution, for $x(s) = x_s$ (Arnold [1, p. 130]),

$$x(t) = x_s e^{A(t-s)} + a\frac{e^{A(t-s)} - 1}{A} + b\int_s^t e^{A(t-u)}\,dw(u) \tag{3.2.2}$$

so that, for $t = p\delta$ and $s = p\delta - \delta$ (Sargan [54]),

$$x_p = e^{A\delta} x_{p-1} + a\frac{e^{A\delta} - 1}{A} + b\int_{p\delta-\delta}^{p\delta} e^{A(p\delta-u)}\,dw(u) = c_1 x_{p-1} + c_0 + u_p \tag{3.2.3a}$$

where

$$c_1 = e^{A\delta}, \qquad c_0 = a\frac{e^{A\delta} - 1}{A}, \qquad u_p \sim \mathcal{N}(0, v^2), \qquad v^2 = b^2\frac{e^{2A\delta} - 1}{2A}, \tag{3.2.3b}$$

---

[26] Marsh and Rosenfeld [44, p. 639], observe that in most markets data are not generated in real time so that as the interval δ shrinks observation/missing errors increase. Non-trading time is a leading example.

[27] See McShane [48, p. 66]. In the case of stochastic integrals, the limit of trapezoidal approximations is used in the definition of the Stratonovich integral.

that is

$$A = \frac{\log c_1}{\delta}, \qquad a = \frac{c_0 \log c_1}{\delta(c_1 - 1)}, \qquad b^2 = v^2 \frac{2 \log c_1}{\delta(c_1^2 - 1)}.$$

The error term is normal i.i.d. and the ordinary least squares (OLS) estimator has optimal properties in large as well as small samples. For a nonlinear stochastic differential equation such as (3.1.12), a popular estimation procedure (Fischer and Zechner [24], Barone and Cesari [4], De Felice and Moriconi [19]) is to linearize the drift around a mean value $\bar{y}$ obtaining a linear stochastic differential equation (approximation of (3.1.12)) for which the exact discrete equivalent (3.2.3) is known. Linearizing $1/y$, the following stochastic differential equation is obtained:

$$
\begin{aligned}
dy(t) &\simeq \left[\left(-\frac{(4\kappa\theta - \sigma^2)}{8\bar{y}^2} - \frac{\kappa}{2} \right) y + \frac{4\kappa\theta - \sigma^2}{4\bar{y}} \right] dt + \frac{\sigma}{2} dw(t) \\
&\equiv [Ay + a] \, dt + b \, dw(t)
\end{aligned}
\tag{3.2.4}
$$

and the above procedure can be applied giving

$$\widehat{\sigma}^2 = 4\widehat{b}^2, \qquad \widehat{\kappa} = -2\widehat{A} - \frac{\widehat{a}}{\bar{y}}, \qquad \widehat{\theta} = \frac{4\widehat{a}\bar{y} + \widehat{\sigma}^2}{4\widehat{\kappa}}. \tag{3.2.5}$$

3.3 Conditional-mean estimation. Using the conditional expected value in (2.5) we can write:

$$
\begin{aligned}
r_p &= \theta(1 - e^{-\kappa\delta})e^{-\kappa\delta}r_{p-1} + \varepsilon_p, \\
\mathrm{E}(\varepsilon_p) &= 0, \qquad \mathrm{E}(\varepsilon_p^2 | r_{p-1}) \simeq r_{p-1}\sigma^2\delta
\end{aligned}
\tag{3.3.1}
$$

where the innovation term is not normal (its conditional distribution is in fact a non-central χ^2) and its variance, from (2.5), is approximately $r_{p-1}\sigma^2\delta$ so that a weighted least squares transformation gives[28]

$$
\begin{aligned}
\frac{r_p}{\sqrt{r_{p-1}}} &= \theta(1 - e^{-\kappa\delta})\frac{1}{\sqrt{r_{p-1}}} + e^{-\kappa\delta}\sqrt{r_{p-1}} + u_p \\
&\equiv A\frac{1}{\sqrt{r_{p-1}}} + B\sqrt{r_{p-1}} + u_p, \\
\mathrm{E}(u_p) &= 0, \qquad \mathrm{E}(u_p^2) \simeq \sigma^2\delta \\
\widehat{k} &= -\frac{\log \widehat{B}}{\delta}, \qquad \widehat{\theta} = \frac{\widehat{A}}{1 - \widehat{B}}, \qquad \widehat{\sigma}^2 = \frac{\mathrm{Var}(\widehat{u})}{\delta}.
\end{aligned}
\tag{3.3.2}
$$

---

[28]The parameter δ represents the observation frequency in terms of the chosen time unit. For example, for annual rates (i.e. time unit = 1 year) and monthly observations $\delta = 1/12$; for monthly rates (time unit = 1 month) and monthly observations $\delta = 1$.

3.4 OLS estimation of crude discretization of SDE. Let us consider an approximate discrete-time specification of the basic stochastic differential equation (2.1)

$$r_p - r_{p-1} = (\alpha + \beta r_{p-1})\delta + \varepsilon_p, \tag{3.4.1}$$

$$\alpha \simeq \kappa\theta, \qquad \beta \simeq -\kappa, \qquad E(\varepsilon_p) = 0, \qquad E(\varepsilon_p^2 | r_{p-1}) = \sigma^2 r_{p-1}\delta$$

where the disturbance term is not necessarily normal. Using a weighted least squares transformation we obtain (see for example Brennan and Schwartz [11])

$$\frac{r_p - r_{p-1}}{\sqrt{r_{p-1}}} = \frac{\alpha\delta}{\sqrt{r_{p-1}}} + \beta\delta\sqrt{r_{p-1}} + u_p \equiv A\frac{1}{\sqrt{r_{p-1}}} + B\sqrt{r_{p-1}} + u_p,$$

$$E(u_p) = 0, \qquad E(u_p^2) = \sigma^2\delta,$$

$$\widehat{\kappa} = -\widehat{B}/\delta, \qquad \widehat{\theta} = -\widehat{A}/\widehat{B}, \qquad \widehat{\sigma}^2 = \frac{\text{Var}(\widehat{u})}{\delta}. \tag{3.4.2}$$

Note that the regression coefficients are the linear approximation of the coefficients in the previous case. Moreover the estimates for θ and σ^2 are clearly numerically equivalent.

3.5 GMM estimation of crude discretization of SDE. Under the assumption of stationary ergodic processes, Hansen's [30] generalized method of moments (GMM) can be used to estimate the above given crude discretization

$$r_p - r_{p-1} = (\alpha + \beta r_{p-1})\delta + \varepsilon_p, \tag{3.5.1}$$

$$\alpha \simeq \kappa\theta, \qquad \beta \simeq -\kappa, \qquad E(\varepsilon_p) = 0, \qquad E(\varepsilon_p^2 - \sigma^2 r_{p-1}\delta) = 0.$$

The idea under the GMM procedure is to impose a set of orthogonality conditions on the variables involved, choosing as estimates the values that minimize (in a certain optimal metric) the relevant distance. Various forms of least squares, quasi-maximum-likelihood and nonlinear instrumental variables (IV) estimators can be viewed as special case of GMM. In our case (see Chan, Karolyi, Longstaff and Sanders [14]) the orthogonality condition is[29]

$$E(f_p(\zeta)) = 0, \quad \text{where } f_p(\zeta) = \begin{bmatrix} \varepsilon_p \\ \varepsilon_p r_{p-1} \\ \varepsilon_p^2 - \sigma^2 r_{p-1}\delta \\ (\varepsilon_p^2 - \sigma^2 r_{p-1}\delta)r_{p-1} \end{bmatrix}, \tag{3.5.2}$$

$$\zeta \equiv (\alpha\delta, \beta\delta, \sigma^2\delta),$$

---

[29]The number of orthogonality conditions must be not less than the number of parameters. As Hansen [30, p. 1048] observes, his result "is limited in that it takes the specification of the orthogonality conditions as given and does not discuss how to construct optimally orthogonality conditions".

and the GMM estimator is obtained as

$$\zeta_n = \arg\min_{\zeta} g_n'(\zeta) D_n g_n(\zeta)$$

where

$$g_n(\zeta) \equiv \frac{1}{n} \sum_{p=1}^{n} f_p(\zeta),$$

i. e.,

$$\frac{\partial g_n(\zeta)}{\partial \zeta} D_n g_n(\zeta) = 0 \quad \text{f. o. c.} \tag{3.5.3}$$

D_n being a positive semi-definite weighting matrix. The optimal GMM estimator (Hansen [30, Theorem 3.2]) is obtained using $D_n = \widehat{S}_n^{-1}$ where $\widehat{S}_n$ is a consistent estimator of the covariance matrix $S(\zeta) = \mathrm{E}(f_p(\zeta) f_p'(\zeta))$ so that

$$\sqrt{n}(\widehat{\zeta}_n - \zeta) \overset{aD}{\sim} \mathcal{N}\left(0, \left[\frac{\partial g_n}{\partial \zeta} S_n^{-1} \frac{\partial g_n}{\partial \zeta} \right]^{-1} \right) \quad \text{as } T \to \infty. \tag{3.5.4}$$

Newey and West [51] give a consistent, positive semi-definite estimator of $S(\zeta)$:

$$\widehat{S}_n = \widehat{\Omega}_0 + \sum_{j=1}^{m} \frac{m+1-j}{m+1} [\widehat{\Omega}_j + \widehat{\Omega}_j'] \quad \text{with} \quad \widehat{\Omega}_j = \frac{1}{n} \sum_{p=j+1}^{n} \widehat{f}_p \widehat{f}_{p-j}' \tag{3.5.5}$$

where m is the number of nonzero autocorrelations of $f_p(\zeta)$.[30] A goodness-of-fit test for the model is obtained by the result of Hansen [30, Lemma 4.2]:

$$n g_n'(\widehat{\zeta}) \widehat{S}_n^{-1} g_n(\widehat{\zeta}) \overset{aD}{\sim} \chi_v^2 \quad \text{as } T \to \infty, \quad v = \dim(f(\zeta)) - \dim(\zeta). \tag{3.5.6}$$

3.6 Discrete-time maximum-likelihood estimation. Let us consider a set of n observations at times $t_1, \ldots, t_n$, possibly not equally spaced,[31] of the process $r(t)$, i. e. $(r(t_1), r(t_2), \ldots, r(t_p), \ldots, r(t_n))$. The likelihood function, given an initial value $r(t_0)$, is

$$
\begin{aligned}
L\big(r(t_1), r(t_2), &\ldots, r(t_n) \,|\, r(t_0), \zeta\big) \\
&= p\big(r(t_1) \,|\, r(t_0), \zeta\big) \, p\big(r(t_2) \,|\, r(t_1), r(t_0), \zeta\big) \cdots \\
&\quad \times p\big(r(t_n) \,|\, r(t_{n-1}), r(t_{n-2}), \ldots, r(t_0), \zeta\big) \tag{3.6.1} \\
&= \prod_{p=1}^{n} p\big(r(t_p) \,|\, r(t_{p-1}), \zeta\big)
\end{aligned}
$$

---

[30] In general (Newey and West [51, Theorem 2]) $m(n)$ may be a function of n going to infinity with n more slowly than $n^{0.25}$. In our Monte Carlo experiments we used the largest integer less than $n^{0.24}$.

[31] See Marsh and Rosenfeld [44], Lo [42, 43], Robinson [53]. When the observation interval is small (day, hour, etc.) not equally spaced data is often due to non-trading time which generates missing observations.

where the last equality comes from Markov's property and $p(r(t_p)|r(t_{p-1}))$ is the transition density from $r(t_{p-1})$ to $r(t_p)$, in our case a non-central χ^2 (see (2.3)).

It is well known that the maximum-likelihood estimator has optimal asymptotic properties. It is given by

$$\widehat{\zeta}_n \equiv \arg\max_\zeta L(\zeta)$$

$$= \arg\max_\zeta \prod_{p=1}^n c e^{-(u_{p-1}+v_p)/2} \left(\frac{v_p}{u_{p-1}}\right)^{(q-1)/2} I_{q-1}\left(\sqrt{u_{p-1}v_p}\right)$$

where

$$\zeta = (\kappa, \theta, \sigma^2), \qquad c = \frac{2\kappa}{\sigma^2(1-e^{-\kappa\delta_p})}, \qquad u_{p-1} = 2cr(t_{p-1})e^{-\kappa\delta_p},$$

$$v_p = 2cr(t_p), \qquad q = \frac{2\kappa\theta}{\sigma^2}, \qquad \delta_p = t_p - t_{p-1}, \qquad (3.6.2)$$

and it is consistent and asymptotically normal:[32]

$$\text{P-}\lim_{n\to\infty} \widehat{\zeta}_n = \zeta \quad \text{(true value)},$$

$$\sqrt{n}(\widehat{\zeta}_n - \zeta) \overset{aD}{\sim} \mathcal{N}(0, I^{-1}(\zeta)) \quad \text{as } n \to \infty,$$

$$I(\zeta) = \lim_{n\to\infty} \frac{1}{n} \sum_{p=1}^n - E_{p-1}\left[\frac{\partial^2 \log p(r(t_p)|r(t_{p-1}), \zeta)}{\partial\zeta^2}\right]$$

$$= \lim_{n\to\infty} \frac{1}{n} \sum_{p=1}^n E_{p-1}\left[\frac{\partial \log p(r(t_p)|r(t_{p-1}), \zeta)}{\partial\zeta}\right]^2. \qquad (3.6.3)$$

For the discount bond price $P(r, t, T, \zeta, \lambda)$, if λ were known,

$$\widehat{P}_n = P(r, t, T, \widehat{\zeta}_n, \lambda) \qquad (3.6.4)$$

is the maximum-likelihood estimator and

$$\sqrt{n}(\widehat{P}_n - P) \overset{aD}{\sim} \mathcal{N}\left(0, \frac{\partial P}{\partial\zeta} I^{-1}(\zeta) \frac{\partial P}{\partial\zeta}\right) \quad \text{as } n \to \infty. \qquad (3.6.5)$$

---

[32]See Bar–Shalom [7] and Bhat [8], where the result is obtained for the general case of dependent observations.

§4 Monte Carlo simulations of continuous processes

4.1 Introduction. Quoting Talay [58], "the numerical analysis of stochastic differential systems is at its very beginning". A major critical point is the fact that the concept of white noise (posing no problem in discrete time) in continuous-time means a process independent at every time point t with respect to any other time point s however close to t:

$$\text{Cov}(\xi(t), \xi(s)) = \begin{cases} \sigma^2 & \text{for } t = s, \\ 0 & \text{for } t \neq s. \end{cases}$$

This discontinuity of $\xi(t)$ (for example in the mean-square sense) suggests the definition of stochastic processes of a more general type (see for example Yaglom [61, p. 210]) for which a generalized differential and a generalized calculus can be defined. These general processes are an abstract concept with no direct real counterpart[33] and this impiles a second trouble whenever the theory has to be compared with data, real or empirically simulated. Roughly speaking, one could say that real data imply only discrete, colored noise and the link with a model driven by continuous white noise is not clear.

A trade-off, however, comes out due to the fact that continuous time is theoretically very appealing and manageable and the white noise hypothesis (if not a theoretical result)[34] is a very economic parametrization of the model.

There are many possible combinations between real data, simulated data and the theoretical or model data. In a performance analysis, if the real data-generating mechanism (real DGM) was known (its features, not its parameters) the simulation task would be just to replicate it "in vitro". However, the real DGM is unknown and the question is whether the student has to follow his or her "a priori" on it or, instead, just to accurately replicate the theoretical DGM assumed in the model, or also to take an average position between the two alternatives. According to the taken position, (more near to the "a priori" real DGM or to the model DGM) the distance between simulated and theoretical points may change significantly.[35]

Moreover, when the theoretical model is time-continuous, a problem is raised concerning the convergence of discrete (simulated) data to a continuous theoretical process. It is well known, for example, that differential equations of linear interpolations of Brownian motion paths at discrete time-points converge (in mean square) to Stratonovich (not Itô) stochastic differential equations (Stratonovich calculus).

---

[33] "The point is that in practice one must always use some physical device to measure $\xi(t)$, and since the device always has 'inertia' (or 'memory'), corresponding to its nonzero 'time constant', the input process $\xi(t)$ will inevitably be subjected to some time averaging." (Yaglom [61, p. 208]).

[34] A fundamental result of finance is that under the no-arbitrage condition the asset price vector is a martingale (orthogonal increments). See for example Harrison and Kreps [31] and Harrison and Pliska [32]. This is also a property of Itô integrals.

[35] In principle, the best position should be where an isometry exists between simulated and theoretical data on one hand and real and theoretical data on the other.

4.2 Some simulation data-generating mechanisms. Let $[0, T]$ be a time interval and $0 = t_0 < t_1 < \cdots < t_n = T$ a partition of $[0, T]$ such that $\delta_n = t_{p+1} - t_p$ is independent of p (constant step) and refining to zero, i. e., $\lim_{n \to \infty} \delta_n = 0$. For simplicity, take $t_p = pT/n \equiv p\delta_n$ for $p = 0, 1, \ldots, n$ and consider the one-dimensional diffusion $x(t)$ which solves of the Itô stochastic differential equation

$$dx(t) = \alpha(x(t)) \, dt + \sigma(x(t)) \, dw(t) \qquad (4.2.1)$$

or the corresponding equivalent Itô stochastic integral equation

$$x(t) = x(v) + \int_v^t \alpha(x(s)) \, ds + \int_v^t \sigma(x(s)) \, dw(s). \qquad (4.2.2)$$

A simple simulation approach (strong approximation) is to approximate the diffusion $x(t)$ by a discrete process $x_p^\circ = x^\circ(p\delta_n)$ for $p = 0, 1, \ldots, n$ such that, omitting the subscript n for δ_n,

$$\mathrm{E}\left[\sup_p |x(p\delta) - x_p^\circ|\right] \leq c\delta^\gamma \qquad (4.2.3)$$

where γ is the order of strong convergence. A different approach (moment or weak approximation) is suggested by a different intuition: given that $x(t)$ is a stochastic process, its moments or certain quantities depending on the law of $x(t)$ may be more interesting than some particular path (Talay [58]). This implies to approximate the mean of a smooth function g of the process according to the criterion

$$\max_p \left|\mathrm{E}\big(g(x(p\delta))\big) - \mathrm{E}(g(x_p^\circ))\right| \leq c_g \delta^\beta \qquad (4.2.4)$$

where β is the order of weak convergence.

Let us now consider the following simulation schemes.

- *Euler scheme* (Maruyama [45]):

$$x_{p+1}^\circ = x_p^\circ + \alpha(x_p^\circ)\delta + \sigma(x_p^\circ)u_{p+1}, \quad p = 0, 1, \ldots, n-1,$$

  where $u_{p+1} \equiv w(p\delta + \delta) - w(p\delta)$ is normal $\mathcal{N}(0, \delta)$.
- *Milstein* [49] *scheme:*

$$x_{p+1}^\circ = \text{Euler} + \sigma'(x_p^\circ)\sigma(x_p^\circ)\frac{1}{2}[u_{p+1}^2 - \delta]$$

  where a prime means first derivative.
- *Talay* [57] *scheme:*

$$x_{p+1}^\circ = \text{Milstein} + \left[\alpha'\alpha + \frac{1}{2}\alpha''\sigma^2\right]\frac{1}{2}\delta^2 + \left[\alpha'\sigma + \sigma'\alpha + \frac{1}{2}\sigma''\sigma^2\right]\frac{1}{2}u_{p+1}\delta$$

  where a double prime means second derivative and, as before, the functions are evaluated at x_p°.

The stochastic foundation of the schemes and their approximation errors are sketched in Appendix A, using a stochastic Taylor formula. It can be shown (Milstein [49]) that the Euler scheme is of order $\gamma = 0.5$ of strong convergence and the Milstein scheme is of order $\gamma = 1$. In terms of weak convergence, they are both of order $\beta = 1$ while the Talay scheme is of order $\beta = 2$.[36]

4.3 The Monte Carlo experiment. The first point to stress is that in actual markets the instantaneous interest rate $r(t)$ is not observable. In fact, the rate $r(t)$ is defined as the interest rate on a loan obtained at time t and maturing in the next instant, i.e. at time $t + dt$. It is, therefore, a limit concept (see (2.16)), not observable in real markets. What is in fact observable is a particular, short maturity interest rate $R(t, \tau^\circ) \equiv R^\circ(t)$, for τ° fixed. More precisely, at any given time t, the term structure is known at a small number of maturities, for example 1 day to maturity (so called overnight rate), 2, 3 days, 1, 2, 3 weeks, 1, 3, 6, 12 months.

We tried to replicate this situation in which only interest rates with positive maturities are available at discrete-time observation intervals. The basic set-up is given as follows. We assumed that the time unit is equal to 1 year. The instantaneous rate $r(t)$ has been generated using the three given data-generating mechanisms assuming δ equal to 1 hour ($\delta = 1/8760$) and parameter values $\kappa = 0.3$, $\theta = 0.1$, $\sigma = 0.06$. The observable process $R(t)$ has been obtained using definition (2.13), $\lambda = -0.03$, and assuming a positive fixed maturity (τ° equal to 3 months). Finally, the data have been sampled at fixed interval length Δ (one observation every month). The parameter values are similar, by and large, to real data estimates.[37] The starting value $r(0)$ is set to θ, the long-run mean. The usual estimation procedures do not recognize that available observations concern discount bonds with a given non-zero maturity τ° and use these data as an approximation of the instantaneous rate ($\tau = 0$). Our experiment evaluates the effects of this approximation. We considered 200 simulations over a sample time-span up to 200 years (2, 5, 10, 20, 50, 100, 200 years).[38] A few examples of generated interest rates $R(t, \tau^\circ)$ over 20 years of sample time-span, for $\tau^\circ = 3$ months, are given in Figure 1. The straight line is the long-run mean θ of the instantaneous rate $r(t)$. A real time-series of three months Italian treasury bill rates over the last 20 years is displayed in Figure 2 and compared with a similar simulated path (Figure 2b).

---

[36]See Kloeden and Platen [37, p. 465 and Chapters 10–15] for more refined strong and weak approximation schemes.

[37]Fournie and Talay [26] use $\delta = 1/365$ (one day), $\kappa = 0.75$, $\theta = 0.1$ (10%), $\sigma = 0.105$ and T up to 110 years ($365 \times 110 = 40150$ simulated data.) They are interested only in the large sample properties of estimators and simulate just one sample path. Barone and Cesari [4] and De Felice and Moriconi [19] use 3-months Treasury bills and estimate for the period 8106–8411 (monthly data), $\kappa = 0.39341$, $\theta = 0.146844$, $\sigma = 0.05611$ and $\lambda = 0.065296$. Barone, Cuoco and Zautzik [6] for daily data 8312–8903 find $\kappa = 0.25118$, $\theta = 0.1117$, $\sigma = 0.0627$ and $\lambda = -0.00783$ (calculated under the assumption $\theta = R(\infty)$).

[38]In the case of small samples (2 years of monthly observations) 13 simulations were rejected because of the influence of large outliers.

C. BIANCHI, R. CESARI AND L. PANATTONI

Figure 1: Simulated series of interest rates
(monthly data)

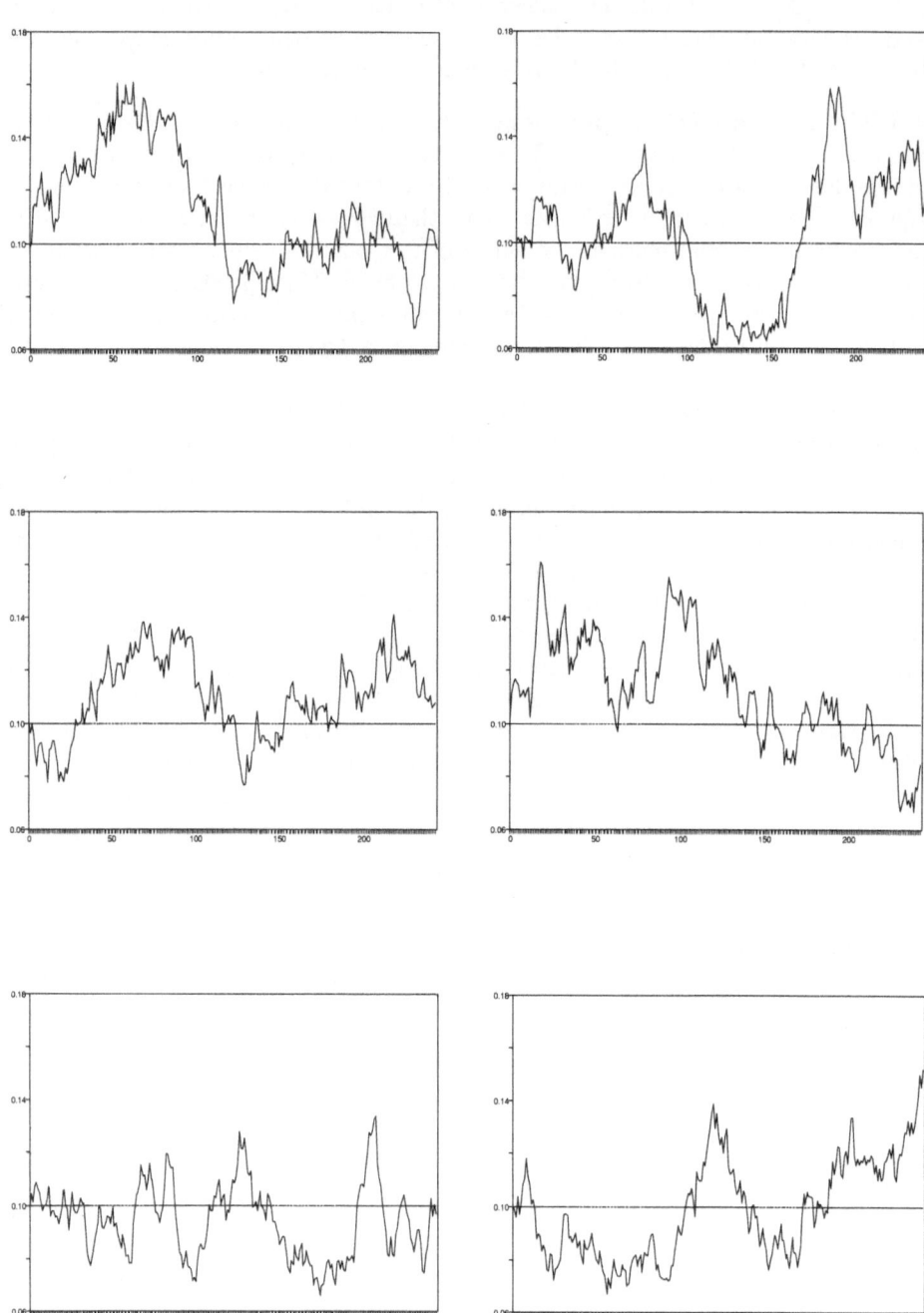

Figure 2: Three-months Italian treasury bill rates
(1973–1993)

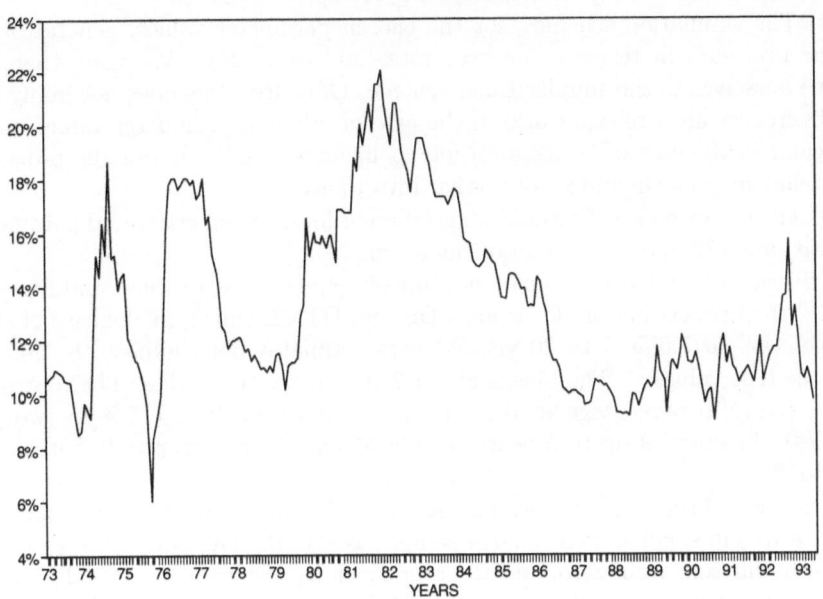

Figure 2b: Simulated three-months treasury bill rates
(20 years of monthly data)

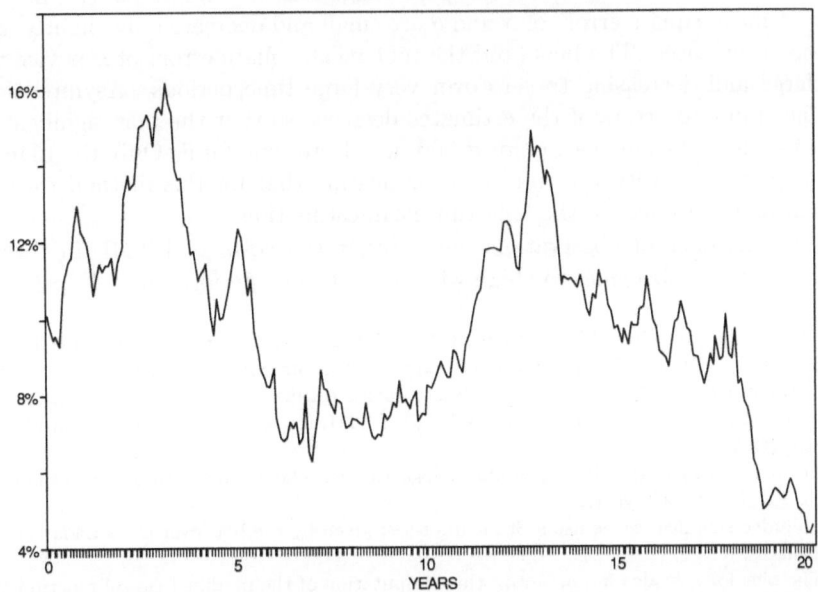

§5 MONTE CARLO RESULTS

The main results of the given Monte Carlo set-up are the following:[39]

(A) The simulation schemes, for the chosen parameter values, generates very similar processes in terms of interest rates and estimates. We may, therefore, confine ourselves to the simpler Euler scheme. Of course, this does not imply that the differences are irrelevant also in the case of different parameter values and in particular in the case of larger diffusion coefficients σ and non-ergodic processes. This point may be the object of further investigation.

(B) The performance of estimators is different for different estimated parameters (κ, θ, σ) and different time-span of the sample.

In the case of our basic set-up of monthly observations (see Table 1 and Figures 3 to 5),[40] the best estimators for σ are LDE and DTML which, in the case of small samples (in particular 5 to 20 years), gives estimates not significantly different from the true values.[41] The bias is about 2–3 % of the true value. The root mean square error (in percentage of the true value) is below 10 % (16 % in two year samples). In samples up to 5 years the GMM for σ performs poorly with a bias over 10 %.

In the case of the long-run mean θ the best estimator (in terms of the root mean square error) in small as well as large samples is CTML. The bias is 1.3 %, not significantly different from zero, and the root mean square error is 13 %. The GMM has a statistically not significant bias of 2.7 % and it is the second-best estimator.

For the parameter κ, all the estimators are grossly upward biased for small samples and converge to the true values only for very large sample time-spans: for 50 years of monthly observations the estimators reach from above the first decimal figure of the true value (0.3), with a percentage bias as large as 30 %. The biases and root mean square errors of σ and θ are small and decrease only slightly as the time-span increases. The bias (and the root mean square error) of κ is vice versa very large and decreasing to zero over very large time periods. Asymptotically, also the standard errors of the estimates decrease so that the bias significance is decreasing for κ but increasing for σ and non-decreasing for θ. Only the LDE bias for θ is increasing with the time-span, indicating that for this method there is a cumulating error effect in the long-run mean estimation.

As the number of observations increases,[42] the costs of DTML in terms of computation time becomes too high with respect to its performance. In particular,

---

[39] We use the following abbreviations: CTML for continuous-time maximum-likelihood estimator (§3.1), LDE for linearized discrete equivalent estimator (§3.2), CME for conditional mean estimator (§3.3), OLS for ordinary least squares estimator of crude discretization (§3.4), GMM for generalized method of moments (§3.5), DTML for discrete-time maximum-likelihood estimator (§3.6).

[40] In Figures 3 to 11 the time-span axis is logarithmic. The displayed values correspond to 1, 2, 5, 10, 20, 50, 100, 200 years.

[41] To judge significance we use a Student's t-test given by the bias over the standard error of the estimates.

[42] The bias for κ is also important in the computation of the modified Bessel function (2.4).

Table 1

Percentage relative bias, root mean square error and Student's t-statistic: Simulated monthly data for three-months treasury bill rates

years=2 187		bias σ	κ	θ	rmse σ	κ	θ	Student t σ	κ	θ
CTML	-5.76	732.69	1.34	14.25	842.39	13.14	-0.44	1.76	0.10
LDE	2.16	843.91	5.72	15.95	1034.72	36.84	0.14	1.41	0.16
CME	-8.68	845.49	4.92	15.76	1036.55	33.65	-0.66	1.41	0.15
OLS	-8.68	707.28	4.92	15.76	846.88	33.65	-0.66	1.52	0.15
GMM	-14.05	723.62	2.71	19.65	859.88	26.72	-1.02	1.56	0.10
DTML	-2.04	848.20	5.34	15.25	1031.80	35.20	-0.13	1.44	0.15
years=5 199		**bias** σ	κ	θ	**rmse** σ	κ	θ	**Student t** σ	κ	θ
CTML	-6.38	366.95	1.74	10.20	465.64	15.82	-0.80	1.28	0.11
LDE	-2.18	397.27	3.15	8.95	535.04	31.36	-0.25	1.11	0.10
CME	-7.85	396.32	2.47	11.23	533.98	30.03	-0.98	1.11	0.08
OLS	-7.85	353.23	2.47	11.23	462.86	30.03	-0.98	1.18	0.08
GMM	-10.53	370.27	2.15	13.58	483.36	25.64	-1.23	1.19	0.08
DTML	-3.81	395.92	3.14	9.33	533.36	30.20	-0.45	1.11	0.10
years=10 198		**bias** σ	κ	θ	**rmse** σ	κ	θ	**Student t** σ	κ	θ
CTML	-4.89	181.39	0.64	8.24	236.86	14.34	-0.74	1.19	0.04
LDE	-2.40	177.30	2.97	7.26	247.01	25.33	-0.35	1.03	0.12
CME	-5.60	177.27	1.81	8.60	246.57	21.49	-0.86	1.03	0.08
OLS	-5.60	164.61	1.81	8.60	227.36	21.49	-0.86	1.05	0.08
GMM	-6.88	172.07	1.56	9.62	235.80	19.81	-1.02	1.07	0.08
DTML	-3.21	177.23	1.96	7.52	246.95	22.85	-0.47	1.03	0.09
years=20 200		**bias** σ	κ	θ	**rmse** σ	κ	θ	**Student t** σ	κ	θ
CTML	-4.79	82.91	1.34	6.56	122.94	12.51	-1.07	0.91	0.11
LDE	-3.36	80.30	2.90	5.53	127.46	17.07	-0.77	0.81	0.17
CME	-5.43	79.57	1.82	6.92	127.07	16.32	-1.27	0.80	0.11
OLS	-5.43	74.51	1.82	6.92	119.06	16.32	-1.27	0.80	0.11
GMM	-6.13	81.75	1.84	7.52	126.41	18.26	-1.41	0.85	0.10
DTML	-3.75	79.79	1.80	5.76	127.23	16.50	-0.86	0.81	0.11
years=50 200		**bias** σ	κ	θ	**rmse** σ	κ	θ	**Student t** σ	κ	θ
CTML	-4.44	31.66	0.48	5.24	51.77	8.80	-1.59	0.77	0.05
LDE	-3.53	31.05	1.86	4.47	52.60	9.46	-1.28	0.73	0.20
CME	-5.00	29.58	0.69	5.70	51.53	9.18	-1.83	0.70	0.08
OLS	-5.00	27.32	0.69	5.70	49.09	9.18	-1.83	0.67	0.08
GMM	-5.38	32.64	0.51	6.09	54.29	9.26	-1.89	0.75	0.05
DTML	-3.69	29.69	0.69	4.59	51.68	9.18	-1.34	0.70	0.08
years=100 200		**bias** σ	κ	θ	**rmse** σ	κ	θ	**Student t** σ	κ	θ
CTML	-4.16	14.52	0.45	4.63	33.01	6.59	-2.04	0.49	0.07
LDE	-3.52	14.74	1.81	4.07	34.54	7.09	-1.73	0.47	0.26
CME	-4.82	13.84	0.53	5.22	33.55	6.79	-2.41	0.45	0.08
OLS	-4.82	12.14	0.53	5.22	32.01	6.79	-2.41	0.41	0.08
GMM	-4.98	15.41	0.28	5.38	35.37	6.79	-2.47	0.48	0.04
DTML									
years=200 200		**bias** σ	κ	θ	**rmse** σ	κ	θ	**Student t** σ	κ	θ
CTML	-3.93	5.50	0.74	4.17	18.78	4.66	-2.79	0.31	0.16
LDE	-3.39	6.01	2.09	3.67	19.65	5.17	-2.42	0.32	0.44
CME	-4.60	5.49	0.74	4.79	19.36	4.73	-3.40	0.30	0.16
OLS	-4.60	4.09	0.74	4.79	18.54	4.73	-3.40	0.23	0.16
GMM	-4.68	6.24	0.51	4.87	19.90	4.63	-3.43	0.33	0.11
DTML									

Figure 3: Mean estimated of the diffusion coefficient σ

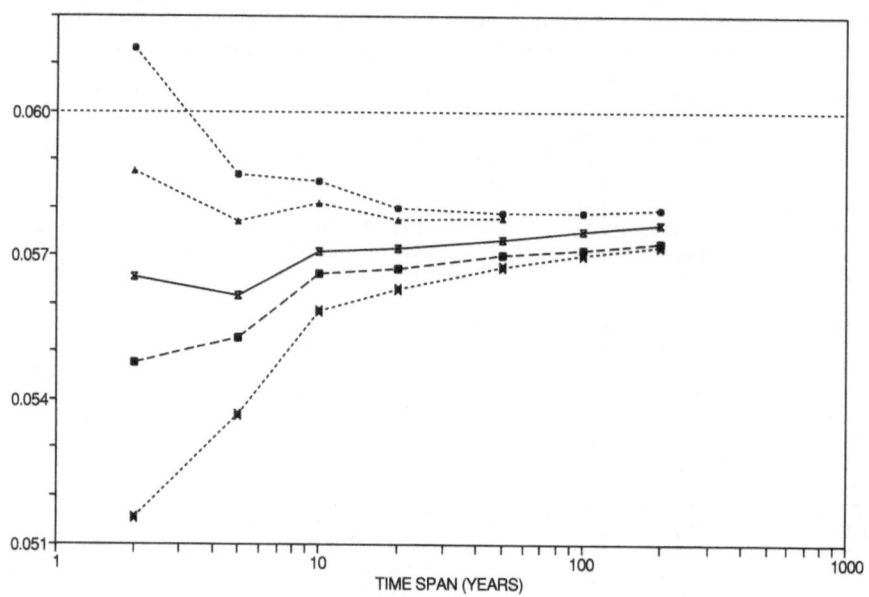

Standard deviation

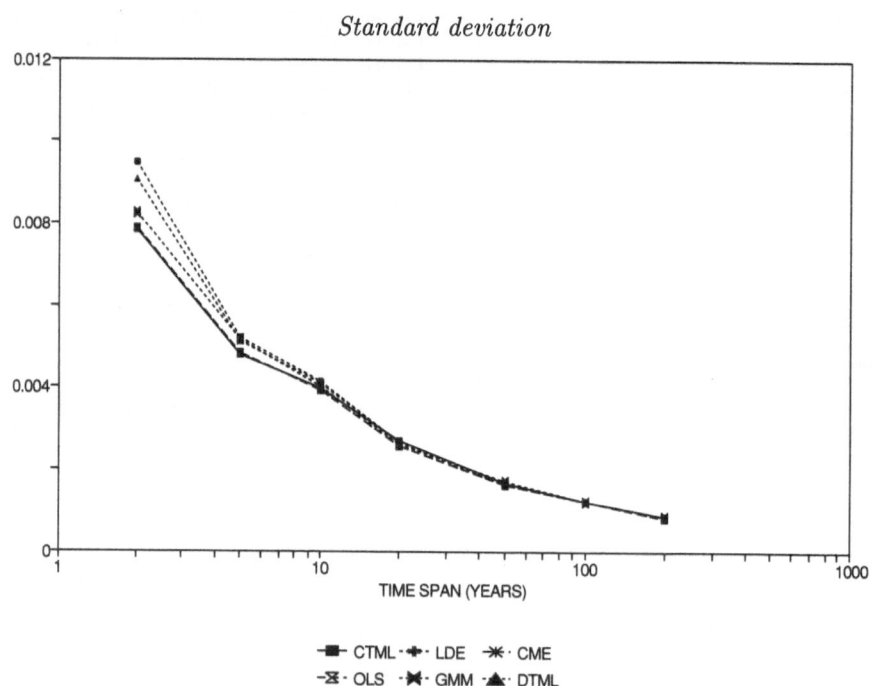

Figure 4: Mean estimated value of the speed adjustment κ

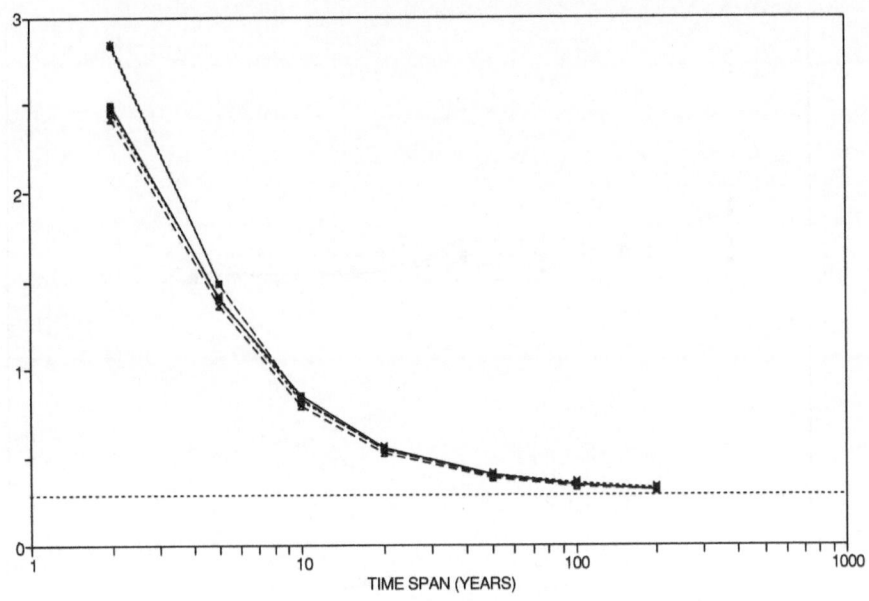

Standard deviation

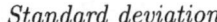

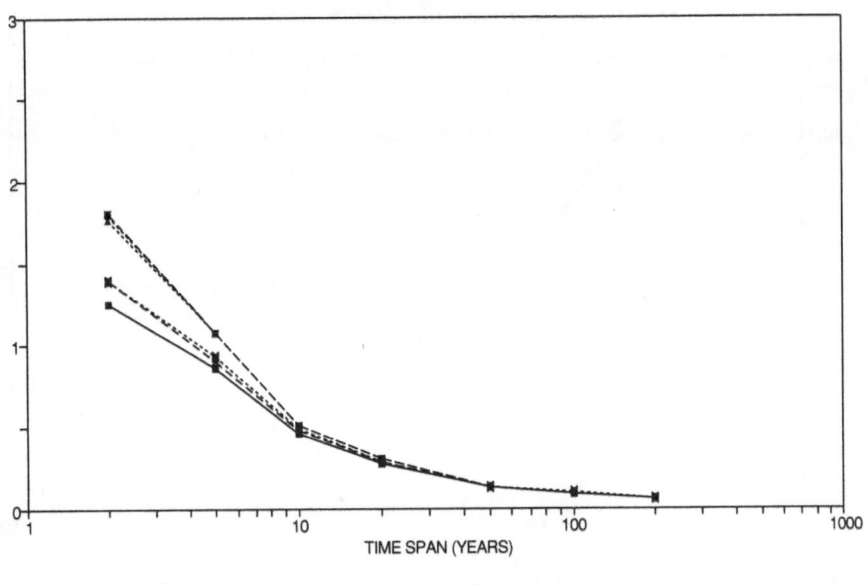

Figure 5: Mean estimated value of the long-run mean θ

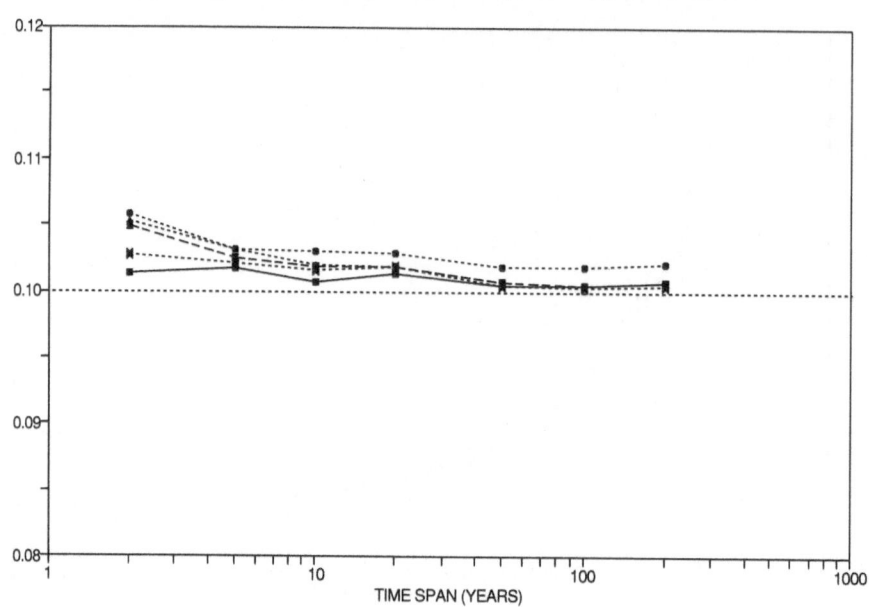

Standard deviation

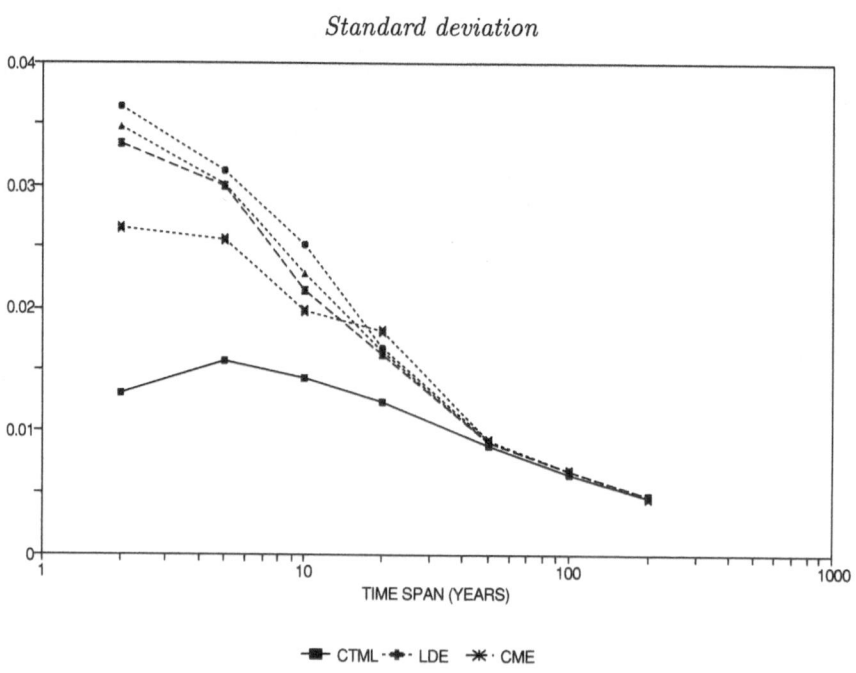

for 100 replications of a 20-year sample of monthly data the DTML requires 7530 seconds of cpu-time (on a IBM 3090) with respect to 2/100 seconds for CTML, 4/100 seconds for LDE, 7/100 seconds for CME, and 1.38 seconds for GMM. For large samples this method has been dropped.

(C) The increased frequency of observations per unit time (from monthly to daily data) for a given time-span and a given maturity τ° implies a general reduction of the bias for σ to 3–4 % as well as a reduced dispersion of results for different methods in small samples. In particular, LDE displays a slight increase of the bias (from 2.2 % to 3.4 % in samples up to 5 years) while GMM presents a large bias reduction from above 10 % to below 4 % (see Table 2 and Figures 6 to 8). LDE is still the best estimator for σ (DTML is too costly with daily data) but now the larger number of observations leads to smaller standard errors and to significant bias already in 5-year samples. Moreover, the difference with the other estimators is negligible. The best estimator for θ is again CTML, with a not significant bias of 1.6 % in 2-year samples, half as large as the bias (3.2 %) of the GMM. This one and OLS are the second-best estimators, while LDE displays again an increasing significant bias (2.8 % for a 20-year sample against 1.3 % for CTML). The effect on κ of higher frequency observations is even a worsening of the finite sample bias.

(D) In order to assess the effect of approximating $r(t)$ with $R(t, \tau^\circ)$, we have directly used monthly data of the instantaneous rate $r(t)$ (see Table 3 and Figures 9 to 11). For σ the best estimator is now DTML with a bias of about 2 % for 2-year samples and practically negligible ($-0.2\,\%$) from 5-year samples onward. Quite good results are obtained with CTML and, when samples are not too short (10 years or more), with LDE. In any case the decreasing bias is never significant. For κ all the estimators display a large bias in small samples. The same results for CTML can be found in Fournie and Talay [26]. The best estimator for θ is CTML with a bias of 1 % and a root mean square error of 14 % (2 years of data). LDE has a persistent bias (2.6 % in 20-year samples). GMM and OLS have similar performance in large samples (the root mean square error gets 9.7 % for 50- year samples). If, instead of monthly data, we use high frequency observations for $r(t)$, for example hourly observations, with a step Δ (1 hour) equal to the generation step δ, we obtain, for a 2-year sample, a root mean square error of 0.55 % for σ using LDE and of 13.8 % for θ using CTML. Even in this case, the bias and standard error for κ are quite large for every method, showing definitely that the relevant information to efficiently estimate the speed of adjustment κ can be obtained only from a large sample time-span (the range $[0, T]$ in continuous time), almost independently from the frequency of observation.

(E) To better appreciate the effect of the approximation $R(\tau^\circ) \approx r$ we have used different values for τ°: 0, 1 day, 1 week, 1 month, 3 months, 6 months, 1 year. The results for the various methods (DTML excluded) are displayed in Fig. 12.[43] For τ° equal to one month or greater, the estimation bias is increasing for σ and θ. For κ, CTML is sensitive to an increasing τ° while the other estimators are not.

---

[43]The exercise consists of 200 replications of monthly data over 10 years of time-span.

Table 2
Percentage relative bias, root mean square error and Student's t-statistic:
Simulated daily data for three-months treasury bill rates

years=1		bias			rmse			Student t	
196	σ	κ	θ	σ	κ	θ	σ	κ	θ
CTML	-3.76	1823.80	1.38	4.90	2248.97	9.88	-1.19	1.39	0.14
LDE	-3.26	1845.73	2.61	4.52	2306.06	20.15	-1.04	1.32	0.13
CME	-4.03	1844.59	2.58	5.12	2313.97	20.43	-1.27	1.32	0.13
OLS	-4.03	1821.31	2.58	5.12	2277.08	20.43	-1.27	1.33	0.13
GMM	-4.61	1888.54	3.31	5.70	2361.90	23.37	-1.37	1.33	0.14
DTML									

years=2		bias			rmse			Student t	
193	σ	κ	θ	σ	κ	θ	σ	κ	θ
CTML	-3.62	909.39	1.65	4.36	1123.21	12.70	-1.49	1.38	0.13
LDE	-3.37	934.59	4.61	4.15	1179.29	26.46	-1.39	1.30	0.18
CME	-3.78	934.22	4.31	4.49	1179.48	26.55	-1.57	1.30	0.16
OLS	-3.78	927.74	4.31	4.49	1168.77	26.55	-1.57	1.31	0.16
GMM	-4.09	963.45	3.18	4.77	1210.49	22.33	-1.66	1.31	0.14
DTML									

years=5		bias			rmse			Student t	
198	σ	κ	θ	σ	κ	θ	σ	κ	θ
CTML	-3.53	401.30	1.66	3.86	518.73	15.85	-2.27	1.22	0.11
LDE	-3.41	400.22	3.24	3.75	530.86	27.75	-2.16	1.15	0.12
CME	-3.60	398.59	2.63	3.93	529.97	26.32	-2.29	1.14	0.10
OLS	-3.60	397.07	2.63	3.93	527.39	26.32	-2.29	1.14	0.10
GMM	-3.77	425.81	2.51	4.09	549.63	23.21	-2.39	1.16	0.11
DTML									

years=10		bias			rmse			Student t	
199	σ	κ	θ	σ	κ	θ	σ	κ	θ
CTML	-3.62	187.10	0.35	3.78	245.86	14.33	-3.34	1.17	0.02
LDE	-3.61	171.39	1.69	3.76	238.35	19.75	-3.38	1.03	0.09
CME	-3.71	170.80	0.94	3.86	237.92	19.46	-3.50	1.03	0.05
OLS	-3.71	170.38	0.94	3.86	237.29	19.46	-3.50	1.03	0.05
GMM	-3.78	177.87	1.14	3.93	246.40	20.06	-3.53	1.04	0.06
DTML									

years=20		bias			rmse			Student t	
200	σ	κ	θ	σ	κ	θ	σ	κ	θ
CTML	-3.56	88.27	1.31	3.66	129.88	12.51	-4.24	0.93	0.11
LDE	-3.59	80.71	2.84	3.68	126.81	15.96	-4.37	0.83	0.18
CME	-3.66	79.90	1.83	3.75	126.39	15.47	-4.46	0.82	0.12
OLS	-3.66	79.72	1.83	3.75	126.11	15.47	-4.46	0.82	0.12
GMM	-3.71	84.40	1.73	3.79	130.70	15.17	-4.52	0.85	0.11
DTML									

years=50		bias			rmse			Student t	
200	σ	κ	θ	σ	κ	θ	σ	κ	θ
CTML	-3.49	31.87	0.42	3.53	54.10	9.04	-6.61	0.73	0.05
LDE	-3.55	28.41	1.70	3.59	52.54	9.56	-6.94	0.64	0.18
CME	-3.60	27.48	0.51	3.64	52.04	9.31	-7.06	0.62	0.06
OLS	-3.60	27.40	0.51	3.64	51.96	9.31	-7.06	0.62	0.06
GMM	-3.61	30.42	0.45	3.65	54.53	9.35	-7.03	0.67	0.05
DTML									

years=100		bias			rmse			Student t	
200	σ	κ	θ	σ	κ	θ	σ	κ	θ
CTML	-3.49	16.24	0.43	3.51	34.21	6.56	-9.02	0.54	0.07
LDE	-3.56	14.82	1.80	3.58	34.19	7.06	-9.62	0.48	0.26
CME	-3.60	13.94	0.52	3.62	33.12	6.76	-9.74	0.46	0.08
OLS	-3.60	13.89	0.52	3.62	33.07	6.76	-9.74	0.46	0.08
GMM	-3.60	15.50	0.49	3.62	35.45	6.74	-9.81	0.49	0.07
DTML									

Figure 6: Mean estimated value of the diffusion coefficient σ

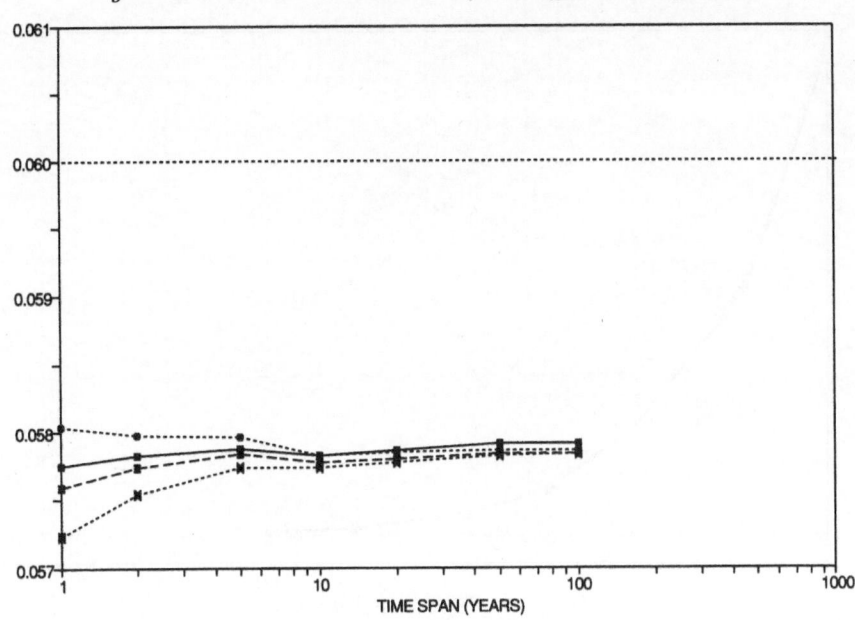

Standard deviation

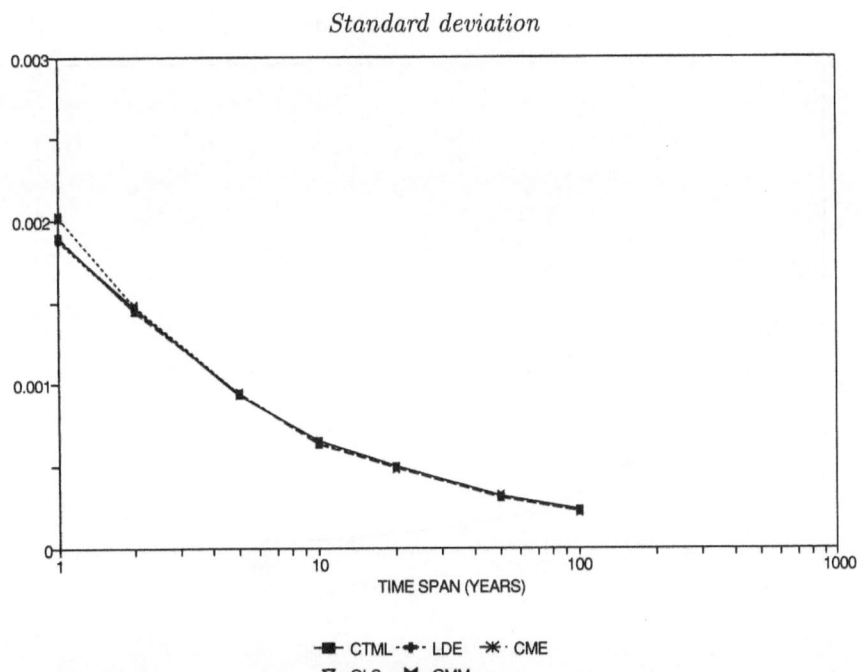

Figure 7: Mean estimated value of the speed adjustment κ

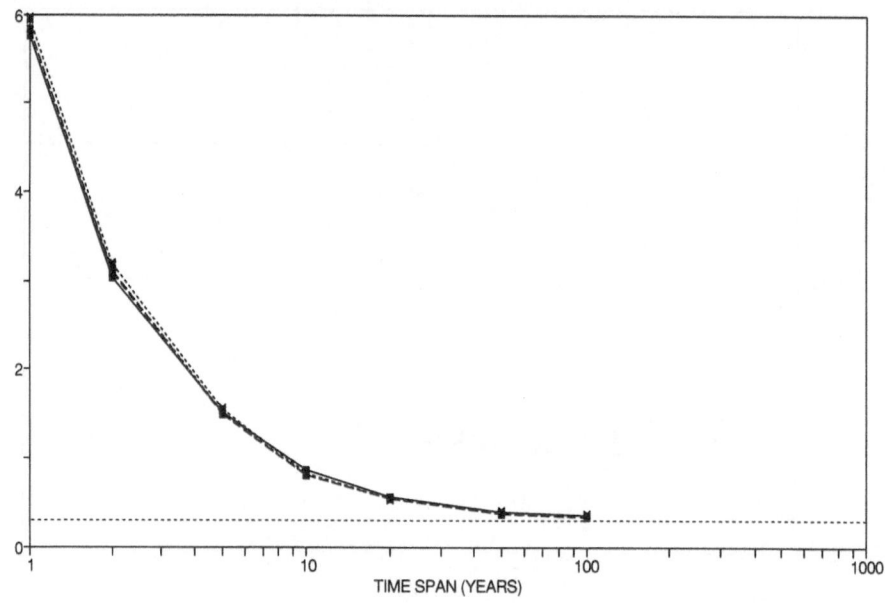

Standard deviation

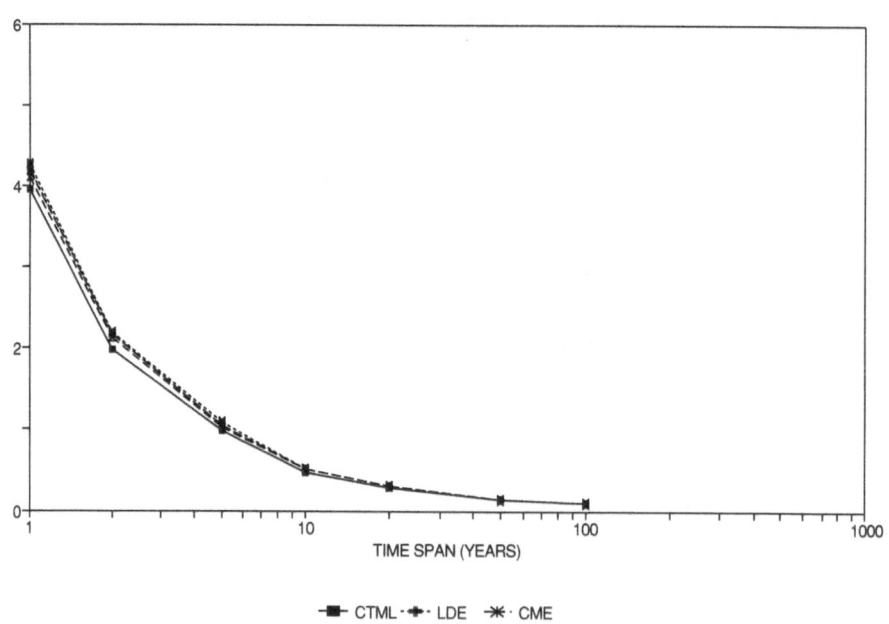

Figure 8: Mean estimated value of the long-run mean θ

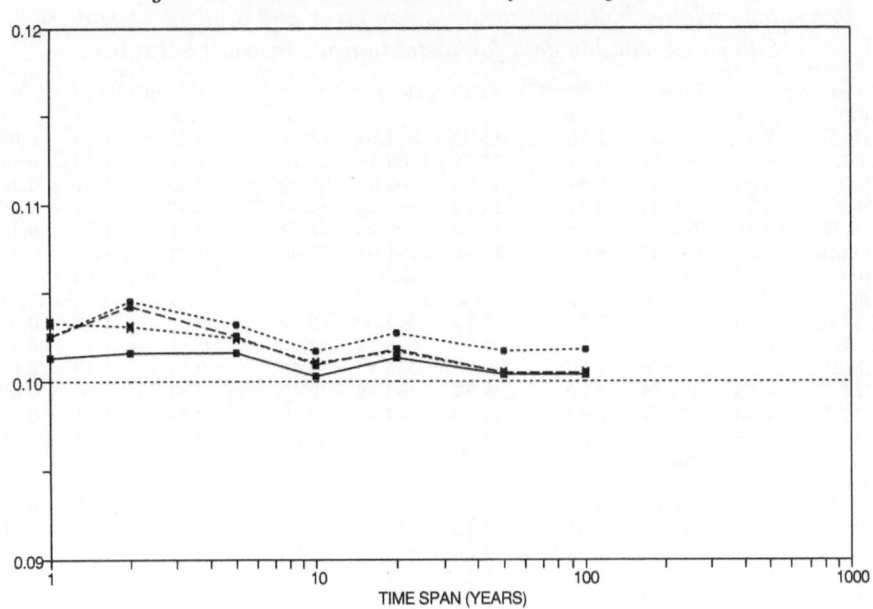

Standard deviation

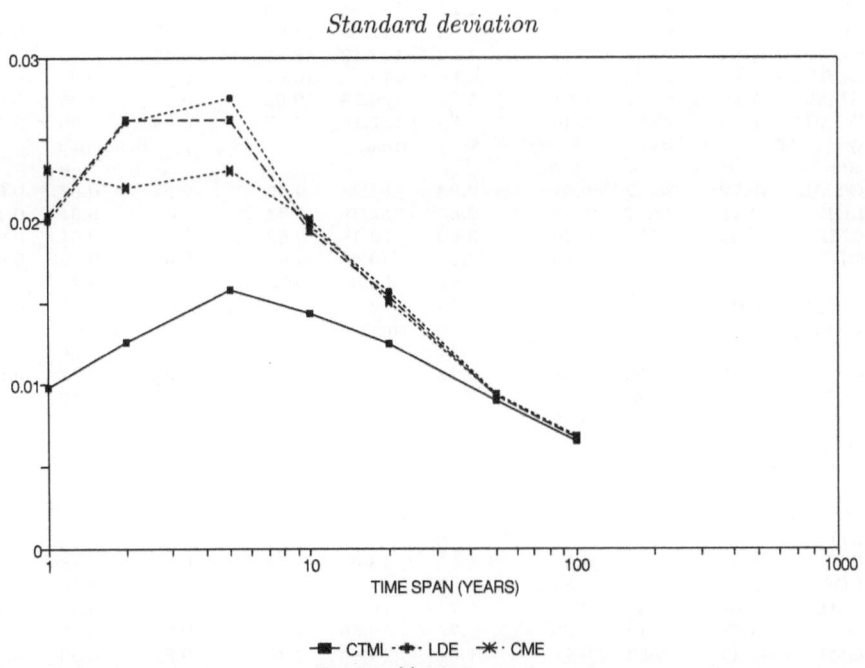

　　　　　C. BIANCHI, R. CESARI AND L. PANATTONI

Table 3
Percentage relative bias, root mean square error and Student's t-statistic: Simulated monthly data for instantaneous treasury bill rates

years=2		bias			rmse			Student t	
187	σ	κ	θ	σ	κ	θ	σ	κ	θ
CTML	-2.03	751.30	1.10	13.35	877.61	13.84	-0.15	1.66	0.08
LDE	6.70	901.56	6.27	17.79	1103.18	38.16	0.41	1.42	0.17
CME	-5.30	903.48	5.47	14.29	1105.92	35.39	-0.40	1.42	0.16
OLS	-5.30	749.39	5.47	14.29	888.34	35.39	-0.40	1.57	0.16
GMM	-10.91	764.95	3.46	17.72	899.99	27.22	-0.78	1.61	0.13
DTML	2.21	902.37	5.21	15.86	1101.91	37.04	0.14	1.43	0.14
years=5		bias			rmse			Student t	
199	σ	κ	θ	σ	κ	θ	σ	κ	θ
CTML	-2.96	363.20	1.51	8.72	462.62	16.62	-0.36	1.27	0.09
LDE	1.48	400.72	3.50	9.03	536.14	32.06	0.17	1.13	0.11
CME	-4.43	399.79	2.79	9.37	535.12	30.49	-0.54	1.12	0.09
OLS	-4.43	356.48	2.79	9.37	463.79	30.49	-0.54	1.20	0.09
GMM	-7.28	373.91	2.43	11.44	484.30	25.49	-0.82	1.21	0.10
DTML	-0.23	398.54	2.89	8.75	534.43	31.31	-0.03	1.12	0.09
years=10		bias			rmse			Student t	
198	σ	κ	θ	σ	κ	θ	σ	κ	θ
CTML	-1.40	179.09	0.13	6.96	234.73	14.92	-0.20	1.18	0.01
LDE	1.24	177.95	2.60	7.16	247.20	25.92	0.18	1.04	0.10
CME	-2.08	177.99	1.39	7.04	246.79	22.03	-0.31	1.04	0.06
OLS	-2.08	165.31	1.39	7.04	227.57	22.03	-0.31	1.06	0.06
GMM	-3.47	172.73	1.10	7.75	236.01	20.40	-0.50	1.07	0.05
DTML	0.41	177.74	1.54	7.01	247.02	23.45	0.06	1.04	0.07
years=20		bias			rmse			Student t	
200	σ	κ	θ	σ	κ	θ	σ	κ	θ
CTML	-1.35	81.67	0.98	4.84	121.94	12.93	-0.29	0.90	0.08
LDE	0.21	80.33	2.65	4.56	127.49	17.62	0.05	0.81	0.15
CME	-1.94	79.66	1.49	4.86	127.13	16.81	-0.43	0.80	0.09
OLS	-1.94	74.60	1.49	4.86	119.12	16.81	-0.43	0.80	0.09
GMM	-2.69	81.97	1.51	5.28	126.75	19.06	-0.59	0.85	0.08
DTML	-0.20	79.65	1.46	4.55	127.16	17.07	-0.04	0.80	0.09
years=50		bias			rmse			Student t	
200	σ	κ	θ	σ	κ	θ	σ	κ	θ
CTML	-0.72	28.92	0.04	2.93	51.73	9.35	-0.25	0.67	0.00
LDE	0.21	28.37	1.44	2.85	52.76	9.83	0.07	0.64	0.15
CME	-1.30	27.30	0.15	3.08	52.14	9.62	-0.46	0.61	0.02
OLS	-1.30	25.08	0.15	3.08	49.67	9.62	-0.46	0.59	0.02
GMM	-1.70	31.28	-0.16	3.42	54.53	9.70	-0.57	0.70	-0.02
DTML	0.05	27.45	0.16	2.85	52.37	9.62	0.02	0.62	0.02
years=100		bias			rmse			Student t	
200	σ	κ	θ	σ	κ	θ	σ	κ	θ
CTML	-0.70	13.96	0.08	2.22	32.70	6.80	-0.33	0.47	0.01
LDE	0.07	14.79	1.55	2.12	34.53	7.26	0.03	0.47	0.22
CME	-1.29	13.86	0.17	2.44	33.52	7.01	-0.62	0.45	0.02
OLS	-1.18	12.17	0.17	2.44	31.99	7.01	-0.62	0.41	0.02
GMM	-1.49	15.47	0.14	2.57	35.44	7.02	-0.71	0.49	-0.02
DTML									
years=200		bias			rmse			Student t	
200	σ	κ	θ	σ	κ	θ	σ	κ	θ
CTML	-0.47	5.01	0.38	1.55	18.65	4.78	-0.32	0.28	0.08
LDE	0.19	6.07	1.84	1.48	19.66	5.22	0.13	0.32	0.38
CME	-1.06	5.53	0.39	1.77	19.38	4.85	-0.75	0.30	0.08
OLS	-1.06	4.13	0.39	1.77	18.56	4.85	-0.75	0.23	0.08
GMM	-1.17	6.23	0.11	1.84	19.83	4.75	-0.82	0.33	0.02
DTML									

Figure 9: Mean estimated value of the diffusion coefficient σ

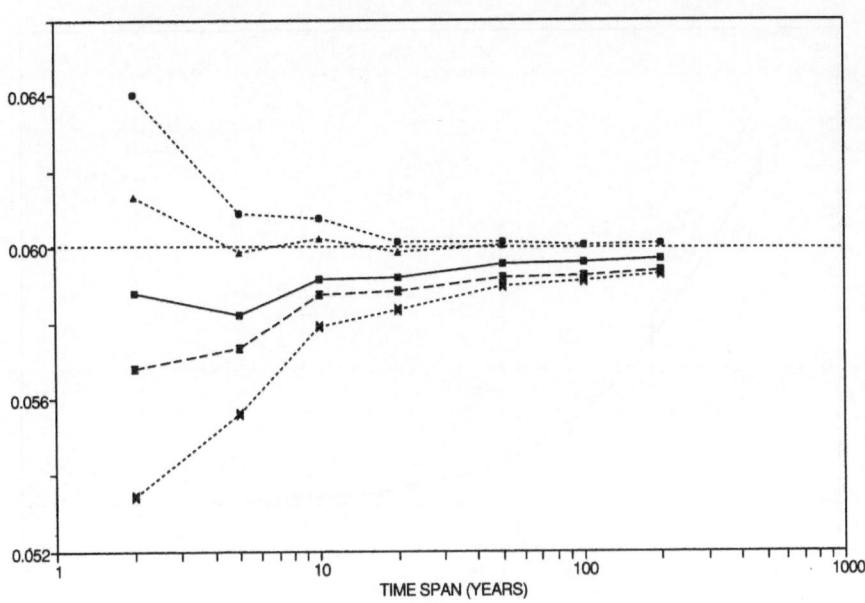

Standard deviation

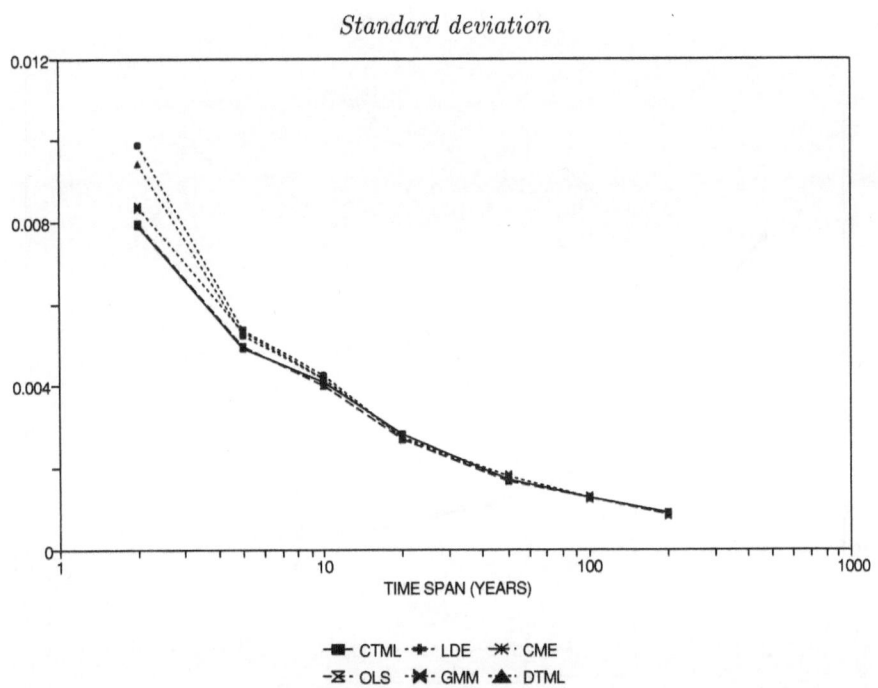

Figure 10: Mean estimated value of the speed adjustment κ

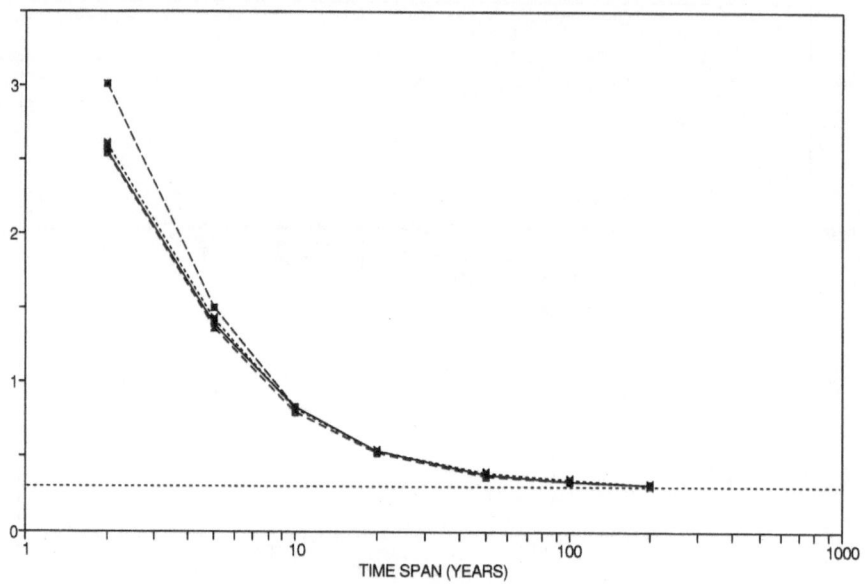

Standard deviation

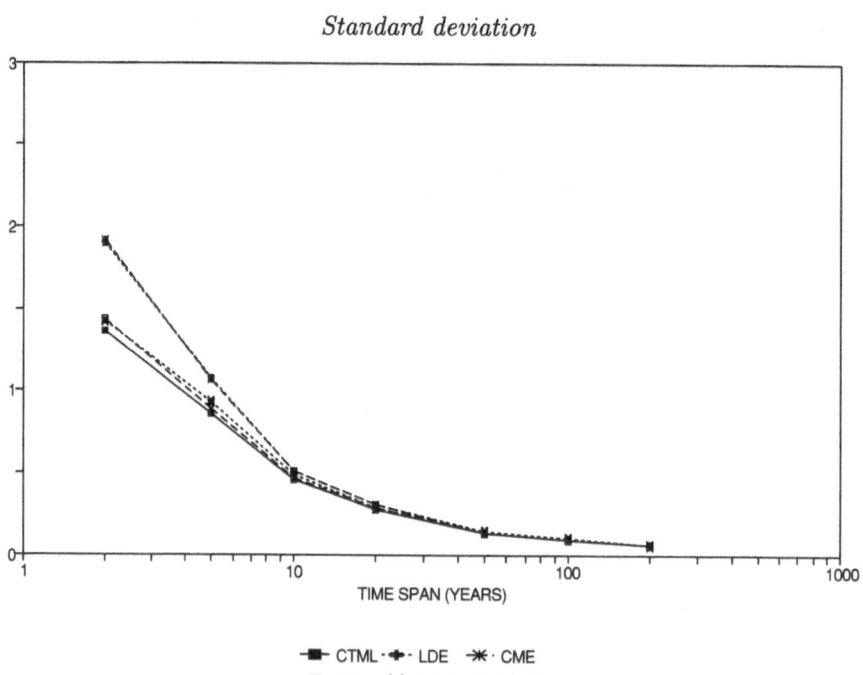

Figure 11: Mean estimated value of the long-run mean

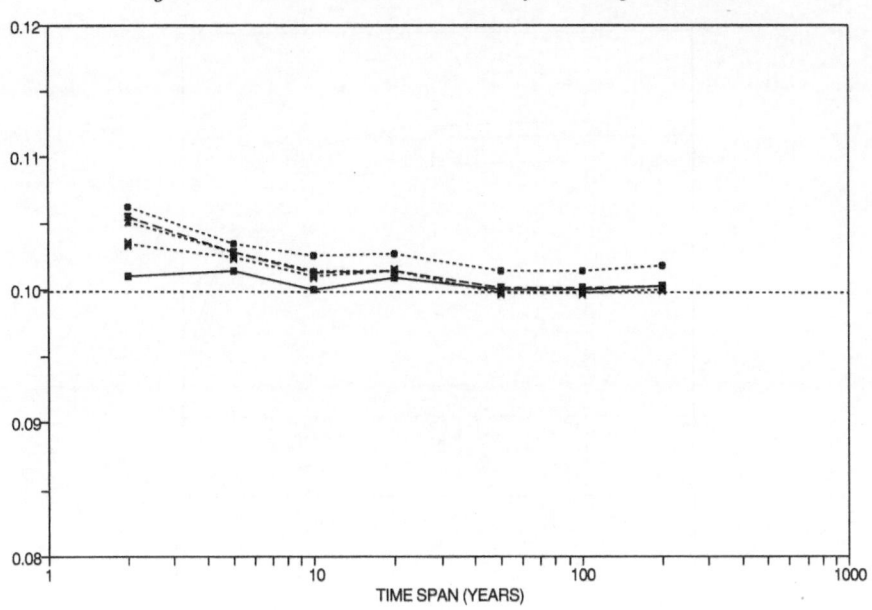

Standard deviation

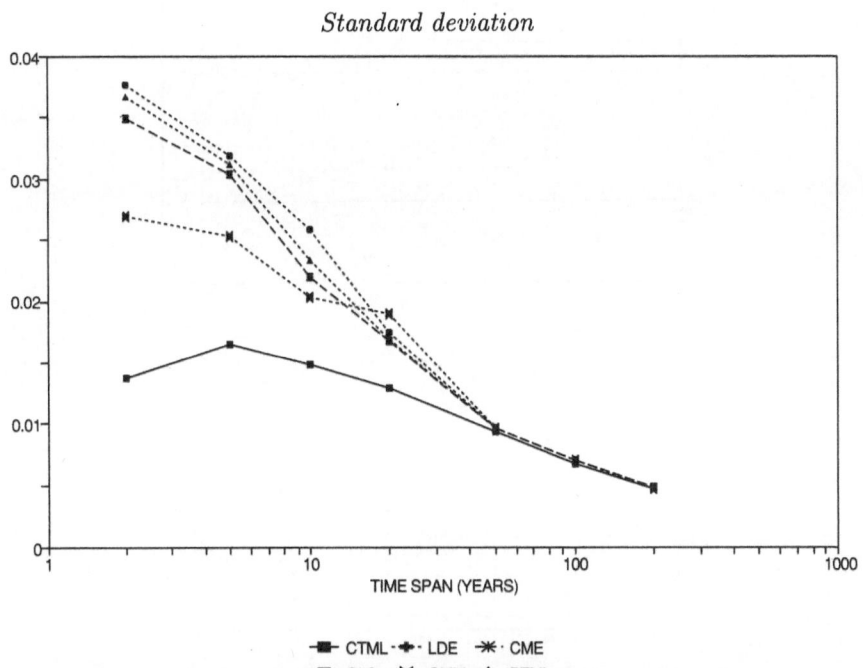

Figure 12: Effect of different maturity terms τ°

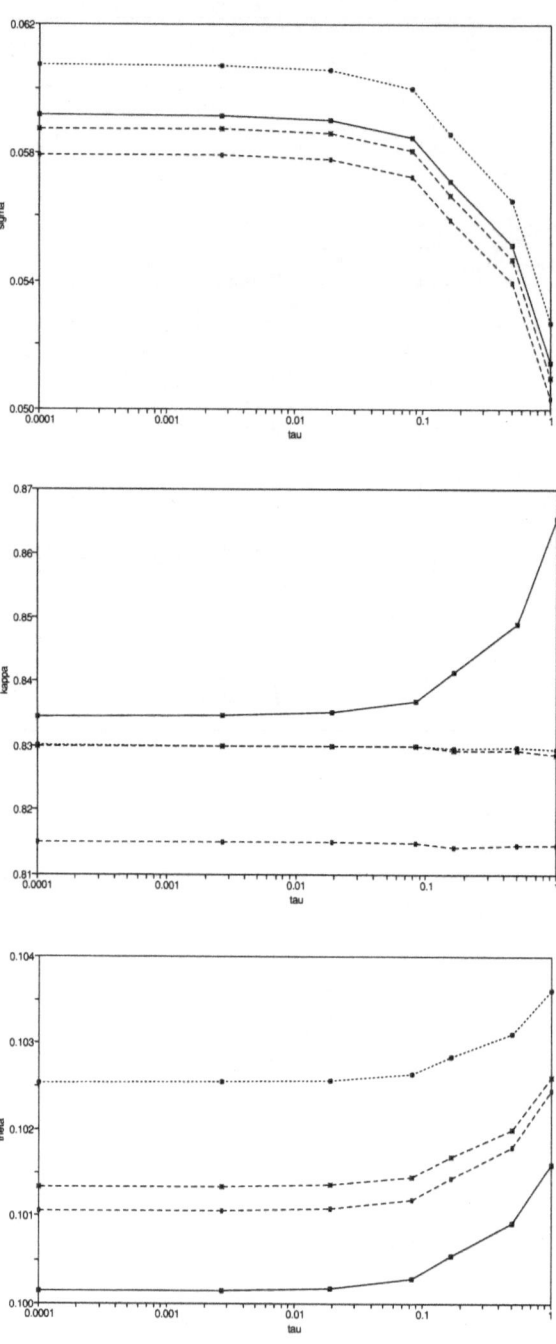

A similar exercise was made to assess the effect of the "price of risk" parameter λ (see equation (2.8)). For $\tau°$ equal to three months and $\lambda = -0.1, -0.03, 0, 0.03$ we obtained the results in Figure 13. We can see that an increasing λ has small reducing effect (i. e. increasing bias in this case) on the estimates for σ, almost no effect for κ and sensible reducing effect for θ. The simultaneous effects of $\tau°$ and λ are one-sided and reciprocally reinforcing for σ and κ while they tend to compensate for θ when the maturity $\tau°$ is relatively large.

(F) As expected, the starting point $r(0)$ has important effects in small samples. A starting point 25 % away from the long-run value implies a bias for θ as large as 18 % in 2-year samples and 12 % in 5-year samples. The estimators are not equally affected. CTML is no longer the best estimator when the starting point is out of equilibrium. For this method the bias is still at 8 % when the sample has been increased to 10 years of data. Better estimators are LDE and OLS. We also note that the effects of initial values above or below the long-run value are not symmetric. The bias is larger and more persistent starting from above θ (1.25θ) than from below (0.75θ): for LDE with a 2-year sample we have in the former case a bias of 17.3 % and in the latter case a bias of $-12{,}7$ %; with a 5-year sample the two biases are 11.8 % and -4.8 % respectively; with a 10-year sample they reach 4.1 % and -0.9 %. The estimation for σ is not affected by $r(0)$ while the identification of the speed of adjustment κ, always very poor in finite samples, seems slightly improved by a staring value out of the long-run equilibrium.

(G) Finally, given the distorsion of the considered estimators, we have investigated the effect of simulating the bond price dynamics using wrong parameters, in order to assess the financial effects (for example in a portfolio strategy) of inferential problems. We have, therefore, simulated the bond price (2.10) for a single maturity τ equal to three months, using alternatively the "true" parameter set (control series) and a different set with biased parameter values (biased series).[44] According to our previous results in the case of a 20-year time-span, σ has been reduced by -5 %, κ increased by $+100$ % and θ increased by $+2$ %. The price difference, calculated as biased price less control price, results to be small and not significant for σ and κ: after 1 month it is about 0.0000013 (1.3 \$ per million); after 12 months it reaches 0.0000096 for σ and 0.000031 (31 \$ per million) for κ. In the case of θ the difference is larger and stable over time at about -0.00049 (490 \$ per million). The negative sign indicates that increasing θ reduces the bond price. The long-run mean θ seems therefore a crucial parameter to estimate, while biased values for σ and κ have minor effects on the asset price.

§6 CONCLUSIONS

We have investigated the relative performance of alternative estimators of a popular diffusion model of the term structure of interest rates, characterized by a

---

[44]We have run 500 replications using hourly data over one year of time-span. Note that simulation is required instead of partial derivatives $\partial P/\partial\alpha$ because $r(t)$ is an endogenous function of the basic parameters.

Figure 13: Effect of the "price of risk" parameter λ

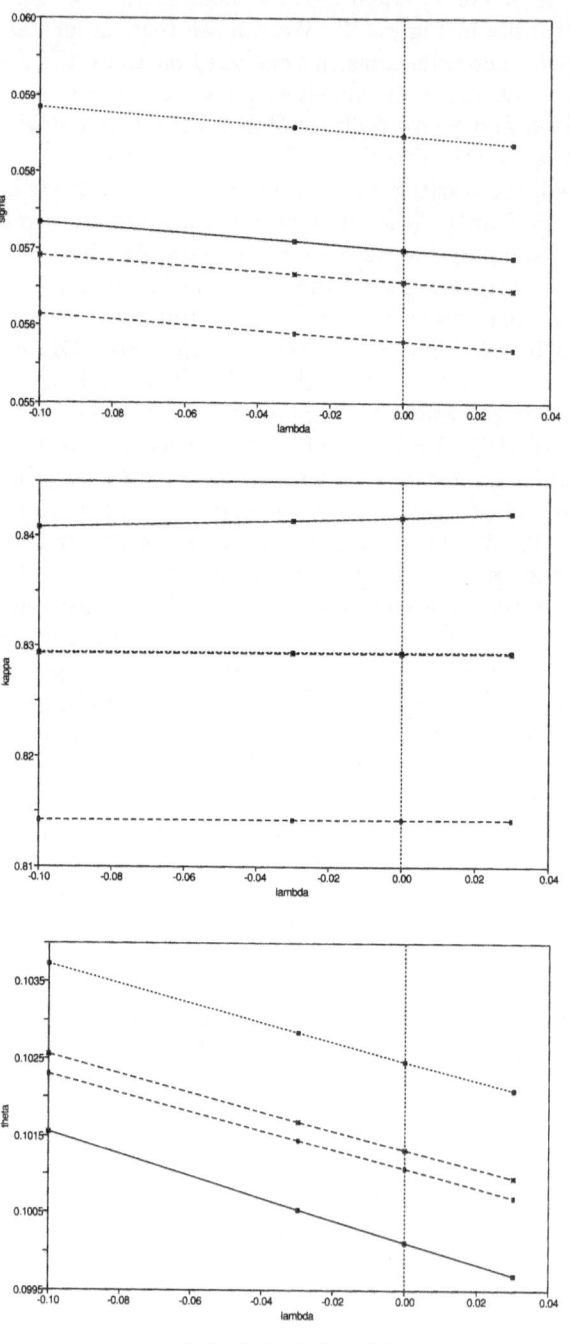

single state-variable with Feller (square root) dynamics (see Section 2). A number of different estimators, reviewed in Section 3, has been proposed in the empirical literature: the continuous-time maximum-likelihood estimator (CTML), the linearized discrete equivalent (LDE), the conditional mean estimator (CME) and the estimator of the crude discretization of the stochastic differential equation (OLS), the generalized method of moments estimator (GMM), and the discrete-time maximum-likelihood estimator (DTML). However, no general evaluation of their econometric performance and relative behaviour, particularly with small samples of data, was available.

In order to attack this problem we built a set of Monte Carlo experiments (see Section 4) designed to simulate continuous-time (diffusion) processes and replicate the typical situation of empirical applications. In particular, we set the time unit to 1 year, the generation time interval to 1 hour, the observation time frequency from 1 to 30 days, the time-span from 1 to 200 years. The state-variable parameters (κ, θ, σ) were set to values approximately equal to published empirical estimates. Different simulation schemes (Euler–Maruyama, Milstein [49], Talay [57]) were used without finding important differences in the estimators results. Particular attention has been paid to the unobservability of the state variable, the instantaneous (zero maturity) interest rate $r(t)$, and the consequences of using proxy, observable rates $R^\circ(t)$, for a given positive maturity τ°. For σ, the diffusion parameter, we have found a general downward distortion of the estimators. This asymptotic bias is due to the approximating rate R° and it would disappear if the (unobservable) instantaneous rate $r(t)$ (or a very short maturity rate: see Fig. 12) were used. Using data of higher frequency (e. g. from monthly to daily) the estimator variance is reduced, not the bias. LDE and DTML appear to be the best estimators, but the latter is computationally more demanding. The worst performance, in small samples, comes from GMM. For κ, the speed of adjustment parameter, there is no bias asymptotically but large upward bias in finite samples. No estimator clearly outperforms the others. Increasing the frequency of the observations has a worsening effect on the already bad finite sample results. It appears that to confidently estimate the speed of adjustment, a time series is needed much longer than usually available data. The parameter θ, the steady-state mean of the instantaneous rate, is in general estimated with no significant bias. The best estimator is CTML and the worst is LDE. This result, however, depends on the starting point $r(0) = \theta$. Starting values out of equilibrium reduce the relative performance of CTML. Moreover, starting values below equilibrium are more quickly absorbed than starting values above θ.

Finally, we have investigated the financial effect (in terms of bond price) of using biased parameters. Considering the typical distortion found in the case of data over 20 years of time-span, we obtain no significant effect from σ and κ but a more sensible price gap from even small bias (2 %) of θ. This parameter is therefore a crucial one to estimate but it is also estimated, by various methods, with the smaller relative bias even in small samples: a bad news and a good news as often is the case.

APPENDIX: A STOCHASTIC TAYLOR FORMULA

In this appendix we shall give an heuristic derivation of a stochastic Itô–Taylor formula (Platen and Wagner [52], Kloeden and Platen [37, Chapter 5]) which generalizes Itô's formula and is useful to understand the discrete data schemes used in the text (see §4.2).

Let us consider the one-dimensional diffusion $x(t)$ which solves the Itô stochastic integral equation

$$x(t) = x(v) + \int_v^t \alpha(x(s)) \, ds + \int_v^t \sigma(x(s)) \, dw(s). \tag{A.1}$$

Let f be a real-valued function with continuous partial derivatives f', f'', etc. Writing x_s for $x(s)$, α_s for $\alpha(x_s)$, and σ_s for $\sigma(x_s)$, Itô's formula in integral form is:

$$f(x_t) = f(x_v) + \int_v^t \left[f'(x_s)\alpha_s + \frac{1}{2} f''(x_s)\sigma_s^2 \right] ds + \int_v^t f'(x_s)\sigma_s \, dw_s$$
$$\equiv I_{t,x}(f_v). \tag{A.2}$$

For $f(x) = x$, (A.2) reduces to (A.1). For $f(x) = \alpha(x)$ and $f(x) = \sigma(x)$, we obtain two expressions for the drift and diffusion function:

$$\alpha_s = \alpha_v + \int_v^s \left[\alpha_u'\alpha_u + \frac{1}{2}\alpha_u''\sigma_u^2 \right] du + \int_v^s \alpha_u'\sigma_u \, dw_u \equiv I_{s,x}(\alpha_v),$$
$$\sigma_s = \sigma_v + \int_v^s \left[\sigma_u'\alpha_u + \frac{1}{2}\sigma_u''\sigma_u^2 \right] du + \int_v^s \sigma_u'\sigma_u \, dw_u \equiv I_{s,x}(\sigma_v) \tag{A.3}$$

so that, substituting in (A.1),

$$x_t = x_v + \alpha_v \int_v^t ds + \int_v^t \int_v^s \left[\alpha_u'\alpha_u + \frac{1}{2}\alpha_u''\sigma_u^2 \right] du \, ds + \int_v^t \int_v^s \alpha_u'\sigma_u \, dw_u \, ds$$
$$+ \sigma_v \int_v^t dw_s + \int_v^t \int_v^s \left[\sigma_u'\alpha_u + \frac{1}{2}\sigma_u''\sigma_u^2 \right] du \, dw_s$$
$$+ \int_v^t \int_v^s \sigma_u'\sigma_u \, dw_u \, dw_s. \tag{A.4}$$

This is the simplest stochastic Taylor formula for $x(t)$ and could be written as

$$x_t = x_v + \alpha_v \int_v^t ds + \sigma_v \int_v^t dw_s + R_1 \tag{A.5}$$

where R_1 is the remainder. This formula suggests the Euler scheme given in the text. More elaborate expressions can be obtained by substituting again the integrand functions in (A.4) with their Itô expansions $I_{u,x}$, obtaining expressions with deterministic and stochastic triple integrals. In particular:

$$
\begin{aligned}
x_t = v_x + \alpha_v \int_v^t ds &+ \sigma_v \int_v^t dw_s + \sigma_v' \sigma_v \int_v^t \int_v^s dw_u \, dw_s \\
&+ \left[\alpha_v' \alpha_v + \frac{1}{2} \alpha_v'' \sigma_v^2 \right] \int_v^t \int_v^s du \, ds \\
&+ \alpha_v' \sigma_v \int_v^t \int_v^s dw_u \, ds + \left[\sigma_v' \alpha_v + \frac{1}{2} \sigma_v'' \sigma_v^2 \right] \int_v^t \int_v^s du \, dw_s + R_2.
\end{aligned}
\tag{A.6}
$$

Using Itô's formula for w_t^2, we notice that

$$
\int_v^t \int_v^s dw_u \, dw_s = \frac{(w_t - w_v)^2 - (t - v)}{2}
\tag{A.7}
$$

which is the term in the Milstein scheme. Moreover, using Itô's formula for tw_t,

$$
tw_t = vw_v + \int_v^t w_s \, ds + \int_v^t s \, dw_s,
\tag{A.8}
$$

we have

$$
\int_v^t \int_v^s dw_u \, ds = (t - v) \int_v^t dw_s - \int_v^t \int_v^s du \, dw_s
\tag{A.9}
$$

with

$$
\begin{aligned}
\mathrm{E}\left(\int_v^t \int_v^s du \, dw_s \right) &= 0, \\
\mathrm{Var}\left(\int_v^t \int_v^s du \, dw_s \right) &= \frac{1}{3}(t - v)^3, \\
\mathrm{Cov}\left(\int_v^t dw_s, \int_v^t \int_v^s du \, dw_s \right) &= \frac{1}{2}(t - v)^2,
\end{aligned}
\tag{A.10}
$$

so that the two double integrals have the same mean, variance and covariance with the Wiener increments. The last two integrals in (A.6) justify the last component in the Talay scheme. The remainder of the stochastic Taylor formula allows to assess the order of strong and weak convergence of different approximation schemes. Using the same token to (A.2), we obtain a stochastic Taylor formula for $f(x_t)$ as an extension of Itô's formula.

References

1. L. Arnold, *Stochastic Differential Equations: Theory and Applications*, Wiley, New York, 1974.
2. G. Bamberg and K. Spremann (eds.), *Risk and Capital*, Lecture Notes in Economics and Mathematical Systems **227** (1984), Springer, Berlin.
3. G. Banon, *Nonparametric identification for diffusion processes*, SIAM Journal of Control and Optimization **16, 3** (1978), 380–395.
4. E. Barone and R. Cesari, *Rischio e rendimento dei titoli a tasso fisso e a tasso variabile in un modello stocastico univariato*, Temi di discussione, vol. 73, Banca d'Italia, 1986.
5. E. Barone and D. Cuoco, *La Valutazione delle Obbligazioni e delle Opzioni su Obbligazioni*, Quaderni di Ricerca N.10, LUISS, Roma, 1991.
6. E. Barone, D. Cuoco, and E. Zautzik, *La struttura dei rendimenti per scadenza secondo il modello di Cox, Ingersoll e Ross: una verifica empirica*, Temi di discussione, vol. 128, Banca d'Italia, 1989.
7. Y. Bar-Shalom, *On the asymptotic properties of the maximum-likelihood estimate obtained from dependent observations*, Journal of the Royal Statistical Society, series B **33, 1** (1971), 72–77.
8. B. R. Bhat, *On the method of maximum-likelihood for dependent observations*, Journal of the Royal Statistical Society, series B **36, 1** (1974), 48–53.
9. F. Black and M. Scholes, *The pricing of options and corporate liabilities*, Journal of Political Economy **781, 3** (1973), 637–654.
10. M. J. Brennan and E. S. Schwartz, *A continuous-time approach to the pricing of bonds*, Journal of Banking and Finance **3** (1979), 133–155.
11. M. J. Brennan and E. S. Schwartz, *An equilibrium model of bond pricing and a test of market efficiency*, Journal of Financial and Quantitative Analysis **17, 3** (1982), 301–329.
12. R. H. Brown and S. M. Schaefer, *The term structure of real interest rates and the Cox, Ingersoll & Ross model*, London Business School, mimeo, 1988.
13. S. J. Brown and P. H. Dybvig, *The empirical implications of the Cox, Ingersoll, Ross theory of the term structure of interest rates*, Journal of Finance **41, 3** (1986), 617–630.
14. K. C. Chan, G. A. Karolyi, F. A. Longstaff, and A. B. Sanders, *An empirical comparison of alternative models of the short-term interest rate*, Journal of Finance **47, 3** (1992), 1209–1227.
15. J. C. Cox, J. E. Ingersoll, and S. A. Ross, *Duration and the measurement of basis risk*, Journal of Business **52, 1** (1979), 51–61.
16. J. C. Cox, J. E. Ingersoll, and S. A. Ross, *A reexamination of traditional hypotheses about the term structure of interest rates*, Journal of Finance **36, 4** (1981), 769–799.
17. J. C. Cox, J. E. Ingersoll, and S. A. Ross, *An intertemporal general equilibrium model of asset prices*, Econometrica **53, 2** (1985), 363–384.
18. J. C. Cox, J. E. Ingersoll, and S. A. Ross, *A theory of the term structure of interest rates*, Econometrica **53, 2** (1985), 385–407.
19. M. De Felice and F. Moriconi, *La teoria dell'immunizzazione finanziaria. Modelli e strategie*, Il Mulino, Bologna, 1991.
20. S. Dobson, R. Sutch, and D. Vanderford, *An evaluation of alternative empirical models of the term structure of interest rates*, Journal of Finance **31** (1976), 1035–1065.
21. L. U. Dothan, *On the term structure of interest rates*, Journal of Financial Economics **6** (1978), 59–69.
22. P. D. Feigin, *Maximum-likelihood estimation for continuous-time stochastic processes*, Advances in Applied Probability **8** (1976), 712–736.
23. W. Feller, *Two singular diffusion problems*, Annals of Mathematics **54, 1** (1951), 173–182.
24. E. O. Fischer and J. Zechner, *Diffusion process specifications for interest rates*, Bamberg and Spremann (eds.), 1984, pp. 64–73.
25. I. Fisher, *The Theory of Interest*, Macmillan, London, 1930.

26. E. Fournie and D. Talay, *Application de la statistique des diffusions à un modèle de taux d'intérêt*, Inria, mimeo (1992) (to appear in Finance).

27. A. Friedman, *Stochastic Differential Equations and Applications*, vol. 1, Academic Press, New York, 1975.

28. B. M. Friedman and F. H. Hahn (eds.), *Handbook of Monetary Economics*, North-Holland, 2 voll., Amsterdam, 1990.

29. M. R. Gibbons and K. Ramaswamy, *The term structure of interest rates: empirical evidence*, The Wharton School, mimeo, 1986.

30. L. P. Hansen, *Large sample properties of generalized method of moments estimators*, Econometrica **50**, 4 (1982), 1029–1054.

31. J. M. Harrison and D. M. Kreps, *Martingale and arbitrage in multiperiod securities markets*, Journal of Economic Theory **20** (1979), 381–408.

32. J. M. Harrison and S. R. Pliska, *Martingales and stochastic integrals in the theory of continuous trading*, Stochastic Processes and their Applications **11** (1981), 215–260.

33. J. R. Hicks, *Value and Capital*, Oxford University Press, Oxford, 1939 and 1946.

34. N. Ikeda and S. Watanabe, *Stochastic Differential Equations and Diffusion Processes*, North-Holland, Amsterdam, 1981.

35. S. Karlin and J. McGregor, *Classical diffusion processes and total positivity*, Journal of Mathematical Analysis and Applications **1** (1960), 163–183.

36. S. Karlin and H. M. Taylor, *A Second Course in Stochastic Processes*, Academic Press, San Diego, 1981.

37. P. E. Kloeden and E. Platen, *Numerical Solutions of Stochastic Differential Equations*, Springer, Berlin, 1992.

38. H. Korezlioglu, G. Mazziotto and J. Szpirglas (eds.), *Filtering and Control of Random Processes*, Lecture Notes in Control and Information Sciences, 61, Springer, Berlin, 1984.

39. A. Yu Kutoyants, *Estimation of the trend parameter of a diffusion process in the smooth case*, transl. by Durri-Hamdani, Theory of Probability and its Applications **23**, 3 (1978), 399–405.

40. A. Le Breton, *On continuous and discrete sampling for parameter estimation in diffusion type processes*, Mathematical Programming Study **5** (1976), 124–144.

41. R. S. Liptser and A. N. Shiryayev, *Statistics of Random Processes, I, II*, transl. (1978), Springer, Berlin, 1974.

42. A. W. Lo, *Statistical tests of contingent claims asset-pricing models: a new methodology*, Journal of Financial Economics **17** (1986), 143–174.

43. A. W. Lo, *Maximum-likelihood estimation of generalized Itô processes with discretely sampled data*, Econometric Theory **4** (1988), 231–247.

44. T. A. Marsh and E. R. Rosenfeld, *Stochastic processes for interest rates and equilibrium bond prices*, Journal of Finance **38**, 2 (1983), 635–646.

45. G. Maruyama, *Continuous Markov processes and stochastic equations*, Rendiconti del Circolo Matematico di Palermo **4** (1955), 48–90.

46. R. Masera, *The term structure of interest rates. An expectations model tested on post-war Italian data*, Oxford University Press, Oxford, 1972.

47. I. W. McKeague, *Estimation for diffusion processes under misspecified models*, Journal of Applied Probability **21** (1984), 511–520.

48. E. J. McShane, *Unified Integration*, Academic Press, New York, 1983.

49. G. N. Milstein, *Approximate integration of stochastic differential equations*, Theory of Probability and its Applications **19** (1974), 557–562.

50. F. Modigliani and R. Sutch, *Innovations in interest rate policy*, American Economic Review **56**, 2 (1966), 178–197.

51. W. K. Newey and K. D. West, *A simple, positive semi-definite, heteroskedasticity and autocorrelation consistent covariance matrix*, Econometrica **55**, 3 (1987), 703–708.

52. E. Platen and W. Wagner, *On a Taylor formula for a class of Itô processes*, Probability and Mathematical Statistics **3** (1982), 37–51.

53. P. M. Robinson, *Estimation of a time series model from unequally spaced data*, Stochastic Processes and their Applications **6** (1977), 9–24.

54. J. D. Sargan, *Some discrete approximations to continuous time stochastic models*, Journal of the Royal Statistical Society, Series B **36** (1974), 74–90.

55. R. J. Shiller, *The term structure of interest rates*, in Friedman and Hahn (eds.), Ch. 13, 1990.

56. A. N. Shiryayev, *Martingales: recent developments, results, and applications*, International Statistical Review **49** (1981), 199–233.

57. D. Talay, *Efficient numerical schemes for the approximation of expectations of functionals of stochastic differential equations*, in Korezlioglu, Mazziotto, and Szpirglas (eds.), 1984, pp. 294–313.

58. D. Talay, *Simulation and numerical analysis of stochastic differential systems: a review*, Inria, Report 1313, mimeo, 1990.

59. O. A. Vasicek, *An equilibrium characterization of the term structure*, Journal of Financial Economics **5** (1977), 177–188.

60. E. Wong and B. Hajek, *Stochastic Processes in Engineering Systems*, Springer, New York, 1985.

61. A. M. Yaglom, *An Introduction to the Theory of Stationary Random Functions,y* 1952[1], translated from Russian and edited by R. A. Silverman, Dover, New York, 1962.

Progress in Probability, Vol. 36
© 1995 Birkhäuser Verlag Basel/Switzerland

BACKWARD STOCHASTIC DIFFERENTIAL EQUATIONS.
OPTION HEDGING UNDER ADDITIONAL COST

RAINER BUCKDAHN

ABSTRACT. In a general incomplete model of a financial market we hedge contingent claims by using trading strategies with a small riskyness and consider additional cost which can come, e. g., from greater interests for borrowing and taxes on wealth. The explicit form of the optimal strategy minimizing locally riskyness (in Schweizer's sense) and the associated price of the contingent claim is determined in the Markovian case, first for a small investor, and then for a greater investor who influences the stock price.

§1 INTRODUCTION

In this paper we study a general stochastic model of continuous trading. The stock price process is a martingale and the model of the financial market is incomplete. Our aim is to hedge contingent claims by using trading strategies with a small riskyness, where the cost process accociated to the trading strategies includes, e. g., greater interests for borrowing, taxes on wealth and the possibility of payments of dividends. This paper is strongly motivated by the works of Pardoux/Peng [10], El Karoui/Quenez [5] and Schweizer [12] and bases on common work of the author with E. Pardoux.

In Section 1 we introduce our general model for option trading, recall the notation and terminology and review previous results on the existence and uniqueness of the optimal trading strategy and the associated price process of the contingent claim for preparing the following sections.

In Section 2, reviewing briefly the main result of Buckdahn/Pardoux [3] for a Markovian model of a financial market, we consider a stock price process given as the solution of a stochastic differential equation driven by a Wiener process and a Poisson random measure, and we derive the explicit form of the optimal trading strategy and the associated price of the contingent claim.

While in Section 2 the investor is assumed to be "small", i. e., he is supposed not to influence the stock price process, he is allowed in Section 3 to influence the stock price process via the value of his portfolio. The computation of the

1991 *Mathematics Subject Classification.* 60H10, 60H30, 60J60.

Key words and phrases. Backward stochastic differential equation, forward-backward stochastic differential equation, option hedging, incomplete market, contingent claim, locally risk-minimizing trading strategy.

optimal trading strategy and the associated price processes for the stock and for the contingent claim lead to forward–backward stochastic differential equations (FBSDE). Existence and uniqueness of nodal solutions of such equations are shown under Lipschitz conditions on the coefficients of the model. The technique of the proof we present works only in the one-dimensional case, but, on the other hand, since we do not need a priori the solution of the parabolic partial differential equation associated to the FBSDE in order to construct the nodal solution, we have not to require smoothness of the coefficients of the model assumed in works on the multi-dimensional case by Ma/Protter/Yong [8] and Duffie/Ma/Yong [4].

§2 The basic model

This section has two purposes. First of all we describe our model for option trading, recall the notation and the terminology and, after this, we review a previous result in order to prepare the subsequent development.

We consider a model of a financial market whose basic securities consist of two assets, the only assets available to agents for trading. One of these two assets is non-risky (the bond), the other asset is risky (the stock). Assuming discountation of the price processes with respect to the interest rate paid on the bond, we have a price per unit of bond which is identically equal to one. Let the *price* $X = (X_t)_{0 \le t \le T}$ $(T \in \mathbb{R}_+)$ of one share *of the stock* be modelled by a square-integrable martingale defined over some complete probability space $(\Omega, \mathcal{F}, P)$ and adapted to some filtration $(\mathcal{F}_t)_{0 \le t \le T}$ satisfying the usual conditions on right-continuity and completeness, cf. [7], [12].

A *trading strategy* φ is a pair of processes $\xi = (\xi_t)_{0 \le t \le T}$ and $\eta = (\eta_t)_{0 \le t \le T}$ with the following properties:

(i) ξ is predictable and $\mathrm{E}\left[\int_0^T \xi_s^2 \, d\langle X \rangle_s\right] < \infty$,

(ii) η is an adapted process and such that

$$V_t^\varphi = \xi_t \cdot X_t + \eta_t, \qquad 0 \le t \le T,$$

is a right-continuous process with $V_t^\varphi \in L^2(\Omega, \mathcal{F}, P)_{0 \le t \le T}$.

In accordance with the usual terminology, the process $V^\varphi = (V_t^\varphi)_{0 \le t \le T}$ will be called the *value of the portfolio* (of an investor using the trading strategy φ) and the process

$$C_t^\varphi = V_t^\varphi - \int_0^t \xi_s \, dX_s + \int_0^t g_s(V_s, \xi_s, R_s(C^\varphi)) \, dA_s, \qquad 0 \le t \le T, \qquad (1.1)$$

the *cost process* counting up the cost of the trading strategy φ till time t. Here

(iii) $g : [0, T] \times \Omega \times \mathbb{R}^3 \to \mathbb{R}$ is an optional function which is Lipschitz continuous in (v, z, r), uniformly with respect to $(t, \omega) \in [0, T] \times \Omega$, and such that $g(\cdot, \cdot, 0, 0, 0)$ is bounded,

(iv) A is a finite variation process which is absolutely continuous with respect to $[X] + \langle X \rangle$ and such that the process

$$\left(\frac{dA_t}{d([X]_t + \langle X \rangle_t)} \right)_{0 \le t \le T}$$

is bounded, and

(v) $R_t(C^\varphi)$ denotes the measure of the risk associated to the trading strategy φ,

$$R_t(C^\varphi) = (\mathrm{E}[(C_T^\varphi - C_t^\varphi)^2 \,|\, \mathcal{F}_t])^{1/2}, \quad 0 \le t \le T,$$

introduced by Föllmer/Sondermann [7].

Formula (1.1) expresses the fact that the cumulative cost up to time t decomposes into the sum of the current value of the portfolio V_t^φ minus the accumulated gains

$$\int_0^t \xi_s \, dX_s$$

and additional cost coming, for instance, from greater interests for borrowing, taxes on wealth or payments of dividends depending on the value of the portfolio and the risk of the used trading strategy.

Finally, let $B \in L^2(\Omega, \mathcal{F}, P)$ be a *contingent claim*, i.e., a random payoff at maturity date $T > 0$. A trading strategy duplicating the contingent claim B till time T, $V_T^\varphi = B$, is called mean-self-financing, if the cost process C^φ is a square-integrable martingale ($C^\varphi \in \mathcal{M}(P)$). We say that such a mean-self-financing trading strategy φ is *optimal*, if $\langle C^\varphi, X \rangle = 0$. Concerning the interpretation, cf. [7] and [12]. In particular, if essentially the assumptions (X.1)–(X.5) in [12] are satisfied for $X \in \mathcal{M}(P)$ and the finite variation process A, then the trading strategy φ is optimal, if and only if it minimizes locally the risk – we refer to [12] for the definition of the local risk minimization and the proof of Proposition 2.3, whose arguments work also in our situation here.

We are interested in the determination of the *optimal price* $\pi(B)$ which is to be paid at time $t = 0$ for the contingent claim B, i.e., that initial value of a portfolio for which there is an optimal trading strategy φ duplicating B till maturity time T. Using equation (1.1) we get the backward stochastic differential equation (BSDE)

$$dV_t = -g_t(V_t, \xi_t, R_t(C)) \, dA_t + \xi_t \cdot dX_t + dC_t, \quad 0 \le t \le T,$$
$$V_T = B. \tag{1.2}$$

A triple (V, ξ, C) is said to be a solution of the BSDE (1.2), if V is a semimartingale with

$$\sup_{0 \le t \le T} |V_t| \in L^2(\Omega, \mathcal{F}, P),$$

ξ is a predictable process with $\mathrm{E}[\int_0^T \xi_s^2 \, d\langle X \rangle_s] < \infty$, and $C \in \mathcal{M}^2(P)$ with $\langle X, C \rangle = 0$.

Proposition 1. *(cf. [2], Theorem 2.1): Suppose that, in addition to the assumptions on g, A, X, and B made above, it holds*

  (i) $[X]_T \in L^\infty(\Omega, \mathcal{F}, P)$, *or*

  (ii) $dA_t \ll d\langle X\rangle_t$ *and* $\langle X\rangle_T \in L^\infty(\Omega, \mathcal{F}, P)$.

Then the BSDE (1.2) *has a unique solution* (V, ξ, C).

Finally, having the unique solution (V, ξ, C) of the BSDE (1.2), we easily see that

$$\varphi_t = (\xi_t, \eta_t), \quad \eta_t = V_t - \xi_t \cdot X_t, \quad \pi(B) = V_0$$

define the optimal trading strategy and the optimal price of the contingent claim we have looked for, and C is the cost process associated to the strategy φ.

§3 THE MARKOVIAN MODEL FOR A SMALL INVESTOR

The aim of this section is to compute explicitly the unique solution (V, ξ, C) of the BSDE (1.2) in the Markovian case, and so the optimal price $\pi(B)$ and the optimal trading strategy φ for a small investor, i. e., for a holder of the portfolio who does not influence the stock price by his trading.

Let $W = (W_t)_{0 \le t \le T}$ be a Wiener process and $\mu = \{\mu(dt\,dy), (t, y) \in [0, T] \times E\}$, $E = \mathbb{R} \setminus \{0\}$, be a Poisson random measure defined over the complete probability space $(\Omega, \mathcal{F}, P)$, and suppose that, for a certain square-integrable Borel measure λ on E, the compensator ν of μ has the form $\nu(dt\,dy) = dt\,\lambda(dy)$. By $(\mathcal{F}_t)_{0 \le t \le T}$ we denote the smallest filtration with the usual conditions on right-continuity and completeness to which W and μ are adapted. Let $\sigma \in C^3_{\ell, b}(\mathbb{R})$ and $\beta : \mathbb{R} \times E \to \mathbb{R}$ be a measurable function with $\beta(\cdot, y) \in C^3_{\ell, b}(\mathbb{R})$ for every $y \in E$, which is such that, for some real K,

$$|\beta(0, y)| \le K(1 \wedge |y|), \quad y \in E,$$

$$\left|\frac{\partial^k}{\partial x^k}\beta(x, y)\right| \le K(1 \wedge |y|), \quad x \in \mathbb{R}, y \in E, 1 \le k \le 3,$$

and assume that, for some $\varepsilon > 0$,

$$\sigma(x)^2 + \int_E \beta(x, y)^2 \lambda(dy) \ge \varepsilon, \quad \text{for all } x \in \mathbb{R}.$$

Suppose that the price $X = (X_t)$ of one share of the stock can be modelled by the stochastic differential equation

$$dX_t = \sigma(X_t)\,dW_t + \int_E \beta(X_{t-}, y)\,\tilde{\mu}(dt\,dy), \quad 0 \le t \le T,$$

$$X_0 = x \in \mathbb{R}, \tag{2.1}$$

driven by the Wiener process and the compensated Poisson random measure $\tilde{\mu} = \mu - \nu$. Concerning the proof of existence and uniqueness of the solution of this equation as well as its properties, we refer to [6].

Assume that the function $g_t(\omega, v, z, r)$ introduced in Section 1 has the form

$$g_t(\omega, v, z, r) = f(X_t(\omega), v, z), \quad f \in C^3_{\ell,b}(\mathbb{R}^3),$$

and consider a contingent claim $B = h(X_t)$, $h \in C^3_p(\mathbb{R})$. Following [3] we can state

Proposition 2. *Let* $u \in C^{1,2}([0,T] \times \mathbb{R})$ *be the unique solution of the partial differential equation (PDE)*

$$\frac{\partial}{\partial t} u_t(x) + L u_t(x) + f(x, u_t(x), k(u_t))(x) = 0, \quad 0 \le t \le T,$$
$$u_T(x) = h(x), \quad x \in \mathbb{R}, \tag{2.2}$$

where, for every $\varphi \in C^2_{\ell,b}(\mathbb{R})$,

$$L\varphi(x) = \frac{1}{2}\sigma(x)^2\varphi''(x) + \int_E \left(\varphi(x + \beta(x,y)) - \varphi(x) - \varphi'(x)\beta(x,y)\right)\lambda(dy),$$

$$k(\varphi)(x) = \frac{\sigma(x)^2\varphi'(x) + \int_E(\varphi(x + \beta(x,y)) - \varphi(x))\beta(x,y)\,\lambda(dy)}{\sigma(x)^2 E + \int_E \beta(x,y)^2\lambda(dy)}.$$

Then the unique solution of the BSDE (1.2) is given by

$$V_t = u_t(X_t),$$
$$\xi_t = k(u_t)(X_{t-}),$$
$$c_t = \int_0^t \left(u_s'(X_s) - k(u_s)(X_s)\right)\sigma(X_s)\,dW_s +$$
$$+ \int_0^t \int_E \left(u_{s-}(X_{s-} + \beta(X_{s-}, y))\right. \tag{2.3}$$
$$\left. - u_s(X_{s-}) - k(u_s)(X_{s-})\right)\beta(X_{s-}, y)\,\tilde{\mu}(ds\,dy), \quad 0 \le t \le T.$$

The proof of Proposition 2 can be found in [3], where the result is extended to a Lipschitz function $f : \mathbb{R}^3 \to \mathbb{R}$ and a Borel function $h : \mathbb{R} \to \mathbb{R}$ with at most linear growth by passing to the limit via a sequence of smooth functions f_n, h_n. Of course, for the limit we do not have relation (2.2).

Remark 1. Note that once having the existence of a solution $u \in C^{1,2}([0,t] \times \mathbb{R})$ of equation (2.2), we can get relation (2.3) quite easily: We apply the Itô formula to $u_t(X_t)$ and use (2.2) to see hat

$$u_t(X_t) = h(X_T) - \int_t^T \int_E \left(u_s(X_{s-} + \beta(X_{s-}, y)) - u_s(X_{s-})\right)\tilde{\mu}(ds\,dy)$$
$$- \int_t^T u_s'(X_s)\,dX_s^c + \int_t^T f(X_s, u_s(X_s), k(u_s)(X_s))\,ds,$$

and decomposing the martingale part of the semimartingale $u_t(X_t)$ into a martingale of the subspace of $\mathcal{M}^2(P)$ generated by X and a martingale orthogonal to X, we obtain

$$
\begin{aligned}
u_t(X_t) = h(X_T) &- \int_t^T k(u_s)(X_{s-})\, dX_s - \int_t^T \left(u_s'(X_s) - k(u_s)(X_s) \right) \sigma(X_s)\, dW_s \\
&+ \int_t^T \int_E \left(u_s(X_{s-} + \beta(X_{s-}, y)) - u_s(X_{s-}) \right. \\
&\qquad\qquad \left. - k(u_s)(X_{s-})\beta(X_{s-}, y) \right) \tilde{\mu}(ds\,dy) \\
&+ \int_t^T f\left(X_s, u(X_s), k(u_s)(X_s) \right) ds, \quad 0 \le t \le T.
\end{aligned}
$$

Finally, taking into account the uniqueness of the solution of the BSDE (1.2), we see (2.3).

Remark 2. As explained in Section 1, together with the unique solution of the BSDE (1.2) we also have the optimal trading strategy φ which hedges the contingent claim B and the optimal price for B. A straight-forward calculation following [12] shows that the optimal trading strategy φ is the locally risk minimizing one.

§4 THE MARKOVIAN MODEL FOR A GREATER INVESTOR

In this section we use essentially the same model as in Section 1, but we now allow the investor to influence the stock price via the value of his portfolio. More precisely, we assume that $\sigma : \mathbb{R}^2 \to \mathbb{R}$ is a Lipschitz function, $\beta : \mathbb{R}^2 \times E \to \mathbb{R}$ is a measurable function with

$$
|\beta(x, v, y) - \beta(x', v', y)| \le K(1 \wedge |y|)(|x - x'| + |v - v'|),
$$

$$
|\beta(x, v, y)| \le \frac{1}{2} \wedge (K|y|), \quad x, x', v, v' \in \mathbb{R},\ y \in E,
$$

$$
\sigma(x, v)^2 + \int_E \beta(x, v, y)^2\, \lambda(dy) \ge \varepsilon > 0, \quad x, v \in \mathbb{R},
$$

for some reals K and ε. Using Lipschitz functions $f : \mathbb{R}^2 \to \mathbb{R}$ and $h : \mathbb{R} \to \mathbb{R}$, we model the interaction between the stock price process X and the value of the portfolio V of the investor by the FBSDE

$$
\begin{aligned}
dX_t &= \sigma(X_t, V_t)\, dW_t + \int_E \beta(X_{t-}, V_{t-}, y)\, \tilde{\mu}(dt\,dy), \\
dV &= -f(X_t, V_t)\, dt + \xi_t\, dX_t + dC_t, \quad 0 \le t \le T, \\
X_0 &= x, \\
V_T &= h(X_T).
\end{aligned}
\tag{3.1}
$$

A quadruple (X, V, ξ, C) is called a solution of the FBSDE (3.1), if V is a square-integrable semimartingale, X and C are square-integrable martingales orthogonal to each other, and ξ is a predictable process with

$$\mathrm{E}\left[\int_0^T \xi_s^2 \, d\langle X\rangle_s\right] < +\infty.$$

Following [9], [8], and [4], we say that a solution (X, V, ξ, C) of the FBSDE (3.1) is *nodal*, if there is a measurable function $u : [0, t] \times \mathbb{R} \to \mathbb{R}$, Lipschitz continuous in $x \in \mathbb{R}$, uniformly with respect to $t \in [0, t]$, and such that $V_t = u_t(X_t)$, $0 \leq t \leq T$.

Proposition 3. *Under the assumptions made in this section there is a unique nodal solution of the FBSDE* (3.1).

Remark 3. Problem (3.1) is studied in [4] for $\beta = 0$, $f(x, v) = 1 - c(x)v$, C_b^1-functions σ, c and $h \in C^{2+\alpha}(\mathbb{R})$, for some $\alpha > 0$, in the multi-dimensional case and with drift in the first equation.

Proof of Proposition 3. For reasons of convenience we restrict the proof to the special case $\beta = 0$, but we emphasize that the proof works in the general case quite analogously. Let $u_t^0(x) = 0$, $(t, x) \in [0, T] \times \mathbb{R}$, and assume that, for some natural $n \geq 1$, we have certain measurable function $u^{n-1} : [0, T] \times \mathbb{R} \to \mathbb{R}$ which is Lipschitz continuous in $x \in \mathbb{R}$, uniformly with respect to $t \in [0, T]$. Then denote by $(X_s^n(t, x))_{t \leq s \leq T}$ the unique solution of the equation

$$dX_s^n(t, x) = \sigma\big(X_s^n(t, x), u_s^{n-1}(X_s^n(t, x))\big) \, dW_s, \quad t \leq s \leq T,$$
$$X_t^n(t, x) = x, \tag{3.2}$$

and by $(V_s^n(t, x), \xi_s^n(t, x), C_s^n(t, x))_{t \leq s \leq T}$ the unique solution of the BSDE

$$dV_s^n(t, x) = -f(X_s^n(t, x), V_s^n(t, x)) \, ds + \xi_s^n(t, x) \, dX_s^n(t, x) + dC_s^n(t, s), \quad t \leq s \leq T,$$
$$V_T^n(t, x) = h(X_T^n(t, x)), \quad x \in \mathbb{R}. \tag{3.3}$$

Recall that the existence and uniqueness of this solution is guaranteed by Proposition 1, note that $V_t^n(t, x)$ is deterministic and introduce the function $u_t^n(x) = V_t^n(t, x)$, $(t, x) \in [0, T] \times \mathbb{R}$. Using the Markov property of $\{X_s^n(t, x), 0 \leq t \leq s \leq T, x \in \mathbb{R}\}$ we obtain

$$V_s^n(t, x) = u_s^n(X_s^n(t, x)), \quad t \leq s \leq T, \, x \in \mathbb{R}.$$

Let us show that $u^n : [0, t] \times \mathbb{R} \to \mathbb{R}$ is a measurable function for which there exists a real K not depending on u^{n-1} or n such that

$$|u_t^n(x)| \leq K, \tag{3.4}$$
$$|u_t^n(x) - u_t^n(x')| \leq K|x - x'|, \quad x, x' \in \mathbb{R}, \, 0 \leq t \leq T. \tag{3.5}$$

For this note that

$$E[|X_s^n(t,x) - X_s^n(t,x')|] = |x - x'| \quad x, x' \in \mathbb{R},\ t \le s \le T,$$

and setting

$$a \oplus b = \begin{cases} b/a, & \text{if } a \ne 0, \\ 0, & \text{otherwise}, \end{cases} \quad a, b \in \mathbb{R},$$

we define the processes

$$\alpha_s^n(t,x,x') = (V_s^n(t,x) - V_s^n(t,x'))^\oplus \{g(X_s^n(t,x), V_s^n(t,x)) \\ - g^s(X_s^n(t,x), V_s^n(t,x'))\},$$

$$\gamma_s^n(t,x,x') = (X_s^n(t,x) - X_s^n(t,x'))^\oplus \{g(X_s^n(t,x), V_s^n(t,x')) \\ - g(X_s^n(t,x'), V_s^n(t,x'))\},$$

which are bounded by a real number not depending on s, t, x, x', n. Then we can transform (3.3) as follows

$$(V_s^n(t,x) - V_s^n(t,x'))$$

$$= \exp\left\{ \int_s^T \alpha_r^n(t,x,x')\,dr \right\} \{h(X_T^n(t,x)) - h(X_T^n(t,x'))\}$$

$$+ \int_s^T \exp\left\{ \int_r^T \alpha_v^n(t,x,x')\,dv \right\} \{\gamma_r^n(t,x,x')(X_r^n(t,x) - X_r^n(t,x'))\,dr$$

$$- (\xi_r^n(t,x) - \xi_r^n(t,x'))\,dW_r\}, \qquad t \le s \le T,$$

i. e., we can find a real K' not depending on t and n, such that

$$|u_t^n(x) - u_t^n(x')| = |E[V_t^n(t,x) - V_t^n(t,x')]|$$

$$\le K'\,E[|X_T^n(t,x) - X_T^n(t,x')|] + K' \int_t^T E[|X_r^n(t,x) - X_r^n(t,x')|]\,dr$$

$$= K'(1 + (T - t))|x - x'|, \qquad 0 \le t \le T,\ x, x' \in \mathbb{R}.$$

Hence, (3.5) is proved. A similar argument allows us to show that (3.4) holds. Consequently, by iteration we obtain a sequence of measurable functions $(u^n)_{n \ge 0}$, $u^n : [0,T] \times \mathbb{R} \to \mathbb{R}$, which are uniformly bounded and Lipschitz continuous in $x \in \mathbb{R}$, uniformly with respect to $t \in [0,T]$, $n \ge 0$. Now, taking (3.5) into account we can use standard estimates to derive from (3.2)

$$E\left[\sup_{t \le r \le s} |X_r^{n+1}(t,x) - X_r^n(t,x)|^2 \right] \le K_1 \int_t^s \|u_r^n - u_r^{n-1}\|_{L^\infty(\mathbb{R})}^2\,dr,$$

$$0 \le t \le s \le T,\ n \ge 1,$$

for some real K_1, and a procedure analogous to the proof of (3.5) then provides

$$\|u_t^{n+1} - u_t^n\|_{L^\infty(\mathbb{R})}^2 \leq K_2 \int_t^T \|u_r^n - u_r^{n-1}\|_{L^\infty(\mathbb{R})}^2 \, dr, \quad 0 \leq t \leq T, \ n \geq 1. \quad (3.6)$$

Therefore, there is a bounded measurable function $u : [0,T] \times \mathbb{R} \to \mathbb{R}$, Lipschitz continuous in $x \in \mathbb{R}$, uniformly with respect to $t \in [0,T]$, which is the limit of the sequence $(u^n)_{n \geq 0}$ in $L^\infty([0,T] \times \mathbb{R})$.

Let now $(X_s^t(x))_{t \leq s \leq T}$ be the solution of equation

$$dX_s^t(x) = \sigma(X_s^t(x), u_s(X_s^t(x))) \, dW_s, \quad t \leq s \leq T,$$
$$X_t^t(x) = x, \quad x \in \mathbb{R}, \quad (3.7)$$

and $(V_s^t(x), \xi_s^t(x), C_s^t(x))$ the solution of the BSDE

$$dV_s^t(x) = -f(X_s^t(x), V_s^t(x)) \, ds + \xi_s^t(x) \, dX_s^t(x) + dC_s^t(x), \quad t \leq s \leq T,$$
$$V_T^t(x) = h(X_T^t(x)), \quad x \in \mathbb{R}. \quad (3.8)$$

Clearly, for all $p \geq 2$, we have

$$\mathrm{E}\left[\sup_{t \leq s \leq T} |X_s^t(x) - X^n(t,x)|^p \right] + \mathrm{E}\left[\sup_{0 \leq s \leq T} |V_s^t(x) - V^n(t,x)|^p \right] \to 0, \quad n \to \infty.$$

Since $V_s^n(t,x) = u_s^n(X_s^n(t,x))$, $n \geq 1$, we see that also $V_s^t(x) = u_s(X_s^t(x))$ for $0 \leq t \leq s \leq T$ and $x \in \mathbb{R}$. Consequently,

$$(X, V, \xi, C) = (X^0(x), V^0(x), \xi^0(x), C^0(x))$$

is a nodal solution of the FBSDE (3.1).

In order to complete the proof, note that the uniqueness can be proved, too, by the arguments presented above. $\square$

Finally, we remark

Corollary. *If, in addition to the assumptions introduced above, the functions σ and f are bounded and*

$$h \in W_p^2(\mathbb{R}, \rho_\alpha(dx)) \cap L^\infty(\mathbb{R}), \quad \rho_\alpha(dx) = \exp\{-\alpha(1+x^2)^{1/2}\} \, dx,$$

for some $p \geq 2$ and $\alpha > 0$, then the function $u : [0,T] \times \mathbb{R} \to \mathbb{R}$ constructed in the proof of Proposition 3 is the unique solution of the PDE

$$\frac{\partial}{\partial t} u_t(x) + \frac{1}{2}\sigma(x, u_t(x))^2 u_t''(x) + f(x, u_t(x)) = 0, \quad 0 \leq t \leq T,$$
$$u_T(x) = h(x), \quad x \in \mathbb{R}, \quad (3.9)$$

where $u \in W_p^{1,2}([0,T] \times \mathbb{R}, dt\,\rho_\alpha(dx)) \cap L^\infty([0,T] \times \mathbb{R})$ and u is Lipschitz continuous in $x \in \mathbb{R}$, uniformly with respect to $t \in [0,T]$.

Proof. Let $v \in W_p^{1,2}([0,T] \times \mathbb{R}, dt\,\rho_\alpha(dx)) \cap L^\infty([0,T] \times \mathbb{R})$ be the unique solution of the PDE

$$\frac{\partial}{\partial t} v_t(x) + \frac{1}{2}\sigma(x, u_t(x))^2 v_t''(x) + f(x, u_t(x)) = 0, \qquad 0 \le t \le T,$$
$$v_T(x) = h(x), \qquad x \in \mathbb{R}.$$

It is explicitly given by

$$v_t(x) = \mathrm{E}\left[h(X_T^t(x)) + \int_t^T f(X_s^t(x), u_s(X_s^t(x)))\,ds\right], \quad 0 \le t \le T,\ x \in \mathbb{R}, \quad (3.10)$$

where $(X_s^t(x))$ is defined by (3.7), cf. [1], Theorem 8.1. On the other hand, from (3.8) we conclude that the right-hand side of (3.10) coincides with $V_t^t(x) = u_t(x)$. Hence,

$$u = v \in W_p^{1,2}([0,T] \times \mathbb{R}, dt\,\rho_\alpha(dx)) \cap L^\infty([0,T] \times \mathbb{R})$$

and u solves the PDE (3.9).

If there is another solution $\tilde{u} \in W_p^{1,2}([0,T] \times \mathbb{R}, dt\,\rho_\alpha(dx)) \cap L^\infty([0,T] \times \mathbb{R})$, Lipschitz continuous in $x \in \mathbb{R}$, uniformly with respect to $t \in [0,T]$, then

$$\tilde{u}_t(x) = \mathrm{E}\left[h(\tilde{X}_T^t(x)) + \int_t^T f(\tilde{X}_s^t(x), \tilde{u}_s(\tilde{X}_s^t(x)))\,ds\right], \quad 0 \le t \le T,\ x \in \mathbb{R},$$

where $(\tilde{X}_s^t(x))$ is the solution of the equation

$$d\tilde{X}_s^t(x) = \sigma(\tilde{X}_s^t(x), \tilde{u}_s(\tilde{X}_s^t(x)))\,dW_s, \qquad t \le s \le T,$$
$$\tilde{X}_t^t(x) = x, \qquad x \in \mathbb{R}.$$

Fix any $\tau \in [0,T]$ and $x \in \mathbb{R}$. Then, using the Markov property of $(\tilde{X}_s^t(x))$ and the notations $X_s = X_s^\tau(x)$, $\tilde{X}_s = \tilde{X}_s^\tau(x)$, we get

$$\tilde{u}_t(\tilde{X}_t) = \mathrm{E}\left[h(\tilde{X}_T) + \int_t^T f(\tilde{X}_s, \tilde{u}_s(\tilde{X}_s))\,ds \;\middle|\; \mathcal{F}_t\right], \quad \tau \le t \le T,\ x \in \mathbb{R},$$

i.e.,

$$M_t = \tilde{u}_t(\tilde{X}_t) + \int_\tau^t f(\tilde{X}_s, \tilde{u}_s(\tilde{X}_s))\,ds, \quad \tau \le t \le T,$$

is a square-integrable martingale, and decomposing $M \in \mathcal{M}^2(P)$,

$$dM_t = \tilde{\xi}_t\,d\tilde{X}_t + d\tilde{C}_t, \quad \tau \le t \le T,$$

where $\tilde{\xi}$ is a predictable process with $E[\int_\tau^T \tilde{\xi}_s^2 \, d\langle X \rangle_s] < \infty$, and $\tilde{C} \in \mathcal{M}^2(P)$ with $\langle \tilde{X}, \tilde{C} \rangle = 0$, we get a nodal solution $(\tilde{X}_t, \tilde{u}_t(\tilde{X}_t), \tilde{\xi}_t, \tilde{C}_t)$ of the FBSDE (3.1) (with $\beta = 0$) starting at time τ. From the uniqueness of the nodal solution it follows that

$$X_t = \tilde{X}_t, \quad u_t(X_t) = \tilde{u}_t(\tilde{X}_t) = \tilde{u}_t(X_t), \quad \text{i.e.,} \quad u_\tau(x) = \tilde{u}_\tau(x).$$

This shows the uniqueness of the solution of PDE (3.9) and completes the proof. $\square$

Finally, let us remark that once having the unique nodal solution (X, V, ξ, C) of the FBSDE (3.1), we obtain also here the optimal trading strategy and the associated optimal price of the contingent claim $B = h(X_T)$ by setting

$$\varphi = (\xi, \eta),$$
$$\eta_t = V_t - \xi_t \cdot X_t,$$
$$\pi(B) = V_0.$$

Moreover, if the assumptions on σ, β, and f are such that the solution u of the PDE (3.9) even belongs to $C^{1,2}([0,T] \times \mathbb{R})$ (cf. [8], [4] for $\beta = 0$), then from the argument provided in Remark 1 it follows that V, ξ, and C are given by (2.3) after replacing $\sigma(x)$ by $\sigma(x, u_t(x))$ and $\beta(x, y)$ by $\beta(x, u_t(x), y)$, and so we have the explicit form of φ and $\pi(B)$.

REFERENCES

1. Bensoussan, A., Lions, J. L., *Contrôle Impulsionel et Inéquations quasi variationelles*, Bordas, Paris, 1982.
2. Buckdahn, R., *Backward stochastic differential equations driven by a martingale*, Prépublication 93-05, URA 225 Université de Provence, Marseille.
3. Buckdahn, R., Pardoux, E., *Backward stochastic differential equations. Application to finance*, in preparation.
4. Duffie, D., Ma, J., Yong, J., *Black's consol rate conjecture*, Preprint, 1993.
5. El Karoui, N., Quenez, M.-C., *Dynamic Programming and Pricing of Contingent Claims in an Incomplete Market*, Thèse de Doctorat de l'Université Paris VI présentée par M.-C. Quenez, 1993.
6. Fujiwara, T., Kunita, H., *Stochastic differential equation of jump type and Lévy processes in diffeomorphisms group*, J. Math. Kyoto Univ. **25-1** (1985), 71–106.
7. Föllmer, H., Sondermann, D., *Hedging of non-redundant contingent claims*, in: Contributions to Mathematical Economics (W. Hildenbrand and A. Mas-Colell, eds.), 1986, pp. 205–223.
8. Ma, J., Protter, P., Yong, J., *Solving forward-backward stochastic differential equations explicitly – A four step scheme*, preprint, 1993.
9. Ma, J., Yong, J., *Solving forward-backward SDEs and the nodal set of Hamilton–Jacobi–Bellman equations*, preprint, 1993.
10. Pardoux, E., Peng, S., *Adapted solution of a backward stochastic differential equation*, Systems and Control Letters **14** (1990), 55–61.
11. Pardoux, E., Peng, S., *Backward stochastic differential equations and quasilinear parabolic partial differential equations*, Prépublication 92-03, URA 225, Université de Provence, Marseille.

12. Schweizer, M., *Option hedging for semimartingales*, Stochastic Processes and their Applications **37** (1991), 339–363.

RAINER BUCKDAHN, FB MATHEMATIK, HUMBOLDT-UNIVERSITÄT ZU BERLIN, UNTER DEN LINDEN 6, 10099 BERLIN

Progress in Probability, Vol. 36
© 1995 Birkhäuser Verlag Basel/Switzerland

COMPONENTWISE AND VECTOR STOCHASTIC INTEGRATION WITH RESPECT TO CERTAIN MULTI-DIMENSIONAL CONTINUOUS LOCAL MARTINGALES

MICHEL CHATELAIN AND CHRISTOPHE STRICKER

ABSTRACT. In a previous paper we provided a condition under which the componentwise and the vector stochastic integration with respect to a given continuous $\mathbb{R}^d$-local martingale are the same notions. Here, we improve this result, taking more general volatility matrices as the previous ones which were non-singular.

§1 INTRODUCTION

To construct a multi-dimensional stochastic integral with respect to a given process (for instance an $\mathbb{R}^d$-valued special semimartingale or continuous local martingale), a first idea is to take the sum of the stochastic integrals with respect to each component, i. e. the "componentwise integral". This is the natural notion if one integrates with respect to a process of bounded variation. On the other hand the stochastic integral for a continuous local martingale M has to be constructed using an isometry as in the one-dimensional case in order to obtain good properties, for instance the property of predictable representation (see for instance [2]). Therefore one has to consider a larger class of predictable processes than the one of the componentwise integrable processes. We refer the reader to the carefully presented study of this theory in [6].

Let us give some definitions and results in the case where the given process M is an $\mathbb{R}^d$-valued continuous local martingale on a filtered probability space $(\Omega, \mathcal{F}, (\mathcal{F}_t)_{0 \leq t \leq 1}, P)$ satisfying the usual conditions. There exist a continuous increasing process B (for instance $B := \sum_i \langle M_i, M_i \rangle$) and a predictable symmetric non-negative matrix $C = (C_{ij})_{1 \leq i \leq d, 1 \leq j \leq d}$ such that $d\langle M_i, M_j \rangle = C_{ij} \, dB$ P-a. s. A predictable $\mathbb{R}^d$-valued process H is componentwise integrable with respect to M if, for $i = 1$ to d, we have

$$\int_{[0,1]} H_i^2 C_{ii} \, dB < +\infty \quad P\text{-a. s.}$$

1991 *Mathematics Subject Classification.* 60H05 90A09.

Key words and phrases. Predictable processes, componentwise and vector stochastic integration, equivalent local martingale measures, complete or incomplete markets.

If the resulting local martingale is in fact a square-integrable martingale, an easy calculation shows that

$$E((H \cdot M)_1^2) = E\left(\int_{[0,1]} H^*CH \, dB \right).$$

Thus, in order to construct the vector stochastic integral by isometry, the vector integrability condition for a predictable process H is

$$\int_{[0,1]} H^*CH \, dB < +\infty \quad P\text{-a. s.}$$

Note that the two types of integrals coincide clearly if for instance C is a diagonal matrix, or if one deals only with locally bounded H. We recall the fact that a stochastic integral does not depend on the chosen probability Q, as long as Q is equivalent to P.

In financial theory, we have a certain number of processes representing the discounted prices of the primary traded securities of the market. These prices are often supposed to have a "good" form, for instance they follow Itô processes (see the models in [9]) or more generally they are special semimartingales with an equivalent local martingale measure. By this we understand a probability measure Q equivalent to P under which the discounted prices are local martingales. The existence of such a measure arises from "No Arbitrage" conditions (we refer the reader to [1], [3], [4], [5], and [10]). This is why we deal only with local martingales in the present study. In our previous paper we took d continuous local martingales

$$M_i := \sum_{j=1}^{n} \sigma_{ij} \cdot N_j,$$

the N_i being d continuous mutually orthogonal local martingales and

$$\Sigma = (\sigma_{ij})_{1 \leq i \leq d, 1 \leq j \leq d}$$

a predictable non-singular $d \times d$ volatility matrix. The non-singularity of Σ is necessary to obtain a "complete market". We then provided a condition under which the two types of stochastic integration with respect to M are the same notions. Roughly speaking it consisted in the local boundedness of the processes $a_{kj}\sigma_{ji}$ where $\Sigma^{-1} = A = (a_{ij})$.

D. W. Stroock recently asked us what could be said about the more general case of a $d \times n$ volatility matrix. Indeed such a question is quite natural because it corresponds to the frequent case of an "incomplete market", where the number of primary traded securities is smaller than the number of "uncertainties", i. e. for instance the number of independent Brownian motions generating the filtration. Thus in the present paper we take a local martingale M written

$$M_i := \sum_{j=1}^{n} \sigma_{ij} \cdot W_j$$

for $i = 1$ to d, where the volatility matrix is a $d \times n$ predictable matrix, and W is an n-dimensional Brownian motion. We will simply choose $dB_s = ds$, and C will be the "instantaneous return covariance" matrix $\Sigma\Sigma^*$ ($*$ denotes the transpose), which already appears in [7]. There, the authors supposed Σ to be a $d \times d$ non-singular matrix. The main purpose of our study is to extend our previous theorem to the present case where Σ is more general, using similar methods and working on $\Sigma\Sigma^*$.

Let us recall a definition and a result obtained in [2] that will be used in what follows: a process on Ω is "locally bounded" if there exists an increasing sequence of stopping times that converges stationarily to 1 such that the corresponding stopped processes are bounded. In our paper "On componentwise and vector stochastic integration" we gave the following result:

Lemma 1.1. *Let m and D be two right-continuous predictable processes on $(\Omega, \mathcal{F}, (\mathcal{F}_t)_{0 \leq t \leq 1}, P)$, D being increasing and vanishing at zero. We suppose that for any nonnegative predictable process H satisfying*

$$\int_{[0,1]} H_s \, dD_s < +\infty \quad P\text{-a. s.},$$

we have

$$\int_{[0,1]} H_s |m_s| \, dD_s < +\infty \quad P\text{-a. s.}$$

Then m admits a locally bounded version m', i. e. a locally bounded predictable process such that

$$\int_{[0,1]} \mathbf{1}_{\{m_s \neq m'_s\}} \, dD_s = 0 \quad P\text{-a. s.}$$

§2 THE MAIN RESULT

We start with a remark: if we look at the two integrability conditions, we see that they remain the same if we replace Σ and H by $\widehat{\Sigma}$ and $\widehat{H}$ defined as follows: we cancel a line of Σ whenever it is zero and do the same thing with the corresponding coordinate of H. Note that the obtained $\widehat{\Sigma}$ and $\widehat{H}$ have a variable size, function of ω and t, but that $\widehat{\Sigma}$ and $\widehat{H}$ remain predictable.

Theorem 2.1. *The following conditions are equivalent:*

(A1) $\widehat{\Sigma}\widehat{\Sigma}^*$ *is $dP \otimes ds$-a. s. non-singular and the processes $(\widehat{\Sigma}\widehat{\Sigma}^*)_{ii}(\widehat{\Sigma}\widehat{\Sigma}^*)_{ii}^{-1}$ admit locally bounded versions.*

(A2) *The two types of stochastic integration with respect to M are the same.*

(A3) $\widehat{\Sigma}\widehat{\Sigma}^*$ *is $dP \otimes ds$-a. s. non-singular. There exists a predictable non-singular matrix $A = (a_{ij})_{i,j}$ satisfying $(\widehat{\Sigma}\widehat{\Sigma}^*)^{-1} = A^*A$ and the processes $a_{ij}\widehat{\sigma}_{jk}$ admit locally bounded versions.*

Remark 2.2. If condition (A3) holds for a given A, then it holds for every predictable non-singular matrix A' satisfying $(\widehat{\Sigma}\widehat{\Sigma}^*)^{-1} = A'^* A'$. For instance, in case Σ is non-singular, taking $A := \Sigma^{-1}$, (A3) corresponds to the condition we found in our previous paper [2].

The proof of the theorem is an adaptation of the one in [2]. We first need the following results:

Lemma 2.3. *If $\widehat{\Sigma}\widehat{\Sigma}^*$ is not $dP \otimes ds$-a. s. non-singular, then condition (A2) does not hold.*

Proof. We proceed by contradiction. So we pick in the kernel of $\widehat{\Sigma}\widehat{\Sigma}^*$ a predictable process $\widehat{H}$ which is not identically zero. For any predictable real-valued process α, the predictable process $\alpha\widehat{H}$ is trivially vector integrable. There exists some i such that the following predictable set

$$D := \left\{ (\omega, t) \; ; \; \sum_{j=1}^n \widehat{\sigma}_{ij}^2 \widehat{H}_i^2 \neq 0 \right\}$$

is not negligible. Thus there exist $\varepsilon > 0$ and $\beta > 0$ such that

$$P\left(\int_{[0,1]} \mathbf{1}_D \, ds \geq \beta \right) \geq \varepsilon.$$

Now we define the following increasing sequence of stopping times: $T_0 := 0$ and for $i \geq 1$,

$$T_i := \inf\left\{ t \in [0,1] \; ; \; \int_{[0,t]} \mathbf{1}_D \, ds \geq \sum_{k=1}^i \beta \, 2^{-k} \right\} \qquad .$$

and $T_i = 1$ if the infimum is taken over $\varnothing$. We then define the predictable process

$$\alpha := \left[\mathbf{1}_D \left(\sum_{i=0}^\infty 2^{2i} \mathbf{1}_{]T_i, T_{i+1}]} \right) \frac{1}{\sum_{j=1}^n \widehat{H}_i^2 \widehat{\sigma}_{ij}^2} \right]^{1/2}$$

If we set $\widehat{K} := \alpha\widehat{H}$, a straightforward calculation shows that

$$\int_{[0,1]} \sum_{j=1}^n \widehat{\sigma}_{ij}^2 \widehat{K}_i^2 \, ds = +\infty$$

on the non-negligible set $\{ \omega \; ; \; \int_{[0,1]} \mathbf{1}_D \, ds \geq \beta \}$. This is a contradiction to (A2) and the lemma is proved. $\square$

Remark 2.4. For a symmetric nonnegative matrix S there exists a unique nonnegative symmetric matrix R such that $S = RR$, and R is a continuous function of S. Thus there exists a non-singular predictable matrix Q such that $\widehat{\Sigma}\widehat{\Sigma}^* = QQ^*$

when $\widehat{\Sigma}\widehat{\Sigma}^*$ is non-singular (for instance the symmetric one obtained above). Let $A := Q^{-1}$ and denote by a_{ij} and q_{ij} the entries of A and Q.

Proof of the theorem. Suppose that (A2) holds. Lemma 2.2 shows that $\widehat{\Sigma}\widehat{\Sigma}^*$ is non-singular. Take any predictable real-valued process η_k and call $\widehat{H}$ the process $[(0,\dots,\eta_k,0,\dots)A]^*$. We have $\widehat{H}^*\widehat{\Sigma}\widehat{\Sigma}^*\widehat{H} = \eta_k^2$ whereas $\widehat{H}_i^2\widehat{\sigma}_{ij}^2 a_{ki}^2$. If we remark that

$$(\widehat{\Sigma}\widehat{\Sigma}^*)_{ii} = \sum_j \widehat{\sigma}_{ij}^2 \quad \text{and} \quad (\widehat{\Sigma}\widehat{\Sigma}^*)_{ii}^{-1} = \sum_k a_{ki}^2,$$

we conclude, using Lemma 1.1, that condition (A1) holds. Conversely suppose that (A1) holds and take any vector integrable predictable process H. The vector integrability condition can be written

$$\int_{[0,1]} \|Q^*\widehat{H}\|^2 \, ds < +\infty \quad P\text{-a.s.}$$

or equivalently

$$\int_{[0,1]} \sum_i \left(\sum_j q_{ji}\widehat{H}_j\right)^2 ds < +\infty \quad P\text{-a.s.}$$

We have to show that H is componentwise integrable too. We have

$$\widehat{\sigma}_{ij}^2 \widehat{H}_i^2 = \widehat{\sigma}_{ij}^2 \left(\sum_k \widehat{H}_k \sum_\ell q_{k\ell}a_{\ell i}\right)^2$$

$$= \widehat{\sigma}_{ij}^2 \left(\sum_\ell \left(\sum_k \widehat{H}_k q_{k\ell}\right)a_{\ell i}\right)^2$$

$$\leq d\sum_\ell \left(a_{\ell i}^2\widehat{\sigma}_{ij}^2\left(\sum_k \widehat{H}_k q_{k\ell}\right)^2\right).$$

The processes $a_{\ell i}^2\widehat{\sigma}_{ij}^2$ admit locally bounded versions since $(\widehat{\Sigma}\widehat{\Sigma}^*)_{ii} = \sum_j \widehat{\sigma}_{ij}^2$, $(\widehat{\Sigma}\widehat{\Sigma}^*)_{ii}^{-1} = \Sigma a_{ki}^2$ and the processes $(\widehat{\Sigma}\widehat{\Sigma}^*)_{ii}(\widehat{\Sigma}\widehat{\Sigma}^*)_{ii}^{-1}$ have locally bounded versions. This combined with

$$\int_{[0,1]} \left(\sum_k \widehat{H}_k q_{k\ell}\right)^2 ds < +\infty \quad P\text{-a.s.}$$

yields the desired integrability condition. Hence (A2) holds. When $\widehat{\Sigma}\widehat{\Sigma}^*$ is non-singular, the equalities

$$(\widehat{\Sigma}\widehat{\Sigma}^*)_{ii} = \sum_j \widehat{\sigma}_{ij}^2 \quad \text{and} \quad (\widehat{\Sigma}\widehat{\Sigma}^*)_{ii}^{-1} = \sum_k a_{ki}^2$$

hold for any non-singular matrix A such that

$$(\widehat{\Sigma}\widehat{\Sigma}^*)^{-1} = A^*A.$$

This remark, and a closer look at the preceding proof of the equivalence between (A1) and (A2), show that (A1) and (A3) are equivalent. The theorem is proved. $\square$

§3 APPLICATION TO FINANCIAL THEORY

We consider a standard model for a financial market with continuous prices. We note S_i, $i = 1, \dots, d$, the d positive continuous processes representing the discounted prices of the primary traded securities of the market. We suppose that the S_i satisfy an equation of the type: for $i = 1$ to d,

$$\frac{dS_i}{S_i} = \sum_{j=1}^{n} \sigma_{ij} \, dW_j = b_j \, dt \quad \text{a.\,s.,}$$

where the b_j are predictable processes, $\Sigma = (\sigma_{ij})_{1 \le i \le d, 1 \le j \le n}$ is a predictable volatility matrix, and W is an n-dimensional Brownian motion. Set $dX_i := \frac{dS_i}{S_i}$. Starting from the economically meaningful assumption that S does not allow arbitrage profits, the fundamental theorem of asset pricing [5] states the existence of an equivalent local martingale measure Q for the discounted price process S. In fact there exists a new Brownian motion $\widetilde{W}$ under Q such that

$$dX_i = \sum_{j=1}^{n} \sigma_{ij} \, d\widetilde{W}_j \quad \text{a.s.}$$

Thus we can use our Theorem 2.1 for the X_i and we get the following:

Theorem 3.1. *The following conditions are equivalent:*

(A1) $\widehat{\Sigma}\widehat{\Sigma}^*$ *is* $dQ \otimes ds$-*a.\,s. non-singular and the processes* $(\widehat{\Sigma}\widehat{\Sigma}^*)_{ii} (\widehat{\Sigma}\widehat{\Sigma}^*)_{ii}^{-1}$ *admit locally bounded versions.*

(A2) *The two types of stochastic integrals with respect to S are the same.*

(A3) $\widehat{\Sigma}\widehat{\Sigma}^*$ *is* $dQ \otimes ds$-*a.\,s. non-singular. There exists a predictable non-singular matrix* $A = (a_{ij})_{i,j}$ *satisfying* $(\widehat{\Sigma}\widehat{\Sigma}^*)^{-1} = A^*A$ *and the processes* $a_{ij}\widehat{\sigma}_{jk}$ *admit locally bounded versions.*

Proof. The theorem is obvious if we make the following remark: a process $H := (H_1, \dots, H_d)^*$ is predictable if and only if the process

$$\widetilde{H} := \left(\frac{H_1}{S_1}, \dots, \frac{H_d}{S_d} \right)^*$$

is predictable, and H is componentwise (resp. vector) integrable with respect to S if and only if $\widetilde{H}$ is componentwise (resp. vector) integrable with respect to X). $\square$

Remark 3.2. A model with a non-singular volatility matrix Σ corresponds to a "complete" financial market without arbitrage opportunities. If the volatility matrix is rectangular, it corresponds to an "incomplete" market. This is for instance the case in [8] where the authors suppose that Σ has full rank – or equivalently that $\Sigma\Sigma^*$ is non-singular. Under this assumption the first part of condition (A1) is fulfilled. (The non-singularity of $\Sigma\Sigma^*$ allows them to complete the market with "fictitious securities" and to obtain results on utility maximizations). Certain models (see [9]) even satisfy a stronger condition: Σ is bounded and there exists a positive ε such that, for each vector H, we have $H^*\Sigma\Sigma^*H \geq \varepsilon H^*H$ (the "strong non-degeneracy" condition). This yields the non-degeneracy of $\Sigma\Sigma^*$, and the boundedness of $\Sigma\Sigma^*$ and $(\Sigma\Sigma^*)^{-1}$. In particular, condition (A1) holds in such models. Therefore the two types of integration are the same in this context.

REFERENCES

1. Ansel, J. P., Stricker, C., *Lois de martingale, densités et décomposition de Föllmer-Schweizer*, Ann. Inst. Henri Poincaré, **28 (3)** (1992), 375–392.
2. Chatelain, M., Stricker, C., *On componentwise and vector stochastic integration* (to appear in Mathematical Finance).
3. Dalang, R. C., Morton, A., Willinger, W., *Equivalent martingale measures and no-arbitrage in stochastic securities market models*, Stochastics and Stochastics Reports **29** (1990), 185–201.
4. Delbaen, F., *Representing martingale measures when asset prices are continuous and bounded*, Mathematical Finance **2** (1992), 107–130.
5. Delbaen, F., Schachermayer, W., *A general version of the fundamental theorem of asset pricing* (to appear).
6. Jacod, J., *Calcul stochastique et problèmes de martingales*, Lectures Notes in Math., vol. 714, Springer Verlag, New York, 1979.
7. Jarrow, R. A., Madan, D. B., *A characterization of complete security markets on a Brownian filtration*, Mathematical Finance **1 (3)** (1991), 31–43.
8. Karatzas, I., Lehoczky, J., Shreve, S., Xu, G., *Martingale and duality methods for utility maximization in an incomplete market*, SIAM J. Control and Optimization **29 (3)** (1991), 702–730.
9. Karatzas, I., Shreve, S., *Brownian Motion and Stochastic Calculus*, Springer, Berlin-Heidelberg-New-York, 1988.
10. Stricker, C., *Arbitrage et lois de martingale*, Ann. Inst. Henri Poincaré **26 (3)** (1990), 451–460.

MICHEL CHATELAIN AND CHRISTOPHE STRICKER, UNIVERSITÉ DE FRANCHE-COMTÉ, URA CNRS 741, 25030 BESANÇON CEDEX - FRANCE

Progress in Probability, Vol. 36
© 1995 Birkhäuser Verlag Basel/Switzerland

STOCK PRICE RETURNS AND THE JOSEPH EFFECT:
A FRACTIONAL VERSION
OF THE BLACK–SCHOLES MODEL

NIGEL J. CUTLAND, P. EKKEHARD KOPP
AND WALTER WILLINGER

ABSTRACT. In mathematical finance, semimartingales are traditionally viewed as the largest class of stochastic processes which are economically reasonable models for stock price movements. This is mainly because stochastic integrals play a crucial role in the modern theory of finance, and semimartingales represent the largest class of stochastic processes for which a general theory of stochastic integration exists. However, some empirical evidence from actual stock price data suggests stochastic models that are not covered by the semimartingale setting.

In this paper, we discuss the empirical evidence that suggests that *long-range dependence* (also called the *Joseph effect*) be accounted for when modeling stock price movements. We present a *fractional* version of the Black–Scholes model that (i) is based on fractional Brownian motion, (ii) accounts for long-range dependence and is therefore inconsistent with the semimartingale framework (except in the case of ordinary Brownian motion) and (iii) yields the ordinary (independent) Black–Scholes model as a special case. Mathematical problems of practical importance to finance (e. g., completeness, equivalent martingale measures, arbitrage, financial gains) for this class of fractional Black–Scholes models are dealt with in an elementary fashion, namely using the (hyperfinite) *fractional* versions of the corresponding Cox–Ross–Rubinstein models.

§1 INTRODUCTION

In mathematical finance, semimartingales are typically viewed as a class of stochastic processes that is large enough to model a great variety of 'realistic' stock price behaviors. In addition, since stochastic integrals play a crucial role in the modern theory of finance and semimartingales are, in some sense (see Protter [63]), the largest class of stochastic processes for which a general stochastic integration theory exists, the semimartingale assumption has become a generally accepted modeling framework in the mathematical finance literature. In contrast

1991 *Mathematics Subject Classification.* 60F17, 60H05, 62M07, 90A09.

Key words and phrases. Long-range dependence, fractional Brownian motion, efficient markets, arbitrage opportunities.

to this theory-oriented approach to stock price modeling, there have been numerous papers in the financial economics and econometrics literature that pursue a data-oriented approach and try to answer the question "How do currently used models for stock price movements (e. g., Black and Scholes's [8] geometric Brownian motion) compare to actual stock price data?" In this paper, we illustrate how this data-oriented approach to stock price modeling can lead to price processes that are not semimartingales and, nevertheless, result in interesting and intriguing problems in mathematical finance.

In general, when trying to fit a model to a given empirical record (time series), the modeler is concerned with adequately capturing properties in the data that are related to (i) the marginal distribution and (ii) the correlation structure. In the particular context of stock price data, aspects of the marginal distribution have played an important role in the early empirical studies of asset returns (see for example [51], [22], [23], [52], [67]) but are dealt with in this paper the conventional way; that is, we assume that the stock price returns are Gaussian. Instead, we concentrate in this paper on properties related to the autocorrelation function of asset returns and summarize the findings of relevant empirical studies, including the works of Mandelbrot [53], Greene and Fielitz [31], and Lo [48]. In particular, we discuss the arguments put forward in these papers in favor of or against the presence of *long-range dependence* (or the *Joseph effect*) in empirical records of stock price returns. Intuitively, stock price returns exhibit the Joseph effect if their autocorrelations decay very slowly (e. g., like a power); so slowly, in fact, that the sum of the autocorrelations (when summed over all lags) diverges. This behavior is in stark contrast to the traditional models for stock returns where the autocorrelations decay exponentially fast and their sum converges ("short-range dependence"). In this sense, the well-known Black–Scholes model is an extreme case for it results in a degenerate autocorrelation function, i. e., the single-period returns are Gaussian and uncorrelated, hence independent.

The presence of the Joseph effect is widespread in nature (see [54]) and well-documented in hydrology, meteorology, geophysics (e. g., [57], [58], [59]) and, more recently, in modern telecommunications (see [44]) and the analysis of DNA sequences (see [61]). On the other hand, empirical studies that conclude that economic time series do or do not exhibit the Joseph effect (e. g., [53], [31], [48]) are often the source of major controversies and are usually perceived with more skepticism than similar studies in other fields of science. While a typical plot of an economic time series displays long-run nonperiodic cyclical patterns (the longest being roughly of the order of the length of the available empirical record) that are suggestive of the Joseph effect, statistical techniques used in the past to test for the presence of long-range dependence in economic time series provide a less coherent picture and have resulted in often contradictory conclusions. At the same time, it is undisputed (e. g., see [53], [50], [43]) that the presence of the Joseph effect in asset returns would have important practical and theoretical implications for many problems in modern financial economics, e. g., the pricing of derivative securities using martingale methods.

Given this inconclusive evidence in favor of or against the presence of the Joseph effect in asset returns, the purpose of this paper is to propose and study a parametric class of "Black–Scholes–like" models that (i) includes the ordinary Black–Scholes model as a special case and (ii) is able to account for the possibility of long-run nonperiodic statistical dependence in stock price returns. We call the resulting models *fractional Black–Scholes models* for they are obtained by replacing the ordinary Brownian motion as a driving process in the Black–Scholes model by fractional Brownian motion with parameter $1/2 \le H < 1$. Fractional Brownian motion with $H \in [0.5, 1)$ is a Gaussian process that was originally introduced in Mandelbrot and Van Ness [56] and whose increment process exhibits long-range dependence. Fractional Black–Scholes or closely related models have also been considered in [50] and [43], and a number of the results below have already been established by these authors. However, in contrast to their work, our approach to constructing and studying fractional Black–Scholes models is elementary in the sense that we analyze fractional Black–Scholes models via their discrete counterparts, the so-called *fractional Cox–Ross–Rubinstein models*, in the same way we can analyze the geometric Brownian motion model of Black–Scholes by investigating the corresponding ordinary Cox–Ross–Rubinstein [14] binomial geometric random walk model. In fact, the approach in this paper parallels the method developed in Cutland et al. [17] where we proposed using techniques from nonstandard analysis for studying the (continuous) Black–Scholes option pricing model via a built-in version of the (discrete) Cox–Ross–Rubinstein option pricing model. Here we give a nonstandard construction of fractional Brownian motion that yields directly a (standard) Donsker–type invariance principle result for the *geometric fractional Brownian motion model* in terms of *geometric fractional random walk models*. Subsequently, this construction allows for an elementary treatment of many interesting mathematical finance problems related to the class of fractional Black–Scholes models; e. g., the semimartingale property and the Joseph effect, long-range dependence and the existence of equivalent martingale measures, completeness, and arbitrage-pricing.

The rest of the paper is organized as follows. In Section 2, we provide a critical discussion of the literature relevant to the question of the Joseph effect in empirical records of asset returns. In particular, note that this paper contains *no* statistical analysis of actual data of asset returns; instead, we merely comment on the statistical methods used in the various empirical studies reported in the literature. Section 3 contains the Donsker–type invariance principle result for fractional Brownian motion and, subsequently, for the fractional Black–Scholes model. Properties of the class of fractional Black–Scholes models are discussed in Section 4; we give elementary arguments for why – with the exception of the ordinary Black–Scholes model – the proposed price processes (i) are not semimartingales, (ii) cannot be turned into a martingale by an equivalent change of measure, (iii) are complete, and (iv) allow for arbitrage opportunities. Some practical aspects of modeling stock prices in terms of geometric fractional Brownian motion and some open problems are discussed in Section 5.

§2 ON THE JOSEPH EFFECT IN STOCK PRICE RETURNS

In this section, we give the definition of long-range dependence (or of the Joseph effect) and illustrate some of its implications, especially with regard to problems of statistical inference for long-range dependent processes. We also discuss a number of statistical methods for checking for the Joseph effect in a given empirical record and summarize the findings of some of the major empirical studies reported in the literature about long-range dependence in asset returns. Recall, that in this paper, we do not provide any statistical analysis of actual stock price data but merely summarize the state-of-the-art of the current understanding of the question "Is there long-range dependence in stock price returns?"

2.1 Definitions, Properties and Examples. Let $X = (X_t : t = 0, 1, \dots)$ be a *covariance stationary* (sometimes called *wide-sense stationary*) stochastic process with mean μ, variance σ^2, and autocorrelation function $r(k)$, $k \geq 0$. In particular, we assume that X has an autocorrelation function of the form

$$r(k) \sim k^{-\beta} L_1(k), \quad \text{as } k \to \infty, \tag{1}$$

where $0 < \beta < 1$ and L_1 is a function which is slowly varying at infinity, that is, $\lim_{t \to \infty} L_1(tx)/L_1(t) = 1$ for all $x > 0$ (examples of such slowly varying functions are $L_1(t) = \text{const}$ and $L_1(t) = \log(t)$). A stochastic process satisfying relation (1) is said to exhibit *long-range dependence* (see, for example, [4], [69], [41], or [13]). In Mandelbrot's terminology, long-range dependence is also called the *Joseph effect* in reference to the Old Testament prophet who foretold of the "seven fat years and seven lean years" that Egypt was to experience. Thus, processes with long-range dependence are characterized by an autocorrelation function that decays hyperbolically as the lag increases. Moreover, it is easy to see that (1) implies $\sum_k r(k) = \infty$. This non-summability of the correlations captures the intuition behind long-range dependence, namely that while high-lag correlations are all individually small, their cumulative effect is of importance and gives rise to features which are drastically different from those of the more conventional, i. e., short-range dependent processes. Short-range dependent processes are characterized by an exponential decay of the autocorrelations, i. e., $r(k) \sim \rho^k$ as $k \to \infty$ for some $\rho \in (0, 1)$, resulting in a summable autocorrelation function $0 < \sum_k r(k) < \infty$. Heuristically, long-range dependence manifests itself in the presence of cycles of all frequencies and orders of magnitude and, as a result, displays features that are commonly attributed to nonstationarity.

When working in the frequency domain, long-range dependence manifests itself in a spectral density that obeys a power-law behavior near the origin. In fact, equivalently to (1) (under weak regularity conditions on the slowly varying function L_1), there is long-range dependence in X if

$$f(\lambda) \sim \lambda^{-\gamma} L_2(\lambda), \quad \text{as } \lambda \to 0, \tag{2}$$

where $0 < \gamma < 1$, L_2 is slowly varying at the origin, and $f(\lambda) = \sum_k r(k) \exp(ik\lambda)$ denotes the spectral density function. Thus, from the point of view of spectral analysis, long-range dependence implies that $f(0) = \sum_k r(k) = \infty$, that is, it requires a spectral density which tends to infinity as the frequency λ approaches zero ("$1/f$-noise"). On the other hand, short-range dependence is characterized by a spectral density function $f(\lambda)$ which is positive and finite for $\lambda = 0$.

Historically, the importance of the Joseph effect lies in the fact that it provides an elegant explanation and interpretation of an empirical law that is commonly referred to as *Hurst's law* or the *Hurst effect*. Briefly, for a given set of observations $(X_k : k = 1, 2, \ldots, n)$ with sample mean $\bar{X}(n)$ and sample variance $S^2(n)$, define the *rescaled adjusted range* or the *R/S-statistic* by

$$R(n)/S(n) = (\max\{0, W_1, W_2, \ldots, W_n\} - \min\{0, W_1, W_2, \ldots, W_n\})/S(n), \quad (3)$$

with $W_k = (X_1 + X_2 + \cdots + X_k) - k\bar{X}(n), 1 \leq k \leq n$. Hurst (e. g., see [38]) found that many naturally occurring time series appear to be well represented by the relation

$$\mathrm{E}[R(n)/S(n)] \sim cn^H, \quad \text{as } n \to \infty, \quad (4)$$

with *Hurst parameter* $0.5 < H \leq 1.0$, and c a finite positive constant that does not dependent on n. If the observations X_k come from a short-range dependent model, then it is shown in [56] that

$$\mathrm{E}[R(n)/S(n)] \sim dn^{0.5}, \quad \text{as } n \to \infty, \quad (5)$$

(d a finite positive constant, independent of n). This discrepancy between relations (4) and (5) is generally referred to as the *Hurst effect* or the *Hurst phenomenon*.

From a statistical point of view, the most salient feature of long-range dependent processes is that the variance of the arithmetic mean decreases more slowly than the reciprocal of the sample size; that is, it behaves like $n^{-\beta}$ for some $\beta \in (0, 1)$, instead of like n^{-1} for processes with short-range dependence. For our discussion below, we assume for simplicity that the slowly varying functions L_1 and L_2 in (1) and (2), respectively, are asymptotically constant. Moreover, for each $m \geq 1$, let $X^{(m)} = (X_k^{(m)} : k \geq 1)$ denote a new covariance stationary process obtained by averaging the original time series X over non-overlapping blocks of size m; i.e., $X_k^{(m)} = m^{-1}(X_{km-m+1} + \cdots + X_{km}), k \geq 1, m \geq 1$. It is shown in [13] that a specification of the autocorrelation function satisfying (1) (or equivalently, of the spectral density function satisfying (2)) is the same as a specification of the sequence $(\mathrm{var}(X^{(m)}) : m \geq 1)$ with the property

$$\mathrm{var}(X^{(m)}) \sim am^{-\beta}, \quad \text{as } m \to \infty, \quad (6)$$

where a is a finite positive constant independent of m, and $0 < \beta < 1$; in fact, the parameter β is the same as in (1) and is related to the parameter γ in (2) by $\beta = 1 - \gamma$. On the other hand, for covariance stationary processes that exhibit short-range dependence, it is easy to see that the sequence $(\mathrm{var}(X^{(m)}) : m \geq 1)$ satisfies

$$\mathrm{var}(X^{(m)}) \sim bm^{-1}, \quad \text{as } m \to \infty, \quad (7)$$

where b is a finite positive constant independent of m. The consequences of the slowly decaying variances $\mathrm{var}(X^{(m)})$ for classical statistical tests and confidence or prediction intervals can be disastrous (e. g., see [4], [32], [42]), since usual standard errors (derived for conventional models) are wrong by a factor that tends to infinity as the sample size increases!

The two best known examples of stationary stochastic models with long-range dependence are certain *fractional Gaussian noise models* and *fractional ARIMA processes*. Fractional Gaussian noise models are Gaussian processes with correlation function $r(k) = \frac{1}{2}(|k+1|^{2H} - |2k|^{2H} + |k-1|^{2H})$, $k \geq 0$, $H \in (0,1]$; that is, they exhibit long-range dependence if $1/2 < H \leq 1$ and are independent for $H = 1/2$. The case $0 < H < 1/2$ where $\sum_k r(k) = 0$ is somewhat artificial and less interesting for modeling purposes and statistical applications. Fractional Gaussian noise is the increment process of *fractional Brownian motion* $b_H = (b_H(t), t \in \mathbb{R})$ with parameter $0 < H \leq 1$, i. e., a Gaussian process with mean zero and covariance $\mathrm{E}[b_H(s)b_H(t)] = \frac{1}{2}(|s|^{2H} + |t|^{2H} - |t-s|^{2H})$. Fractional Brownian motion was originally introduced in [40] and brought to the attention of probabilists and statisticians by Mandelbrot and his co-workers (see [56], [57], [58], [59], or [55]). Fractional Brownian motion will play a crucial role in Section 3 for generalizing the ordinary Black–Scholes option pricing model in order to allow for the possibility of the Joseph effect in stock prices. Finally, fractional ARIMA models were introduced in [30] and [36] and extend in a natural way the standard ARIMA(p, d, q) models defined in [11] to allow for non-integer values of the parameter d (for more details, see [3] and [36]).

2.2 Statistical Inference Methods. With large data sets, trying to fit traditional (i. e., short-range dependent) models to "truly" long-range dependent data is equivalent to approximating a hyperbolically decaying autocorrelation function by a sum of exponentials. Although always possible, the number of parameters needed will tend to infinity as the sample size increases. Thus, long-range dependent processes become a necessity from the point of view of *parsimonious* modeling of large data sets that exhibit the Joseph effect. On the other hand, given a finite set of data, it is in principle not possible to decide whether the asymptotic relationships (1), (2), (4), etc. hold or not. For short-range dependent processes, the correlations will eventually decrease exponentially, continuity of the spectral density function at the origin will eventually show up, the variances of the aggregated processes $X^{(m)}$ will eventually decrease as m^{-1}, and the rescaled adjusted range will eventually increase as $n^{0.5}$. For finite sample sizes, distinguishing between these asymptotics and the ones corresponding to long-range dependent processes is, in general, problematic and empirical studies of one and the same data set can result, therefore, in very different conclusions, depending on the statistical methods used (see Section 2.3 below).

However, there has been considerable progress recently in developing a theory of statistical inference for long-range dependent processes. While many problems remain unsolved, for some commonly encountered situations, suitable statistical methods are now sufficiently well understood to be used by a larger class of data analysts. Below we review three such methods; namely (i) the *variance–time function method* related to the slowly decaying variance property (6), (ii) the *R/S-analysis* related to the Hurst effect (4), and (iii) a periodogram-based approximate maximum likelihood type method *(Whittle's estimate)* related to property (2) of the spectral density. While methods (i) and (ii) are heuristic in nature and result in point estimates of the Hurst parameter H, method (iii) is more rigorous and more adequate for a refined data analysis, including confidence intervals for H.

We have seen in Section 2.1 that long-range dependence gives rise to variances of the aggregated processes $X^{(m)}$, $m \geq 1$, that decrease linearly (for large m) in log–log plots against m with slopes arbitrarily flatter than -1 (see (6)). On the other hand, none of the short-range dependent processes commonly considered in the finance literature yield a power-law for the variances of the form (6); it can be approximated for some transient period of time by short-range dependent models with a large number of parameters, but the variance of $X^{(m)}$ will eventually decrease linearly in log–log plots against m with a slope equal to -1 (see (7)). The so-called *variance–time plots* are obtained by plotting $\log(\text{var}(X^{(m)}))$ against $\log(m)$ ("time") and by fitting a simple least-squares line through the resulting points in the plane, ignoring the small values for m. Values of the estimate $\hat{\beta}$ of the asymptotic slope between -1 and 0 indicate long-range dependence, and an estimate for the degree of long-range dependence is given by $\hat{H} = 1 - \hat{\beta}/2$. Clearly, variance–time plots are not reliable for empirical records with small sample sizes. However, for reasonably large empirical records, "eyeball tests" such as the variance–time plots become highly useful and give a rather accurate picture about the presence or absence of the Joseph effect in the underlying time series and, in the former case, about the degree of long-range dependence as measured by the Hurst parameter.

The objective of the R/S-analysis of an empirical record is to infer the value of the Hurst parameter in relation (4) for the long-range dependent process that presumably generated the record under consideration. In practice, R/S-analysis is based on a heuristic graphical approach (originally described in detail in [59], see also [55]) that tries to exploit as fully as possible the information in a given record. The following graphical method has been used extensively in the past. Given a sample of N observations $(X_k \colon k = 1, 2, \ldots, N)$, one subdivides the whole sample into K non-overlapping blocks and computes the rescaled adjusted range $R(t_i, n)/S(t_i, n)$ for each of the new "starting points" $t_1 = 1$, $t_2 = N/K + 1$, $t_3 = 2N/K + 1,\ldots$ which satisfy $(t_i - 1) + n \leq N$. Here, the R/S-statistic $R(t_i, n)/S(t_i, n)$ is defined as in (3) with W_k replaced by $W_{t_i+k} - W_{t_i}$ and $S^2(t_i, n)$ is the sample variance of $X_{t_i+1}, X_{t_i+2}, \ldots, X_{t_i+n}$. Thus, for a given value ("lag") of n, one obtains many samples of R/S, as many as K for small n and as few as one when n is close to the total sample size N. Next, one takes logarithmically spaced values of n, starting with about $n = 10$. Plotting $\log(R(t_i, n)/S(t_i, n))$ versus $\log(n)$ results in the *rescaled adjusted-range plot* (also called the *pox diagram of R/S*). When the parameter H in relation (4) is well defined, a typical rescaled adjusted-range plot starts with a transient zone representing the nature of short-range dependence in the sample, but eventually settles down and fluctuates in a straight "street" of a certain slope. Graphical R/S-analysis is used to determine whether such asymptotic behavior appears supported by the data. In the affirmative case, an estimate $\hat{H}$ of the Hurst parameter H is given by the street's asymptotic slope (typically obtained by a simple least-squares fit) which can take any value between $1/2$ and 1. With regard to the effectiveness of R/S-analysis as a function of the sample size, similar comments as above apply. Further shortcomings will be discussed in Section 2.3 below. For practical purposes, the most useful and attractive feature of the R/S-analysis is its relative robustness against changes of the marginal distribution. This feature allows for practically separate investigations of the Joseph effect in a given empirical record and of its distributional characteristics.

The absence of any results for the limit laws of the Hurst parameter estimates obtained via the variance–time plot method or the R/S-analysis makes these methods inadequate for a more refined data analysis requiring, for example, confidence intervals for $\hat{H}$, model selection criteria, and goodness of fit tests. In contrast, a more refined data analysis is possible with maximum likelihood type estimates and related estimation methods based on the periodogram. In particular, for Gaussian processes $X = (X_k : k \geq 0)$, Whittle's approximate maximum likelihood estimator has been studied extensively (see [72], [3], [27], and [18]) and is defined as follows. Let $f(x; \theta) = \sigma_\epsilon^2 f(x; (1, \eta))$ be the spectral density of X with $\theta = (\sigma_\epsilon^2, \eta) = (\sigma_\epsilon^2, H, \theta_3, \ldots, \theta_k)$, where $H = (\gamma + 1)/2$ (with γ as in (2)) is the Hurst parameter and $\theta_3, \ldots, \theta_k$ model the short-range dependence structure of the process. As the scale parameter, we use the variance σ_ϵ^2 of the innovation ϵ in the infinite autoregressive representation of the process, i.e., $X_j = \sum_{i \geq 1} \alpha_i X_{j-i} + \epsilon_j$, with $\sigma_\epsilon^2 = \operatorname{var}(\epsilon_j)$. Note that this implies $\int \log(f(x; (1, \eta))) \, dx = 0$. The Whittle estimator $\hat{\eta}$ of η minimizes

$$Q(\eta) = \int_{-\pi}^{\pi} \frac{I(x)}{f(x; (1, \eta))} \, dx, \tag{8}$$

where $I(\cdot)$ denotes the *periodogram* of X; i.e.,

$$I(x) = \frac{1}{2\pi n} \left| \sum_{j=1}^{n} X_j \exp(ijx) \right|^2,$$

and the estimate of σ_ϵ^2 is given by $\hat{\sigma}_\epsilon^2 = \int_{-\pi}^{\pi} I(x)/f(x; (1, \hat{\eta})) \, dx$. Then $\sqrt{n}(\hat{\theta} - \theta)$ is asymptotically normally distributed if X can be written as an infinite moving average process. For Gaussian processes, $\hat{\theta}$ has the same asymptotic distribution as the maximum likelihood estimator and is asymptotically efficient. Note that in the context of periodogram-based methods, loss of efficiency is often due to (i) deviations from Gaussianity or (ii) deviations from the assumed model spectrum.

Transforming the data so as to obtain approximately the desired marginal (normal) distribution is generally considered a viable method to overcome (i). For problem (ii), there are several proposals in the literature, including estimating H from the periodogram ordinates at low frequencies only or to bound the influence of $I(x)$ at high frequencies (see for example [29], [29], [64]). Finally, for large data sets, a powerful alternative and more direct method for tackling (ii) uses Whittle's estimate in combination with the method of aggregation (see [44]).

2.3 Empirical Studies of Asset Returns: A Survey. The presence of the Joseph effect is well documented in many economic time series (e. g., see [9] for gold prices, [10] for foreign exchange rates, [35] for future contracts, [21] for U.S. post-war quarterly real gross national product (GNP) and other macroeconomic time series, [35] for future prices for soybeans, soybean oil, and soybean meal), and its existence is commonly accepted in the economics literature. In contrast, the question of whether or not returns on common stocks also exhibit the Joseph effect has been typically met with more skepticism and the literature on this problem accurately reflects this attitude. Early investigations into the dependence structure of asset returns (see [23], [24]) used simple significance tests for the first few autocorrelation coefficients and – finding them to be very close to zero – concluded that successive returns can be assumed to be independent, i. e., stock price returns follow a random walk or, equivalently, stock prices follow a geometric random walk. Notice that this observation, together with the assumed Gaussian nature of the marginal distributions, implies a stock price process that has become synonymous with modern finance theory, namely the Black–Scholes model introduced in [8] and [60].

Among the first to have considered the possibility of long-run statistical dependence in asset returns was Mandelbrot (see [53]). Since then, a number of studies have provided empirical evidence in favor of Mandelbrot's arguments and against the random walk model. For example, in [31] Greene and Fielitz use the R/S-analysis to study 200 daily stock return series of securities listed on the New York Stock Exchange, and they claim to have found significant long-range dependence in many of these series. More recent results about long-run swings in long-horizon stock returns reported in [25] and [62] can be interpreted as further evidence in favor of the Joseph effect in returns on common stocks. More recently, the random walk model has again received intense scrutiny. For example, Lo and MacKinlay [49] analyze market returns from a 25-year period (1962–1987) – the same data considered in [31] – and strongly reject the random walk hypothesis, not only for the whole period but also for all subperiods and for a number of aggregate-returns indices and size-sorted portfolios. Having observed such substantial short-range dependence in the data, in [48] Lo then re-examines the question of long-range dependence in asset returns and – using a modified R/S-analysis (see below) – reports, contrary to the findings in [31] and others, that there is no statistical evidence of long-range dependence in the data once the effects of short-range dependence have been accounted for. In particular, Lo claims that the results in [31] are due to the fact that the standard R/S-analysis is sensitive to the presence of short-range dependence. In view of a recent convergence result obtained in [43], Lo's findings suggest that the Black–Scholes model may be a good approximation to a broad range of time series models for the discrete asset return processes, including models with substantial short-range dependences. On the other hand, combining the results in [43] with the claims made by Greene and Fielitz [31] shows that the Black–Scholes model would not be an adequate process for stock prices but should be replaced by a model that uses the fractional Brownian motion as driving process for the stock returns.

While the empirical studies mentioned above all rely on a R/S-analysis-based approach for checking for the presence of the Joseph effect in stock return data, they nevertheless give rise to very different conclusions with quite opposite implications – often for the same data! To illustrate this, note that Greene and Fielitz [31] use the method for R/S-analysis originally introduced in [38], refined in [59] and briefly described in Section 3.2. This method has been found to be largely insensitive to changes in the marginal distribution; data analysts are especially attracted by its reported robustness with respect to infinite variance distributions such as stable laws or Pareto distributions (see [58] and [55]). Moreover, the method is generally viewed to discriminate reliably between long-range and short-range dependence, given that a large enough amount of data is available. Clearly, the size of the data set is crucial for the latter observation to hold because graphical R/S-analysis as proposed in [59] can only be successful when the asymptotic behavior of the rescaled adjusted-range statistic is clearly evident. For small to moderately large data sets, a slope estimate of the pox plot of R/S between 0.5 and 1 is often not an indication of long-range dependence in the data but is simply due to the fact that the R/S-statistic has not yet reached the "asymptotic regime". In this case, the presence of a substantial amount of short-range dependence is generally responsible for an observed H-value above 0.5. From this perspective, it is certainly plausible to re-interpret the findings in [31] along the lines suggested in [48], especially given the only moderately large time series considered and the observed H-values which are typically only slightly above 0.5.

In order to overcome the sensitivity of the originally proposed R/S-method, Lo [48] introduced a *modified R/S-statistic* that is obtained by replacing the denominator S (sample standard deviation) by a consistent estimator of the square root of the partial sum's variance. Note that in the case of short-range dependence, this variance is not simply the sum of the variances of the individual terms but also includes autocovariances up to some lag q where the truncation lag q has to be chosen with some consideration of the data at hand. Lo provides distributional results for this modified test statistic and claims that it has good robustness properties with respect to short-range dependence. Moreover, Monte Carlo experiments in [48] indicate that the modified R/S-test has reasonable power against certain long-range dependence models. In this sense, Lo's conclusion that there is no evidence of the Joseph effect in daily or monthly stock return indices suggests that the dynamic behavior of asset returns may be adequately described by more traditional (i. e., short-range dependent) time series models.

However, although Lo's modified R/S-statistic is a definitive improvement over the original version for which no distributional results exist, his method does not appear to provide the "ultimate" test for long-range dependence. A number of problems remain and the results using his method still have to be interpreted with caution. For example, in [12] it is observed that the modified R/S-test is sensitive to the presence of a large negative moving-average component and to infrequent shifts in the mean. In a different context, one of the authors (W. W.) has observed interesting differences between the results obtained using the modified R/S-statistic and those obtained using a combination of the original R/S-statistic and the aggregation method (see [44]). Notice that the combined R/S-aggregation method has the same objective as Lo's modification of R/S, namely to make the test more robust with respect to short-range dependence. More precisely, we frequently encountered the following situation when dealing with large data sets (100,000 observations and more) where even after aggregating the original time series over blocks of size 1000 or more, the aggregated sequence still contains a few hundred observations: while an elaborate (standard) R/S-analysis of all the aggregated time series provided a convincing argument for the presence of long-range dependence in the original time series, a direct application of the modified R/S-statistic to the original series showed that there is no statistical evidence for long-range dependence in the data. Obviously, these problems are due to the presence and choice of the truncation lag q in the modified R/S-statistic; a large q often means that too much of the dependence due to the Joseph effect is being discounted which makes the modified R/S-test statistic clearly inadequate for detecting "true" long-range dependence (see also [49] for problems when q is large relative to the sample size).

To summarize, a survey of the literature on the Joseph effect in stock returns shows that there is inconclusive statistical evidence in favor for or against the presence of long-range dependence in returns from common stocks. Put differently, if there is indeed some form of long-range dependence present in stock return data, it will not be easily detected by relying on a single test method such as the R/S-based tests and variations thereof. What is needed is a "portfolio" of different statistical methods for testing for long-range dependence, including R/S-based methods, variance–time techniques and periodogram-based approaches, preferably in combination with the powerful aggregation method and suitable transformation of the data. In the absence of such a "full-fledged" empirical study of the Joseph effect in stock prices, stochastic modeling of stock returns or stock prices should allow for the possibility of long-range dependence; that is, it should include long-range dependent processes as well as the traditional short-range dependent models. Currently used models typically are not able to capture long-run statistical dependences and are therefore inadequate for investigating potential implications of the Joseph effect for the modern theory of finance.

2.4 Modern Finance Theory and the Joseph effect. One of the first to have considered and realized the implications of long-range dependence in asset returns was Mandelbrot [53]. In particular, he showed that the question whether or not the Joseph effect is present in returns is directly related to the question whether or not markets are *efficient* (see [31] for an short discussion). Briefly, a securities market is efficient if every price reflects all the available and relevant information; newly arrived information is promptly arbitraged away. In other words, an arbitraged price, if implemented, must follow a martingale (see also [24], [66]) or – using the terminology of modern finance theory – there exists an *equivalent martingale measure* for the securities price process. One of Mandelbrot's [53] results states that in the presence of the Joseph effect, perfect arbitraging does not lead to a martingale process, i. e., the price process cannot be changed into a martingale via an equivalent change of measure. Clearly, this causes problems for the pricing of derivative securities (such as options and futures) using modern martingale methods. At a more basic level, it has been shown recently (see for example [50]) that Mandelbrot's observation when cast in modern finance language means that long-range dependence is inconsistent with the semimartingale framework, commonly assumed in the modern theory of finance. As noted in [50], since much of the modern theory of stochastic calculus is restricted to semimartingales, price processes with long-range dependence represent examples where assuming that asset prices follow semimartingales is a substantive restriction and not a mere regularity condition.

In addition to these implications of the Joseph effect for many of the paradigms used in modern finance theory, Lo [48] also observed that (i) traditional statistical tests of the capital asset pricing model and the arbitrage pricing theory are no longer valid because the statistical inference methods used do not apply to time series with long-range dependence, and (ii) conclusions of more recent tests of efficient markets hypotheses or stock market rationality depend crucially on the presence or absence of the Joseph effect.

§3 THE FRACTIONAL BLACK–SCHOLES MODEL

In this section, we present an explicit construction of a class of stock price processes which allows for the possibility of long-range dependence, while satisfying dynamics analogous to those of the well-known Black–Scholes model. The basic ingredient of our construction is *fractional Brownian motion* with parameter $0 < H < 1$ as introduced in [56]. Since we are mainly interested in capturing the Joseph effect, our presentation below concentrates on the case $1/2 < H < 1$; with slightly more work, the definitions and results below also apply to the case $0 < H < 1/2$ which is, however, of less importance from a modeling point of view. Our approach relies on methods from *nonstandard analysis* as presented, for example, in [1], [15], [45], or [37]. Note that for the statement of our results, it is not necessary to know any nonstandard analysis. However, some of the notions we work with (such as Donsker–type invariance principles) have a very natural equivalent formulation in nonstandard terms, which as well as being of intrinsic interest provide straightforward proofs of many of our results.

3.1 Preliminaries and Notation. We assume familiarity with the basics of nonstandard analysis – see for example [1, 2, 15, 16, 17, 37, 39, 45]. For an infinite integer N, set $\Delta t = 1/N$ and consider the *hyperfinite time line* $\mathbb{T}_N = \{0, \Delta t, 2\Delta t, \dots, N\Delta t = 1\}$ with the uniform counting measure P_N. Let $\Omega_N^0 = \{-1, 1\}^{\mathbb{T}_N}$ and let $\boldsymbol{\Omega}_N^0 = (\Omega_N^0, \mathcal{A}_N^0, P_N^0)$ denote the hyperfinite probability space corresponding to the coin toss experiment on $\mathbb{T}_N$. Anderson's construction [2] of Brownian motion $b = (b(t): 0 \leq t \leq 1)$ on the Loeb space $(\Omega_N^0, L(\mathcal{A}_N^0), L(P_N^0))$ gives b as the standard part $^\circ B_N^0$ of the hyperfinite random walk $B_N^0(\omega, t) = \sum_{s<t} \omega(s)\sqrt{\Delta t}$ for $(\omega, t) \in \Omega_N^0 \times \mathbb{T}_N$. Anderson showed that almost all paths of B_N^0 are S-continuous, and his construction provides a rigorous basis for the intuition that Brownian motion contains a built-in version of a random walk with infinitesimal steps. In standard language, Anderson's result provides an easy proof of Donsker's invariance principle (see [7]):

$$B_n^0 \to b \qquad \text{weakly, as } n \to \infty.$$

In the following we will need Anderson's construction of Brownian motion on $\mathbb{R}$, which uses the hyperfinite time line $\mathbb{T}_N^\# = \{k\Delta t: -N^2 \leq k \leq N^2\}$ and the corresponding hyperfinite coin toss space $\boldsymbol{\Omega}_N = (\{-1, 1\}^{\mathbb{T}_N^\#}, \mathcal{A}_N, P_N)$. For $t \in \mathbb{T}_N^\#$ with $t < 0$ we define $B_N(\omega, t) = \sum_{t \leq s < 0} \omega(s)\sqrt{\Delta t}$ so that for all $s < t \in \mathbb{T}_N^\#$ we have

$$B_N(\omega, t) - B_N(\omega, s) = \sum_{s \leq u < t} \omega(u)\sqrt{\Delta t} = \sum_{s \leq u < t} \Delta B_N(\omega, u)$$

where $\Delta B_N(\omega, u) = B_N(\omega, u + \Delta t) - B_N(\omega, u) = \omega(u)\sqrt{\Delta t}$.

3.2 Fractional Brownian Motion as the Limit of Fractional Random Walks. Fractional Brownian motion $b_H = (b_H(t): 0 \leq t \leq 1)$ on $[0, 1]$ with parameter H $(1/2 < H < 1)$, normalised so that $\mathrm{E}[b_H^2(1)] = 1$, is a Gaussian process with continuous sample paths and covariance function

$$R(s, t) = \tfrac{1}{2}(s^{2H} + t^{2H} - (t - s)^{2H}), \qquad 0 \leq s < t \leq 1. \tag{9}$$

It is easy to see that the increments $b_H(t) - b_H(t - 1)$ of b_H form a stationary process with autocorrelation function $r(k)$ such that $r(k) \sim c(H)k^{2H-2}$ as $k \to \infty$, where $c(H)$ is a positive constant independent of k. Thus, $r(k)$ satisfies (1) as long as $1/2 < H < 1$.

Using Anderson's construction of the Wiener integral, a nonstandard construction of b_H is obtained as follows. We regard $N \in {}^*\mathbb{N} \setminus \mathbb{N}$ as fixed for the moment, so we shall drop the suffix N and write $B = B_N$, $\mathbb{T} = \mathbb{T}_N$, $\Omega = \Omega_N$, etc. Set $\alpha = H - \tfrac{1}{2}$, and consider the internal (hyperfinite) process $B_H : \Omega \times \mathbb{T} \to {}^*\mathbb{R}$ defined by

$$B_H(\omega, t) = c_H \sum_{-N \leq s < t} F(t, s)\Delta B(\omega, s) \tag{10}$$

where c_H is a standard positive constant, defined below, that depends only on H, and where the internal function $F : \mathbb{T} \times \mathbb{T}^{\#} \to {}^{*}\mathbb{R}$ is given by

$$F(t,s) = \begin{cases} (t-s)^{\alpha} - (-s)^{\alpha} 1_{\{s \leq 0\}}(s), & \text{if } s \leq t, \\ 0, & \text{otherwise.} \end{cases} \tag{11}$$

We call B_H a *hyperfinite fractional random walk* with parameter $1/2 < H < 1$ on $(\Omega, \mathcal{A}, P)$, a terminology that is justified below. The formulation (10) is motivated by the well-known integral representation for b_H [54], namely

$$b_H(t) = c_H \int_{-\infty}^{t} \left[(t-s)^{\alpha} - (-s)^{\alpha} 1_{\{s \leq 0\}}(s) \right] db(s) \tag{12}$$

$$= c_H \int_{-\infty}^{t} f(t,s) \, db(\omega, s) \tag{13}$$

say, where $f = {}^{\circ}F$ and

$$c_H = \left(\int_{-\infty}^{0} [(1-u)^{\alpha} - (-u)^{\alpha}]^2 \, du + \int_{0}^{1} (1-u)^{2\alpha} \, du \right)^{1/2}.$$

We now have the following Donsker-type invariance principle for fractional random walks. We write $B_{H,n}$ to denote the fractional random walk defined as in (10) but with n in place of N, for all $n \in \mathbb{N}$.

Theorem 3.1.

(a) *For almost all ω, B_H is S-continuous and*

$${}^{\circ}B_H(\omega, t) = b_H(\omega, {}^{\circ}t) \text{ for all } t \in \mathbb{T}.$$

(b) $B_{H,n} \to b_H$ *weakly, as $n \to \infty$.*

Proof. (a) From Anderson's hyperfinite formulation [2] of the Itô integral, and the fact that for fixed t, $F(t,s)$ is an SL^2 lifting of $f(t,s)$, it follows that ${}^{\circ}B_H(\omega, t)$ is an SL^2 lifting of $b_H(\omega, {}^{\circ}t)$ and that $\mathrm{E}[(B_H(\omega, t') - B_H(\omega, t))^2] = (t' - t)^{2H}$. From this, using the "Kolmogorov continuity" theorem [1, Theorem 4.8.4], it follows that B_H is S-continuous for almost all ω.

(b) As in the case of Brownian motion (see [2]), the weak convergence result follows immediately from (a). ☐

Remarks.

(1) Our construction of fractional Brownian motion as a hyperfinite fractional random walk parallels Anderson's construction of ordinary Brownian motion as a hyperfinite random walk (with infinitesimal steps). For a slightly different approach see Lindstrøm [46], where white noise integrals are used to construct *fractional Brownian fields* (with fractional Brownian motion

as the special case of dimension one). This approach can be used to give an alternative nonstandard construction rather similar to ours.

(2) Theorem 3.1 is a special case of a more general result proved without nonstandard methods in Davydov [19]. Davydov gives conditions under which linear time series models (such as (10)) converge weakly to b_H.

3.3 The Fractional Black–Scholes Model and an Invariance Result.

The classical Black–Scholes option pricing model [8] assumes that the price of a stock is represented by a process $s(\omega, t)$ satisfying the differential equation

$$ds(\omega, t) = s(\omega, t)(\sigma\, db(\omega, t) + \mu\, dt)$$

which has the solution

$$s(\omega, t) = s_0 \exp\left(\sigma b(\omega, t) + (\mu - \frac{1}{2}\sigma^2)t\right). \qquad (14)$$

The question arises as to what is the corresponding *fractional* version of the Black–Scholes option pricing model. By analogy, we expect a price process s_H obeying a differential equation

$$ds_H(\omega, t) = s_H(\omega, t)(\sigma\, db_H(\omega, t) + \mu\, dt)$$

but since b_H is not a semimartingale (see Section 4 below) it is hard to see how to solve this. We can, however, at the discrete hyperfinite level, define the hyperfinite *fractional* version of the Cox–Ross–Rubinstein option pricing model [14, henceforth CRR] with parameter $1/2 < H < 1$ by setting

$$S_H(\omega, t) = s_0 \prod_{u<t}\left(1 + \sigma\Delta B_H(\omega, u) + \mu\Delta t\right) \qquad (15)$$

(where $\sigma > 0$, $\mu \in \mathbb{R}$ are standard) so that

$$\Delta S_H(\omega, t) = S_H(\omega, t)\left(\sigma\Delta B_H(\omega, t) + \mu\Delta t\right). \qquad (16)$$

Note that the hyperfinite price process S_H is obtained from the ordinary (hyperfinite) geometric random walk model of CRR [14] (see [17]) by replacing the hyperfinite random walk B as driving process by the hyperfinite fractional random walk B_H. Subsequently, the price process S_H defined by (15) and the corresponding option pricing model are called (hyperfinite) *geometric fractional random walk* and *fractional Cox-Ross-Rubinstein model*, respectively. We can now prove the following main result.

Theorem 3.2. *Let $\frac{1}{2} < H < 1$. For almost all $\omega \in \Omega$, $S_H(\omega, t)$ is S-continuous, and for all $t \in \mathbb{T}$*

$$^{\circ}S_H(\omega, t) = s_0 \exp(\sigma b_H(\omega, {}^{\circ}t) + \mu\, {}^{\circ}t)$$

First we prove the following Lemma.

Lemma 3.3. *There is a finite constant C such that for each $t \in \mathbb{T}$*

$$E[(\Delta B_H(t))^2] = \Delta t^{2H}(1 + C(t))$$

where $C(t) \approx C$. Hence, if $\frac{1}{2} < H < 1$, then

$$E\left[\sum_{0 \le t < 1}(\Delta B_H(t))^2\right] \approx 0.$$

Proof. Elementary calculation shows that

$$E[(\Delta B_H(t))^2] = \sum_{0 < s < N+t}((s + \Delta t)^\alpha - s^\alpha)^2 \Delta t + \Delta t^{2H}. \qquad (17)$$

Now the mean value theorem gives $(s + \Delta t)^\alpha - s^\alpha = \alpha \Delta t(s + \delta_s)^{\alpha-1}$ with $0 < \delta_s < \Delta t$. Writing $s + \delta_s = (k + \varepsilon_k)\Delta t$ (with $0 < \varepsilon_k < 1$) and $M\Delta t = N + t$, the first term on the right of (17) is

$$\sum_{0 < k < M}(\alpha \Delta t)^2(k + \varepsilon_k)^{2\alpha-2}\Delta t^{2\alpha-2}\Delta t = \alpha^2 \Delta t^{2H}\sum_{0 < k < M}(k + \varepsilon_k)^{2H-3}.$$

Let $C = \alpha^2 \sum_{k=1}^{\infty}(k + {}^\circ\varepsilon_k)^{2H-3}$, which is finite since $2H - 3 < -1$, and the result follows. □

Proof of Theorem 3.2. Suppressing the dependence on ω, take any path with $B_H(\cdot)$ S-continuous and with $\sum_{0 \le t < 1}\Delta B_H(t)^2 \approx 0$ (using Lemma 3.3). We have

$$\log S_H(u) - \log s_0 = \sum_{t < u}\log(1 + \sigma \Delta B_H(t) + \mu \Delta t) = \sigma B_H(u) + \mu u + \sum_{t < u}r(t)$$

say, where $r(t) = \log(1 + \sigma \Delta B_H(t) + \mu \Delta t) - (\sigma \Delta B_H(t) + \mu \Delta t)$. Now since $\Delta B_H(t) \approx 0$, we have $|r(t)| \le c(\Delta B_H(t)^2 + \Delta t^2)$ for some finite constant c, and so $|\sum_{t < 1}r(t)| \approx 0$. Thus $\log S_H(t)$ is S-continuous, and

$$\log S_H(t) - \log s_0 \approx \sigma B_H(t) + \mu t \approx \sigma b_H({}^\circ t) + \mu {}^\circ t$$

for all $t \in \mathbb{T}$. Taking exponentials completes the proof. □

Hence we make the following

Definition 3.4. The fractional Black–Scholes price model for a stock is a geometric fractional Brownian motion with parameter $1/2 < H < 1$, i.e.,

$$s_H(\omega, t) = s_0 \exp(\sigma b_H(\omega, t) + \mu t), \quad 0 \le t \le 1.$$

From Theorem 3.2, we obtain immediately the following invariance result for geometric fractional random walks.

Corollary 3.5. *Let $S_{H,n}$ denote the (standard) discrete-time geometric fractional random walk process with parameter $1/2 < H < 1$ defined as in (15) but with $n \in \mathbb{N}$ in place of N (so that $\Delta t = 1/n$); i. e. $S_{H,n}(\omega, t) = s_0 \prod_{u < t}(1 + \sigma \Delta B_{H,n}(\omega, u) + \mu \Delta t)$. Then*

$$S_{H,n} \to s_H \quad \text{weakly, as } n \to \infty$$

Proof. Exactly as the corresponding result for Brownian motion. □

Observe that we obtained the fractional Black–Scholes model from an appropriately built-in version of the fractional Cox–Ross–Rubinstein model in the same way as we constructed in [17] the (standard) geometric Brownian motion model of Black–Scholes from the (hyperfinite) geometric random walk model of CRR. As in [17], we will show in Section 4 below, how such a construction allows one to exploit combinatorial properties that are easy to prove for the discrete object (i. e., the hyperfinite fractional CRR model) in a continuous setting (namely, the fractional Black–Scholes model). Also notice that while our approach directly generalizes the classical Black–Scholes setup at the level of the differential and difference equations (14) and (16), respectively, it fails to do so at the level of the explicit formula for the solution of (14) and (16). We shall return to this problem in a later paper.

§4 PROPERTIES AND IMPLICATIONS

Fractional Black–Scholes–type or related models which are able to capture the Joseph effect in stock prices have also been considered in recent papers by Maheswaran and Sims [50] and Kunitomo [43]. These authors show that the Joseph effect is inconsistent with the semimartingale property of the underlying stock price process and thus also with the modern martingale approach to pricing derivative securities (such as options and futures). In particular, they demonstrate that long-range dependent processes (Kunitomo [43] only considers the special case of fractional Brownian motion) cannot be turned into martingales by an equivalent change of measure. By Stricker's theorem [68], this then implies that option pricing models that exhibit long-range dependence allow for *arbitrage opportunities* or *free lunches*, i. e., trading strategies that "produce something out of nothing".

We give below elementary proofs of these properties for both the hyperfinite fractional Black–Scholes model and the fractional CRR models introduced in Section 3. Our purpose is two-fold: first, we illustrate the ease with which non-standard methods yield these results for the discrete (hyperfinite) fractional CRR model and hence often directly for the continuous fractional Black–Scholes model. This is in stark contrast to some of the sophisticated general-purpose stochastic analysis tools used in [50] and [43] (for the semimartingale problem, see also [47, p. 238] and [65]). Second, we show how our elementary approach provides means for identifying and constructing arbitrage opportunities. While the latter is clearly a problem of very practical importance, it is also a highly nontrivial problem given the absence of the semimartingale framework and, as a result, of a useful stochastic calculus.

4.1 On the Semimartingale Property. Mandelbrot and Van Ness [56] showed that for $0 < H < 1$ almost all sample paths of fractional Brownian motion $b_H = (b_H(t) : 0 \le t \le 1)$ are continuous and nowhere differentiable. Consequently, almost all paths of b_H are of unbounded (total) variation, that is, for almost all $\omega \in \Omega$,

$$\sum_{t \in \mathbb{T}_n} |b_H(\omega, t + \Delta t) - b_H(\omega, t)| \to \infty, \quad \text{as } n \to \infty \tag{18}$$

(see also Verwaat [71]). Concerning the quadratic variation process of b_H, Lemma 3.3 shows that for the S-continuous lifting B_H of b_H, we have for all $\omega \in \Omega$,

$$\mathrm{E}\left[\sum_{t \in \mathbb{T}_N} |B_H(\omega, t + \Delta t) - B_H(\omega, t)|^2 \right] = \gamma N^{1-2H}, \tag{19}$$

where γ is finite. Thus for $1/2 < H < 1$, since $\gamma N^{1-2H} \approx 0$, b_H has zero quadratic variation, because

$$\sum_{t \in \mathbb{T}_n} |b_H(\omega, t + \Delta t) - b_H(\omega, t)|^2 \to {}^{\circ}\!\left(\sum_{t \in \mathbb{T}_N} |B_H(\omega, t + \Delta t) - B_H(\omega, t)|^2 \right) = 0,$$

for almost all ω, by (19). This implies that b_H must be of bounded variation. However, this contradicts (18) and shows that b_H is not a semimartingale for $1/2 < H < 1$. Finally, using Definition 3.4, it is easy to see that the process $\sigma b_H(t) + \mu t$ (which is also not a semimartingale) can be written as a convex function of the geometric fractional Brownian motion process $s_H = (s_H(t) : 0 \le t \le 1)$. Hence (see [20, p. 221]), s_H cannot be a semimartingale for all $1/2 < H < 1$, and we have proved the following result.

Proposition 4.1. *For all $1/2 < H < 1$, the processes b_H and s_H are not semimartingales.*

As mentioned in [50], stock price processes such as geometric fractional Brownian motion represent examples where assuming an underlying semimartingale dynamics for the asset price is a substantive restriction and not a mere regularity condition. However, since much of the modern theory of stochastic calculus is restricted to semimartingales, the consequences of allowing for the possibility of the Joseph effect in stock price returns from a financial engineering perspective appear prohibitive. We will return to this point in Section 4.3 below and in Section 5.

4.2 The Equivalent Martingale Measure Problem. The problem of pricing derivative securities (such as options and futures) using modern martingale techniques relies crucially on the fact that the stock price process can be made into a martingale by an equivalent change of the underlying probability measure. The main result in [43] states that for $0 < H < 1$, $H \ne 1/2$, fractional Brownian motion cannot be turned into a martingale by an equivalent measure change,

i. e., the equivalent martingale measure approach to contingent claim valuation as developed in [33] and [34] does not apply for the fractional Black–Scholes models introduced in Section 3 (see also [50]). Here we show how our understanding of the equivalent martingale measure problem in a discrete setting (see for example [70, Theorem 3.1]), namely in the context of the hyperfinite fractional CRR model, yields an elementary explanation of Kunitomo's result.

To this end, we recall that for a discrete process, constructing an equivalent martingale measure reduces to solving for new transition or conditional probabilities which satisfy the martingale property "locally"; i. e., at every node of the tree associated with the given process. Put differently, there does not exist an equivalent martingale measure if for some nodes of the tree, no new transition probabilities exist that are consistent with the martingale property at that node. In the notation used in [70], this happens if 0 is not contained in the relative interior of the convex hull (denoted by ri(conv(·))) generated by the increments of the process at that node. We now show that for the hyperfinite fractional random walk process $B_H = (B_H(t) : t \in \mathbb{T})$ with parameter $1/2 < H < 1$ as defined in (10), there exit nodes $(\omega, t) \in \Omega \times \mathbb{T}$ such that the increment $\Delta B_H(\omega', t)$ remains positive as ω' ranges through the set $[\omega]_t = \{\omega' \in \Omega : \omega'(s) = \omega(s), \ s < t\}$; that is, $0 \notin \mathrm{ri}(\mathrm{conv}(\Delta B_H(\omega', t) : \omega' \in [\omega]_t))$.

Fix $1/2 < H < 1$ and consider the increment

$$\Delta B_H(\omega, t) = B_H(\omega, t + \Delta t) - B_H(\omega, t) \tag{20}$$

$$= c_H Y(\omega, t) + c_H \omega(t) \Delta t^H, \tag{21}$$

where $Y(\omega, t) = \sum_{-N \le s < t} \{(t + \Delta t - s)^\alpha - (t - s)^\alpha\} \omega(s) \sqrt{\Delta t}$. Fixing $t \in \mathbb{T}$, consider any path ω in the set $A_t = \{\omega : \omega(s) = 1, \text{ all } s < t\}$ and note that A_t has positive probability $P(A_t) = 2^{-N(N-t)-1}$. For $\omega \in A_t$, we have

$$\Delta B_H(\omega, t) = c_H \left(\omega(t) \Delta t^H + \sum_{0 < s < N+t} \{(s + \Delta t)^\alpha - s^\alpha\} \omega(s) \sqrt{\Delta t} \right) \tag{22}$$

$$= c_H \sqrt{\Delta t} (\omega(t) \Delta t^\alpha + (N + t + \Delta t)^\alpha - \Delta t^\alpha), \tag{23}$$

since the sum telescopes if $\omega(s) = 1$ for all $s < t$. Clearly, if $\omega(t) = 1$, the increment $\Delta B_H(\omega, t)$ is positive as $c_H > 0$. For $\omega(t) = -1$ we consider

$$\Delta B_H(\omega, t) = c_H \sqrt{\Delta t} ((N + t + \Delta t)^\alpha - 2\Delta t^\alpha) \tag{24}$$

$$\ge c_H \sqrt{\Delta t} (N^\alpha - 2\Delta t^\alpha) \tag{25}$$

which is positive. Since $\Delta S_H(\omega, t) = S_H(\omega, t) (\mu \Delta t + \sigma \Delta B_H(\omega, t))$ we also have that for $t \in \mathbb{T}$ and $\omega \in A_t$, the increment $\Delta S_H(\omega, t)$ of the fractional geometric random walk process S_H with parameter $1/2 < H < 1$ remains positive irrespective of the value of $\omega(t)$. Thus, we have shown the following:

Proposition 4.2.

(a) *Let $1/2 < H < 1$. For $t \in \mathbb{T}$ and $\omega \in A_t$, the increments $\Delta B_H(\omega, t)$ of the fractional random walk B_H and $\Delta S_H(\omega, t)$ of the geometric fractional random walk S_H remain positive, whatever the value of $\omega(t)$.*

(b) *For all sufficiently large $n \in \mathbb{N}$, the fractional random walk $B_{H,n}$ and geometric fractional random walk $S_{H,n}$ $(1/2 < H < 1)$ admit no equivalent martingale measures.*

While Proposition 4.2(b) follows directly from the underflow principle of non-standard analysis (see for example [1, Proposition 1.2.7]), it does not deal with the continuous fractional Black–Scholes model itself and hence, stops short of providing an alternative proof of Kunitomo's [43] result mentioned earlier. We will show in a forthcoming paper how to achieve this goal in the present setting with only slightly more effort. Also for later reference, we conclude this subsection with the following completeness result for the (hyperfinite or standard) fractional CRR models. Recall (for more details, see for example [34] or [70]) that a market model is said to be *complete* if all contingent claims are attainable or, equivalently, if for all $t \in \mathbb{T}$ and all $[\omega]_t$,

$$K([\omega]_t) = \dim\{\operatorname{span}\{S_H(\omega', t + \Delta t) : \omega' \in [\omega]_t\}\} + 1,$$

where for $t \in \mathbb{T}$, $K([\omega]_t) = \operatorname{cardinality}([\omega]_{t+\Delta t} : [\omega]_{t+\Delta t} \subseteq [\omega]_t)$ is called the *splitting index of* $[\omega]_t$. Proposition 4.3 below follows now immediately from the fact that the tree structure corresponding to the fractional geometric random walk is binomial (just as in the case of the ordinary CRR model).

Proposition 4.3. *Let $1/2 < H < 1$. The fractional CRR model $S_{H,n}$ is complete for all $n \in \mathbb{N}$ (or $n \in {}^*\mathbb{N}$).*

4.3 About Arbitrage Opportunities. Stricker's [68] general result about the equivalence between the existence of equivalent martingale measures for continuous sample paths price processes and the absence of arbitrage opportunities applies directly to the geometric fractional Brownian motion price process s_H with $1/2 < H < 1$. In particular, it states that as a result of [43, Theorem 3] the fractional Black–Scholes option pricing model contains arbitrage opportunities. However, Stricker's result gives no information about identifying or actually constructing some of these arbitrage opportunities. For an illustration about what certain asymptotic arbitrage strategies in the contingent claims market (as the risk goes to zero) and spot market (as the rate of trading increases) might look like, see the discussion in [50]. Here we indicate how our understanding of Stricker's result in the discrete setting of the hyperfinite fractional CRR model yields information about arbitrage strategies that is of practical use.

[70, Theorem 3.1] provides an explicit relationship between the equivalent notions of existence of an equivalent martingale measure for discrete processes such as the hyperfinite or standard fractional geometric random walks and the absence of arbitrage opportunities for discrete option pricing models such as the hyper-finite or standard fractional CRR models. In the present setting, however we can see immediately from Proposition 4.2 how to describe an arbitrage opportunity in the hyperfinite fractional CRR model. This illustrates but does not draw upon the general result [70, Theorem 3.1] (which applies here in connection with Proposition 4.2).

Proposition 4.4.

  (a) *The hyperfinite fractional Cox–Ross–Rubinstein option pricing model S_H $(1/2 < H < 1)$ contains arbitrage opportunities.*
  (b) *For all sufficiently large $n \in \mathbb{N}$, the fractional CRR option pricing models $S_{H,n}$ $(1/2 < H < 1)$ contain arbitrage opportunities.*

Proof. (a) We construct a very simple arbitrage opportunity for the hyperfinite CRR market models following the path used in deriving Proposition 4.2. At time zero, the investor places his total capital in (constant) bonds. Fix a time t (say, $t = \frac{1}{2}$). If $\omega \notin A_t$, he does nothing further throughout the interval $[0, 1]$. If $\omega \in A_t$, he transfers all his funds to stock at time t, and sells all these shares immediately at time $t + \Delta t$, after which point the funds remain in bonds until the investment horizon. Since A_t has positive P-measure, it follows that the investor has made a non-zero riskless profit buy using this strategy.

  (b) This follows immediately by the underflow principle of nonstandard analysis. □

An interesting theoretical problem centres on the fractional Black–Scholes model itself: it should be possible to identify a set of paths of positive measure where similar arbitrage strategies are possible, since s_H is not a semimartingale, and since S_H is a lifting of s_H. The proof of Proposition 4.2 indicates that there is considerable scope for the increments to remain positive, even when a proportion of time points $s \in \mathbb{T}$ for which $\omega(s) = -1$ is significant. This suggests the possibility of identifying a set of positive P-measure (hence including S-continuous paths) where this behavior is retained, and hence of determining a finite time interval on which the price increments remain positive throughout. We shall return to these issues in a later paper.

Thinking of practical implications, Proposition 4.4(b) is likely to provide the most useful result. It states that for sufficiently large $n \in \mathbb{N}$, there exist nodes in the tree corresponding to the fractional geometric random walk process $S_{H,n}$ where the construction of an equivalent martingale measure for $S_{H,n}$ will fail or, equivalently, where arbitrage opportunities can be identified and an arbitrage strategy can be constructed explicitly. For example, this objective can be accomplished in a systematic manner by applying the algorithm presented in [70] for analyzing discrete market models.

§5 Discussion

Empirical studies of asset returns typically concentrate either on the properties of their marginals or on features related to their autocorrelation functions. This paper deals exclusively with the latter (in fact, we assume throughout Gaussian marginals), and we have shown that accounting for the Joseph effect in modeling stock price returns leads directly to price processes that are inconsistent with the traditionally assumed semimartingale setting in modern finance theory. In view of convincing empirical evidence against Gaussian marginals and in favor of stable laws for stock returns (see for example [22], [23], [51], [52], [53], [67]), a more realistic approach to modeling stock price returns should also take into account stable laws (or, using Mandelbrot's terminology, the *Noah effect*) for asset return distributions. Clearly, pricing models that allow for a combination of the Joseph and Noah effects are appealing from a modeling perspective and promise to generate interesting and intriguing problems in mathematical finance.

In addition to the practical implications mentioned in Section 4 of allowing for long-range dependence in stock price returns, there is considerable scope for theoretical work centered around providing a "useful" stochastic calculus for processes such as fractional Brownian motion b_H. Clearly, since modern stochastic calculus is intimately related to the semimartingale framework, stochastic integrals with respect to s_H or stochastic differential equations involving s_H are meaningless per se, unless it is possible to give these objects a well-defined meaning. While there have been numerous attempts to extend today's stochastic calculus theory beyond the semimartingale setting (see for example, [26] and [5], and [6] in a finance-related context), the nonstandard analysis approach pursued in this paper seems to have definite advantages for this particular purpose. Indeed, the results in Sections 3 and 4 suggest that the hyperfinite setup is a natural environment for studying objects that are no longer well-defined on a standard probability space. At the same time, our results also indicate that the translation from the hyperfinite to the standard world ("taking standard parts") is not always obvious (e. g., see Propositions 4.2, 4.3, and 4.4) and typically requires some more delicate arguments. We intend to explore these and related issues in more detail in a forthcoming paper.

REFERENCES

1. S. Albeverio, J.-E. Fenstad, R. Høegh-Krohn, and T. Lindstrøm, *Nonstandard Methods in Stochastic Analysis and Mathematical Physics*, Academic Press, New York, 1986.
2. R. M. Anderson, *A nonstandard representation for Brownian motion and Itô integration*, Israel Math. J. **25** (1976), 15–46.
3. J. Beran, *Estimation, Testing and Prediction for Self-Similar and related Processes*, Ph. D. Dissertation, ETH Zürich, 1986.
4. J. Beran, *Statistical methods for data with long-range dependence*, Statistical Science **7** (1992), 404–427.
5. P. J. Bertoin, *Sur une integrale pour les processus a α-variation bornee*, Ann. Probab. **17** (1989), 1521–1535.
6. A. Bick and W. Willinger, *Dynamic spanning without probabilities* (1994) (to appear in Stoch. Proc. Appl.).
7. P. Billingsley, *Convergence of Probability Measures*, Wiley, New York, 1968.
8. F. Black and M. Scholes, *The pricing of contingent claims and corporate liabilities*, J. Polit. Econom. **81** (1973), 637–654.

9. G. G. Booth and F. R. Kaen, *Gold and silver spot prices and market information efficiency*, Financial Rev. **14** (1979), 21–26.

10. G. G. Booth, F. R. Kaen, and P. E. Koveos, *R/S analysis of foreign exchange rates under two international monetary regimes*, J. Monetary Econom. **10** (1982), 407–415.

11. G. E. P. Box and G. M. Jenkins, *Time Series Analysis: Forecasting and Control*, Holden-Day, San Francisco, 1970.

12. Y.-W. Cheung, *Tests for fractional integration: A Monte Carlo experiment*, preprint, 1991.

13. D. R. Cox, *Long-range dependence: A review*, in: Statistics: An Appraisal (H. A. David and H. T. David, eds.), Iowa State Univ. Press, Iowa, 1984, pp. 55–74.

14. J. Cox, S. Ross, and M. Rubinstein, *Option pricing: A simplified approach*, Financial Econom. **7** (1979), 229–263.

15. N. J. Cutland, *Nonstandard measure theory and its applications*, J. Bull. London Math. Soc. **15** (1983), 529–589.

16. N. J. Cutland (ed.), *Nonstandard Analysis and its Applications*, Cambridge University Press, Cambridge, 1988.

17. N. J. Cutland, P. E. Kopp, and W. Willinger, *A nonstandard approach to option pricing*, Mathematical Finance **1** (1991), 1–38.

18. R. Dahlhaus, *Efficient parameter estimation for self-similar processes*, Ann. Statist. **17** (1989), 1749–1766.

19. Y. A. Davydov, *The invariance principle for stationary processes*, Theory Probab. Appl. **15** (1970), 487–498.

20. C. Dellacherie and P.-A. Meyer, *Probabilities and Potential B*, North-Holland, Amsterdam, 1982.

21. F. X. Diebold and G. D. Rudebusch, *Long memory and persistence in aggregate output*, J. Monetary Econom. **24** (1989), 189–209.

22. E. F. Fama, *Mandelbrot and the stable Paretian hypothesis*, J. Business **36** (1963), 420–429.

23. E. F. Fama, *The behavior of stock-market prices*, J. Business **38** (1965), 34–105.

24. E. F. Fama, *Efficient capital markets: A review of theory and empirical work*, J. Finance **25** (1970), 383–417.

25. E. F. Fama and K. R. French, *Permanent and temporary components of stock prices*, J. Polit. Econom. **96** (1988), 246–273.

26. H. Föllmer, *Calcul d'Itô sans probabilities*, Seminaire de Probabilities XV, Lecture Notes in Math., vol. 850, Springer-Verlag, New York, 1981, pp. 143–150.

27. R. Fox and M. S. Taqqu, *Large sample properties of parameter estimates for strongly dependent stationary Gaussian time series*, Ann. Statist. **14** (1986), 517–532.

28. J. Geweke and S. Porter-Hudak, *The estimation and application of long memory time series models*, J. Time Ser. Anal. **4** (1983), 221–237.

29. H. P. Graf, *Longe-Range Correlations and Estimation of the Self-Similarity Parameter*, Ph. D. Dissertation, ETH Zürich, 1983.

30. C. W. Granger and R. Joyeux, *An introduction to long-range time series models and fractional differencing*, J. Time Ser. Anal. **1** (1980), 15–30.

31. M. T. Greene and B. D. Fielitz, *Long-term dependence in common stock returns*, J. Financial Econom. **4** (1977), 339–349.

32. F. R. Hampel, E. M. Ronchetti, P. J. Rousseeuw, and W. A. Stahel, *Robust Statistics. The Approach Based on Influence Functions*, Wiley, New York, 1986.

33. J. M. Harrison and D. M. Kreps, *Martingales and Arbitrage in Multiperiod Securities Markets*, J. Econom. Theory **20** (1979), 381–408.

34. J. M. Harrison and S. R. Pliska, *Martingales, stochastic integrals and continuous trading*, Stoch. Proc. Appl. **11** (1981), 215–260.

35. B. P. Helms, F. R. Kaen and R. E. Rosenman, *Memory in commodity futures contracts*, J. Futures Markets **4** (1984), 559–567.

36. J. R. M. Hosking, *Fractional differencing*, Biometrica **68** (1981), 165–176.

37. A. E. Hurd and P. A. Loeb, *An Introduction to Nonstandard Real Analysis*, Academic Press, New York, 1985.
38. H. E. Hurst, *Long-term storage capacity of reservoirs*, Trans. Amer. Soc. Civil Engineers **116** (1951), 770–779.
39. H. J. Keisler, *An infinitesimal approach to stochastic analysis*, Mem. Amer. Math. Soc. **297** (1984).
40. A. N. Kolmogorov, *Wiennersche Spiralen und einige andere interessante Kurven in Hilbertschen Raum*, Comptes Rendus (Doklady) Academie des Sciences de l'USSR (N.S.) **26** (1940), 115–118.
41. H. R. Künsch, *Statistical aspects of self-similar processes*, in: Proc. 1st World Congress of the Bernoulli Society 1 (Y. Prohorov and V.V. Sazonov, eds.), VNU Science Press, Utrecht, 1987, pp. 67–74.
42. H. R. Künsch, J. Beran, and F. R. Hampel, *Contrasts under long-range correlations*, Ann. Statist. **21** (1993), 943–964.
43. N. Kunitomo, *Long-term memory and fractional Brownian motion in financial markets*, preprint, 1993.
44. W. E. Leland, M. S. Taqqu, W. Willinger, and D. V. Wilson, *On the self-similar nature of Ethernet traffic*, Proc. ACM/SIGCOMM '93 (1993), 183–193.
45. T. L. Lindstrøm, *An invitation to nonstandard analysis*, in [16], pp. 1–105.
46. T. L. Lindstrøm, *Fractional Brownian fields as integrals of white noise*, Bull. London Math. Soc. **24** (1992).
47. R. S. Liptser and A. N. Shiryaev, *Theory of Martingales*, (in Russian), Nauka, Moscow, 1986.
48. A. W. Lo, *Long-term memory in stock market prices*, Econometrica **59** (1991), 1279–1313.
49. A. W. Lo and A. C. MacKinlay, *Stock market prices do not follow random walks: Evidence from a simple specification test*, Rev. Financial Studies **1** (1988), 41–66.
50. S. Maheswaran and C. A. Sims, *Empirical implications of arbitrage-free asset markets,* (1993) (to appear in: Models, Methods and Applications of Econometrics: The A. R. Bergstrom Festschrift).
51. B. B. Mandelbrot, *The Pareto–Lévy law and the distribution of income*, Int. Econom. Review **1** (196), 79–109.
52. B. B. Mandelbrot, *The variation of some other speculative prices*, J. Business **40** (1967), 393–413.
53. B. B. Mandelbrot, *When can price be arbitraged efficiently? A limit to the validity of the random walk and martingale models*, Rev. Econom. Statist. **53** (1971), 225–236.
54. B. B. Mandelbrot, *The Fractal Geometry of Nature*, Freeman, New York, 1983.
55. B. B. Mandelbrot and M. S. Taqqu, *Robust R/S analysis of long run serial correlation*, Proc. 42nd Session of the ISI **2** (1979), Manila, 69–100.
56. B. B. Mandelbrot and J. W. Van Ness, *Fractional Brownian motions, fractional noises and applications*, SIAM Review **10** (1968), 422–437.
57. B. B. Mandelbrot and J. R. Wallis, *Noah, Joseph and operational hydrology*, Water Resources Research **4** (1968), 909–918.
58. B. B. Mandelbrot and J. R. Wallis, *Computer experiments with fractional Gaussian noises, Parts 1–3*, Water Resources Research **5** (1969), 228–267.
59. B. B. Mandelbrot and J. R. Wallis, *Some long run properties of geophysical records*, Water Resources Research **5** (1969), 321–340.
60. R. C. Merton, *Theory of rational option pricing*, Bell J. Econom. Manag. Sci. **4** (1973), 141–183.
61. C.-K. Peng, S. V. Buldyrev, A. L. Goldberger, S. Havlin, F. Sciortino, M. Simons, and H. E. Stanley, *Long-range correlations in nucleotide sequences*, Nature **356** (1992), 168–170.
62. J. M. Poterba and L. H. Summers, *Mean reversion in stock prices: Evidence and implications*, J. Financial Econom. **22** (1988), 27–59.

63. P. Protter, *Stochastic Integration and Differential Equations*, Springer-Verlag, New York, 1990.
64. P. M. Robinson, *Semiparametric analysis of long-memory time series*, preprint, 1992.
65. L. C. G. Rogers, *personal communication*, 1989.
66. P. A. Samuelson, *Proof that properly anticipated prices fluctuate randomly*, Industrial Manag. Review **6** (1965), 41–49.
67. P. A. Samuelson, *Efficient portfolio selection for Pareto–Lévy investments*, J. Financ. and Qualit. Analysis **2** (1967), 107–122.
68. C. Stricker, *Arbitrage et lois de martingale*, Ann. Inst. Henri Poincare **26** (1990), 451–460.
69. M. S. Taqqu, *Self-similar processes*, in: Encyclopedia of Statistical Science, vol. 8, Wiley, New York, 1988, pp. 352–357.
70. M. S. Taqqu and W. Willinger, *The analysis of finite security markets using martingales*, Adv. Appl. Prob. **18** (1987), 1–25.
71. W. Verwaat, *Properties of general self-similar processes*, Bull. Inst. Internat. Statist. **52** (1987), 199–216.
72. P. Whittle, *Estimation and information in stationary time series*, Ark. Mat. **2** (1953), 423–434.

NIGEL J. CUTLAND AND P. EKKEHARD KOPP, SCHOOL OF MATHEMATICS, UNIVERSITY OF HULL, HULL HU67RX, UNITED KINGDOM

WALTER WILLINGER, BELLCORE, 445 SOUTH STREET, ROOM 2P-372, MORRISTOWN, NJ 07960-6438, USA

CRITICAL PRICE FOR AN AMERICAN OPTION
NEAR MATURITY

DAMIEN LAMBERTON

ABSTRACT. We give an alternate proof of a recent result of Barles, Burdeau, Romano and Samsœn on the behavior of the critical price for American put options near maturity.

§1 INTRODUCTION

In the Black–Scoles model, under the so-called "risk-neutral" probability measure $\mathbb{P}$, the stock price process satisfies the following stochastic differential equation:

$$\frac{dS_t}{S_t} = r\,dt + \sigma\,dW_t,$$

where the interest rate r is a positive constant and $(W_t)_{t\geq 0}$ is a standard Brownian motion under $\mathbb{P}$ (see [6, 8] for a modern presentation of the model). According to this model, the price of a *European* put option on one share, with exercise price K and date of maturity T is given (at time $t < T$) by the conditional expectation of the terminal payoff attached to the option, i.e. $\mathbb{E}(e^{-r(T-t)}(K - S_T)^+|\mathcal{F}_t)$, where $(\mathcal{F}_t)_{t\geq 0}$ is the natural filtration of $(W_t)_{t\geq 0}$. This conditional expectation can be written in the following form:

$$\mathbb{E}\big(e^{-r(T-t)}(K - S_T)^+\big|\mathcal{F}_t\big) = P(t, S_t),$$

where the function $P(t,x)$ is defined, for $(t,x) \in [0,T] \times \mathbb{R}^+$, by

$$P(t,x) = \mathbb{E}\big(e^{-r(T-t)}K - xe^{\sigma W_{T-t} - \sigma^2(T-t)/2}\big)^+.$$

From this formula, it is clear that $P(t,x)$ is a decreasing convex function of x, satisfying $P(t,0) = Ke^{-r(T-t)}$ and $\lim_{x\to+\infty} P(t,x) = 0$. Therefore, the equation

$$P(t,x) = (K - x)$$

1991 *Mathematics Subject Classification.* 60G40, 90A09.
Key words and phrases. American options, critical price, free boundary, optimal stopping.

admits a unique solution in the open interval $(0, K)$, if $t < T$. This solution will be denoted by $s_e(t)$ in the sequel.

The price of an *American* put option, with exercise price K and date of maturity T is given (at time $t < T$) by

$$\operatorname*{ess\,sup}_{\tau \in \mathcal{T}_{t,T}} \mathbb{E}\left(e^{-r(\tau - t)}(K - S_\tau)^+ \big| \mathcal{F}_t\right),$$

where $\mathcal{T}_{t,T}$ is the set of all stopping times with values in $[t, T]$ (see [7, 8]). This price can be written as $P_a(t, S_t)$, where the function $P_a(t, x)$ is defined, for every $(t, x) \in [0, T] \times \mathbb{R}^+$, by

$$P_a(t, x) = \sup_{\tau \in \mathcal{T}_{0, T-t}} \mathbb{E}\left(e^{-r\tau}K - xe^{\sigma W_\tau - \sigma^2 \tau/2}\right)^+.$$

From this formula, it is clear that $P_a(t, x)$ is also a decreasing convex function of x and that it satisfies $P_a(t, 0) = K$ and $\lim_{x \to +\infty} P_a(t, x) = 0$. Moreover,

$$\forall x \in \mathbb{R}^+ \quad P(t, x) \le P_a(t, x) \quad \text{and} \quad (K - x)^+ \le P_a(t, x).$$

For $t \in [0, T)$, define the *critical stock price* at time t as

$$s^*(t) = \sup \left\{ x \ge 0 \mid P_a(t, x) = (K - x)^+ \right\}.$$

It is known that the function $t \mapsto s^*(t)$ is non-decreasing, C^∞ over $[0, T)$ and that $\lim_{t \to T} s^*(t) = K$ (cf. [9, 4]).

Throughout the paper, we will use the following notation. If f and g are two functions defined on $[0, T]$, we will write that $f(t) \sim g(t)$ as t approaches T if $\lim_{t \to T} \frac{f(t)}{g(t)} = 1$. In [1, 2] the following result is proved:

Theorem 1.1. *As t approaches T, the critical price satisfies the following:*

$$K - s^*(t) \sim K\sigma\sqrt{(T - t)\ln(1/(T - t))}.$$

The purpose of this note is to give an alternate proof of this result. The proof of Barles et al. is based on the construction of sub- and supersolutions of the variational inequality satisfied by P_a. Actually, the subsolution they consider is just $P(t, x)$ and they use the inequality $s_e(t) \ge s^*(t)$, which follows from $P \le P_a$. The supersolution is of the form $f(t)P(t, x)$. Our method consists in studying $s_e(t)$ (essentially as in [2]) and in giving an estimate for $s^*(t) - s_e(t)$, without using supersolutions.

§2 BEHAVIOR OF $s_e(t)$ NEAR T

The following proposition is implicit in [2]. The proof given below is slightly different.

Proposition 2.1. *For $t \in [0, T)$, let*

$$\phi(t) = \frac{\ln(K/s_{\mathrm{e}}(t))}{\sigma\sqrt{T-t}}.$$

As t approaches T,

$$\phi(t)^2 \exp\left(\frac{\phi(t)^2}{2}\right) \sim \frac{\sigma}{r\sqrt{2\pi(T-t)}}.$$

It is easy to check that this result implies that

$$K - s_{\mathrm{e}}(t) \sim K\sigma\sqrt{(T-t)\ln(1/(T-t))}$$

as t approaches T. Before proving Proposition 2.1, we have the following lemma:

Lemma 2.2.
$$\lim_{t \to T} \frac{K - s_{\mathrm{e}}(t)}{\sqrt{T-t}} = +\infty.$$

Proof. To our knowledge, this property was first observed by Charretour and Viswanathan (cf. [3]). For completeness, we give a short proof. From the definition of $s_{\mathrm{e}}(t)$, we have

$$K - s_{\mathrm{e}}(t) = \mathbb{E}\big(Ke^{-r(T-t)} - s_{\mathrm{e}}(t)e^{\sigma W_{T-t} - \sigma^2(T-t)/2}\big)^+. \tag{1}$$

Therefore

$$\frac{K - s_{\mathrm{e}}(t)}{\sqrt{T-t}} = \mathbb{E}\left[e^{-r(T-t)}\left(\frac{K - s_{\mathrm{e}}(t)}{\sqrt{T-t}} + s_{\mathrm{e}}(t)\frac{1 - e^{(r-\sigma^2/2)\theta + \sigma\sqrt{\theta}g}}{\sqrt{\theta}}\right)^+\right],$$

where $\theta = T - t$ and g is a standard normal random variable. Using Fatou's lemma and $\lim_{t \to T} s_{\mathrm{e}}(t) = K$, we can state

$$\liminf_{t \to T} \frac{K - s_{\mathrm{e}}(t)}{\sqrt{T-t}} \geq \mathbb{E}\left(\liminf_{t \to T} \frac{K - s_{\mathrm{e}}(t)}{\sqrt{T-t}} - K\sigma g\right)^+.$$

Now, if $\eta \in \mathbb{R}$, then $(\eta - K\sigma g)^+ \geq \eta - K\sigma g$ and the inequality is strict whenever $\eta - K\sigma g < 0$. Hence, by taking expectations, we have $\mathbb{E}(\eta - K\sigma g)^+ > \eta$ for any finite η, which proves that $\liminf_{t \to T} \frac{K - s_{\mathrm{e}}(t)}{\sqrt{T-t}}$ must be infinite. $\square$

Proof of Proposition 2.1. We still denote by g a standard normal random variable. Using the notation $\theta = T - t$ again, we derive from equation (1) that

$$K - s_{\mathrm{e}}(t) = e^{-r\theta}K - s_{\mathrm{e}}(t) + \mathbb{E}\big(Ke^{-r\theta} - s_{\mathrm{e}}(t)e^{\sigma\sqrt{\theta}g - \sigma^2\theta/2}\big)^-$$

and

$$\left(1 - e^{-r\theta}\right)K = \mathbb{E}\left(Ke^{-r\theta} - s_e(t)e^{\sigma\sqrt{\theta}g - \sigma^2\theta/2}\right)^-.$$

Now divide both sides by $s_e(t)$ and let

$$\alpha = \alpha(t) = \frac{1}{\sigma\sqrt{\theta}}\left(\ln\left(\frac{K}{s_e(t)}\right) - \left(r - \frac{\sigma^2}{2}\right)\theta\right).$$

Note that, by Lemma 2.2, $\lim_{t\to T} \alpha(t) = +\infty$ and $\lim_{t\to T} \sqrt{T-t}\,\alpha(t) = 0$. With these notations, we have

$$\left(1 - e^{-r\theta}\right)\frac{K}{s_e(t)} = \mathbb{E}\left[e^{-\sigma^2\theta/2}\left(e^{\alpha\sigma\sqrt{\theta}} - e^{\sigma\sqrt{\theta}g}\right)^-\right].$$

The left-hand side in this equality is equivalent to $r\theta$ as t approaches T. To estimate the right-hand side, let

$$f(\theta) = \mathbb{E}\left[\left(e^{\alpha\sigma\sqrt{\theta}} - e^{\sigma\sqrt{\theta}g}\right)^-\right] = \mathbb{E}\left[\left(e^{\sigma\sqrt{\theta}g} - e^{\alpha\sigma\sqrt{\theta}}\right)\mathbf{1}_{\{g>\alpha\}}\right].$$

Using $|e^x - 1 - x| \le \frac{x^2}{2}e^{|x|}$, we have

$$\left|f(\theta) - \mathbb{E}\left(\sigma\sqrt{\theta}(g - \alpha)\mathbf{1}_{\{g>\alpha\}}\right)\right|$$
$$\le \frac{\sigma^2\theta}{2}\mathbb{E}\left(g^2 e^{\sigma\sqrt{\theta}|g|}\mathbf{1}_{\{g>\alpha\}}\right) + \frac{\sigma^2\theta\alpha^2}{2}e^{\sigma\sqrt{\theta}\alpha}\mathbb{P}(g > \alpha).$$

As t approaches T, θ goes to 0 and α tends to infinity. Therefore,

$$f(\theta) = \sigma\sqrt{\theta}\,\mathbb{E}\left((g - \alpha)\mathbf{1}_{\{g>\alpha\}}\right) + o(\theta).$$

Since we know that $f(\theta) \sim r\theta$, we must have

$$r\theta \sim \sigma\sqrt{\theta}\,\mathbb{E}\left((g - \alpha)\mathbf{1}_{\{g>\alpha\}}\right) = \sigma\sqrt{\theta}\int_\alpha^{+\infty}(y - \alpha)e^{-y^2/2}\frac{dy}{\sqrt{2\pi}}$$
$$= \frac{\sigma\sqrt{\theta}}{\sqrt{2\pi}\alpha^2 e^{\alpha^2/2}}\int_0^{+\infty} xe^{-x-x^2/(2\alpha^2)}\,dx,$$

where, in the last integral, we have set $x = \alpha(y - \alpha)$. Hence

$$\alpha^2(t)e^{\alpha^2(t)/2} \sim \frac{\sigma}{r\sqrt{2\pi(T-t)}}.$$

We now easily obtain Proposition 2.1, since $\phi(t) - \alpha(t) = \left(r - \frac{\sigma^2}{2}\right)\frac{\sqrt{T-t}}{\sigma}$. $\quad\square$

§3 BEHAVIOR OF $s^*(t)$ NEAR T

Theorem 1.1 is an immediate consequence of Proposition 2.1 and the following estimate.

Proposition 3.1. *There exists a constant $C > 0$ such that, for all $t \in [0, T)$,*

$$0 \le s_e(t) - s^*(t) \le C\sqrt{T - t}.$$

Proof. Let $t \in [0, T)$. The inequality $0 \le s_e(t) - s^*(t)$ follows from $P(t, x) \le P_a(t, x)$. Before proving the other inequality, recall that $P(t, \cdot)$ is C^1 on $[0, +\infty)$ and C^∞ on the interval $(s^*(t), +\infty)$ and that the following equality holds for $t < T$ and $x > s^*(t)$:

$$\frac{\partial P_a}{\partial t}(t, x) + \frac{\sigma^2}{2} x^2 \frac{\partial^2 P_a}{\partial x^2}(t, x) + rx \frac{\partial P_a}{\partial x}(t, x) - r P_a(t, x) = 0.$$

Since $s_e(t) \le s^*(t)$, we can apply Taylor's formula and write:

$$P_a(t, s_e(t)) = P_a(t, s^*(t)) + (s_e(t) - s^*(t)) \frac{\partial P_a}{\partial x}(t, s^*(t)) + \frac{(s_e(t) - s^*(t))^2}{2} \frac{\partial^2 P_a}{\partial x^2}(t, \xi),$$

with $s^*(t) < \xi < s_e(t)$. Hence (using $\frac{\partial P_a}{\partial x}(t, s^*(t)) = -1$),

$$
\begin{aligned}
P_a(t, s_e(t)) &= K - s_i^*(t) - (s_e(t) - s^*(t)) + \frac{(s_e(t) - s^*(t))^2}{2} \frac{\partial^2 P_a}{\partial x^2}(t, \xi) \\
&= P(t, s_e(t)) + \frac{(s_e(t) - s^*(t))^2}{2} \frac{\partial^2 P_a}{\partial x^2}(t, \xi).
\end{aligned}
\tag{2}
$$

Now recall the following formula relating P_a and P (cf. [8], Corollary 3.1, or [5]):

$$
\begin{aligned}
&P_a(t, x) - P(t, x) \\
&= rK \int_0^{T-t} e^{-ru} \mathbb{P}\big((r - (\sigma^2/2))u + \sigma W_u < \ln(s^*(t + u)/x)\big) \, du.
\end{aligned}
\tag{3}
$$

This formula clearly implies that $P_a(t, x) - P(t, x) \le rK(T - t)$. Therefore, going back to (2),

$$\frac{(s_e(t) - s^*(t))^2}{2} \frac{\partial^2 P_a}{\partial x^2}(t, \xi) \le rK(T - t).$$

Now, since $\xi > s^*(t)$, we have

$$
\begin{aligned}
\frac{\sigma^2}{2} \xi^2 \frac{\partial^2 P_a}{\partial x^2}(t, \xi) &= -\frac{\partial P_a}{\partial t}(t, \xi) - r\xi \frac{\partial P_a}{\partial x}(t, \xi) + r P_a(t, \xi) \\
&\ge r\xi \left(-\frac{\partial P_a}{\partial x}(t, \xi) \right),
\end{aligned}
$$

since $t \mapsto P_a(t, x)$ is non-increasing. Moreover, it follows from equation (3) that $x \mapsto P_a(t, x) - P(t, x)$ is non-increasing. Hence, $\frac{\partial P_a}{\partial x}(t, \xi) \leq \frac{\partial P}{\partial x}(t, \xi)$. Therefore,

$$\frac{\sigma^2}{2}\xi^2 \frac{\partial^2 P_a}{\partial x^2}(t, \xi) \geq r\xi\left(-\frac{\partial P}{\partial x}(t, \xi)\right).$$

An easy computation shows that

$$-\frac{\partial P}{\partial x}(t, x) = \mathbb{P}\left(g < \frac{\ln(K/x) - (r + \frac{\sigma^2}{2})\theta}{\sigma\sqrt{\theta}}\right) \geq \mathbb{P}\left(g < -\left(r + \frac{\sigma^2}{2}\right)\sqrt{T - t}\right),$$

if $x < K$. Hence, for t close enough to T,

$$-\frac{\partial P}{\partial x}(t, \xi) \geq \frac{1}{4}.$$

Therefore $(s_e(t) - s^*(t))^2 \leq C(T - t)$ for t close to T, which suffices to prove Proposition 3.1. □

Remark. It would be interesting to have an equivalent for $e^{\psi(t)^2/2}$, where $\psi(t) = \frac{\ln(K/s^*(t))}{\sigma\sqrt{T-t}}$. The following estimate is (implicitly) proved in [2]:

$$\limsup_{t \to T} \sqrt{T - t}\, e^{\psi(t)^2/2} \leq \frac{\sigma}{r\sqrt{2\pi}}.$$

It can also be obtained by pushing the methods of the present paper, but we were not able to find a more precise result.

REFERENCES

1. G. Barles, J. Burdeau, M. Romano, N. Sansœn, *Estimation de la frontière libre des options américaines au voisinage de l'échéance*, C. R. Acad Sci. Paris, Série I **316** (1993), 171–174.
2. G. Barles, J. Burdeau, M. Romano, N. Sansœn, *Critical stock price near expiration* (to appear in Mathematical Finance).
3. F. Charretour, R. J. Elliott, R. Myneni, R. Viswanathan, *Paper presented at the Oberwolfach meeting on mathematical finance*, August 1992.
4. A. Friedman, *Parabolic variational inequalities in one space dimension and smoothness of the free boundary*, J. Funct. Anal. **18** (1975), 151–176.
5. S. D. Jacka, *Optimal stopping and the American put price*, Math. Finance **1** (1991), 1–14.
6. I. Karatzas, *Optimization problems in the theory of continuous trading*, SIAM J. Control Optim. **27** (1989), 1221–1259.
7. I. Karatzas, *On the pricing of American options*, Appl. Math. Optim. **17** (1988), 37–60.
8. R. Myneni, *The pricing of the American option*, Ann. Appl. Probab. **2** (1992), 1–23.
9. P. L. J. Van Moerbeke, *On optimal stopping and free boundary problems*, Arch. Rational Mech. Anal **60** (1976), 101–148.

DAMIEN LAMBERTON, EQUIPE D'ANALYSE ET DE MATHÉMATIQUES APPLIQUÉES, UNIVERSITÉ DE MARNE-LA-VALLÉE, 2 RUE DE LA BUTTE VERTE, 93166 NOISY-LE-GRAND CEDEX, FRANCE

HEDGING OF OPTIONS UNDER DISCRETE OBSERVATION ON ASSETS WITH STOCHASTIC VOLATILITY

G. B. DI MASI, E. PLATEN, AND W. J. RUNGGALDIER

ABSTRACT. The paper considers the hedging of contingent claims on assets with stochastic volatilities when the asset price is only observable at discrete time instants. Explicit formulae are given for risk-minimizing hedging strategies.

§1 INTRODUCTION

In most practical cases a trader has to base the hedging of contingent claims on discrete observations of a risky asset whose price is characterized by continuous-time dynamics. In this paper we consider the hedging of a contingent claim on a continuous-time asset observable only at discrete times. Additionally, this asset may have a stochastic volatility.

Discrete-time observations cannot be considered in the framework of [2]; in fact they induce an additional risk. By a modification of the methods of Föllmer and his coauthors in [2] and [3], we shall prove a result that allows the determination of a risk-minimizing hedging strategy also under discrete-time observations.

§2 MODEL AND MAIN RESULT

Let the stochastic process $S = \{ S_t, 0 \leq t \leq T \}$ describe the price of a risky asset (e. g. a stock or a currency) as square integrable solution of the Itô stochastic differential equation

$$dS_t = \sigma_t S_t \, dW_t, \quad 0 \leq t \leq T, \tag{1}$$

with initial condition S_0 and where $W = \{ W_t, 0 \leq t \leq T \}$ is a given Wiener process. The volatility $\sigma = \{ \sigma_t, 0 \leq t \leq T \}$ is supposed to form a positive, càdlàg, square integrable semi-martingale independent of W. At this point we do not further specify the stochastic volatility, but one may think of σ as proposed e. g. in Hull and White [8] or Hofmann, Platen, Schweizer [7]. At the end of the paper we shall discuss an example, where σ represents a discrete-time, finite-state, inhomogeneous Markov chain. We remark that we consider our asset price evolution in a risk-neutral world (see [5], [6]) which we choose to be the one

1991 *Mathematics Subject Classification.* primary 90A09, secondary 60H99, 93E99.

Key words and phrases. Hedging of options, incomplete markets, stochastic volatility, incomplete observations.

corresponding to the minimal equivalent martingale measure proposed in [2] (see also Hofmann, Platen, Schweizer [7]). For simplicity we assume that the riskless interest rate is zero. Besides the risky asset S, let there be given a bond with constant unit value.

We denote the underlying probability space by $(\Omega, \underline{\mathcal{F}}, \mathcal{F}, P)$, where the filtration $\mathcal{F} = (\mathcal{F}_t)_{0 \leq t \leq T}$ is generated by the flow of σ-algebras

$$\mathcal{F}_t = \sigma\{\sigma_r, S_s; 0 \leq r \leq T, 0 \leq s \leq t\},$$

for $t \in [0, T]$. We also set $\underline{\mathcal{F}} = \mathcal{F}_T$. The process $S = \{S_t, 0 \leq t \leq T\}$ is thus an $(\mathcal{F}, P)$-martingale.

Now let us consider the problem of hedging a contingent claim of the form $H = f(S_T)$ on the price of the risky asset at maturity T, which is traded at time $t = 0$.

A *dynamical trading strategy* $\phi = \{\phi_t = (\xi_t, \eta_t), 0 \leq t \leq T\}$ is a strategy to build a portfolio consisting at time t of the amount ξ_t of the risky asset and the amount η_t of the bond. Here ξ_t is assumed to be a predictable $\mathcal{F}_{t-}$-adapted process and η_t is $\mathcal{F}$-adapted.

Given a trading strategy ϕ, we shall define its *value process* $V(\phi) = \{V_t(\phi), 0 \leq t \leq T\}$ by

$$V_t(\phi) = \xi_t S_t + \eta_t \cdot 1 \tag{2}$$

and its *cost process* $C(\phi) = \{C_t(\phi), 0 \leq t \leq T\}$ (see [3]) by

$$C_t(\phi) = V_t(\phi) - \int_0^t \xi_s \, dS_s. \tag{3}$$

Given a contingent claim H at maturity T, we shall say that ϕ *hedges against* H if $V_T(\phi) = H$.

The problem that we address in this paper is that of determining a suitable trading strategy that hedges against H, when the trader does not have full information about the underlying price process. We are going to model lack of information by working with a subfiltration to which S is possibly not adapted. The latter is the case if the asset is observed only at discrete times.

A right-continuous filtration $\mathcal{A} = (\mathcal{A}_t, 0 \leq t \leq T)$ with $\mathcal{A}_t \subseteq \mathcal{F}_t$, $t \in [0, T]$, will be called a *subfiltration* if $\mathcal{A}_0$ measures S_0 and $\mathcal{A}_T$ measures S_T.

Examples for subfiltrations are easily obtained from the σ-algebras generated by the observations S_{τ_n} at some given time instants $0 = \tau_0 < \tau_1 < \cdots < \tau_T = T$, which could represent e. g., daily stock data excluding weekends. We remark that such subfiltrations do not satisfy the assumptions in [2] since the asset S is not adapted with respect to them.

Given a subfiltration $\mathcal{A}$, a trading strategy $\phi = \{\phi_t = (\xi_t, \eta_t), 0 \leq t \leq T\}$ will be called $\mathcal{A}$-*admissible* if it hedges against H, η_t is $\mathcal{A}_t$-adapted, and ξ_t is predictable $\mathcal{A}_{t-}$-adapted with

$$E\left\{ \int_0^T |\xi_t|^2 \sigma_{t-}^2 S_t^2 \, dt \,\bigg|\, \mathcal{A}_0 \right\} < \infty. \tag{4}$$

We denote by $\psi(\mathcal{A})$ the set of $\mathcal{A}$-admissible trading strategies.

A trading strategy ϕ will be called $\mathcal{A}$-*mean self-financing* if it is $\mathcal{A}$-admissible and its *conditional cost process*

$$B(\phi, \mathcal{A}) = \left\{ B_t(\phi, \mathcal{A}) = \mathrm{E}\{C_t(\phi) \,|\, \mathcal{A}_t\}, \, 0 \le t \le T \right\} \tag{5}$$

is an $(\mathcal{A}, P)$-martingale.

We can now adapt a lemma from [11].

Lemma 1. *There exists a bijection between the $\mathcal{A}$-predictable processes $\xi = \{\xi_t, \, 0 \le t \le T\}$ satisfying (4) and the $\mathcal{A}$-mean self-financing strategies $\phi = \{\phi_t = (\xi_t, \eta_t), \, 0 \le t \le T\}$ by putting*

$$\eta_t = \mathrm{E}\{H \,|\, \mathcal{A}_t\} - \xi_t \, \mathrm{E}\{S_t \,|\, \mathcal{A}_t\}. \tag{6}$$

Proof. It is enough to note that with the choice (6) we obtain from (5), (3), and (2) the conditional cost

$$\begin{aligned}
B_t(\phi, \mathcal{A}) &= \mathrm{E}\left\{ V_t(\phi) - \int_0^t \xi_s \, dS_s \,\bigg|\, \mathcal{A}_t \right\} \\
&= \mathrm{E}\left\{ H - \int_0^T \xi_s \, dS_s \,\bigg|\, \mathcal{A}_t \right\} = \mathrm{E}\{B_T(\phi, \mathcal{A}) \,|\, \mathcal{A}_t\}
\end{aligned}$$

as an $(\mathcal{A}, P)$-martingale. $\square$

Given a subfiltration $\mathcal{A}$ and a trading strategy $\phi \in \psi(\mathcal{A})$, we recall from [3] the notion of *remaining risk process* $R(\phi, \mathcal{A}) = \{ R_t(\phi, \mathcal{A}), \, 0 \le t \le T \}$ given by

$$R_t(\phi, \mathcal{A}) = \mathrm{E}\{(C_T(\phi) - C_t(\phi))^2 \,|\, \mathcal{A}_t\}. \tag{7}$$

A trading strategy $\phi^* \in \psi(\mathcal{A})$ will be called $\mathcal{A}$-*risk-minimizing* if for all $\phi \in \psi(\mathcal{A})$ and all $t \in [0, T]$ for which S_t is $\mathcal{A}_t$-measurable

$$R_t(\phi^*, \mathcal{A}) \le R_t(\phi, \mathcal{A}). \tag{8}$$

In this paper we address the following

Problem. *Given any subfiltration $\mathcal{A}$, determine an $\mathcal{A}$-mean self-financing and $\mathcal{A}$-risk-minimizing strategy ϕ^* that hedges against H, together with its value process $V_t(\phi^*)$.*

To obtain an answer to this problem take into account that under our filtration $\mathcal{F}$, using a well-known decomposition ([10]) the contingent claim H can be represented in the form

$$H = \mathrm{E}\{H \,|\, \mathcal{F}_t\} + \int_t^T \mu_s \, dS_s, \tag{9}$$

where $\mu = \{ \mu_\tau, \, t \le \tau \le T \}$ is $\mathcal{F}$-predictable.

Theorem 1. *Given a subfiltration $\mathcal{A}$ and a contingent claim H with $\mathrm{E}\{H^2 \,|\, \mathcal{A}_0\} < \infty$ we consider the $\mathcal{A}$-predictable process $\mu^* = \{\mu_t^*, 0 \le t \le T\}$ with*

$$\mu_t^* = \frac{\mathrm{E}\{\mu_t \sigma_{t-}^2 S_t^2 \,|\, \mathcal{A}_{t-}\}}{\mathrm{E}\{\sigma_{t-}^2 S_t^2 \,|\, \mathcal{A}_{t-}\}}, \tag{10}$$

where in the numerator and denominator we take the predictable versions. Then $\phi = (\xi, \eta) \in \psi(\mathcal{A})$ is an $\mathcal{A}$-risk-minimizing and $\mathcal{A}$-mean self-financing strategy that hedges against H if and only if $\xi_t = \mu_t^*$ with equality holding in $L_2(dP \times d\langle S\rangle)$ and η_t is chosen according to (6) for all $t \in [0, T]$. Furthermore, we have $v_0 = V_0(\phi) = \mathrm{E}\{H \,|\, \mathcal{A}_0\}$.

Proof. Given any $\phi \in \psi(\mathcal{A})$ we can compute the remaining risk for times t for which S_t is $\mathcal{A}_t$-measurable in the form

$$R_t(\phi, \mathcal{A}) = \mathrm{E}\{(C_T(\phi) - C_t(\phi))^2 \,|\, \mathcal{A}_t\}$$

$$= \mathrm{E}\left\{\left(H - \mathrm{E}\{H \,|\, \mathcal{A}_t\} - \int_t^T \xi_s \, dS_s\right)^2 \,\Big|\, \mathcal{A}_t\right\}$$

$$= \mathrm{E}\left\{\left(\int_t^T (\mu_s - \xi_s) \, dS_s + \theta_t\right)^2 \,\Big|\, \mathcal{A}_t\right\}$$

with

$$\theta_t = \mathrm{E}\{H \,|\, \mathcal{F}_t\} - \mathrm{E}\{H \,|\, \mathcal{A}_t\}.$$

We get

$$Z_t := R_t(\phi, \mathcal{A}) - \mathrm{E}\{\theta_t^2 \,|\, \mathcal{A}_t\}$$

$$= \mathrm{E}\left\{\int_t^T (\mu_s - \xi_s)^2 \sigma_{s-}^2 S_s^2 \, ds \,\Big|\, \mathcal{A}_t\right\}$$

$$= \mathrm{E}\left\{\int_t^T \mathrm{E}\{(\mu_s - \mu_s^*)^2 \sigma_{s-}^2 S_s^2 \,|\, \mathcal{A}_{s-}\} \, ds \,\Big|\, \mathcal{A}_t\right\}$$

$$+ \mathrm{E}\left\{\int_t^T \mathrm{E}\{[2(\mu_s - \mu_s^*)(\mu_s^* - \xi_s) + (\mu_s^* - \xi_s)^2] \sigma_{s-}^2 S_s^2 \,|\, \mathcal{A}_{s-}\} \, ds \,\Big|\, \mathcal{A}_t\right\}.$$

Finally, we obtain using (10)

$$Z_t = \mathrm{E}\left\{\int_t^T \mathrm{E}\{[(\mu_s - \mu_s^*)^2 + (\mu_s^* - \xi_s)^2] \sigma_{s-}^2 S_s^2 \,|\, \mathcal{A}_{s-}\} \, ds \,\Big|\, \mathcal{A}_t\right\}.$$

Taking into account that $(\mu_s^* - \xi_s)^2 \sigma_{s-}^2 S_s^2$ is non-negative, it then follows that $R_t(\phi, \mathcal{A})$ is minimized for all $t \in [0, T]$ for which S_t is $\mathcal{A}_t$-measurable if and only if

$$\int_0^T \mathrm{E}\{(\mu_s^* - \xi_s)^2 \sigma_{s-}^2 S_s^2 \,|\, \mathcal{A}_{s-}\} \, ds = 0,$$

which corresponds to the statement of the theorem recalling that by (1)

$$d\langle S\rangle_t = \sigma_{t-}^2 S_t^2\, dt.$$

$\square$

The above proof uses mainly the fact that the stochastic integral $\int (\mu_s - \mu_s^*)\, dS_s$ is orthogonal to other stochastic integrals $\int \alpha_s\, dS_s$ where α_s is $\mathcal{A}$-adapted. Process μ^* appears in this way as a conditional expectation of μ with respect to $dP \times d\langle S\rangle$.

§3 COMPUTATION OF A RISK-MINIMIZING STRATEGY
UNDER DISCRETE OBSERVATION

To get an explicit example we now assume the volatility σ to be a discrete-time finite-state inhomogeneous and right-continuous Markov chain which is sometimes suggested by the analysis of historical volatilities as in Galai [4]. We shall assume that the jump times are the points τ_n, $n = 0, 1, \ldots, N$, at which we also observe the asset price S_{τ_n}. Thus our subfiltration is

$$\mathcal{A}_t = \mathcal{F}_t^* = \sigma\{S_{\tau_n}, 0 \le \tau_n \le t\}.$$

We denote by $J = \{a_1, \ldots, a_k\}$ the finite state space of the Markov chain σ and by

$$p_{\tau_n}(j, i) = P\{\sigma_{\tau_{n+1}} = a_j \mid \sigma_{\tau_n} = a_i\}$$

its transition probabilities at time τ_n. Then we obtain from Theorem 1

$$\xi_t^* = \frac{\mathrm{E}\{\mu_t \sigma_{\tau_n}^2 S_t^2 \mid \mathcal{F}_{\tau_{n-1}}^*\}}{\mathrm{E}\{\sigma_{\tau_n}^2 S_t^2 \mid \mathcal{F}_{\tau_{n-1}}^*\}},$$

for $t \in (\tau_n, \tau_{n+1}]$, $n = 1, \ldots, N-1$, where μ_t is $\mathcal{F}_t$-measurable and follows from a Black–Scholes type formula as described in Hull and White [8].

In principle there is no problem to compute explicitly the values ξ_t^* in the above case. We omit these formulae. They depend on the conditional probabilities $P(\sigma_{\tau_n} = a_j \mid \mathcal{F}_{\tau_{n-1}}^*)$ for $j = 1, \ldots, k$. These probabilities can be estimated by filtering techniques, see e.g. Di Masi and Runggaldier [1].

In many cases, when one finds another specific structure for the volatility, one can use stochastic numerical methods as described in Kloeden, Platen [9] or used in Hofmann, Platen, Schweizer [7] to compute the hedging strategy and the option price.

REFERENCES

1. G. B. Di Masi, W. J. Runggaldier, *On measure transformations for combined filtering and parameter estimation in discrete time*, Systems & Control Letters **2** (1982), 57–62.
2. H. Föllmer, M. Schweizer, *Hedging of contingent claims under incomplete information*, In: Applied Stochastic Analysis, Stochastic Monographs (M. H. A. Davis and R. J. Elliott, eds.), vol. 5, Gordon & Breach, London/New York, 1991, pp. 389–414.

3. H. Föllmer, D. Sondermann, *Hedging of non-redundant contingent claims*, In: Contributions to Mathematical Economics (W. Hildenbrand and A. Mas-Colell, eds.), 1986, pp. 205–223.
4. D. Galai, *Inferring volatility from option prices*, Finance **12** (1991).
5. J. M. Harrison, D. M. Kreps, *Martingales and arbitrage in multiperiod securities markets*, J. of Economic Theory **20** (1979), 381–408.
6. J. M. Harrison, S. R. Pliska, *Martingales and stochastic integrals in the theory of continuous trading*, Stochastic Processes and Their Applications **11** (1981), 215–260.
7. N. Hofmann, E. Platen, M. Schweizer, *Option pricing under incompleteness and stochastic volatility*, Mathematical Finance **2** (1992), 153–187.
8. J. Hull, A. White, *The pricing of options on assets with stochastic volatilities*, J. of Finance **42** (1987), 281–300.
9. P. E. Kloeden, E. Platen, *Numerical solution of stochastic differential equations*, Applications of Mathematics Series, vol. 23, Springer-Verlag, New York/Berlin, 1992.
10. H. Kunita, S. Watanabe, *On square integrable martingales*, Nagoya Math. J. **30** (1967), 209–245.
11. M. Schweizer, *Hedging of Options in a General Semimartingale Model*, Diss., ETHZ n. 8615, Zürich, 1988.

G. B. DI MASI, UNIVERSITÀ DI PADOVA-DIPARTIMENTO DI MATEMATICA PURA ED APPLICATA AND CNR-LADSEB, I-35100 PADOVA, ITALY

E. PLATEN, AUSTRALIAN NATIONAL UNIVERSITY, GPO BOX 4, CANBERRA ACT, AUSTRALIA AND IAAS, PF 1304, D-1086 BERLIN, GERMANY

W. J. RUNGGALDIER, UNIVERSITÀ DI PADOVA-DIPARTIMENTO DI MATEMATICA PURA ED APPLICATA, I-35100 PADOVA, ITALY

Progress in Probability, Vol. 36
© 1995 Birkhäuser Verlag Basel/Switzerland

CONVERGENCE OF OPTION VALUES
UNDER INCOMPLETENESS

WOLFGANG J. RUNGGALDIER AND MARTIN SCHWEIZER*

ABSTRACT. We study the problem of convergence of discrete-time option values to continuous-time option values. While previous papers typically concentrate on the approximation of geometric Brownian motion by a binomial tree, we consider here the case where the model is incomplete in both continuous and discrete time. Option values are defined with respect to the criterion of local risk-minimization and thus computed as expectations under the respective minimal martingale measures. We prove that for a jump-diffusion model with deterministic coefficients, these values converge; this shows that local risk-minimization possesses an inherent stability property under discretization.

§0 INTRODUCTION

A major controversy in the modern theory of option pricing is the debate of *discrete-time versus continuous-time* modelling. Whereas a continuous-time formulation is often more amenable to analysis and thus tends to provide better insight, one also frequently hears the counter-argument that a description of real market behaviour can only be based on discrete observations. In view of these contrasting opinions, it is natural to look for connections between the two competing approaches. In particular, one should like to obtain convergence results as one passes from discrete time to continuous time by a limiting procedure. There is an abundant literature on this question and we mention here only a very few references. One of the starting points is the paper by Cox/Ross/Rubinstein [2] which provides a derivation of the famous *Black–Scholes formula,* first obtained in Black/Scholes [1] and Merton [11], by a passage to the limit from a binomial model. The survey by Willinger/Taqqu [19] contains an excellent overview of several convergence approaches used so far, as well as an extensive list of references. Among the more recent contributions, we mention Eberlein [5], where pathwise approximations of geometric Brownian motion by piecewise constant processes are

1991 *Mathematics Subject Classification.* 60G35, 90A09.

Key words and phrases. option pricing, incomplete markets, convergence, minimal martingale measure, locally risk-minimizing trading strategies, jump-diffusion.

*Financial support by Deutsche Forschungsgemeinschaft, Sonderforschungsbereich 303 at the University of Bonn, is gratefully acknowledged.

constructed, and Duffie/Protter [4], who discuss the convergence of the process of
cumulative gains from trade.

Perhaps the first result that one would like to establish in this context is the
convergence of option prices. There are quite a few results in this direction when
stock prices are given by diffusion processes and the discrete-time models are
binomial or suitable multinomial trees; see for instance Cox/Ross/Rubinstein [2],
He [7], or Duffie/Protter [4]. The reason for this very restrictive choice of model
is the fact that one has *completeness* at the level of both continuous and discrete
time. This allows perfect replication of any contingent claim, and so option prices
are uniquely determined by the assumption of absence of arbitrage.

In this paper, we attack the same question of convergence in an *incomplete*
market. In that case, one is immediately and simultaneously faced with *two* closely
intertwined problems. Not only is there the difficulty of establishing a convergence
result, but it is even not clear in the first place what the appropriate definition
of an option price or *option value* should be. Intuitively, one feels of course that
a reasonable valuation methodology should allow one to deduce convergence. We
show here that the criterion of *local risk-minimization* introduced in Schweizer
[15, 16] possesses this feature, at least for the particular example considered here.
This means that local risk-minimization has an inherent *stability property* under
discretization which may be regarded as an additional argument in favour of this
approach. For other recent results in a similar direction, see also Dengler [3].

More precisely, we study the preceding problem in the case where the price S
of the underlying asset is given by a *jump-diffusion process* with deterministic coefficients; see Merton [12], Jeanblanc-Picqué/Pontier [9], Shirakawa [18], Xue [20]
and Mercurio/Runggaldier [10] for similar models. The discrete-time processes
S^m are obtained by first approximating the coefficient functions by piecewise constant ones and then simply evaluating the resulting continuous-time process at
the given discretization points. Note that this is rather straightforward and does
not require an elaborate construction of the approximating processes S^m as for
instance in Nelson/Ramaswamy [13]. Section 1 contains a detailed description of
the model and the discretization procedure explained above. In the continuous
and in each discrete model, we then apply the criterion of local risk-minimization
to determine option values. By the results of Schweizer [15, 16, 17], this means
that for each S^m, we use the *minimal martingale measure* $\widehat{P}^m$ for S^m to compute
the value of a contingent claim H^m as $\widehat{E}^m[H^m]$; the valuation for H is $\widehat{E}[H]$ with
$\widehat{P}$ corresponding to S. We should like to emphasize that these quantities are not
necessarily option prices in the usual sense of the word. They give the initial
capital required for the construction of a locally risk-minimizing strategy which
duplicates the contingent claim under consideration, but this strategy is typically
not self-financing, and it may well happen that a non-vanishing hedging cost appears. For these reasons, we use the more cautious terminology "value" rather
than "price".

In Section 2, we prove that in our situation, the densities $\frac{d\widehat{P}^m}{dP}$ converge to $\frac{d\widehat{P}}{dP}$
in $\mathcal{L}^p(P)$ for every $p \in [1, \infty)$. Although our method of proof relies crucially on the

jump-diffusion structure and in particular on the assumption of deterministic coefficients, we feel that the theorem itself is likely to hold in more general situations as well. We remark that a related result was obtained by He [7] who proved the weak convergence of the density processes in the case where S is a multi-dimensional diffusion process and S^m is a suitable multinomial process. However, this is not comparable to our result here since he assumed in addition that S as well as each S^m is complete and thus admits a unique equivalent martingale measure. In Section 3, we discuss some applications of our convergence theorem. One immediate consequence is the convergence of the values $\widehat{E}^m[H^m]$ to $\widehat{E}[H]$ if H^m converges to H in $\mathcal{L}^q(P)$ for some $q > 1$, and the last condition is usually easy to verify. Since the values $\widehat{E}^m[H^m]$ correspond to a discrete-time model, they can in principle always be computed, but the computational burden may occasionally become rather heavy. As a by-product of our convergence approach, we also obtain an additional approximation result which in some cases allows an easier computation of the approximating values as simple averages of Black–Scholes–type formulae.

§1 MODEL AND PROBLEM FORMULATION

Let $(\Omega, \mathcal{F}, P)$ be a probability space and $T > 0$ a fixed and finite time horizon. Let $W = (W_t)_{0 \le t \le T}$ be a Brownian motion and $N = (N_t)_{0 \le t \le T}$ a 1-variate point process with deterministic intensity $\nu(t)$. Thus N is a Poisson process, and W and N are independent by [8, Theorem II.6.3]. We shall assume that $\nu(t)$ is bounded away from 0, uniformly in $t \in [0, T]$. Finally, $\mathbb{F} = (\mathcal{F}_t)_{0 \le t \le T}$ denotes the P-augmentation of the filtration generated by W and N.

Now denote by $S = (S_t)_{0 \le t \le T}$ the unique strong solution of the stochastic differential equation

$$dS_t = S_{t-}\big(b(t)\, dt + v(t)\, dW_t + \varphi(t)\, dN_t\big), \qquad S_0 > 0. \tag{1.1}$$

We shall assume that b, v, φ, ν are left-continuous functions with right limits from $[0, T]$ to $\mathbb{R}$ which are bounded uniformly in $t \in [0, T]$. Furthermore, we impose the conditions

$$\varphi(t) > -1 \qquad \text{for all } t \in [0, T]$$

and

$$v^2(t) + \varphi^2(t) \inf_{0 \le s \le T} \nu(s) \ge \varepsilon \qquad \text{for some } \varepsilon > 0, \text{ uniformly in } t \in [0, T]. \tag{1.2}$$

This implies that S is strictly positive and that the function

$$\varrho(t) := \frac{b(t) + \varphi(t)\nu(t)}{v^2(t) + \varphi^2(t)\nu(t)}, \qquad 0 \le t \le T,$$

is also left-continuous with right limits and bounded uniformly in $t \in [0, T]$. Finally we assume that

$$\varphi(t)\varrho(t) \le 1 - \delta \qquad \text{for some } \delta > 0, \text{ uniformly in } t \in [0, T]. \tag{1.3}$$

By Itô's formula, the solution of (1.1) is explicitly given by

$$S_t = S_0 \exp\left(\int_0^t \left(b(s) - \frac{1}{2}v^2(s) \right) ds + \int_0^t v(s)\, dW_s + \int_0^t \log\left(1 + \varphi(s)\right) dN_s \right) \quad (1.4)$$

for $t \in [0, T]$. Due to the boundedness of all coefficients, one can then show that

$$\sup_{0 \le t \le T} |S_t| \in \mathcal{L}^p(P) \qquad \text{for every } p \in [1, \infty);$$

see for instance [20, Lemma III.2.1] or [17, Lemma II.8.1]. In particular, S is a special semimartingale with canonical decomposition $S = S_0 + M + A$, where

$$M_t = \int_0^t S_{u-}\left(v(u)\, dW_u + \varphi(u)\left(dN_u - \nu(u)\, du \right) \right), \qquad 0 \le t \le T,$$

and

$$A_t = \int_0^t S_{u-}\left(b(u) + \varphi(u)\nu(u) \right) du = \int_0^t \alpha_u\, d\langle M \rangle_u, \qquad 0 \le t \le T,$$

with

$$\alpha_t = \frac{1}{S_{t-}} \frac{b(t) + \varphi(t)\nu(t)}{v^2(t) + \varphi^2(t)\nu(t)} = \frac{\varrho(t)}{S_{t-}}, \qquad 0 \le t \le T.$$

Next we define the process $\widehat{Z} = (\widehat{Z}_t)_{0 \le t \le T}$ by

$$\widehat{Z}_t := \exp\left(\int_0^t \left(\varphi(s)\varrho(s)\nu(s) - \frac{1}{2}v^2(s)\varrho^2(s) \right) ds \right.$$
$$\left. - \int_0^t v(s)\varrho(s)\, dW_s + \int_0^t \log\left(1 - \varphi(s)\varrho(s)\right) dN_s \right), \qquad 0 \le t \le T.$$

Similar estimates as for S then show, using (1.3), that

$$\sup_{0 \le t \le T} |\widehat{Z}_t| \in \mathcal{L}^p(P) \qquad \text{for every } p \in [1, \infty). \qquad (1.5)$$

Since $\widehat{Z}$ also solves the stochastic differential equation

$$d\widehat{Z}_t = -\widehat{Z}_{t-}\left(v(t)\varrho(t)\, dW_t + \varphi(t)\varrho(t)\left(dN_t - \nu(t)\, dt \right) \right), \qquad \widehat{Z}_0 = 1,$$

we see that $\widehat{Z}$ is a strictly positive martingale under P, and this allows us to define an equivalent probability measure $\widehat{P}$ on $(\Omega, \mathcal{F})$ by setting $\frac{d\widehat{P}}{dP} := \widehat{Z}_T$. It is easy to check that S is a $(\widehat{P}, \mathbb{F})$-martingale, and since $\widehat{Z}$ can be written as $\mathcal{E}(-\int \alpha\, dM)$, $\widehat{P}$ is in fact the *minimal equivalent martingale measure* for S with respect to $\mathbb{F}$;

this can be proved as in [6, Theorem (3.5)]. Moreover, Girsanov's theorem shows that

$$\widehat{W}_t := W_t + \int_0^t v(s)\varrho(s)\,ds, \qquad 0 \le t \le T,$$

is a Brownian motion under $\widehat{P}$ and that N has intensity $\widehat{\nu}(t) := \nu(t)(1 - \varphi(t)\varrho(t))$ under $\widehat{P}$. Again using [8, Theorem II.6.3], we conclude that N is a Poisson process under $\widehat{P}$ and that $\widehat{W}$ and N are independent under $\widehat{P}$. For future reference, we note that (1.4) can be rewritten as

$$S_t = S_0 \exp\left(-\int_0^t \left(\frac{1}{2}v^2(s) + \varphi(s)\widehat{\nu}(s)\right) ds + \int_0^t v(s)\,d\widehat{W}_s + \int_0^t \log\left(1 + \varphi(s)\right) dN_s\right)$$

for all $t \in [0, T]$, since S satisfies

$$dS_t = S_{t-}\left(v(t)\,d\widehat{W}_t + \varphi(t)\big(dN_t - \widehat{\nu}(t)\,dt\big)\right).$$

Now consider any $\mathcal{F}_T$-measurable random variable H. If $H \in \mathcal{L}^p(P)$ for some $p > 2$, then H admits a decomposition as

$$H = \widehat{\mathbb{E}}[H] + \int_0^T \xi_u^H\,dS_u + L_T^H \qquad \text{P-a. s.,} \tag{1.6}$$

where $\xi^H = (\xi_t^H)_{0 \le t \le T}$ is an $\mathbb{F}$-predictable process satisfying

$$\mathbb{E}\left[\int_0^T \left(\xi_u^H\right)^2 d\langle M\rangle_u + \left(\int_0^T |\xi_u^H \alpha_u|\,d\langle M\rangle_u\right)^2\right] < \infty,$$

and $L^H = (L_t^H)_{0 \le t \le T}$ is a P-square-integrable $(P, \mathbb{F})$-martingale, null at 0, which is strongly P-orthogonal to M. For a proof of this result, we refer to [17, Theorem II.8.3]. Moreover, the argument given there also shows that

$$\widehat{V}_t := \widehat{\mathbb{E}}[H] + \int_0^t \xi_u^H\,dS_u + L_t^H = \widehat{\mathbb{E}}[H \,|\, \mathcal{F}_t], \qquad 0 \le t \le T.$$

If we now interpret S as the discounted price of some risky asset in a financial market where there also exists a riskless asset whose discounted price is identically 1, then the existence of the above decomposition of H implies the existence of a dynamic *trading strategy* which is *H-admissible* and *locally risk-minimizing* in the sense of [16]. More precisely, define an adapted process $\eta^H = (\eta_t^H)_{0 \le t \le T}$ by setting

$$\eta_t^H := \widehat{V}_t - \xi_t^H S_t, \qquad 0 \le t \le T.$$

If we interpret ξ_t^H as the number of shares of S held at time t and η_t^H as the amount invested in the riskless asset, then the value of this portfolio $\varphi^H = (\xi^H, \eta^H)$ is given by

$$V_t(\varphi^H) = \xi_t^H S_t + \eta_t^H = \widehat{V}_t, \qquad 0 \le t \le T,$$

so that $V_T(\varphi^H) = H$ P-a. s. The cumulative costs incurred by using φ^H are given by

$$C_t(\varphi^H) = V_t(\varphi^H) - \int_0^t \xi_u^H \, dS_u = \widehat{\mathbb{E}}[H] + L_t^H, \qquad 0 \le t \le T.$$

Since this is a $(P, \mathbb{F})$-martingale strongly P-orthogonal to M and since it is easy to verify that S satisfies assumptions (X1)–(X5) of [16], it follows from [16, Proposition 2.3] that φ^H is indeed H-admissible and locally risk-minimizing with respect to $\mathbb{F}$. The value process $V(\varphi^H) = \widehat{V}$ can thus be viewed as a *valuation process* for the contingent claim H with respect to the criterion of local risk-minimization. In particular, $\widehat{V}_0 = \widehat{\mathbb{E}}[H]$ can be interpreted as a *valuation* for H at time 0.

What happens now if we use the same criterion to value options along a sequence of discretizations of S? If the above valuation concept is reasonable, then economic intuition suggests that the sequence of discrete-time values should converge to the continuous-time value. However, this is not so clear from a mathematical point of view, since the valuation measures will usually be different in every discretization. The convergence result established in the next section thus shows that the criterion of local risk-minimization has a very appealing stability property; this will be discussed below in more detail.

To be more precise, fix a sequence $(\tau_m)_{m \in \mathbb{N}}$ of partitions of the interval $[0, T]$, i.e., $\tau_m = \{t_0^m, t_1^m, \ldots, t_{n_m}^m\}$ with $0 = t_0^m < t_1^m < \cdots < t_{n_m}^m = T$, whose mesh size $|\tau_m| := \max_{t_i, t_{i+1} \in \tau_m} |t_{i+1} - t_i|$ tends to 0 as $m \to \infty$. For ease of notation, we shall henceforth simply write n instead of n_m. Define piecewise constant functions $\psi^m : [0, T] \to \mathbb{R}$ by setting

$$\psi^m(t) := \psi(0) I_{\{0\}}(t) + \sum_{k=1}^n \psi(t_{k-1}^m) I_{(t_{k-1}^m, t_k^m]}(t) \tag{1.7}$$

for $m \in \mathbb{N}$ and $\psi \in \{b, v, \varphi\}$. Then each ψ^m is clearly left-continuous with right limits, $\psi^m(t)$ is bounded by $\|\psi\|_\infty$ uniformly in m and t, and (1.2) also holds for the corresponding approximating functions v^m and φ^m. Note that the function

$$\varrho^m(t) := \frac{b^m(t) + \varphi^m(t) v(t)}{\left(v^m(t)\right)^2 + \left(\varphi^m(t)\right)^2 v(t)}, \qquad 0 \le t \le T,$$

is also left-continuous with right limits, but in general not piecewise constant. We shall assume that for m large enough, ϱ^m also satisfies the condition (1.3), uniformly in m; this is for instance the case if $v(t)$ is constant or, more generally, if $v(t)$ is continuous.

If we define $X^m = (X_t^m)_{0 \le t \le T}$ as the solution of the stochastic differential equation

$$dX_t^m = X_{t-}^m \left(b^m(t)\, dt + v^m(t)\, dW_t + \varphi^m(t)\, dN_t \right), \qquad X_0^m = S_0,$$

then (1.4) with ψ^m replacing ψ gives the explicit expression for X_t^m, and we get

$$\sup_{m \in \mathbb{N}} \mathrm{E}\left[\sup_{0 \le t \le T} |X_t^m|^p \right] < \infty \qquad \text{for every } p \in [1, \infty).$$

The discrete-time process corresponding to τ_m is now obtained by simply evaluating X^m at all discretization points $t_k^m \in \tau_m$, and we write

$$S_k^m := X_{t_k^m}^m \qquad \text{for } k = 0, 1, \ldots, n.$$

Finally, the discrete-time filtration $\mathbb{F}^m = (\mathcal{F}_k^m)_{k=0,1,\ldots,n}$ is obtained by setting

$$\mathcal{F}_k^m := \mathcal{F}_{t_k^m} \qquad \text{for } k = 0, 1, \ldots, n,$$

so that $\mathcal{F}_k^m \supseteq \sigma(S_0^m, \ldots, S_k^m)$ for every k.

In analogy to the continuous-time case, we now introduce the discrete-time process $\widehat{Z}^m = (\widehat{Z}_k^m)_{k=0,1,\ldots,n}$ defined by

$$\widehat{Z}_k^m := \prod_{j=1}^k \left(1 - \frac{\mathrm{E}\left[\Delta S_j^m \,\middle|\, \mathcal{F}_{j-1}^m \right]}{\mathrm{Var}\left[\Delta S_j^m \,\middle|\, \mathcal{F}_{j-1}^m \right]} \left(\Delta S_j^m - \mathrm{E}\left[\Delta S_j^m \,\middle|\, \mathcal{F}_{j-1}^m \right] \right) \right)$$

for $k = 0, 1, \ldots, n$, where $\Delta S_j^m := S_j^m - S_{j-1}^m$ denotes the increment of S^m between t_{j-1}^m and t_j^m. Using the explicit expression for $\widehat{Z}^m$ provided in (2.11)–(2.14) below, one readily verifies that $\widehat{Z}^m$ is a P-square-integrable $(P, \mathbb{F}^m)$-martingale and that the product $\widehat{Z}^m S^m$ is also a $(P, \mathbb{F}^m)$-martingale. For this reason, we call the signed measure $\widehat{P}^m$ on $(\Omega, \mathcal{F})$ defined by $\frac{d\widehat{P}^m}{dP} := \widehat{Z}_n^m$ the *minimal signed martingale measure* for S^m with respect to $\mathbb{F}^m$. More details can be found for instance in [17].

If H^m is now any $\mathcal{F}_n^m$-measurable random variable and in $\mathcal{L}^2(P)$, then H^m can be written as

$$H^m = \widehat{\mathrm{E}}^m[H^m] + \sum_{j=1}^n \xi_j^m \Delta S_j^m + L_n^m \qquad P\text{-a. s.,} \tag{1.8}$$

where $\xi^m = (\xi_k^m)_{k=1,\ldots,n}$ is an $\mathbb{F}^m$-predictable process with $\xi_k^m \Delta S_k^m \in \mathcal{L}^2(P)$ for every k, and $L^m = (L_k^m)_{k=0,1,\ldots,n}$ is a P-square-integrable $(P, \mathbb{F}^m)$-martingale, null at 0, which is strongly P-orthogonal to the martingale part M^m in the Doob decomposition $S^m = S_0^m + M^m + A^m$ of S^m with respect to $\mathbb{F}^m$. Note that (1.8) is exactly the discrete-time counterpart of (1.6). For a proof of (1.8), see for instance [14, Lemma 4.10] or [17, Proposition I.6.1]. Moreover, [15, Theorem I.9] implies that ξ^m determines a unique H^m-admissible discrete-time strategy $\varphi^m = (\xi^m, \eta^m)$ which is locally risk-minimizing with respect to the discrete-time filtration $\mathbb{F}^m$. Its value process is given by

$$V_k(\varphi^m) = \xi_k^m S_k^m + \eta_k^m = \widehat{E}^m[H^m] + \sum_{j=1}^{k} \xi_j^m \Delta S_j^m + L_k^m =: \widehat{V}_k^m \quad \text{for } k = 0, 1, \ldots, n,$$

and the value of H^m at time 0 with respect to $\mathbb{F}^m$ is $\widehat{V}_0^m = \widehat{E}^m[H^m]$. Furthermore,

$$\widehat{V}_k^m = \widehat{E}^m\left[H^m | \mathcal{F}_k^m\right] \qquad P\text{-a. s. for } k = 0, 1, \ldots, n$$

in the sense that $\widehat{V}_n^m = H^m$ P-a. s. and $\widehat{V}^m \widehat{Z}^m$ is a $(P, \mathbb{F}^m)$-martingale.

Now we can reformulate our question: How do these option values behave if $|\tau_m|$ tends to zero? If for instance $H = (S_T - K)^+$ is a European call option on S and $H^m = (S_T^m - K)^+$ is its discretized version, then we certainly expect H^m to converge to H, and we hope that this will imply the convergence of the values $\widehat{E}^m[H^m]$ to $\widehat{E}[H]$. The conclusion below will be that this is indeed true, and the essential step in the argument will be to prove that $\widehat{Z}_n^m$ converges to $\widehat{Z}_T$ in a sufficiently strong sense.

§2 CONVERGENCE OF THE MINIMAL DENSITIES

In this section, we prove that $\widehat{Z}_n^m$ converges to $\widehat{Z}_T$ as $|\tau_m|$ tends to zero. For that purpose, we first establish an auxiliary result. Define the left-continuous *piecewise constant* functions $\bar{\varrho}^m : [0, T] \to \mathbb{R}$ by

$$\bar{\varrho}^m(t) := \frac{1}{\Delta t_k^m} \int_{t_{k-1}^m}^{t_k^m} \varrho^m(s)\, ds \qquad \text{for } t \in (t_{k-1}^m, t_k^m]$$

and $\bar{\varrho}^m(0) := \varrho(0)$. Then $\bar{\varrho}^m(t)$ is bounded uniformly in m and t, $\bar{\varrho}^m$ satisfies (1.3) whenever ϱ^m does, and

$$\lim_{m \to \infty} \bar{\varrho}^m(t) = \varrho(t) \qquad \text{for almost every } t \in [0, T]. \tag{2.1}$$

Denote by $U^m = (U_t^m)_{0 \le t \le T}$ the process defined by

$$U_t^m := \exp\left(\int_0^t \left(\varphi^m(s) \bar{\varrho}^m(s) \nu(s) - \frac{1}{2} \left(v^m(s) \bar{\varrho}^m(s) \right)^2 \right) ds \right.$$

$$\left. - \int_0^t v^m(s) \bar{\varrho}^m(s)\, dW_s + \int_0^t \log\left(1 - \varphi^m(s) \bar{\varrho}^m(s)\right) dN_s \right) \tag{2.2}$$

for $t \in [0, T]$. The same arguments as for (1.5) then show that

$$\sup_{m \in \mathbb{N}} E\left[\sup_{0 \le t \le T} |U_t^m|^p \right] < \infty \qquad \text{for every } p \in [1, \infty). \tag{2.3}$$

Lemma 1. *As* $|\tau_m|$ *tends to zero,*

$$U_T^m \longrightarrow \widehat{Z}_T \qquad \text{in } \mathcal{L}^p(P) \text{ for every } p \in [1, \infty) \tag{2.4}$$

and

$$X_T^m \longrightarrow S_T \qquad \text{in } \mathcal{L}^p(P) \text{ for every } p \in [1, \infty). \tag{2.5}$$

Proof. By the definition of ψ^m, $\psi^m(t)$ converges to $\psi(t)$ for every $t \in [0, T]$ and for $\psi \in \{b, v, \varphi\}$. Hence we conclude that

$$\int_0^T \left(\varphi^m(s)\bar{\varrho}^m(s)\nu(s) - \frac{1}{2}\left(v^m(s)\bar{\varrho}^m(s)\right)^2 \right) ds$$

$$\longrightarrow \int_0^T \left(\varphi(s)\varrho(s)\nu(s) - \frac{1}{2}v^2(s)\varrho^2(s) \right) ds$$

by (2.1) and the dominated convergence theorem. Furthermore,

$$\int_0^T \log\left(1 - \varphi^m(s)\bar{\varrho}^m(s)\right) dN_s \longrightarrow \int_0^T \log\left(1 - \varphi(s)\varrho(s)\right) dN_s \qquad P\text{-a. s.}$$

by (2.1) and dominated convergence, since $1 - \varphi^m(t)\bar{\varrho}^m(t)$ is bounded away from zero uniformly in m and t for large m by assumption. Finally,

$$\int_0^T \left(v^m(s)\bar{\varrho}^m(s) - v(s)\varrho(s) \right)^2 ds \longrightarrow 0$$

by (2.1) and dominated convergence and thus

$$\int_0^T v^m(s)\bar{\varrho}^m(s) \, dW_s \longrightarrow \int_0^T v(s)\varrho(s) \, dW_s \qquad \text{in } \mathcal{L}^2(P).$$

This implies that U_T^m converges to $\widehat{Z}_T$ in probability as $m \to \infty$, and combining this with (1.5) and (2.3) yields (2.4). The proof of (2.5) is perfectly analogous. $\square$

Now we are ready to state and prove the main result of this section.

Theorem 2. *As* $|\tau_m|$ *tends to zero,*

$$\widehat{Z}_{n_m}^m \longrightarrow \widehat{Z}_T \qquad \text{in } \mathcal{L}^p(P) \text{ for every } p \in [1, \infty). \tag{2.6}$$

Proof. (1) By Lemma 1, it is enough to show that $\widehat{Z}_n^m - U_T^m$ converges to zero in $\mathcal{L}^p(P)$ for every $p \in [1, \infty)$ or even only for every $p \in \mathbb{N}$. Since

$$\left\| \widehat{Z}_n^m - U_T^m \right\|_{\mathcal{L}^p(P)} \le \|U_T^m\|_{\mathcal{L}^{2p}(P)} \left\| \frac{\widehat{Z}_n^m}{U_T^m} - 1 \right\|_{\mathcal{L}^{2p}(P)}$$

and $(\|U_T^m\|_{\mathcal{L}^{2p}(P)})_{m\in\mathbb{N}}$ is bounded due to (2.3), we only need to show that

$$\mathrm{E}\left[\left(\frac{\widehat{Z}_n^m}{U_T^m} - 1\right)^{2p}\right] \longrightarrow 0 \qquad \text{as } m \to \infty \text{ for every } p \in \mathbb{N}.$$

But

$$\mathrm{E}\left[\left(\frac{\widehat{Z}_n^m}{U_T^m} - 1\right)^{2p}\right] = \sum_{\ell=0}^{2p} \binom{2p}{\ell} \mathrm{E}\left[\left(\frac{\widehat{Z}_n^m}{U_T^m}\right)^{\ell}\right](-1)^{2p-\ell},$$

and so (2.6) will be proved once we show that

$$\lim_{m\to\infty} \mathrm{E}\left[\left(\frac{\widehat{Z}_n^m}{U_T^m}\right)^{\ell}\right] = 1 \qquad \text{for every } \ell \in \mathbb{N}_0. \tag{2.7}$$

(2) Now we compute $\widehat{Z}_n^m$. Due to (1.4) and (1.7), S^m can be written recursively as

$$S_k^m = S_{k-1}^m \exp\left(\left(b_k^m - \frac{1}{2}(v_k^m)^2\right)\Delta t_k^m + v_k^m \Delta W_k^m + \Delta N_k^m \log(1 + \varphi_k^m)\right) \tag{2.8}$$

with the shorthand notation

$$\psi_k^m := \psi(t_{k-1}^m) = \psi^m(t_k^m) \qquad \text{for } \psi \in \{b, v, \varphi, W, N\}. \tag{2.9}$$

Using the fact that ΔW_k^m and ΔN_k^m are independent of each other and of $\mathcal{F}_{k-1}^m$ with respective distributions $\mathcal{N}(0, \Delta t_k^m)$ and $\mathcal{P}(\bar{\nu}_k^m \Delta t_k^m)$, where

$$\bar{\nu}_k^m := \frac{1}{\Delta t_k^m} \int_{t_{k-1}^m}^{t_k^m} \nu(s)\, ds, \tag{2.10}$$

we can compute

$$\mathrm{E}\left[\Delta S_k^m | \mathcal{F}_{k-1}^m\right] = S_{k-1}^m \left(\exp\left((b_k^m + \varphi_k^m \bar{\nu}_k^m)\Delta t_k^m\right) - 1\right)$$

and

$$\mathrm{Var}\left[\Delta S_k^m | \mathcal{F}_{k-1}^m\right]$$
$$= (S_{k-1}^m)^2 \exp\left(2(b_k^m + \varphi_k^m \bar{\nu}_k^m)\Delta t_k^m\right)\left(\exp\left(((v_k^m)^2 + (\varphi_k^m)^2 \bar{\nu}_k^m)\Delta t_k^m\right) - 1\right).$$

With the abbreviations

$$q_k^m := \frac{\exp\left((b_k^m + \varphi_k^m \bar{\nu}_k^m)\Delta t_k^m\right) - 1}{\exp\left((b_k^m + \varphi_k^m \bar{\nu}_k^m)\Delta t_k^m\right)\left(\exp\left(((v_k^m)^2 + (\varphi_k^m)^2 \bar{\nu}_k^m)\Delta t_k^m\right) - 1\right)}, \tag{2.11}$$

and

$$\widetilde{R}_k^m := \exp\left(v_k^m \Delta W_k^m + \Delta N_k^m \log(1 + \varphi_k^m) - \left(\frac{1}{2}(v_k^m)^2 + \varphi_k^m \bar{\nu}_k^m\right)\Delta t_k^m\right), \quad (2.12)$$

as well as

$$R_k^m := q_k^m(\widetilde{R}_k^m - 1), \quad (2.13)$$

we thus obtain

$$\widehat{Z}_n^m = \prod_{j=1}^{n}(1 - R_j^m). \quad (2.14)$$

By (2.2), this implies

$$\frac{\widehat{Z}_n^m}{U_T^m} = \prod_{j=1}^{n}(1 - R_j^m)\exp(-L_j^m)$$

with

$$L_k^m := \left(\varphi_k^m \bar{\varrho}_k^m \bar{\nu}_k^m - \frac{1}{2}(v_k^m \bar{\varrho}_k^m)^2\right)\Delta t_k^m - v_k^m \bar{\varrho}_k^m \Delta W_k^m$$
$$+ \Delta N_k^m \log(1 - \varphi_k^m \bar{\varrho}_k^m); \quad (2.15)$$

note that we have used here the fact that φ^m, v^m, and $\bar{\varrho}^m$ are piecewise constant. Since the processes W and N are independent and have independent increments, the random variables $(1 - R_j^m)\exp(-L_j^m)$ are independent for $j = 1, \ldots, n$ and so

$$E\left[\left(\frac{\widehat{Z}_n^m}{U_T^m}\right)^\ell\right] = \prod_{j=1}^{n} E\left[(1 - R_j^m)^\ell \exp(-\ell L_j^m)\right]$$
$$= \exp\left(\sum_{j=1}^{n} \log E\left[(1 - R_j^m)^\ell \exp(-\ell L_j^m)\right]\right).$$

Hence (2.7) will follow if we show that

$$\lim_{m\to\infty} \sum_{j=1}^{n} \log E\left[(1 - R_j^m)^\ell \exp(-\ell L_j^m)\right] = 0 \qquad \text{for every } \ell \in \mathbb{N}_0. \quad (2.16)$$

(3) Now fix $\ell \in \mathbb{N}_0$ and $m \in \mathbb{N}$, and drop the index m for the moment to ease the notation. From (2.13), we get

$$(1 - R_k)^\ell \exp(-\ell L_k) = \sum_{i=0}^{\ell} \binom{\ell}{i}(1 + q_k)^{\ell-i}(-q_k)^i \widetilde{R}_k^i \exp(-\ell L_k),$$

and using (2.12) and (2.15) gives

$$\mathrm{E}\big[\widetilde{R}_k^i \exp(-\ell L_k)\big] = \exp\big(f_k(\ell, i)\Delta t_k\big)$$

with

$$f_k(\ell, i) := \frac{1}{2}v_k^2\big((i + \ell\bar{\varrho}_k)^2 - i\big) - \ell\left(\varphi_k\bar{\varrho}_k\bar{\nu}_k - \frac{1}{2}v_k^2\bar{\varrho}_k^2\right)$$
$$+ \bar{\nu}_k\left(\frac{(1 + \varphi_k)^i}{(1 - \varphi_k\bar{\varrho}_k)^\ell} - 1 - i\varphi_k\right).$$

Expanding e^x into a power series, we get

$$\mathrm{E}\big[\widetilde{R}_k^i \exp(-\ell L_k)\big] = 1 + f_k(\ell, i)\Delta t_k + O\big((\Delta t_k)^2\big),$$

and since we have uniform bounds on all coefficients for large m, the error term is $O\big((\Delta t_k)^2\big)$ uniformly in m for large m, i.e.,

$$\limsup_{m \to \infty} \frac{1}{(\Delta t_k)^2} O\big((\Delta t_k)^2\big) < \infty.$$

Summing over i yields

$$\mathrm{E}\big[(1 - R_k)^\ell \exp(-\ell L_k)\big]$$
$$= \sum_{i=0}^{\ell} \binom{\ell}{i}(1 + q_k)^{\ell-i}(-q_k)^i\left(1 + f_k(\ell, i)\Delta t_k + O\big((\Delta t_k)^2\big)\right)$$
$$= 1 + \Delta t_k \sum_{i=0}^{\ell} \binom{\ell}{i}(1 + q_k)^{\ell-i}(-q_k)^i f_k(\ell, i) + O\big((\Delta t_k)^2\big),$$

and expanding $\log(1 + x)$ into a power series leads to

$$\log \mathrm{E}\big[(1 - R_k)^\ell \exp(-\ell L_k)\big]$$
$$= \Delta t_k \sum_{i=0}^{\ell} \binom{\ell}{i}(1 + q_k)^{\ell-i}(-q_k)^i f_k(\ell, i) + O\big((\Delta t_k)^2\big),$$

again with an error term which is uniform in m for large m; notice that q_k^m is bounded uniformly in m for large m.

(4) Next we compute

$$\sum_{i=0}^{\ell} \binom{\ell}{i} (1+q_k)^{\ell-i}(-q_k)^i f_k(\ell,i)$$

$$= \frac{1}{2}\ell^2 v_k^2 \bar{\varrho}_k^2 - \ell\left(\varphi_k \bar{\varrho}_k \bar{\nu}_k - \frac{1}{2}v_k^2 \bar{\varrho}_k^2\right) - \bar{\nu}_k$$

$$+ \frac{\bar{\nu}_k}{(1-\varphi_k \bar{\varrho}_k)^\ell} \sum_{i=0}^{\ell} \binom{\ell}{i}(1+q_k)^{\ell-i}(-q_k - q_k\varphi_k)^i$$

$$+ (\ell v_k^2 \bar{\varrho}_k - \bar{\nu}_k \varphi_k)\sum_{i=0}^{\ell} i\binom{\ell}{i}(1+q_k)^{\ell-i}(-q_k)^i$$

$$+ \frac{1}{2}v_k^2 \sum_{i=0}^{\ell} i(i-1)\binom{\ell}{i}(1+q_k)^{\ell-i}(-q_k)^i$$

$$= \frac{1}{2}\ell^2 v_k^2 \bar{\varrho}_k^2 - \ell\left(\varphi_k \bar{\varrho}_k \bar{\nu}_k - \frac{1}{2}v_k^2 \bar{\varrho}_k^2\right) - \bar{\nu}_k + \bar{\nu}_k\left(\frac{1-q_k\varphi_k}{1-\varphi_k \bar{\varrho}_k}\right)^\ell$$

$$- \ell q_k(\ell v_k^2 \bar{\varrho}_k - \bar{\nu}_k \varphi_k) + \frac{1}{2}v_k^2 q_k^2 \ell(\ell-1)$$

$$= \frac{1}{2}\ell^2 v_k^2(\bar{\varrho}_k - q_k)^2 + \ell\left(\varphi_k \bar{\nu}_k(q_k - \bar{\varrho}_k) - \frac{1}{2}v_k^2(q_k^2 - \bar{\varrho}_k^2)\right)$$

$$+ \bar{\nu}_k\left(\left(\frac{1-q_k\varphi_k}{1-\varphi_k \bar{\varrho}_k}\right)^\ell - 1\right)$$

$$=: g_k(\ell).$$

(5) Now sum over k and reinstate the index m to obtain

$$\sum_{k=1}^{n_m} \log \mathrm{E}\left[(1-R_k^m)^\ell \exp(-\ell L_k^m)\right] = \sum_{k=1}^{n_m}\left(g_k^m(\ell)\Delta t_k^m + O((\Delta t_k^m)^2)\right).$$

The sum of the error terms is $O(1)|\tau_m|$ and thus tends to zero. Furthermore, (2.11) shows that

$$\lim_{m\to\infty} q_k^m = \frac{b(t) + \varphi(t)\nu(t)}{v^2(t) + \varphi^2(t)\nu(t)} = \varrho(t)$$

for every t, and as $|\tau_m|$ tends to zero,

$$\psi_k^m \longrightarrow \psi(t) \qquad \text{for every } t \text{ and for } \psi \in \{v, \varphi\}$$

by (2.9) and

$$\bar{\psi}_k^m \longrightarrow \psi(t) \qquad \text{for almost every } t \text{ and for } \psi \in \{\nu, \varrho\}$$

by (2.1) and (2.10). Hence we conclude that (2.16) holds, and this completes the proof. $\square$

§3 Applications

As an immediate consequence of Theorem 2, we obtain

Theorem 3. *Suppose H^m is $\mathcal{F}^m_{n_m}$-measurable for every $m \in \mathbb{N}$, H is $\mathcal{F}_T$-measurable and H^m converges to H in $\mathcal{L}^q(P)$ for some $q > 1$. Then*

$$\lim_{m \to \infty} \widehat{E}^m[H^m] = \widehat{E}[H]. \tag{3.1}$$

From the perspective of possible applications, this is the central result of this paper. It tells us that even in incomplete markets, one can get *convergence* of discrete-time option values to continuous-time option values if these values are determined at each step with respect to the criterion of local risk-minimization. We emphasize once more that this is not a trivial result: local risk-minimization is defined with respect to a given filtration, and so we have a different optimization problem in each discretization. Theorem 3 then shows that local risk-minimization has an inherent *stability property* under discretization and thus provides a strong argument in favour of this criterion for valuing options under incompleteness.

Consider now briefly the case where $\varphi \equiv 0$, i.e., S has no jump component. Then S is just geometric Brownian motion with (time-dependent) drift $b(t)$ and volatility $v(t)$, and this implies that S is *complete*. Hence $\widehat{P}$ is the unique equivalent martingale measure for S with respect to $\mathbb{F}$, and $\widehat{E}[H]$ is the unique price for H which is consistent with absence of arbitrage opportunities. If H has a complicated form, then an explicit formula for $\widehat{E}[H]$ is in general not available and so one resorts to approximations by using discrete-time models. In most papers so far, these discrete models are binomial trees, and one major argument for this choice (apart from computational reasons) is the fact that this is the only discrete-time process which, like its continuous-time counterpart S, is complete and thus allows pricing by arbitrage. Theorem 3 shows that this very restrictive choice is not necessary; one can equally well take a simple (incomplete) discretization of S if one then uses the minimal signed martingale measure $\widehat{P}^m$ to compute the value of the approximating claim H^m.

Remark. Although our proof of Theorem 2 relies crucially on the explicit structure of our model and in particular on the assumption of deterministic coefficients, we conjecture that Theorem 3 is valid in more generality. Obviously, (3.1) will hold whenever H^m tends to H in $\mathcal{L}^q(P)$ and

$$\frac{d\widehat{P}^m}{dP} \longrightarrow \frac{d\widehat{P}}{dP} \qquad \text{in } \mathcal{L}^p(P) \tag{3.2}$$

with $\frac{1}{p} + \frac{1}{q} = 1$. It would be interesting to see a proof of (3.2) in a more general situation.

Note that we have assumed in Theorem 3 that H^m converges to H in $\mathcal{L}^q(P)$. In general, this condition is easy to verify; if for instance $(\mathcal{F}^m_n)_{m \in \mathbb{N}}$ increases to $\mathcal{F}_T$, we can always choose

$$H^m := E\left[H \big| \mathcal{F}^m_n\right]$$

by the martingale convergence theorem. In more specific examples, however, other choices of H^m may be more natural.

Example 1. Suppose $H = (S_T - K)^+$ is a European call option on S with strike price K. If we denote by $H^m = (S_n^m - K)^+$ the corresponding call option in the discrete-time model, then Lemma 1 implies that H^m tends to H in $\mathcal{L}^q(P)$ for every $q \in [1, \infty)$ and so we can apply Theorem 3 to deduce the convergence of the corresponding values. More generally, the same arguments work with $H = f(S_T)$ and $H^m = f(S_n^m)$ for every continuous function f which satisfies for instance a polynomial growth condition.

Example 2. If $H = (\frac{1}{T} \int_0^T S_u \, du - K)^+$ is a fixed strike Asian option, its natural discrete-time counterpart is

$$H^m = \left(\frac{1}{T} \sum_{j=1}^n S_{j-1}^m \Delta t_j^m - K \right)^+ .$$

It is then straightforward to check that H^m tends to H in $\mathcal{L}^q(P)$ for every $q \in [1, \infty)$. Hence $\widehat{E}^m[H^m]$ provides an approximation for $\widehat{E}[H]$ by Theorem 3, and this may be useful since the latter expectation is quite difficult to compute.

In general terms, Theorem 3 tells us that an approximation for the value $\widehat{E}[H]$ can be obtained by computing a suitable expectation in a suitable discretization. The great advantage of this lies in the fact that in a discrete-time model, every quantity of interest can in principle be computed explicitly. Let us illustrate this in our situation for a call option of the form $H^m = (S_n^m - K)^+$. By (2.14), $\widehat{E}^m[H^m]$ can be written as

$$\widehat{E}^m[H^m] = S_0 \, E\left[\prod_{j=1}^n (1 - R_j^m) \left(\prod_{k=1}^n \frac{S_k^m}{S_{k-1}^m} - \frac{K}{S_0} \right)^+ \right] . \tag{3.3}$$

Now drop the index m for ease of notation and use (2.8), (2.12), and (2.13) to obtain

$$E\left[\prod_{j=1}^n (1 - R_j) \left(\prod_{k=1}^n \frac{S_k}{S_{k-1}} - \frac{K}{S_0} \right)^+ \right]$$

$$= E\left[\prod_{j=1}^n \left(1 + q_j - q_j \exp\left(v_j \Delta W_j + \Delta N_j \log(1 + \varphi_j) - \left(\frac{1}{2} v_j^2 + \varphi_j \bar{\nu}_j \right) \Delta t_j \right) \right) \right.$$

$$\left. \times \left(\prod_{k=1}^n \exp\left(\left(b_k - \frac{1}{2} v_k^2 \right) \Delta t_k + v_k \Delta W_k + \Delta N_k \log(1 + \varphi_k) \right) - \frac{K}{S_0} \right)^+ \right] .$$

With suitable constants f_j, g_j, this can be expressed as

$$
\mathrm{E}\Bigg[\prod_{j=1}^{n}\Big(f_j + g_j \exp\big(v_j\Delta W_j + \Delta N_j \log(1+\varphi_j)\big)\Big)
$$

$$
\times \Bigg(\prod_{k=1}^{n}\exp\big(v_k\Delta W_k + \Delta N_k \log(1+\varphi_k)\big) - \bar{K}\Bigg)^{+}\Bigg]
$$

$$
= \mathrm{E}\Bigg[\Bigg\{\prod_{j=1}^{n}f_j + \sum_{j=1}^{n}g_j \exp\big(v_j\Delta W_j + \Delta N_j \log(1+\varphi_j)\big)\prod_{k\neq j}f_k
$$

$$
+ \sum_{i,j=1}^{n}g_i g_j \exp\big(v_i\Delta W_i + v_j\Delta W_j + \Delta N_i \log(1+\varphi_i) + \Delta N_j \log(1+\varphi_j)\big)
$$

$$
\times \prod_{k\neq i,j}f_k + \cdots
$$

$$
+ \Bigg(\prod_{j=1}^{n}g_j\Bigg)\exp\Bigg(\sum_{j=1}^{n}\big(v_j\Delta W_j + \Delta N_j \log(1+\varphi_j)\big)\Bigg)\Bigg\}
$$

$$
\times \Bigg(\exp\Bigg(\sum_{j=1}^{n}\big(v_j\Delta W_j + \Delta N_j \log(1+\varphi_j)\big)\Bigg) - \bar{K}\Bigg)^{+}\Bigg]
$$

which, apart from deterministic constants, is a sum of terms of the form

$$
\mathrm{E}\Bigg[\exp\Bigg(\sum_{j=1}^{n}(c_j\Delta W_j + d_j\Delta N_j)\Bigg)\Bigg(\exp\Bigg(\sum_{j=1}^{n}(\bar{c}_j\Delta W_j + \bar{d}_j\Delta N_j)\Bigg) - \tilde{K}\Bigg)^{+}\Bigg]. \quad (3.4)
$$

Since W and N are independent, the expectation can be performed first with respect to the Poisson and then with respect to the Gaussian variables. Because the random variables ΔN_j are independent with respective distributions $\mathcal{P}(\lambda_j)$, where $\lambda_j = \bar{\nu}_j\Delta t_j$, (3.4) then becomes

$$
\sum_{k_1,\ldots,k_n=0}^{\infty}\Bigg(\prod_{j=1}^{n}\frac{\lambda_j^{k_j}}{k_j!}e^{-\lambda_j}\Bigg)\exp\Bigg(\sum_{j=1}^{n}d_j k_j\Bigg)
$$

$$
\mathrm{E}\Bigg[\Bigg(\exp\Bigg(\sum_{j=1}^{n}\big((c_j + \bar{c}_j)\Delta W_j + \bar{d}_j k_j\big)\Bigg) - \tilde{K}\exp\Bigg(\sum_{j=1}^{n}c_j\Delta W_j\Bigg)\Bigg)^{+}\Bigg].
$$

The problem at this point is reduced to the computation of expectations of the form

$$
(3.6)\qquad\qquad \mathrm{E}\big[(e^{G_1} - \tilde{K}e^{G_2})^{+}\big] = \mathrm{E}\big[\mathrm{E}[(e^{G_1} - \tilde{K}e^{G_2})^{+}\,|\,G_2]\big],
$$

where G_1 and G_2 are Gaussian random variables with given means m_1, m_2, variances σ_1^2, σ_2^2 and covariance $R = \varrho\sigma_1\sigma_2$. In (3.5), we have for instance

$$m_1 = \sum_{j=1}^{n} \bar{d}_j k_j, \qquad\qquad m_2 = 0,$$

$$\sigma_1^2 = \sum_{j=1}^{n} (c_j + \bar{c}_j)^2 \Delta t_j, \qquad \sigma_2^2 = \sum_{j=1}^{n} c_j^2 \Delta t_j,$$

$$R = \sum_{j=1}^{n} c_j(c_j + \bar{c}_j) \Delta t_j.$$

The inner expectation on the right-hand side of (3.6) is now given by a Black–Scholes–type formula corresponding to the case where the terminal value of the risky asset is log-normal with mean

$$m_1 + \varrho\frac{\sigma_2}{\sigma_1} G_2 = m_1 + \frac{RG_2}{\sigma_1^2}$$

and variance

$$\sigma_2^2(1 - \varrho^2) = \sigma_2^2 - \frac{R^2}{\sigma_1^2}.$$

The value of the expression in (3.6) is thus an average of Black–Scholes–type formulae over a Gaussian distribution, and this average must in general be computed numerically. To obtain the option value $\hat{E}^m[H^m]$ in (3.3), a further averaging over a Poisson measure is required according to (3.5). In summary, the option value in the discrete-time model can be obtained as an average of Black–Scholes–type formulae, where the averaging is first performed with respect to a Gaussian measure and then with respect to a Poisson measure.

In the case where the claim H is an Asian option, either with fixed strike, i.e.,

$$H = \left(\frac{1}{T} \int_0^T S_u \, du - K\right)^+,$$

or with average strike, i.e.,

$$H = \left(S_T - \frac{K}{T} \int_0^T S_u \, du\right)^+,$$

completely analogous computations can be made. Consider for instance the second case with

$$H^m = \left(S_n^m - \frac{K}{T} \sum_{j=1}^{n} S_{j-1}^m \Delta t_j^m\right)^+.$$

Instead of (3.6), we then obtain

$$
E\left[\left(e^G - \widetilde{K}\sum_{j=1}^n e^{G_j}\right)^+\right] = E\left[E\left[\left(e^G - \widetilde{K}\sum_{j=1}^n e^{G_j}\right)^+ \Big| G_1,\ldots,G_n\right]\right], \qquad (3.7)
$$

where $G, G_1,\ldots,G_n$ are correlated Gaussian random variables with appropriate means and variances. Thus the value in (3.7) can again be obtained by repeated averaging of Black–Scholes–type formulae over Gaussian distributions, or equivalently by a single averaging over a multivariate Gaussian distribution. To obtain from there the option value $\widehat{E}^m[H^m]$ requires a further averaging over a Poisson measure.

Remark. Despite the fact that Theorem 3 allows us to reduce the computation of approximate option values in our framework to a discrete-time problem, it may sometimes be computationally advantageous to use a different approximation. Suppose that H is of the form $H = f(S_T)$ for a continuous function f satisfying a polynomial growth condition. Define the process $\widehat{U}^m = (\widehat{U}_t^m)_{0\leq t\leq T}$ by setting

$$
\widehat{U}_t^m := \exp\Bigg(\int_0^t \left(\varphi^m(s)\varrho^m(s)\nu(s) - \frac{1}{2}\left(v^m(s)\varrho^m(s)\right)^2\right) ds
$$
$$
- \int_0^t v^m(s)\varrho^m(s)\,dW_s + \int_0^t \log\left(1 - \varphi^m(s)\varrho^m(s)\right) dN_s\Bigg)
$$

for all $t \in [0,T]$; note that $\widehat{U}^m$ differs from U^m in (2.2) by the fact that ϱ^m replaces $\bar{\varrho}^m$. The same arguments as in Section 1 show that $\widehat{U}^m$ is the density process of the minimal martingale measure $\widehat{Q}^m$ for X^m with respect to $\mathbb{F}$. Furthermore, $\widehat{U}_T^m$ converges to $\widehat{Z}_T$ in $\mathcal{L}^p(P)$ for every $p \in [1,\infty)$ by an analogous argument as for (2.4), and so Lemma 1 implies that

$$
\widehat{E}[H] = E\left[\widehat{Z}_T f(S_T)\right] = \lim_{m\to\infty} E\left[\widehat{U}_T^m f(X_T^m)\right] = \lim_{m\to\infty} E_{\widehat{Q}^m}\left[f(X_T^m)\right].
$$

Now use again the arguments of Section 1 to conclude that X_T^m is given by

$$
S_0 \exp\Bigg(\int_0^T v^m(s)\,d\widehat{W}_s^m + \int_0^T \log\left(1 + \varphi^m(s)\right) dN_s
$$
$$
- \int_0^T \left(\frac{1}{2}\left(v^m(s)\right)^2 + \varphi^m(s)\widehat{v}^m(s)\right) ds\Bigg)
$$
$$
= S_0 \exp\Bigg(\int_0^T v^m(s)\,d\widehat{W}_s^m - \int_0^T \left(\frac{1}{2}\left(v^m(s)\right)^2 + \varphi^m(s)\widehat{v}^m(s)\right) ds\Bigg)
$$
$$
\times \prod_{j=1}^n (1 + \varphi_j^m)^{\Delta N_j^m},
$$

since φ^m is piecewise constant. But $\widehat{W}^m$ and N are independent under $\widehat{Q}^m$ and N is a Poisson process with $\widehat{Q}^m$-intensity $\widehat{\nu}^m(t) = \nu(t)\big(1 - \varphi^m(t)\varrho^m(t)\big)$, and so we obtain

$$
\begin{aligned}
&\mathrm{E}_{\widehat{Q}^m}[f(X_T^m)] \\
&= \exp\left(-\int_0^T \widehat{\nu}^m(s)\,ds\right) \sum_{k_1,\dots,k_n=0}^{\infty} \left\{ \prod_{j=1}^{n} \frac{(\widehat{\lambda}_j^m)^{k_j}}{k_j\,!} \right. \\
&\quad \times \mathrm{E}\left[f\left(S_0 e^G \prod_{j=1}^{n} (1+\varphi_j^m)^{k_j} \exp\left(-\int_0^T \left(\frac{1}{2}\big(v^m(s)\big)^2 + \varphi^m(s)\widehat{\nu}^m(s)\right) ds\right)\right)\right] \bigg\},
\end{aligned}
\tag{3.8}
$$

where

$$
\widehat{\lambda}_k^m := \int_{t_{k-1}^m}^{t_k^m} \widehat{\nu}^m(s)\,ds
$$

and G has a normal distribution with mean 0 and variance $\int_0^T \big(v^m(s)\big)^2 ds$. In fact, (3.8) is obtained by conditioning on N and using the independence of N and $\widehat{W}^m$ under $\widehat{Q}^m$, exactly as in the proof of [10, Theorem 2.1]. For the case where H is a European call option, i.e., $f(x) = (x - K)^+$, (3.8) simplifies to a Poisson average of Black–Scholes–type formulae as in [10]. Since no further averaging over a Gaussian distribution is required, (3.8) is therefore easier to compute than (3.3).

REFERENCES

1. F. Black and M. Scholes, *The Pricing of Options and Corporate Liabilities*, Journal of Political Economy **81** (1973), 637–659.
2. J. C. Cox, S. A. Ross, and M. Rubinstein, *Option Pricing: A Simplified Approach*, Journal of Financial Economics **7** (1979), 229–263.
3. H. Dengler, *Poisson Approximations to Continuous Security Market Models*, preprint, Cornell University, 1993.
4. D. Duffie and P. Protter, *From Discrete- to Continuous-Time Finance: Weak Convergence of the Financial Gain Process*, Mathematical Finance **2** (1992), 1–15.
5. E. Eberlein, *On Modeling Questions in Security Valuation*, Mathematical Finance **2** (1992), 17–32.
6. H. Föllmer and M. Schweizer, *Hedging of Contingent Claims under Incomplete Information*, in: Applied Stochastic Analysis, Stochastics Monographs (M. H. A. Davis and R. J. Elliott, eds.), vol. 5, Gordon and Breach, London/New York, 1991, pp. 389–414.
7. H. He, *Convergence from Discrete- to Continuous-Time Contingent Claims Prices*, Review of Financial Studies **3** (1990), 523–546.
8. N. Ikeda and S. Watanabe, *Stochastic Differential Equations and Diffusion Processes*, North-Holland, 1981.
9. M. Jeanblanc-Picqué and M. Pontier, *Optimal Portfolio for a Small Investor in a Market with Discontinuous Prices*, Applied Mathematics and Optimization **22** (1990), 287–310.
10. F. Mercurio and W. J. Runggaldier, *Option Pricing for Jump-Diffusions: Approximations and their Interpretation*, Mathematical Finance **3** (1993), 191–200.
11. R. C. Merton, *Theory of Rational Option Pricing*, Bell Journal of Economics and Management Science **4** (1973), 141–183.

12. R. C. Merton, *Option Pricing when Underlying Stock Returns are Discontinuous*, Journal of Financial Economics **3** (1976), 125–144.
13. D. B. Nelson and K. Ramaswamy, *Simple Binomial Processes as Diffusion Approximations in Financial Models*, Review of Financial Studies **3** (1990), 393–430.
14. M. Schäl, *On Quadratic Cost Criteria for Option Hedging*, (to appear in Mathematics of Operations Research), preprint, University of Bonn, 1992.
15. M. Schweizer, *Hedging of Options in a General Semimartingale Model*, Diss. ETHZ No. 8615, 1988.
16. M. Schweizer, *Option Hedging for Semimartingales*, Stochastic Processes and their Applications **37** (1991), 339–363.
17. M. Schweizer, *Approximating Random Variables by Stochastic Integrals, and Applications in Financial Mathematics*, Habilitationsschrift, University of Göttingen, 1993.
18. H. Shirakawa, *Security Market Model with Poisson and Diffusion Type Return Process*, preprint IHSS 90-18, Tokyo Institute of Technology, 1990.
19. W. Willinger and M. S. Taqqu, *Toward a Convergence Theory for Continuous Stochastic Securities Market Models*, Mathematical Finance **1** (1991), 55–99.
20. X.-X. Xue, *Martingale Representation for a Class of Processes with Independent Increments and its Applications*, in: Applied Stochastic Analysis, Proceedings of a US–French Workshop (I. Karatzas and D. Ocone, eds.), Rutgers University, New Brunswick, N.J., Lecture Notes in Control and Information Sciences 177, Springer, 1992, pp. 279–311.

WOLFGANG J. RUNGGALDIER, DIPARTIMENTO DI MATEMATICA PURA ED APPLICATA, UNIVERSITÀ DEGLI STUDI DI PADOVA, VIA BELZONI 7, I-35131 PADOVA, ITALY

MARTIN SCHWEIZER, UNIVERSITÄT GÖTTINGEN, INSTITUT FÜR MATHEMATISCHE STOCHASTIK, LOTZESTRASSE 13, D-37083 GÖTTINGEN, GERMANY

Progress in Probability, Vol. 36
© 1995 Birkhäuser Verlag Basel/Switzerland

PORTFOLIO SELECTION WITH TRANSACTION COSTS

AGNÈS TOURIN AND THALEIA ZARIPHOPOULOU

ABSTRACT. This paper considers an infinite horizon investment-consumption model in which a single agent consumes and distributes his wealth between two assets, a bond and a stock. The problem of maximization of the total utility from consumption is treated; State (amount allocated in assets) and control (consumption, rates of trading) constraints are present. It is shown that the value function is the unique viscosity solution of a variational inequality with gradient constraints. A monotone numerical scheme is then constructed in order to compute both the value function and the location of the free boundaries of the so-called transaction regions.

§1 THE MODEL

In this paper we examine a general investment and consumption decision problem for a single agent. The investor consumes and distributes his wealth between two financial assets. One asset is a bond, i.e. a riskless security with instantaneous rate of return r. The other asset is a stock, whose value is driven by a Wiener process.

The control objective is to maximize, in an infinite horizon, the total expected discounted utility which comes only from consumption. When the investor makes a transaction, he pays transaction fees which are assumed to be proportional to the amount transacted.

The price P_t^0 of the bond is given by

$$dP_t^0 = rP_t^0 \, dt, \qquad P_0^0 = p_0,$$

where $r > 0$ is the interest rate. The price P_t of the stock satisfies

$$dP_t = bP_t \, dt + \sigma \, dw_t, \qquad P_0 = p,$$

where b is the mean rate of return, σ is the dispersion coefficient and the process w, which represents the source of uncertainty in the market, is a standard Brownian motion defined on the underlying probability space (Ω, F, P). We will denote by F_t the augmentation under P of $F_t^w = \sigma(w_s : 0 < s \le t)$ for $0 < t < +\infty$.

1991 *Mathematics Subject Classification.* 65P05, 90A09.

Key words and phrases. Viscosity solutions, Hamilton–Jacobi–Bellman equation, dynamic programming approximation scheme, infinite horizon investment-consumption model.

The market coefficients r, b, and σ are assumed to be constant with $\sigma \neq 0$ and $b > r > 0$.

The amount of wealth x_t and y_t, invested at time t in bond and stock respectively, are the state variables and they evolve according to the equations

$$
\begin{aligned}
dx_t &= (rx_t - C_t)\, dt - (1 + \lambda)\, dM_t + (1 - \mu)\, dN_t, & x_0 &= x, \\
dy_t &= by_t\, dt + \sigma y_t\, dw_t, & y_0 &= y,
\end{aligned}
\tag{1}
$$

where (x, y) is the initial endowment of the investor. For simplicity we assume here that all financial charges are paid from the holdings in bond. The control processes are the consumption rate C_t and the processes M_t and N_t which represent the cumulative purchases and sales of stock respectively. The controls (C_t, M_t, N_t) are admissible if:

(i) The consumption rate C_t is F_t-measurable, $C_t \geq 0$ almost everywhere for all $t \geq 0$, and $\int_0^t e^{-rs} C_s\, ds < +\infty$ almost surely for all $t \geq 0$.

(ii) M_t, N_t are F_t-measurable, right-continuous and nondecreasing processes.

(iii) If x_t and y_t are the state trajectories given by (1) when the controls M_t and N_t are used, then, for all $t \geq 0$,

$$x_t + (1 + \lambda)y_t \geq 0, \quad \text{a.\,s. if } y_t \geq 0, \quad \text{and } x_t + (1 - \mu)y_t \geq 0 \text{ a.\,s. if } y_t \leq 0.$$

We denote by $\mathcal{A}(x, y)$ the set of admissible policies. The total expected discounted utility J coming from consumption is given by

$$
J(x, y, C, M, N) = \mathrm{E}\left(\int_0^{+\infty} e^{-\beta t} U(C_t)\, dt \right)
$$

with $(C, M, N) \in \mathcal{A}(x, y)$ and $(x, y) \in \bar{\Omega}$ where

$$
\Omega = \{ (x, y) \in \mathbb{R} \times \mathbb{R} : x + (1 + \lambda)y > 0 \text{ if } y < 0 \text{ and } x + (1 - \mu)y \geq 0 \text{ if } y \geq 0 \};
$$

the utility function $U : [0, +\infty) \to [0, +\infty)$ is assumed to have the following properties:

$$
\begin{cases}
U \text{ is a strictly increasing, concave } C^2(0, +\infty)\text{-function}, \\
U(c) \leq K(1 + c)^\gamma \text{ with } 0 < \gamma < 1 \text{ and } K > 0, \\
U(0) \geq 0, \ \lim\limits_{c \to 0} U'(c) = +\infty, \ \lim\limits_{c \to +\infty} U'(c) = 0.
\end{cases}
$$

The value function u is given by

$$
u(x, y) = \sup_{\mathcal{A}(x, y)} \mathrm{E}\left(\int_0^{+\infty} e^{-\beta t} U(C_t)\, dt \right),
$$

where β denotes the discount factor. To guarantee that the value function is well defined when U is unbounded, we assume that

$$
\beta > r\gamma + \gamma(b - r)/\sigma^2(1 - \gamma).
$$

The above condition yields that the value function which corresponds to $\lambda = \mu = 0$ and $U(c) = M(1+c)^\gamma$, and thereby all value functions, is finite for $0 < \lambda$ and $\mu < 1$.

Let us remark that Davis and Norman (see [4]) solve explicitly the problem in the H. A. R. A. case (Hyperbolic Absolute Risk Aversion) $U(c) = \frac{1}{\gamma}c^\gamma$, $0 < \gamma < 1$, or $U(c) = \log c$ for $\gamma = 0$ thanks to the homothetic property satisfied by the corresponding value function.

First, we derive the Hamilton–Jacobi–Bellman equation associated with this control problem and characterize u as its unique solution. It turns out that the Bellman equation here is a variational inequality with gradient constraints.

Theorem 1. *The value function u is the unique constrained viscosity solution of*

$$
\min\Big[(1+\lambda)u_x - u_y, -(1-\mu)u_x + u_y,
$$
$$
-\frac{1}{2}\sigma^2 y^2 u_{yy} - rxu_x - byu_y + \beta u - \max_{c \geq 0}\big(-cu_x + U(c)\big)\Big] = 0, \tag{2}
$$

for all $(x, y) \in \Omega$, in the class of concave and uniformly continuous functions.

§2 THE NUMERICAL SCHEME

The approach which relies on the dynamic programming principle yields monotone, stable and consistent schemes whose convergence was proved in some situations by Barles and Souganidis [2] and recently proved for parabolic equations arising in finance theory by Barles, Daher, and Romano in [1].

In the scheme constructed here, the first-order operators are approximated by monotone finite difference schemes. As far as the second-order operator is concerned, the first-order part is approximated by a monotone explicit scheme based on the dynamic programming principle whereas the second-order term is approximated by an implicit Cranck–Nickolson scheme. Thus, splitting into two half-iterations allows one to choose a time step of the same order as the mesh size. This method is known as the time-splitting method or method of fractional steps.

Finally, let us mention a different approach, based on Kushner's Markov chain ideas, which also yields a monotone consistent and stable scheme and which has been developped in the few last years. In particular, Fitzpatrick and Fleming (see [5]) proposed a discretization for an optimal investment–consumption model.

We next present the monotone scheme which approximates the solution of the variational inequality (2). To this end, we first write it in the form:

$$
\min\{L_0(x, y, u, u_x, u_y, u_{yy}), L_1(u_x, u_y), L_2(u_x, u_y)\} = 0 \tag{3}
$$

where

$$
L_0(x, y, u, u_x, u_y, u_{yy}) = -\frac{1}{2}\sigma^2 y^2 u_{yy} - rxu_x - byu_y + \beta u - \max_{c \geq 0}\{-cu_x + U(c)\}
$$

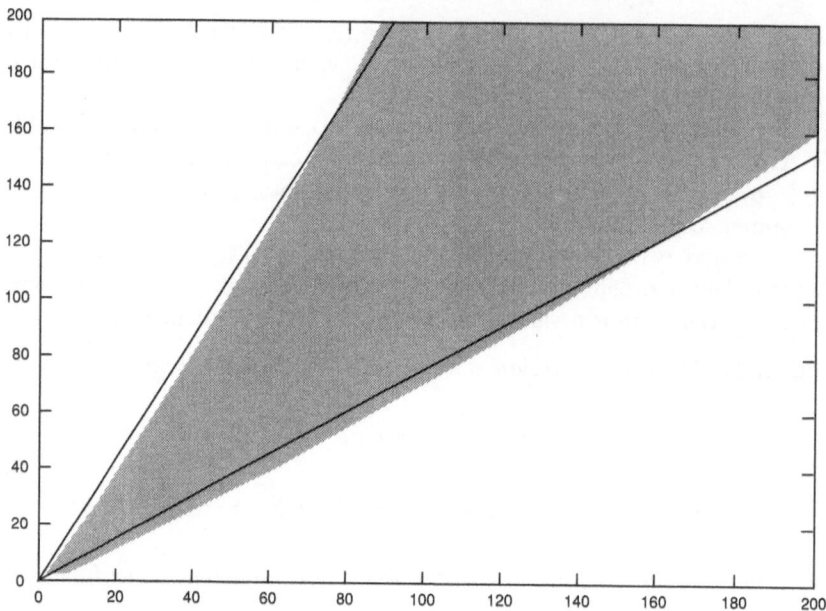

Figure 1: Non-transaction region for the utility function $u(c) = 2\sqrt{c}$.

denotes the second-order operator and

$$L_1(u_x, u_y) = (1 + \lambda)u_x - u_y \qquad \text{and} \qquad L_2(u_x, u_y) = -(1 - \mu)u_x + u_y$$

denote the first-order operators which correspond to the gradient constraints.

To simplify the following computation, we restrict ourselves to the rectangular domain $[0, (M - 1)\Delta x] \times [0, (L - 1)\Delta y]$ of $\mathbb{R}^2$ where M and L denote the number of grid points respectively on the x and y-axis and Δx, $\Delta y > 0$ are the mesh sizes; the value of our numerical approximation at $((i - 1)\Delta x, (j - 1)\Delta y)$ will be denoted by $V_{i,j}$ for $i \in \{1, \ldots, M - 1\}$ and $j \in \{1, \ldots, L - 1\}$. We then make the following usual definitions:

$$D_x^+ V_{i,j} = \frac{V_{i+1,j} - V_{i,j}}{\Delta x}, \qquad D_y^+ V_{i,j} = \frac{V_{i,j+1} - V_{i,j}}{\Delta y},$$

$$D_x^- V_{i,j} = \frac{V_{i,j} - V_{i-1,j}}{\Delta x}, \qquad D_y^- V_{i,j} = \frac{V_{i,j} - V_{i,j-1}}{\Delta y}.$$

Inside the domain, the first-order operators are approximated in a monotone way, using the appropriate backward and forward finite differences:

$$g_1(D_x^- V_{i,j}, D_y^+ V_{i,j}) = (1 + \lambda)\frac{V_{i,j} - V_{i-1,j}}{\Delta x} - \frac{V_{i,j+1} - V_{i,j}}{\Delta y}, \qquad (4)$$

$$g_2(D_x^+ V_{i,j}, D_y^- V_{i,j}) = -(1 - \mu)\frac{V_{i+1,j} - V_{i,j}}{\Delta x} + \frac{V_{i,j} - V_{i,j-1}}{\Delta y}. \qquad (5)$$

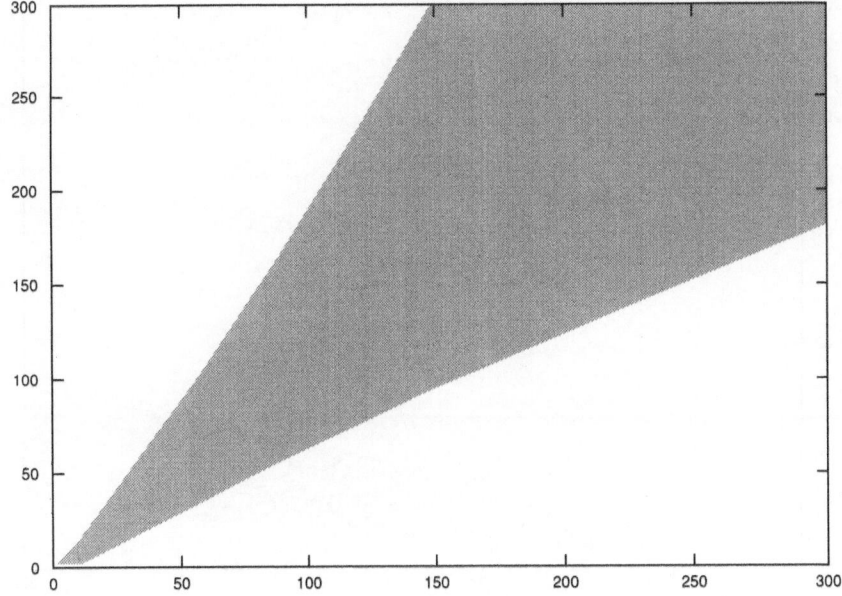

Figure 2: Non-transaction region for the utility function $u(c) = 3c^{1/3} + \frac{3}{2}c^{2/3}$.

We then consider the first-order operator L_0' obtained by removing the second order term from L_0. The solution of the corresponding equation may be rewritten as the value function of a deterministic optimal control problem. In order to get a monotone scheme, we apply the dynamic programming principle to this particular control problem:

$$u(x, y) = \max_{0 \le c(\cdot)} \left\{ \int_0^T e^{-\beta t} U(c_t) \, dt + u(x(T), y(T)) e^{-\beta T} \right\} \quad \text{for all } T > 0.$$

One may choose $T = \Delta t$ arbitrarily small and assume that the control remains constant on the interval $[0, T]$. The dynamic programming then yields

$$\sup_{0 \le c} \left(U(c) + \frac{u(x + \Delta t\,(rx - c), y + \Delta t\,by) - u(x, y)}{\Delta t} e^{-\beta \Delta t} + u(x, y) \frac{e^{-\beta \Delta t} - 1}{\Delta t} \right) = 0.$$

More precisely, in order to approximate the operator L_0', one has to find an explicit formulation for the following optimum:

$$\max \left\{ \sup_{0 \le c \le r(i-1)\Delta x} \left(U(c) - \beta V_{i,j} + D_x^+ V_{i,j}(r(i-1)\Delta x - c) + D_y^+ V_{i,j} b(j-1)\Delta y \right), \right.$$

$$\left. \sup_{0 \le c,\, r(i-1)\Delta x \le c} \left(U(c) - \beta V_{i,j} + D_x^- V_{i,j}(r(i-1)\Delta x - c) + D_y^+ V_{i,j} b(j-1)\Delta y \right) \right\}.$$

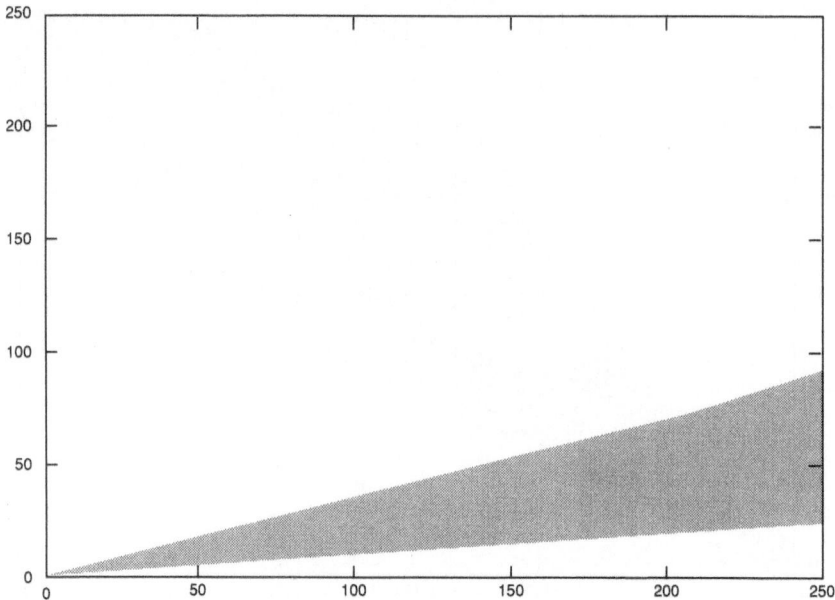

Figure 3: Non-transaction region for the utility function $u(c) = 100 - 1/\sqrt{c}$.

Besides, one has to construct an approximation for the points located on the boundaries of the discrete domain. We first consider the x and y-axis that actually do not correspond to the real boundaries of the solvability region. To simplify, we suppose that the non-transaction region is strictly included in $\mathbb{R}^+ \times \mathbb{R}^+$. Under this assumption, the x and y-axis are respectively included in the sell and buy region, that is to say, on these axis, the value function is the respective solution of $L_1(u_x, u_y) = 0$ and $L_2(u_x, u_y) = 0$.

Under the restrictive above assumption, at the points located on the y-axis, a numerical approximation $V_{1,j}$ of the value function will satisfy, for $1 < j \leq L$, $g_2(D_x^+ V_{1,j}, D_y^- V_{1,j}) = 0$ where g_2 is given by formula (5). At the points located on the x-axis, a numerical approximation $V_{i,1}$ of the value function will satisfy $g_1(D_x^- V_{i,1}, D_y^+ V_{i,1}) = 0$ for $1 < i \leq M$, where g_1 is given by formula (4). Finally, at the point $i = 1$, $j = 1$, we impose the Dirichlet boundary condition $V_{1,1} = 0$.

Besides, in direction to infinity, at the points located on $x = (M - 1)\Delta x$ and $y = (L - 1)\Delta y$, we impose Neumann conditions. The location of the free boundaries is rather sensitive to the given normal derivatives but the error is essentially concentrated near the boundary. Consequently, we take a sufficiently large discrete domain and compute the corresponding value function and the free boundaries. Then, we only take into account the results inside the domain.

In the second half-iteration, we solve the one-dimensional heat equation thanks to a Cranck–Nickolson scheme. Such a scheme requires boundary conditions; on the x-axis we choose Dirichlet boundary conditions whose values are provided by formula (4). At the points located on $y = (L - 1)\Delta y$, we have already imposed

Neumann conditions. Thus, the second half-iteration consists of inverting a tridi-agonal matrix.

Finally, we present some numerical experiments corresponding to three different classes of utility functions. For the H. A. R. A. utility function, an approximation of the value function is computed, on one hand using the algorithm proposed by Davis and Norman, on the other hand, thanks to the numerical scheme described in the preceding section. One may compare the corresponding results, especially the location of the free boundaries. Figures 1–3 show the computed non-transaction regions corresponding to the utility functions $u(c) = 2\sqrt{c}$, $u(c) = 3c^{1/3} + \frac{3}{2}c^{2/3}$ and $u(c) = 100 - 1/\sqrt{c}$.

References

1. G. Barles, C. Daher and M. Romano, *Convergence of numerical schemes for parabolic equations arising in finance theory*, Caisse Autonome de Refinancement, 1991.
2. G. Barles and P. E. Souganidis, *Convergence of approximation schemes for fully nonlinear second order equations* (to appear in Asymptotic Analysis).
3. M. G. Crandall and P. L. Lions, *Viscosity solutions of Hamilton–Jacobi equations*, Trans. Amer. Math. Soc. **277** (1983), 1–42.
4. M. H. A. Davis and A. R. Norman, *Portfolio selection with transaction costs*, Math. Op. Res. **15** (1990), 676–713.
5. B. G. Fitzpatrick and W. H. Fleming, *Numerical methods for an optimal investment/consumption model*, Math. Op. Res. (1991).
6. W. H. Fleming, S. Grossman, J. L. Vila, and T. Zariphopoulou, *Optimal portfolio rebalancing with transaction costs*, submitted to Econometrica.
7. A. Tourin, *Thèse de Doctorat*, Université Paris IX-Dauphine, Januar 1992.

AGNÈS TOURIN, CEREMADE, UNIVERSITÉ DE PARIS IX-DAUPHINE, PLACE DE LATTRE-DE-TASSIGNY, 75775 PARIS CEDEX 16

THALEIA ZARIPHOPOULOU, DEPARTMENT OF MATHEMATICS AND BUSINESS SCHOOL, UNIVERSITY OF WISCONSIN, MADISON, WI 53706 USA

PPS/PP - Progress in Probability

Edited by
Th. M. Liggett / Ch. Newman / L. Pitt

Progress in Probability is designed for the publication of workshops, seminars and conference proceedings on all aspects of probability theory and stochastic processes, as well as their connections with and applications to other areas such as mathematical statistics and statistical physics.

PA - Probability and its Applications

Edited by
Th. M. Liggett / Ch. Newman / L. Pitt

Probability and its Applications publishes research-level monographs and advanced graduate texts dealing with all aspects of probability theory and stochastic processes, as
well as their connections with and applications to other areas such as mathematical statistics and statistical physics.

R. Carmona/J. LaCroix
Spectral Theory of Random Schrödinger Operators
ISBN 3-7643-3486-X

K.L. Chung/R.J. Williams
Introduction to Stochastic Integration
ISBN 3-7643-3386-3

G.F. Lawler
Intersections of Random Walks
ISBN 3-7643-3557-2

R.K. Getoor
Excessive Measures
ISBN 3-7643-3492-4

S. Kwapien/W. Woyczynski
Random Series and Stochastic Integrals: Single and Multiple
ISBN 3-7643-3572-6

R.M. Blumenthal
Excursions of Markov Processes
ISBN 3-7643-3575-0

N. Madras/G. Slade
The Self-Avoiding Walk
ISBN 3-7643-3589-0

LM – Lectures in Mathematics - ETH Zürich

Some Aspects of Brownian Motion
Part I: Some Special Functionals

M. Yor, Université Pierre et Marie Curie, Paris, France

1992. 148 pages. Softcover
ISBN 3-7643-2807-X

Roughly, half of the text consists of new results; hence these notes may be placed midway between an advanced crash course on Brownian motion, and a complement to existing texts, to which precise references are given throughout. This volume will be of interest to researchers either in probability theory or in more applied fields, such as polymer physics or mathematical finance.

"...the aim of the book ... to calculate as explicitly as possible the laws of Brownian functionals, ... has been magnificently achieved.
... the work will be a highly valuable resource for afficianados of Brownian motion."

F. Knight, Univ. of Urbana, Illinois
SIAM Review

"The book presents a great variety of explicit formulae for Brownian motion functionals. ... The book is very well written. Both the style of presentation of the subject matter and typesetting make it very easy to read. The book is addressed to researchers and advanced graduate students as it takes stochastic integration and excursion theory for granted."

Metrika, 3/4 1994